Natural Product Chemistry

A mechanistic, biosynthetic and
ecological approach
by
Kurt B. G. Torssell

APOTEKARSOCIETETEN
SWEDISH PHARMACEUTICAL SOCIETY

The cover shows a panoramic view of the rainforest,
workshop for the production of an astonishing variety
of natural compounds.
The insert on the front cover depicts leafcutting ants in
action, and on the back cover, the collecting of plant
material for chemical investigations.
Kurt B. G. Torssell is Swedish and held a professorship
in the Department of Chemistry at the University of
Aarhus, Denmark from 1969 to 1993.

Photos: Naturbild, Stockholm

Natural Product Chemistry, by Kurt B. G. Torssell

© 1997 Kurt B. G. Torssell and
Apotekarsocieteten—Swedish Pharmaceutical Society, Swedish Pharmaceutical
Press, P.O. Box 1136, S-111 81, Stockholm, Sweden

Project manager: Maud Sundén

ISBN 91 8627 463 5

Kristianstads Boktryckeri AB, Sweden 1997

Contents

Preface of the first edition

The chemistry of natural products has made great progress during the last decades as a result of our better comprehension of enzymatic processes, and the development of biogenetic and biosynthetic theories which logically classify and link together an immense variety of compounds. The endless compilation of compounds could eventually be substituted by biosynthetic schemes easy to survey. Natural product chemistry gradually changes from being traditionally descriptive in nature to being mechanistic and predictable. An enzymatic reaction is no longer just a transformation of reactant into product with unknown transition, but can be or should be understood in the same way as any other analogous reaction carried out non-enzymatically in the test tube. In the present text I have tried to present the formation and reactions of natural products using the vocabulary of ordinary organic reaction mechanisms. It is important to understand the technique nature uses to bring about its multitude of astonishlingly elaborate compounds. The underlying biosynthetic principles and enzyme mechanisms have had a great impact on the imagination of many organic chemists, and imitation of bio-organic processes in the laboratory has led to important advances in synthetic methods. It is my belief that the further development of natural product chemistry will become more biologically oriented. Secondary metabolites shall no longer be regarded as waste products, we just do not know yet how they function. We are at present in a transition of re-evaluation. They certainly play some role necessary for the existence of the organism in its environment. Fruitful research on biochemical interactions in ecosystems is ahead of us and must involve cooperation between chemists and biologists. Recent developments in the area of chemical communication among insects is a stunning example of the opening of new frontiers in science and of the importance of multidisciplinary teamwork.

I have not restricted myself to present established facts only in this text. A textbook ought to be at the frontier. I have tried to demonstrate the dynamic state of mechanistic biosynthetic research by topically following the current developments up to the end of 1981 and 'occasionally' I have preferred to give my own mechanistic interpretation of the events. The critical reader will certainly be able to differentiate the speculative from the accepted. There is often a rich flora of mechanistic proposals to choose among.

I have grouped together the nitrogen containing compounds—the amino acids, alkaloids and *N*-heteroaromatics—in Chapters 7, 8 and 9. However, it is recommended to read Chapter 6, and particularly the pyridoxal catalysed reactions (section 7.3), after the carbohydrates (Chapter 3) because they provide the basis for some of the reactions discussed in the other chapters.

I wish to express my thanks to Professors R. H. Thomson, Aberdeen, and A. Kjaer, Lyngby, for reading the whole manuscript and for their valuable criticism. Not the least are my thanks due to Mrs. Ella Larsen for her indefatigable typing work and for drawing all the figures. I appreciate any comments on errors and suggestions for improvements of the text.

KURT TORSSELL
Aarhus 1981

Preface of the second edition

Many new, important results have been published in the area of natural products chemistry since the first edition appeared in 1983. Proper selection has been difficult. I have focused on two important fields: 1. Biological oxidation and 2. Chemical ecology, which I also used to call Darwinian chemistry. The reasons are first that the various oxidation processes are fundamental to synthesis in the living tissue (Chapter 4) and second, that we have a better understanding of the role of secondary metabolites in the environment and of the role of the compounds in evolutionary processes (Chapter 2). We are about to leave the golden era of structural elucidation, which for a century has engaged and still engages so many distinguished organic chemists. The chemical community is no longer as impressed by structural elucidations of compounds as in the thirties and fourties when the structures of e.g. steroids and penicillins were determined. Generally, the activities have moved towards border zones between chemistry and biology and a fascinating area has emerged concerning the function of metabolites in the ecosystem – in short, chemical ecology. This is defined as the science describing chemical interactions between organisms in the environment, controlling coexistence and coevolution of species.

There has been an efficient cross-fertilization between biology and chemistry as evident from the progress in this area.

In certain dynamic areas such as mechanistic enzyme chemistry, nucleic acid chemistry, gene technology and structural biochemistry the development has been rapid. I feel that the last three areas reside within the domains of biochemistry and are adequately treated in textbooks in that field. Other areas show a more consolidated picture such as the main streams of biosynthesis, which also have a strong impact on three-dimensional synthetic design by organic chemists. Many elegant total syntheses, e.g. of terpenoids, were inspired by principles of biosynthesis.

The placing of the chapter on chemical ecology in the beginning of the book caused some pedagogic problems. Since the secondary metabolites are of paramount evolutionary importance one could very well start the course by discussing the various functions of the compounds in the environment without knowing much about structures and biosynthesis. However, many lecturers prefer to discuss chemical ecology separately at the end of the course when the different pathways have been presented, thus emphasizing biosynthesis. Halfway through the biosynthetic part of the textbook I begin introducing the students to the condensed chapter on chemical ecology supplemented with relevant articles and selected chapters from monographs in the field intermixed with mechanisms of biosynthesis.

All chapters have been updated to 1994–95 and corrected for misprints and mistakes. I wish to thank colleagues, reviewers and students for making suggestions for improvements of the text. I wish to express my thanks to Dr. Steven Lucas for linguistic corrections and to Miss Anna Berggrund for excellent art work.

KURT TORSSELL
Stockholm 1995

Chapter 1

Introduction and general considerations

1.1 The literature

In addition to current journals several comprehensive series are issued containing reviews on various topics of natural product chemistry. Some of these are issued annually or regularly at longer intervals. A great many monographs on special topics are published. The following condensed bibliographic list represents the main sources consulted by the author for the present textbook and is recommended for further studies. Every chapter is supplied with a list of references documenting recent developments. Regrettably, some references must be chosen by chance and consequently they are not always fully representative.

Textbooks

Organic Chemistry
March, J. *Advanced Organic Chemistry. Reactions, Mechanisms and Structure*, 4th Edn. McGraw-Hill, New York, 1992

Biochemistry
Stryer, L. *Biochemistry*, 3rd Edn. Freeman, San Francisco, 1988.
Walsh, C. *Enzymatic Reaction Mechanism*, Freeman, San Francisco, 1979

Natural Product Chemistry
Haslam, E. *Shikimic Acid. Metabolites and Metabolism*. J. Wiley, Chichester, 1993.
Herbert, R. B. *Biosynthesis of Secondary Metabolites*. Chapman and Hall, London, 2nd. Ed. 1989.
Luckner, M. *Secondary Metabolites in Microorganisms, Plants and Animals*, 2nd Edn. Springer Verlag, Berlin, 1985.
Mann, J. *Secondary Metabolism*, 2nd Edn. Clarendon Press, Oxford, 1987.
Mann, J., Davidson, R. S., Hobbs, J. B., Banthorpe, D. V. and Harborne, J. B. *Natural Products. Their Chemistry and Biological Significance*. Longman, Sci. and Tech. 1994.

Thomson, R. H. (Ed.) *The Chemistry of Natural Products*, Blackie, Glasgow, 1985; 2nd Edn. 1993.

Chemical Ecology
Harborne, J. B. *Introduction to Ecological Biochemistry*, 3rd. Edn. Academic Press, London, 1988.

Comprehensive Series, Monographs and Review Articles
Bailey, J. A. and Mansfield, J. W. *Phytoalexins*, Blackie, Glasgow, 1982.
Comprehensive Organic Chemistry, Vol. 5, *Biological Compounds*, Haslam, E. (Ed.), Pergamon, Oxford, 1979.
Connolly, J. D. and Hill, R. A. *Dictionary of Terpenoids*. Vol. I–III. Chapman and Hall, London 1991.
Glasby, J. S. *Encyclopedia of the Terpenes*, J. Wiley, Chichester, 1982.
Dictionary of Natural Products. Vol. **1–7**, Chapman and Hall, London, 1994.
Devon, T. K. and Scott. A. I. *Handbook of Naturally Occurring Compounds* **1–2** (1972–75). Academic Press, New York.
Harborne, J. B. *Recent Advances in Chemical Ecology*. Nat. Prod. Rep. **3** (1986) 323; **6** (1989) 85; **10** (1993) 327.
Hegnauer, R., *Chemotaxonomie der Pflanzen*, **1–6** (1962–1973). *Ergänzungsband* **7–9** (1986–1990). Birkhäuser, Basel.
Karrer, W. *Konstitution und Vorkommen der organischen Pflanzenstoffe*, (1957), Karrer, W., Cherbuliez. E. and Eugster, C. H. *Ergänzungsband* **1** (1977). Birkhäuser Basel.
Marine Natural Products, **1–4** (1978–1981). Scheuer, P. J. (Ed.), Academic Press, New York.
Progress in the Chemistry of Organic Natural Products, **1–61** (1938–1993). Founded by Zechmeister, L., Springer, Vienna.
Advances in Carbohydrate Chemistry and Biochemistry **1–50** (1945–1993), Academic Press, New York.
Carbohydrate Chemistry. Specialist Periodical Reports, **1–24** (1968–1993), Guithrie, R. D. and Brimacombe, J. S. (Eds.), The Chemical Society, London.
Harborne, J. B., Mabry, T. J. and Mabry, H. (Eds.), *The Flavonoids*. Chapman and Hall, London, 1975.
Harborne, J. B. and Tomas-Barberan, F. A. (Eds.), *Ecological Chemistry and Biochemistry of Plant Terpenoids*. Clarendon Press, Oxford 1991.
Haslam, E., *Plant Polyphenols*. Cambridge University Press, Cambridge, 1989.
Thomson, R. H., *Naturally Occurring Quinones*, Academic Press, London, 1971.
Naturally Occurring Quinones, III. Recent Advances, Chapman and Hall, London, 1987.

Amino Acids, Peptides and Proteins, Specialist Periodical Reports **1–23** (1969–1993). Young, G. T. and Sheppard, R. C. (Eds.), The Chemical Society, London.

Recent Advances in Phytochemistry, **10**, *Biochemical Interaction between Plants and Insects*. Wallace, J. W. and Mansell, R. L. (Eds.), Plenum Press, New York, 1975.

Rice, E. L. *Allelopathy*. 2nd Edn. Academic Press, 1984.

Roitberg, B. D. and Isman, M. B., (Eds.), *Insect Chemical Ecology*. Chapman and Hall, New York, 1992.

Rosenthal, G. A., Janzen, D. H. and Berenbaum, M. R. (Eds.), *Herbivores. Their Interaction with Secondary Plant Metabolites*. Academic Press, 1979; 2nd Edn. Vol **1**, 1991, Vol **2** 1992.

The Alkaloids, **1–45**. (1950–1994), Manske, R. H. F., Holmes, H. L., Rodrigo, R. G. A. and Brossi, A. (Eds.), Academic Press, New York.

Dalton, D. R. *The Alkaloids*, M. Dekker, New York, 1979.

The Porphyrins, I–VII (1978–1979). Dolphin, D. (Ed.), Academic Press, New York.

The Total Synthesis of Natural Products, **1–7**, (1973–1988), ApSimon, J. (Ed.), Wiley-Interscience.

  Journals

Journal of Chemical Ecology, [Vol **20**, (1994)].

Journal of Ethnopharmacology, [Vol. **40**, (1994)]. Elsevier.

Journal of Natural Products, (Lloydia), [Vol. **57**, (1994)]. The American Society of Pharmacognosy.

Natural Product Reports, [Vol. **11**, (1994)], Pattenden, G. and Simpson, T. J. (Eds.), The Royal Society of Chemistry, London.

Phytochemistry, [Vol. **37**, (1994)]. *Phytochemistry Review Articles*, **1–88**, (1984–1994), Pergamon Press.

Planta Medica, [Vol. **60**, (1994)]. G. Thieme Verlag.

1.2 Background

Natural product chemistry is in its different aspects an ancient science. The preparation of foodstuffs, colouring matters, fibers, toxins, medicinals and stimulants are examples of activities as old as mankind. When chemists in the late eighteenth century took the final jump from the world of myths into modern science, the true properties of extracts obtained from nature aroused great curiosity amongst scientists. They began to separate, purify, and finally analyse the compounds produced in living cells. Separation methods were developed and without doubt natural product chemisty has brought great stimuli to the development of the refined techniques we have today, such as the various analytical and preparative chromotographic methods: column chromatography (CC), GC, TLC, HPLC, paper chromatography, electrophoresis, ion exchange, etc. These methods have made it possible to isolate compounds present in extremely small quantities. Structural elucidation was typically carried out by degradation to smaller fragments of known structure combined with investigations of the reactivity pattern and elementary analysis of the compounds. These earlier works led in many instances to discoveries of new valuable reactions and rearrangements. However, without spectroscopy in its service, natural product chemistry would never have attained the status and refinement it has today. UV, IR, NMR, MS, ESR, CD and ORD changed working methods and habits considerably through the years. A large number of spectroscopic data, correlating spectral properties with structure, was collected. These data give us sometimes so much information about the structure that pure chemical transformations can be reduced to a minimum. This enables us now to solve structures with much less material than was needed earlier, but as a consequence of these developments, no doubt, valuable information about the chemistry escapes discovery. When the amount of sample available is very limited and the structure too complicated, X-ray crystallography is an ultimate resource. Programmes are now available that make the structural elucidation almost a routine job for the specialist. The classical cumbersome, degradative work is almost turned into a separatory and X-ray crystallographic problem—provided we can obtain suitable crystals.

1.3 Synthesis and biosynthesis

Two other important aspects of structural determination are synthesis and biosynthesis. A structure was earlier not considered to be rigorously proved unless the compound was also synthesized. The structure of fragments was often confirmed by an unambiguous synthesis. The motivation nowadays for total synthesis is not so much the question of confirmation as the challenge inherent in synthesizing intriguing structures in much the same way as the mountaineer is challenged by the precipices of the mountain ridge. Practical and commercial interests often require alternative preparative procedures. These efforts have brought considerable elegance, creativity and efficiency to the art of synthesis. New methods have been designed and worked out specifically for the purpose so that natural product chemistry has stimulated the development of organic synthesis.

The recognition of biosynthetic principles is the most significant development in natural product chemistry. During the last century a great number of new structures was determined. In the earlier phase organic chemists were just content to solve the structure of natural products and group them according to origin, pharmacological activity or structure, but rather soon the mass of information suggested the need for a more coherent view of biogenesis. The ingenuity and intuition of a few chemists led the painstaking search for the hidden biosynthetic pathways in the right direction. It became evident rather early to Wallach and Ruzicka that the terpenes had a common building block, the isoprene unit, the true origin of which was concealed for a long time, and Winterstein and Trier suggested on good grounds that alkaloids were formed from α-amino acids, ideas which were further developed by Schöpf and Robinson. A peep-hole into the fascinating synthetic workshop of the living cell was gradually opened. The biosynthetic routes of the various classes of compounds were mapped out and precursors and intermediates identified. The enzyme catalysed reactions in the cell have their *in vitro* counterparts and their mechanisms can be correlated with known organic reaction mechanisms.

Biosynthetic principles are now continually applied in the process of structural determination of natural products. The structures of the carbon skeletons, the heterocyclic units and the substitution patterns are defined by such considerations, and consequently many "unnatural" structures violating these principles can be discarded in advance as incorrect. Biomimetic studies have led to novel and elegant synthetic procedures and the principles have been applied successfully in total synthesis.

1.4 Primary and secondary metabolism

It is customary to distinguish between primary and secondary metabolism. The former refers to the photosynthetic processes producing simple and widely distributed low molecular weight carboxylic acids of the Krebs cycle, α-amino acids, carbohydrates, fats, proteins and nucleic acids involved in the life processes. These compounds are commonly regarded as the domain of biochemists. We will follow this classification and discuss them in their capacity as starting materials—the precursors—of the secondary metabolites. The secondary metabolites are, in principle, non-essential to life but they definitely contribute to the species' fitness of survival. They are more characteristic for the particular biological group, such as family or genus, and apparently the synthetic machinery involved here is related to the mechanism of evolution of species (see Chapter 2). The specific pattern of constituents in species has, in fact, been used for systematic determination. This field of research, called biochemical systematics or chemotaxonomy, has gained increasing interest during the last decades. Natural product chemistry as defined today concerns mainly the formation, structure and properties of the secondary metabolites. There is no sharp division line between the primary 'biochemical' metabolites and the secondary metabolites. The common sugars glucose, fructose, mannose, the function and chemistry of which has been intensively studied by biochemists, are ranged in the first group, whereas closely related rare sugars, such as chalcose, streptose, mycaminose, discovered as constituents of antibiotics and investigated by organic chemists, are ranged as secondary metabolites. The essential amino acid proline is regarded as a primary metabolite but the likewise widely distributed 6-membered analogue pipecolic acid is classified as a secondary metabolite or an alkaloid (Fig. 1).

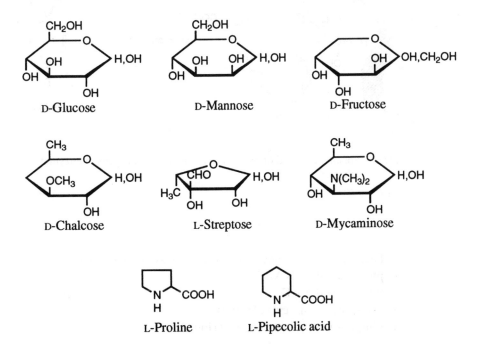

Fig. 1   Primary metabolites and structurally closely related secondary metabolites

1.5 Biochemical reactions and organic reaction mechanisms

Biochemical processes are basically the same as other organic reactions, familiar to us from organic chemistry both as far as thermodynamic and mechanistic considerations are concerned. The stage for the enzyme catalysed reactions is the three-dimensional asymmetric surface of a protein. As a result of the chiral environment the products become asymmetric. A special case is depicted in Fig. 2. The two methylene protons of ethanol are different with respect to their orientation towards the oxidant $NAD^{\oplus}$, and H_S is selectively removed. The two 'faces' of $NADH_S$ are different, since free rotation around the N–R bond is inhibited and from NADH one of the protons is specifically transferred in the reversed reduction. This has been demonstrated by specific labelling. The methylene carbon is said to be prochiral.

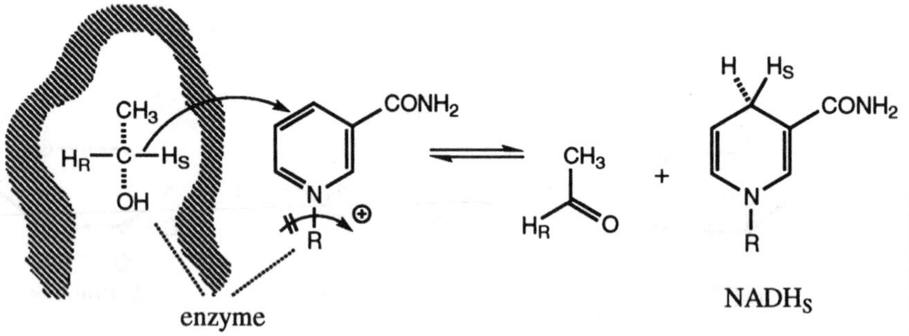

Fig. 2   Specific removal of one of the methylene hydrogens in enzymatic oxidation of ethanol

There are several electron transferring systems known:
1. pyridine based, $NAD^{\oplus}$
2. flavin based, FAD
3. iron-porphyrins, cytochromes
4. ubiquinones or coenzymes Q (Fig. 3)

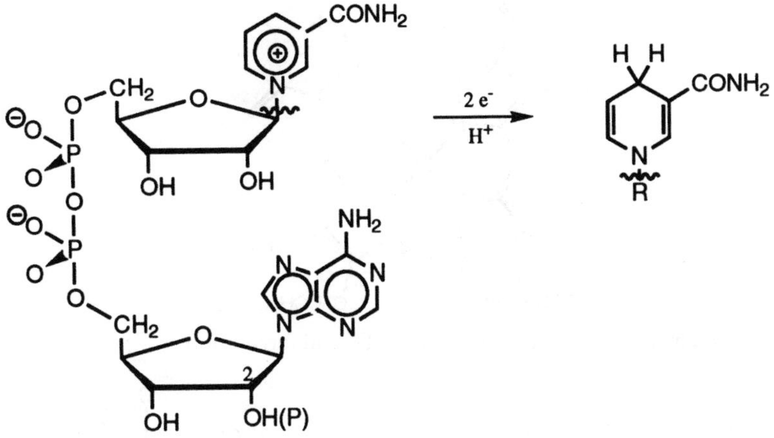

NAD⁺, Nicotinamide adenine dinucleotide
NADP⁺ = NAD⁺-2-phosphate

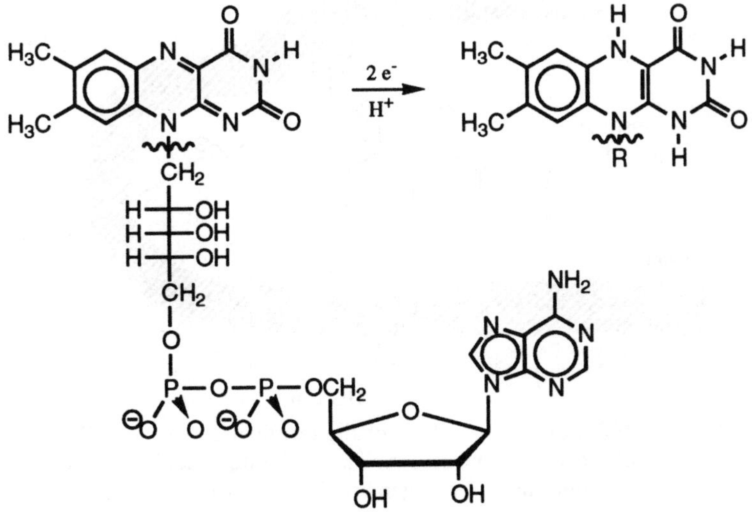

FAD, Flavin adenine dinucleotide, oxidized and reduced forms

Fig. 3 (continued on page 20)

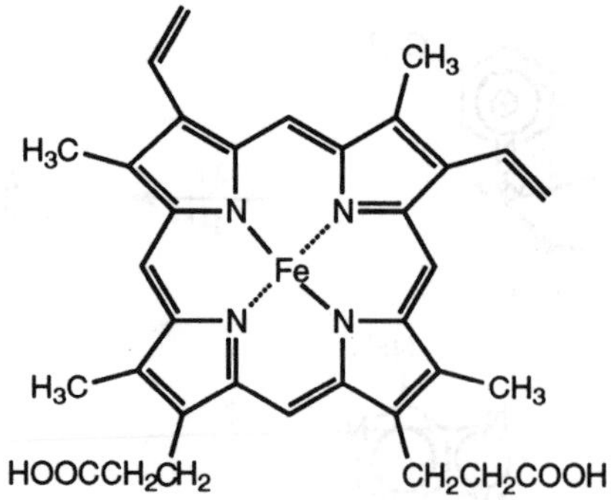

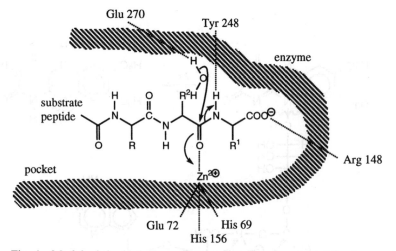

Haem, complex of protoporphyrin IX and ferrous ion

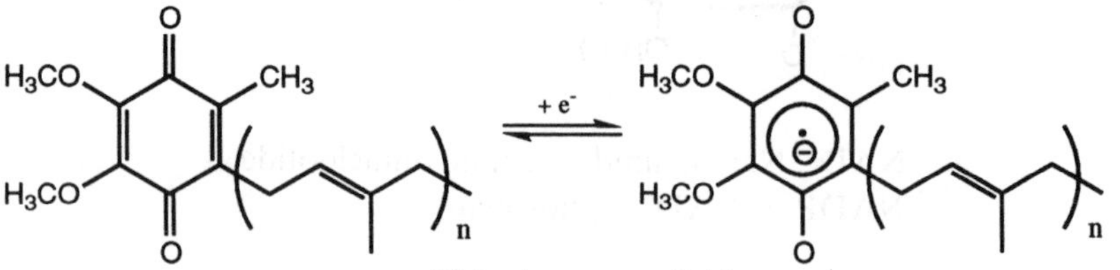

Ubiquinone, n = 6–12

Fig. 3   Structures of electron transferring cofactors

Fig. 4   Model of the hydrolysis of the terminal amino acid, R^1, of a peptide at the active site of carboxypeptidase A. The substrate is held in proper conformation in a pocket of the enzyme by hydrogen bonding. The carbonyl group is associated with the essential zinc ion of the enzyme and the terminal carboxyl and amino groups with arginine and tyrosine residues. Carboxypeptidase A contains a single polypeptide chain with 307 amino acids. The efficiency of the peptidase depends on its ability to carry out three steps simultaneously: (a) enhancement of the dipolar character of the carbonyl bond by its association as ligand to the zinc ion; (b) introduction of a molecule of water close to the carbonyl group; (c) protonation of the amino function thus facilitating the cleavage of the peptide bond

NAD$^\oplus$ or nicotinamide adenine dinucleotide is loosely bounded to the dehydrogenase protein and serves as a hydride transfer agent (Fig. 3). The whole sequence of events can also be formulated as two one-electron transfers intercepted by intermediate proton abstraction (section 5.8).

The enzyme catalysed reactions proceed much faster ($\leq 10^{10}$ times) than the corresponding uncatalysed *in vitro* reactions. Specific adsorption on the protein ensures that the components are brought together and are oriented favourably for the reaction as exemplified for the peptidase reaction (Fig. 4). Several enzymes are produced commercially and used as catalysts, e.g. peptidase, which are added to detergents in order to accelerate the hydrolysis of proteins in stains. Enzymatic reduction of ketones is used in organic synthesis for preparation of asymmetric alcohols.

We can distinguish several types of reactions, here systematized in the terminology of organic reaction mechanisms.

1. Carbon–carbon couplings of Claisen (1) and Michael (2) type guided by principles of polarization and ionization. In the living cell the Claisen reaction corresponds to enzyme promoted acylation with thioesters,

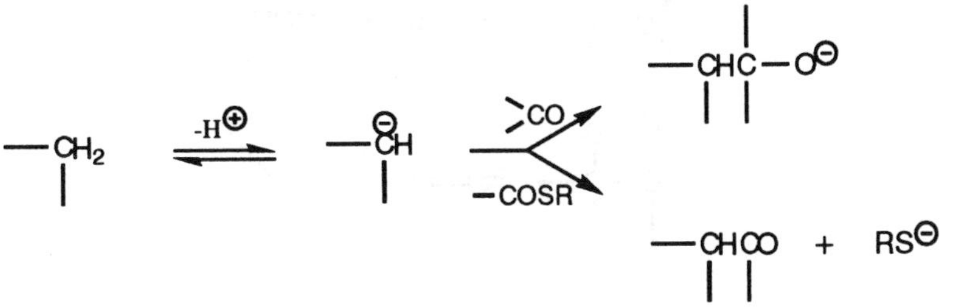

(1)

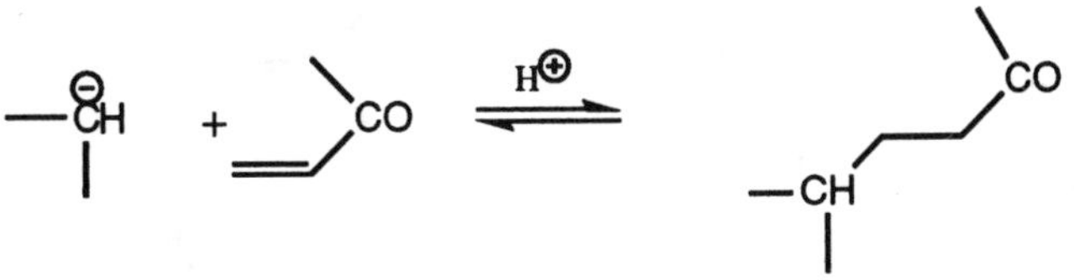

(2)

2. Electrophilic substitutions. *C*-, *N*- and *O*-alkylations with *S*-adenosyl methionine and phosphates,

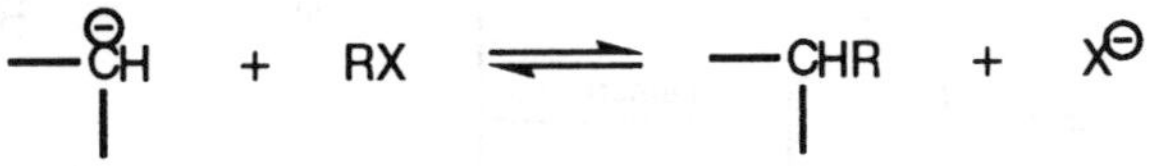

(3)

$$R^1OH \ + \ RX \ \rightleftharpoons \ R^1OR \ + \ HX$$

(4)

(5)

$$X = R^1R^2S^\oplus, \ H_2PO_4(PO)$$

3. Eliminations. In biological systems phosphate, water and ammonia are the leaving groups *par excellence* and B is a nucleophilic group of an enzyme -OH, -NH$_2$, or SH (hydroxy, amino, or mercapto groups),

$$\text{B:} \quad \rightleftharpoons \quad + \quad BH^\oplus \quad + \quad PO^\ominus \tag{6}$$

4. Oxidation, reduction, dehydrogenation. The exact mechanism is uncertain. They can be formulated either as hydride transfer, as one-electron transfer followed by hydrogen abstraction, or as a two-electron transfer. They are promoted by NAD$^\oplus$, FAD and cytochrome enzyme systems.

$$-\overset{\overset{\text{H}}{|}}{\underset{|}{\text{C}}}-\text{OH} \quad \rightleftharpoons \quad \text{C}=\text{O} \tag{7a}$$

$$\overset{\text{H}}{\underset{}{}}\text{C}=\text{O} \quad \rightleftharpoons \quad -\text{COOH} \tag{7b}$$

$$-\overset{\overset{\text{H}}{|}}{\underset{|}{\text{C}}}-\overset{\overset{\text{H}}{|}}{\underset{|}{\text{C}}}- \quad \rightleftharpoons \quad \text{C}=\text{C} \tag{7c}$$

$$-\overset{\overset{\text{H}}{|}}{\underset{|}{\text{C}}}-\text{NH}_2 \quad \rightleftharpoons \quad \text{C}=\text{NH} \tag{7d}$$

$$\tag{7e}$$

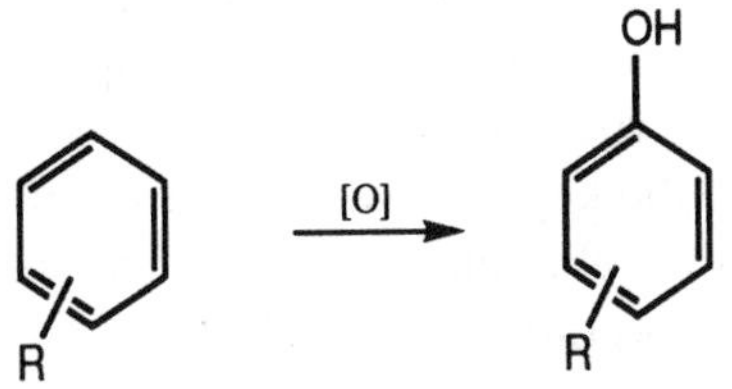

$$(7f)$$

$$\text{AlkH} \xrightarrow{\ [O]\ } \text{AlkOH} \qquad (7g)$$

An important type of oxidation is biological hydroxylation, i.e. insertion of molecular oxygen into the carbon-hydrogen bond to yield alcohols (7f, g). This is promoted by cytochromes or Fe-S clusters which serve as oxygen atom transferring agents. Reaction (7d) is catalysed by peroxidases and can be simulated *in vitro* by Fe^{3+} ions. Reaction (7e) represents phenol oxidation.

5. Wagner-Meerwein or carbonium ion rearrangements occur frequently, especially in terpene biosynthesis, and account for structures which formally do not seem to obey the 'isoprene rule'.

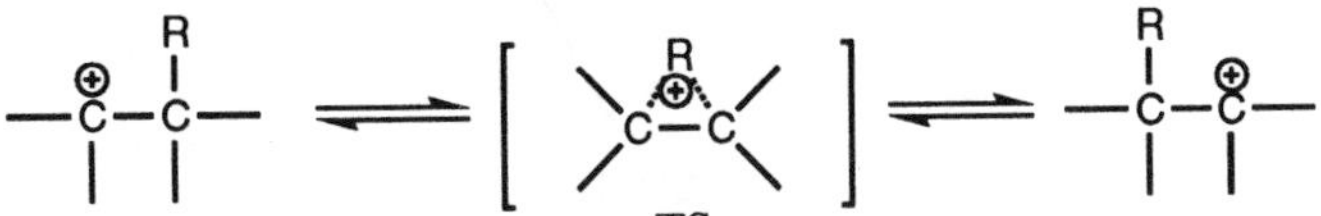

$$(8)$$

TS

R = H, alkyl, aryl

6. Carboxylations and decarboxylations.

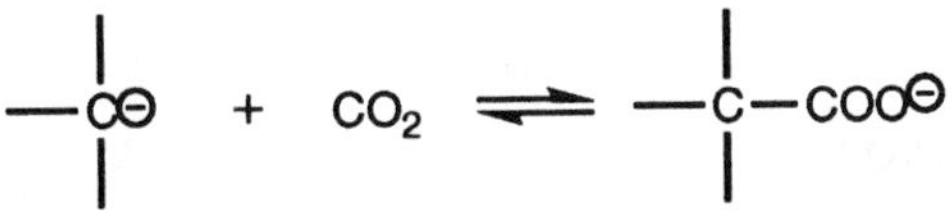

$$(9)$$

This "Grignard"-like reaction is an essential step, e.g. in fatty acid synthesis when acetate is converted into malonate. The reaction is reversible.

1.6 Principal pathways

The main streams of secondary metabolism are outlined in Fig. 5. One remarkable feature is that most metabolites originate from a very limited number of precursors. They are the link to primary metabolism in which they also play an important role. *Acetic acid* has a central position in the form of its thioester, acetyl CoA, or acetyl coenzyme A (Fig. 6). It is produced in the cell from pyruvic acid or fatty acids, or it may be directly formed from acetate and coenzyme A with ATP (Fig. 8) as mediator.

From acetic acid *mevalonic acid* is derived, from which, via 3,3-dimethyl-allyl pyrophosphate and the isomeric isopentenyl pyrophosphate—the iso-prene unit—the terpenoids are formed. From carbohydrates *shikimic acid* is derived which is the key to a wealth of aromatics. Finally, it is worth pointing out the importance of amino acids as precursors of the great variety of nitrogen containing compounds.

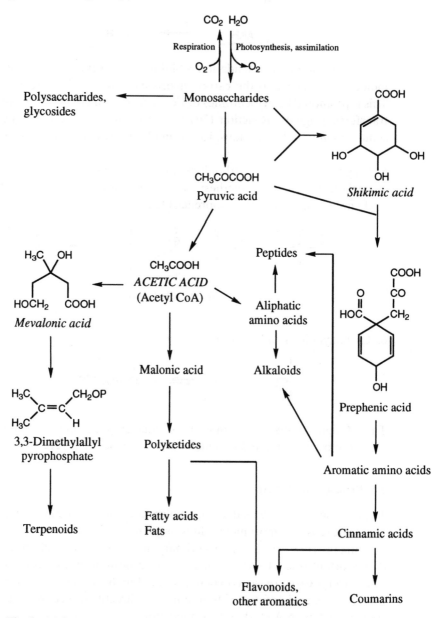

Fig. 5 Main streams of secondary metabolism

Several groups of metabolites have mixed biogenesis, i.e. an intermediate or metabolite from one principal pathway acts as a substrate for another metabolite from a different pathway. Thus, flavonoids are derived from a polyketide (three acetate units) and a cinnamic acid (shikimic acid). The indole alkaloids come from shikimate and a monoterpene (loganin).

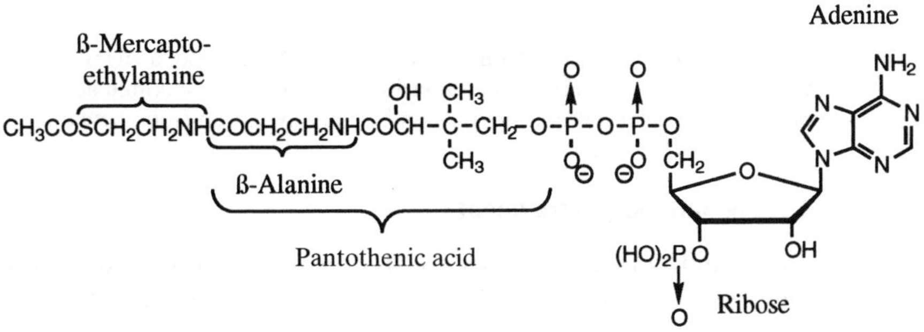

Fig. 6   Coenzyme A acetate

The microorganism *Streptomyces spheroides* has thrust all its synthetic skill into the formation of the complex antibiotic novobiocin (Fig 7). Several other cases of mixed metabolism will be discussed in the following chapters.

Natural products were classified in the past according to structure or biological origin: fatty acids, carbohydrates, terpenes, aromatics, mould metabolites, etc. The biosynthetic scheme groups the compounds according to the synthetic route employed by the cell. There is, of course, overlap between the two systems but as a result of our understanding of biosynthesis we now are able to survey, systematize, and correlate this very great number of diverse natural compounds in a pleasing and logical way. Even if the details of all the intriguing enzymatic processes are still obscure, the results can be rationalized by the principles of organic reaction mechanisms.

The metabolic processes of the three principal pathways: shikimic, polyketide, and mevalonic pathways are discussed in Chapter 4, 5 and 6, respectively. It has been found appropriate to give first a short outline of the constitution and properties of sugars as the final products of photosynthesis and the storehouse of organic matter and energy from which all the other componds derive (Chapter 3). Chapters 7–9 are devoted to the nitrogenous compounds: amino acids, alkaloids, pyrroles, nucleosides, etc.

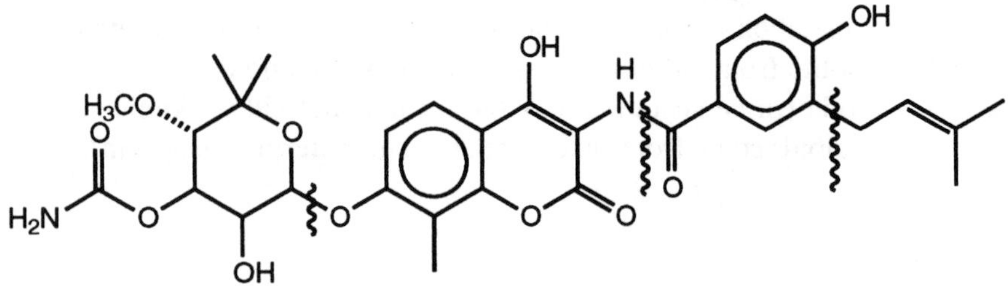

Fig. 7　Structure of the antibiotic novobiocin. A, noviose, a sugar from glucose + a $CONH_2$ group + 2 CH_3; B, a coumarin derivative from shikimic acid; C, *p*-hydroxy-benzoic acid from shikimic acid; D, an isopentenyl group from mevalonic acid

1.7 The one carbon fragment

C-, *N*- and *O*-methyl groups are frequently found in natural products, and they cannot be accounted for by any of the already mentioned pathways. Most biological methylation in animals, plants, and bacteria involves methionine, the methyl group of which is activated by *S*-adenosylation with ATP

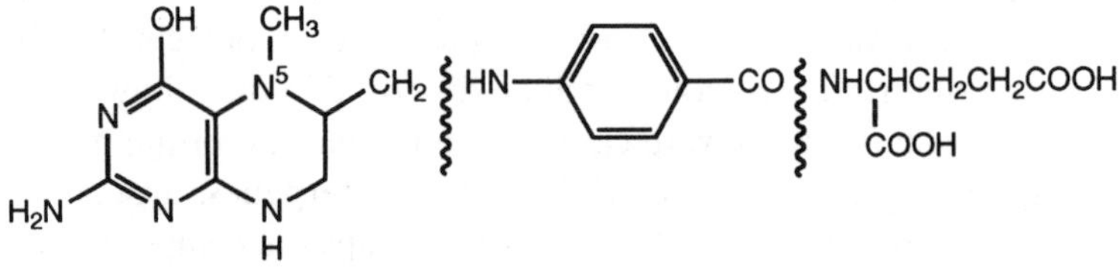

2-Amino-4-hydroxy-　　　　　　*p*-Aminobenzoic　　　Glutamic acid
6-methylaminopteridine　　　　　　acid

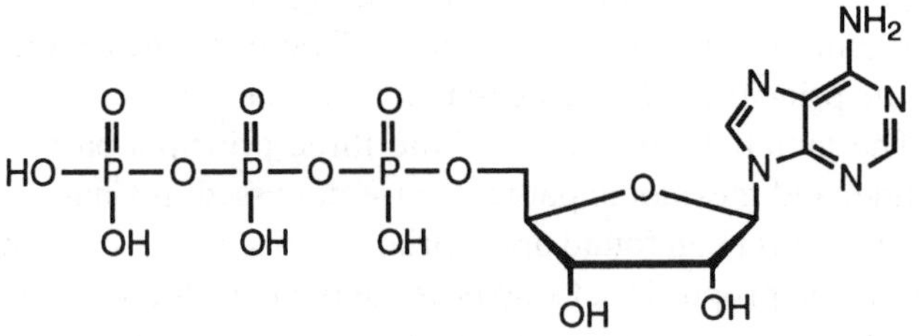

Fig. 8　N^5-Methyltetrahydrofolic acid, I, and adenosine triphosphate, ATP, II

$$
\text{Methionine} \xrightarrow[-3\,H_3PO_4]{ATP} \textit{S-Adenosyl methionine} \tag{10}
$$

(Fig. 8; equation 10). Methionine obtains in its turn its methyl from N^5-methyltetrahydrofolic acid (Fig 8). The ultimate source of the C^1 unit is formate, formaldehyde, serine, or glycine, all of which have been shown by isotopic labelling to give N-formyl derivatives. The formyl group is stepwise and reversibly reduced to a methyl group (equations 11, 12).

$$
\text{5,6,7,8-Tetrahydrofolic acid} \quad + \quad \text{Homocysteine} \xrightarrow{B_{12}} \text{Methionine} \tag{11}
$$

$$
\underset{}{>}N^5\!-\!CHO \underset{NAD^\oplus}{\overset{NADH}{\rightleftharpoons}} >N\!-\!CH_2OH \underset{NAD^\oplus}{\overset{NADH}{\rightleftharpoons}} >N\!-\!CH_3 \tag{12}
$$

1.8 Elucidation of metabolic sequences

For millions of years nature has refined her synthetic skill; no wonder that chemists are eager to learn how nature constructs and degrades its molecules. It has been a painstaking but exciting task to trace out the pathway and the mechanism of each step in a sequence. In the beginning accidental results contributed to the elucidation of intermediate steps. Knoop postulated as early as 1904 that degradation of fatty acids occurred via β-oxidation and presumably produced acetic acid, i.e. the chain is chopped down by two carbons at a time. ω-Phenyl-even-carbon fatty acids were metabolized and extracted in the urine of the test animal as phenylacetic acid, whereas ω-phenyl-odd-carbon fatty acids gave benzoic acid. Collie hypothesized at about the same time (1907) that the reversed reaction, the acetate condensation of the Claisen type, is the origin of many naturally occurring phenolics.

Normally the intermediates are present in a low steady state concentration but specific accumulation can occur in the organism during illness. Thus symptomatic for diabetes is that large amounts of acetoacetate appear in blood or urine, indicating that it is a degradation product from fatty acid oxidation. Nutritional studies on intact organisms established the beginning and the end of the metabolic sequences, but did not disclose so much about the intermediates. The requirements differ very much from one organism to another: *Escherichia coli* is a flexible bacterium able to use glucose, glycerol or acetates as sole carbon source, i.e. it adjusts itself to its environment and it has enzymatic systems active or latent, that are able to interconvert nutrients and manufacture essential metabolites. The lactic acid forming bacterium *Leuconostoc mesenteroides* and most vertebrates including man, on the other hand, require amino acids in their diet.

The breakthrough came with the work on genetically defective organisms, so-called mutants, and with isotopically labelled compounds. Accumulation of intermediates, diversion of intermediates to other products and altered nutritional requirements were studied extensively to map out metabolic sequences. Mutants can arise spontaneously or can be produced by the actions of chemicals or irradiation (X-rays, UV). Provided that the damage is not lethal, it often happens that only one gene is damaged, i.e. the cell is unable to produce one specific enzyme. Damage in the primary metabolism results in inhibited growth. It is more difficult to trace damage to the secondary metabolism since its products are not essential for the development of the organism. However, the underlying principles are the same. Suppose that we are studying the essential sequence A → E (Fig. 9). The normal organism can carry out all steps and A is required for growth. In mutant 1 D → E is blocked. D will accumulate and the growth can be sustained only by adding E to the nutrient. In mutant 2 C will accumulate and addition of D or E will restore normal activity of the organism. In mutant 3 B → C is blocked, but B will not necessarily accumulate, because B can be used as substrate for another enzyme and this special strain will produce a somewhat different spectrum of metabolites, F, G. The filtrate of 1 will restore the growth of 2 but 2 cannot support 1. The shikimic acid pathway was mainly elucidated by Davis and associates using a number of mutants of *E. coli*.

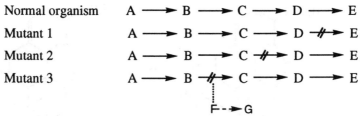

Fig. 9 Mutants with defective enzymes at different points along a metabolic chain

The most powerful method of establishing a metabolic sequence involves the use of isotopes. Table 1 shows the properties of some isotopes used as tracers. The fundamental investigations were run with the radioactive isotopes, ^3H, ^{14}C, and ^{32}P, but in recent years with the advent of pulse Fourier-transform ^{13}C NMR spectroscopy biosynthetic studies have witnessed a new explosive development. Radiotracing and mass spectrometry are the most sensitive methods. Radiotracing is preferably used in cases where quantitative data of incorporation are needed, but where information of location of label in the molecule is less important. The high sensitivity is of advantage if the efficiency of incorporation or conversion is low.

Table 1 Properties of isotopes used as tracers

Isotope	Relative natural abundance (%)	Radiation	Half-life	Spin
^2H	0.015		Stable	1
^3H		β	12.1 years	1/2
^{13}C	1.1		Stable	1/2
^{14}C		β	5700 years	
^{15}N	0.37		Stable	1/2
^{18}O	0.20		Stable	
^{32}P		β	14.3 days	

If the label can be excised from the molecule by some simple well-defined degradative method, the radiotracer technique is the method of choice. Suppose that in a metabolic sequence A–E (Fig. 9) we know the compounds fairly well but we do not know with certainty their exact order of formation, we could apply the radiotracer technique in a primitive and simple way. If radioactive carbon dioxide is administered, it will first appear in A and then in B etc. The measurement of the relative activity of the isolated compounds after a certain period of time would provide us with the correct order. This

Thebaine

Codeine

(13)

Morphine

has been done in the morphine series.[1] Radioactivity appeared first in thebaine, then in codeine and last in morphine. Thus, the methylation is not, as one possibly would guess, the final step of this sequence (13).

Generally, the technique of elucidating metabolic sequences involves feeding likely precursors to the organism and observing whether they are incorporated into the product in an intact state. In this way morphine was proved to be formed from two moles of 3-^{14}C-tyrosine, and demonstrated labelling at the anticipated positions. Today such an experiment could expediently be carried out by using the corresponding ^{13}C-labelled tyrosine. Observation of the enhanced intensities and comparison with the assigned ^{13}C NMR spectrum of morphine will show the mode of incorporation and it is not necessary to degrade the alkaloid (see section 8.3).

In order to locate exactly the labelled atoms in a metabolite it has to be degraded in an unambiguous way. The activity of each fragment, isolated as CO_2, acetic acid or other well-defined small organic molecule, is then measured. Controlled degradation of a big molecule is a very difficult and time-consuming undertaking. This methodology was skilfully demonstrated by Bloch, Lynen, Popjak and Cornforth in the biosynthesis of cholesterol as being ultimately derived from acetate (14).

Calvin used radiotracers for the elucidation of the mechanism of carbon dioxide fixation in green plants, i.e. photosynthesis. The green alga *Chlorella pyrenoidosa* was found to incorporate [14]CO_2 extremely rapidly. Within seconds the label was observed in glycine, alanine, aspartic acid, malic acid, citric acid, triose phosphates, hexose phosphates, glyceric acid-3-phosphate,

$$CH_3\text{---}\overset{*}{C}OOH \longrightarrow \qquad (14)$$

Acetic acid

Cholesterol skeleton

etc. By reducing the time of the exposure, glyceric acid-3-phosphate was recognized as the first stable intermediate in carbon dioxide fixation. This type of investigation called for a sensitive and rapid analytical method. The metabolism was quenched by hot alcohol and the metabolites were separated and identified by paper chromatography. The radioactive spots were localized by scanning the paper with a Geiger-Müller counter or by autoradiography. In the latter method the chromatogram is brought into close contact with a photographic film, sensitive to the short-range β-irradiation of ^{14}C. The active spots develop as black areas on the film and can be identified by the use of reference compounds. This same method is also used for studies of uptake, transport and accumulation of radioactive metabolites in plants. The whole plant is rapidly pressed in frozen condition and applied to a photographic film.

A mass spectrometric determination requires only a few micrograms of the sample, but in order to be able to locate the exact position of the isotope, the fragmentation pattern of the compound must be well understood. Mass spectrometry has the advantage that no previous degradation is necessary; the instrument takes care of that part. ^{2}H, ^{13}C, ^{15}N, and ^{18}O isotopes have been used successfully in structural determinations but scrambling has to be considered when ^{2}H is used.[2,3]

^{13}C NMR spectroscopy along with ^{1}H NMR is today the most efficient tool that the chemist possesses for structural determination of organic molecules. One routinely determines the intensity, shift, and ^{13}C-^{1}H multiplicities from which assignments can be made for each carbon in most low molecular compounds. More complex molecules with a number of very similar carbon atoms are exceptions, e.g. carotenes and long chain fatty acids. A label will be recognized by a change in intensity and its location in the molecule by its shift. Quantitative measurements are difficult because the intensities of ^{13}C peaks of different carbon atoms vary considerably due to differences in relaxation times, NOE effects and instrumental fluctuations during recordings. As a rule the intensity of ^{13}C peaks increases with the number of hydrogen atoms attached, i.e. the intensities increase in the order $CO < C^q < CH < CH_2 < CH_3$.

For a well resolved spectrum we need approximately 5 mg of the metabolite and an enrichment of label of ca. 50 per cent, i.e. for a 90 per cent enriched precursor one can accept a dilution to ca. 1:200 in the metabolic process. These experimental requirements set the limit of applicability of the method.[4] Very important is the use of doubly labelled $1,2$-$^{13}C_2$-acetate enabling detailed investigations of bonds formed and broken in the polyketide pathway and also of metabolites formed in the mevalonic acid pathway. The NMR spectrum of a metabolite will only show couplings between adjacent ^{13}C atoms derived from intact acetate units and inform us about cleavages, rearrangements, type of hybridization, coiling of the polyketide chain and number of individual chains in the molecule. Multiple labelled ^2H-^{13}C compounds allow studies of the integrity of the H-C bond in metabolic processes.

The prominence of spectroscopy for the elucidation of pathways is demonstrated in the biosynthesis of hyoscyamine **4** from the deadly nightshade *Datura stramonium*, Fig. 10. Feeding L-phenyl[1,3-$^{13}C_2$]alanine **1** to the plant gave **4** with $1,2$-^{13}C coupling in the NMR spectrum of the tropate moiety, showing that the carboxylate group had migrated from C^2 to C^3 in **1**.[5] By stereospecific hydrogen labelling of the C^3 methylene group it was demonstrated that the rearrangement occurred with retention of the configuration at C^3 and that H_S shifted intramolecularly from C^3 to C^2. A study with doubly radiolabelled (R,S)-DL-phenyl[1-^{14}C,2-^3H]lactic acid and a stable isotope study with (R,S)-D,L-phenyl[2-^{13}C,^2H]lactic acid **3a** showed that the ratios ^{14}C: ^3H (radioactivity measurements) and ^{13}C: ^2H (NMR measurements) respectively, remained approximately the same in the tropate moiety formed, which indicated that **3a**, **b** are true intermediates of the pathway and that they are incorporated intact. Later it was demonstrated that the tropine ester **3b**, littorine, was the immediate precursor for **4**. Quintuply labelled (R,S)phenyl[1,3-$^{13}C^2$]-lactyl-[N-methyl-$^2H_3$]tropine **3b** was predominantly incorporated intact into **4**, implying no requisite of a preceding hydrolysis. It is also of interest to establish which enantiomer is the preferred mutase substrate. Because of the equilibrium with phenyl pyruvic acid **2**, doubly labelled (R)-D- and (S)-L-[2-^{13}C,^2H]-**3a** were synthesized and submitted to the action of the mutase enzyme. A rapidly established equilibrium with **2** would result in complete loss of ^2H in **3**. (R)-**3a** gave **4** with an enriched peak at δ 64.05 corresponding to the -CH_2OH group and a stronger 1:1:1 triplet at δ 63.7 originating from the -CHD(OH) group with the expected deuterium induced α-shift and ^{13}C-^2H coupling, Fig. 11. The ^{13}C-^2H bond is thus intact in (R)-**3a** during the rearrangement catalysed by mutase. This triplet at δ 63.7 is missing in **4** produced from the (S)-**3a** enantiomer indicating deuterium wash out. Therefore (R)-D-phenyllactate **3a** is the enantiomer processed by mutase in hyoscyamine biosynthesis. GCMS analysis of **4** confirmed the NMR results and allowed a quantitative determination of the isotopic enrichments by studying the M+1 and M+2 peaks. This analysis also revealed the presence of the minor alkaloids

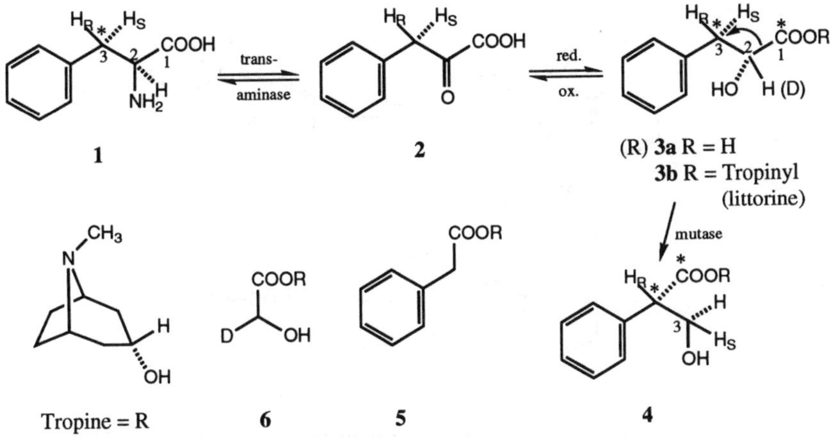

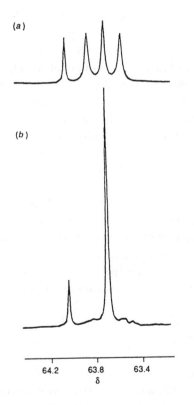

Fig. 10   Biosynthesis of hyoscyamine.

Fig. 11   The $^{13}C^3$-peak of **4**,-CHDOH, (a) showing deuterium induced α-shift, and $^{13}C-^2H$ coupling. (b) The triplet collapses to a singlet by ^2H decoupling. Reproduced with the permision of The Royal Society of Chemistry, UK.

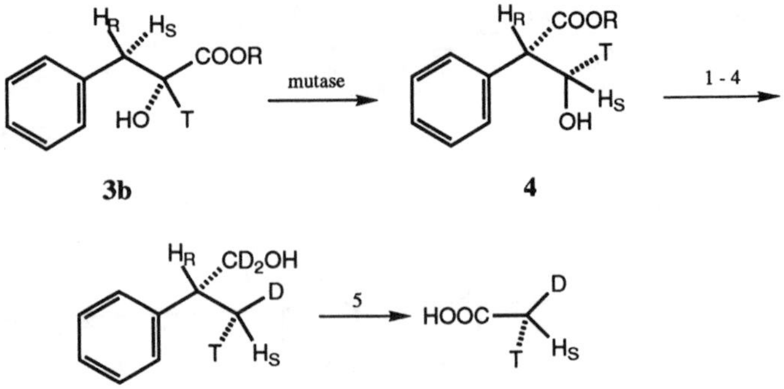

7 (R)-Acetic acid

Fig. 12   Transformation of **4** to chiral acetic acid. 1. OH⁻; 2. CH₂N₂; 3. Mesylation; 4. LiAlD₄; 5. KIO₄–KMnO₄.

5 and **6** not observable by ^{13}C NMR. The alkaloid **5** contains no deuterium and is formed by a retro-Claisen reaction of **4**. The mass spectrum of **6** obtained from (R)-**3a** showed a strong M + 2 peak indicating an intact ^{13}C-^{2}H bond. This alkaloid is presumably formed by C²-C³ scission of littorine **3b**.

Finally, the stereochemical course of the H$_S$ migration could be solved by application of the chiral methyl group assay (the malate synthase/fumarase procedure), (see section 6.2). Littorine **3b** was transformed by conventional methods to chiral acetic acid, **7**, which was analysed. It was found that acetic acid formed had the (R)-configuration (96 % ee) as depicted in Fig. 12, i.e. H$_S$ was introduced with inversion of the configuration at C³ by the mutase enzyme during hyoscyamine biosynthesis.[6]

^{2}H NMR spectroscopy has the advantages of being inexpensive and unlike ^{3}H it does not require special handling. Due to the short relaxation time of ^{2}H and the absence of NOE effect minimizing saturation, it integrates accurately. ^{2}H has a low natural abundance (0.016 per cent) and therefore incorporation of ^{2}H can be detected at a very high dilution: *ca.* 1:10 000 for 50 per cent enrichment. Poor resolution and spectral crowding are the main disadvantages.[7]

^{3}H NMR spectroscopy has gained interest in recent time. It is a sensitive method. ^{3}H has the nuclear spin 1/2 and gives proton-like spectra with narrow line widths and small NOE effects. The spectra can therefore be integrated and, as a result of its low natural abundance, enrichment of label can be detected at extremely low level.[7]

Applications of ^{1}H, ^{2}H, ^{3}H, and ^{13}C NMR spectroscopy is demonstrated in section 5.10 in conjunction with structural elucidations.[8]

A variety of feeding methods have been practised:

1. Uptake by the root from the aqueous medium. Microorganisms are usually grown in a culture medium containing the precursor.

2. Injection into the stem of the plant. Intramuscular or intravenous injection in animals.

3. Feeding through a wick with one end dipped into the medium containing the precursor and the other end inserted into a cut in the plant stem.

4. Oral administration through feeding of the animal or insect.

5. Application of a solution to the leaves of the plant.

6. Suspension of excised parts of an organism with intact metabolism in a solution of the precursor.

Administration of labelled precursors to living organisms often presents several kinds of problems. The precursor added to the nutrient may have difficulty in diffusing through the cell wall of the microorganism. If higher plants are studied, the precursor may not be absorbed by the roots, or if absorbed it may not be transported or may be degraded before it reaches the tissue where the metabolism occurs. A similar fate may also be shared by precursors applied by injection, spreading on leaves or via a wick through the stem. Higher concentrations of a normal plant constituent may have undesirable toxic effects. Dilution presents a serious problem since the added precursor has to compete with the normal pool of metabolites in the cell during the experiment. The cellular reactions are reversible: a constant degradation and rebuilding occur that causes a slow dispersion of the original label. Several of these problems have been circumvented in cell-free preparations with intact enzymatic systems. In fact, the fundamental steps of the mevalonic acid pathway were elucidated by this technique.

Tissue cultures (undifferentiated callus cells) seem to offer several advantages provided that the cells have retained the synthetic power of the parent organism to produce secondary metabolites, actually a serious restriction. Translocation and permeability problems are reduced and there are no seasonal variations. Callus cells have a simpler organization and they are easier to reproduce. It is necessary to work under aseptic conditions but this ensures, on the other hand, that the metabolites originate from the plant and not from symbiotic bacteria. By proper choice of problem the utilization of this technique will undoubtedly be extended in the future.[9]

1.9 Prebiotic Chemistry

In 1953 considerable interest and activity was stimulated among chemists when Urey and Miller demonstrated that electrical discharge in a primitive mixture of water, methane, ammonia and nitrogen gave rise to a complicated mixture of amino acids, e.g. glycine, D,L-alanine and D,L-aspartic acid in a yield of *ca.* 1 %.[10] This work started an array of simulation experiments, subjecting various compositions of simple compounds, such as methane, ethane, formaldehyde, carbon monoxide, carbon dioxide, hydrogen cyanide, acetylene, cyano acetylene, acetonitrile, ammonia, hydrogen sulphide, nitrogen, hydrogen etc. to discharge, heat, shock waves or UV irradiation. It was hypothesized that the atmosphere of the primordial Earth was composed of some of these compounds. Addition of hydrogen sulphide produced amino acids containing sulphur. Equations (15)–(18) describe a reasonable route to methionine.[11]

$$CH_4 + H_2O \xrightarrow{\text{Discharge}} CH_2=CHCHO \tag{15}$$

$$CH_4 + H_2S \xrightarrow{\text{Discharge}} CH_3SH \tag{16}$$

$$CH_3SH + CH_2=CHCHO \xrightarrow{\text{Addition}} CH_3SCH_2CH_2CHO \tag{17}$$

$$CH_3SCH_2CH_2CHO + HCN + NH_3 \xrightarrow{\text{Hydrolysis}} CH_3SCH_2CH_2CH(NH_2)COOH$$
$$\text{Methionine} \tag{18}$$

The occurrence of simple saturated hydrocarbons, alcohols, aldehydes, acids, amines, nitriles and sulphides in interstellar space and on other planets has been spectroscopically detected, adding credibility to the contention that compounds of this kind were present in the atmosphere of primitive Earth. It has been demonstrated that carbon vapour (C_1 or C_2) reacts with ammonia at $-196\,°C$ and low pressure to give a mixture of simple amino acids, a reaction that may play a role in the interstellar formation of amino acids.[12] Amino acids can be formed in low yielding processes in a more primitive system composed principally of water, nitrogen and carbonate. By the action of UV light and transition metal catalysis, water is split into hydrogen and oxygen, and nitrogen is reduced to ammonia. Fe^{2+} ions are reported to photoreduce carbon dioxide to formaldehyde, a reaction which can be considered a very primitive form of photosynthesis.[13] However, carbon dioxide as the sole source of carbon is considered unlikely and free oxygen is not assumed to be present in the atmosphere, which essentially had a reducing character. The temperature at the surface of the Earth had to be in the range of 0–100 °C and the pH about neutral, in view of the stability of the compounds formed and the volatility of the reactants. Adsorption processes on the surface of minerals and clays played an important role in catalysis and in assembling the reactants. The energy required by all these chemical processes was delivered by the sun.

The study of prebiotic chemistry took a new turn when it was found that a number of first stage compounds was formed by oligomerization of one single starting material, hydrocyanic acid, that offered a strikingly unified picture to the formation of amino acids, **6**, pyrimidines, **7,8**, purines, **4,5** and imidazoles, **3**, albeit sometimes in low yields. Furthermore, derivates of hydrocyanic acid, e.g. cyanamide (a tautomer of carbodiimide), cyanoguanidine, diaminomaleonitrile **2** and cyanate, can serve as condensing agents for the synthesis of second stage compounds, such as esters, phosphates, peptides, **9**,

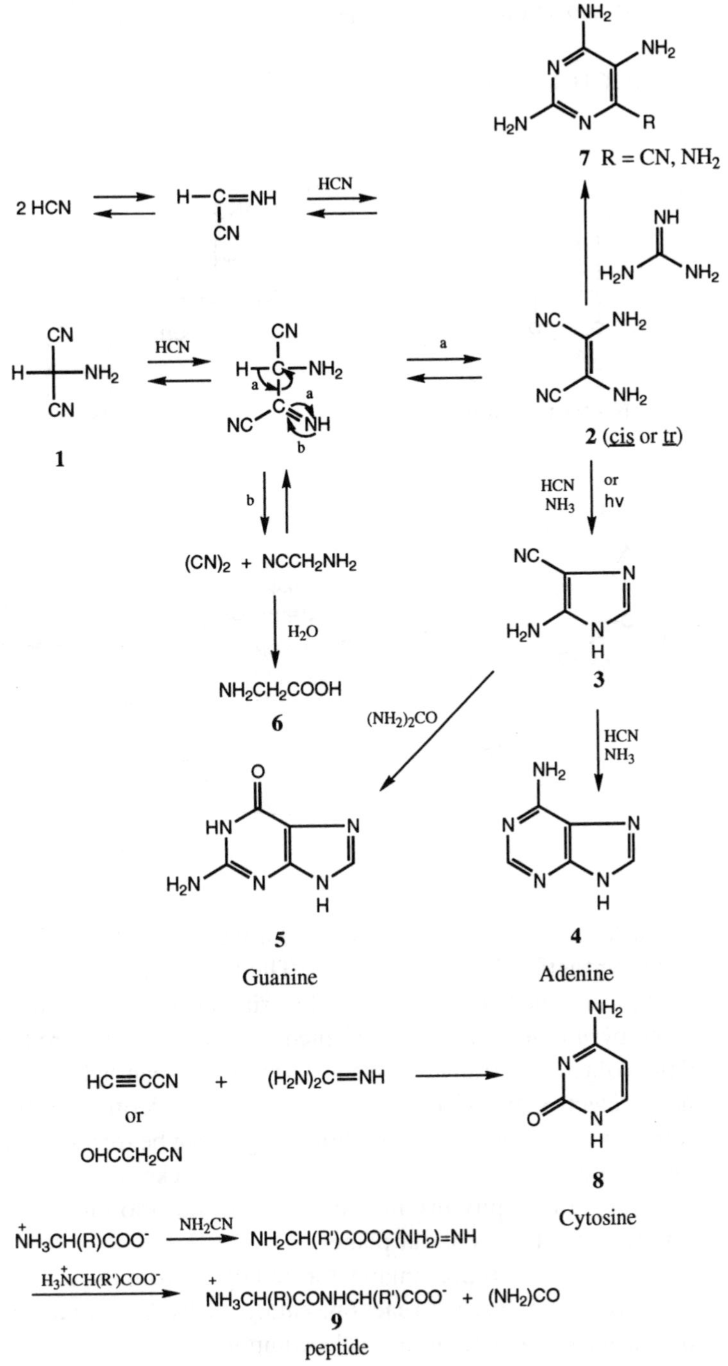

Fig. 13 Synthesis of first stage prebiotic compounds from hydrocyanic acid

$$\text{CH}_2=\text{O} + \text{POCH}_2\text{CHO} \rightleftharpoons \text{HOCH}_2\text{CH(OP)CHO} \overset{\text{POCH}_2\text{CHO}}{\rightleftharpoons}$$

10

$$\text{HOCH}_2\text{CH(OP)CH(OH)CH(OP)CHO}$$

11

(19)

(20)

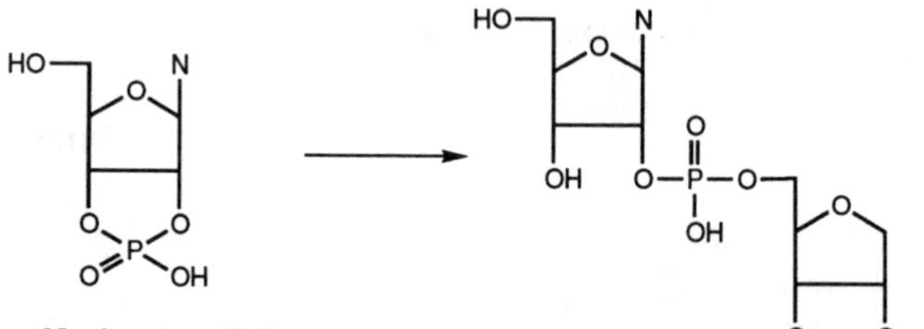

N = heterocyclic base

12   **12**

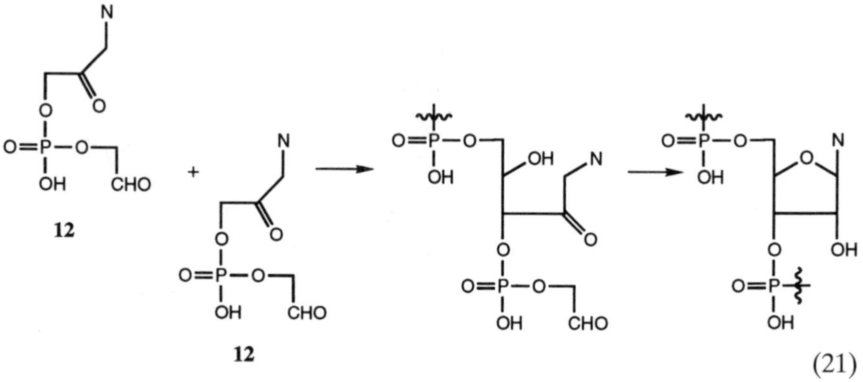

(21)

and nucleotides, [14,15] Fig. 13. Prebiotic formation of adenine **7**, is reminiscent of the biosynthetic scheme, section 9.2. The functionalized imidazole ring is formed first and the annulated pyrimidine ring by the reaction with formamidine.

The pyrimidine structure **7** is encountered in the heteroaromatic systems of the cofactors tetrahydrofolic acid, mediating methyl transfer, section 1.7, and the electron transferring flavin adenine dinucleotide, section 1.5. Furthermore, it is conceivable that porphobilinogen can be formed in a primitive process using glycine and succinate as building blocks, as depicted in section 8.3. Complexes of porphyrins and corrins with transition metal ions give us the first organic catalysts of our planet.

Prebiotic ribose is presumably formed by the formose reaction, also known as the Butlerow-Fischer reaction, catalysed by base and divalent ions, involving formaldehyde as the sole component. Views have been expressed against this reaction as the origin of ribose because of its complexity and the low yield of D,L-ribose, *ca.* 3 %. The formose reaction has been slightly modified by using glycolaldehyde phosphate **10** together with formaldehyde as starting materials, which diastereoselectively form racemic aldopentose 2,4-diphosphates **11** in reasonable yields (19).[16] Sugar phosphates as such occur naturally as essential components in contemporary biosynthetic schemes, thus strengthening the contention that ribose comes from aldolization of formaldehyde.

Considerable progress has thus been made in the simulation studies of the synthesis of monomeric building blocks, which can be seen as the first phase of prebiotic chemistry. Emphasis has shifted to studies of biological polymers, membrane chemistry, cellular structures, catalysis, replication mechanisms, mutability etc.

Molecules do not orient themselves to one other at random. They are influenced by weak dipolar forces, van der Waals forces, hydrogen bonding, steric hindance, and coiling, which limit their freedom and influence the product composition. Formation of nucleotides is a matter of self-organization. It requires first of all that ribose, the purine, inorganic phosphate and catalysts (Zn^{2+} Mg^{2+} Mn^{2+} solid phases) accidentally are together in the solution. Dry heating of such a mixture, simulating evaporation of a primordial pool, indeed gave mixtures of nucleotides, some of which have the suitable substitution pattern. One of the components was the $2',3'$-cyclic phosphate, which by heating, oligomerized and formed predominantly 2,5,-linked oligonucleotides (20).[17] Still more remarkable was the finding that the 2-methylimidazolide of guanosine-$5'$-phosphate regiospecifically gave the $3',5'$-linked oligomer containing up to 50 units. The reaction was carried out on a polycytidylic acid template, which forms Watson-Crick type hydrogen bonds with this substrate. When a non-Watson-Crick complementary nucleotide was reacted under the same conditions the oligomerization ran poorly. The reaction gives us insight into the mechanism of replication in the prebiotic era and also hints at the possibility that biotic catalysis does not necessarily have to be enacted on the protein, which previously was a biochemical dogma. This implies that RNAs could have been formed prior to proteins, and that they both have informational as well as catalytic functions, suggestions that have been substantiated by the discovery of the so-called ribozymes.[18]

It has been suggested that oligomerization occurred as a result of a Claisen condensation followed by an Amadori rearrangement of the bifunctional diphosphate **12** (21).[19] The ribose entity is thus formed in the final step, an idea that requires experimental substantiation. It is not clear how **12** is formed in the prebiotic brew.

The progress in peptide synthesis has been slower. Oligomers are formed by heating amino acids under suitable clay mineral-catalysed conditions. The more simple formation of polynucleotides has bearing on peptide synthesis. A strand of coded polynucleotide could conceivably attach the complementary amino acid to a growing peptide chain.

The mode of formation of lipids, essential to cell membranes, is poorly understood. Long chain fatty acids are virtually absent in the products from Miller's spark experiments.

Finally, it is justified to discuss another fundamental question still awaiting its solution, namely the origin of chirality. All the reactions described give rise to racemates or diastereomeric mixtures whereas biosynthesis produces stereospecific enantiomers. At one evolutionary stage chiral compounds began to be formed; but how and why was the L-form produced in preference to the D-form? Several theories have been advanced for the explanation of asymmetry, including selective destruction of one form by circularly polarized light. Forces originating from Earth's rotation in a magnetic field and chiral β-radiation from radioactive decay have been suggested to induce chirality but these effects seem to be marginal. If the reaction between the organic compounds occurs on the asymmetric surface of a crystalline mineral of layered structure the product will be asymmetric. This is equivalent to contemporary enzyme-catalysed reactions and it has also been experimentally confirmed. However, formation of just one enantiomer presupposes a local occurrence of chiefly one chiral form of the crystalline mineral catalyst. Such an event is conceivable since selective cystallization of D- and L-forms occurs in nature, as Pasteur observed in the tartaric acid case. Once one enantiomer was formed, asymmetry could be induced at adjacent centres of the reacting molecule.

At a symposium in 1983 it was concluded, in the absence of anything better, that we have to accept that chirality evolved through a chance event.

Bibliography

1 Parker, H. I., Blaschke, G. and Rapoport, H., *J. Am Chem. Soc.* **94** (1972) 1276.
2. Budzikiewicz, H., Djerassi, C. and Williams, D. H. *Mass Spectrometry of Organic Compounds*, Holden-Day, Inc., San Francisco, 1967.
3. Grostic, M. F. and Rinehart, Jr., K. L. in *Mass Spectrometry: Techniques and Applications*, Milne, G. W. A. (Ed.), J. Wiley, New York, 1971, p. 217.
4. Simpson, T. J., *Chem. Soc. Revs.* **4** (1975) 479.
5. Leete, E. *Can. J. Chem.* **65** (1987) 226.
6. (a) Chesters, N. C. J. E., O'Hagan, D. and Robins, R. J. *J. Chem. Soc. Chem. Commun.* **1995** 127. (b) Chesters, N. C. J. E., O'Hagan, D., Robins, R. J., Kastelle, A. and Floss, H. G. *J. Chem. Soc. Chem. Commun.* **1995** 129.
7. Garson, M. J. and Staunton, J. *Chem. Soc. Revs.* **8** (1979) 539.
8. Further reading on applied spectroscopy: (a) Evans, E. A., Warrell, D. C. Elvidge, J. A. and Jones, J. R. *Handbook of Tritium NMR Spectroscopy and Applications*, J. Wiley, New York, 1985 (b) Wehrli, F. W. and Nishida, T. *The Use of Carbon-13 Nuclear Magnetic Resonance Spectroscopy in Natural Products Chemistry*, in *Progr. Chem. Org. Nat. Prod.* **36** (1979) 1. (c) Levy, G. C. and Lichter, R. L. *Nitrogen-15 Nuclear Magnetic Resonance Spectroscopy*, J. Wiley, New York, 1979. (d) Bellamy, L. J. *The Infrared Spectra of Complex Molecules*, 3rd Ed. L. Methuen and Co LTD, London, 1975. (e) Gillam, A. E. and Stern, E. S. *An Introduction to*

Electronic Absorption Spectroscopy in Organic Chemistry, Arnold, London, 1954. (f) Williams, D. H. and Fleming, I. *Spectroscopic Methods in Organic Chemistry*, 4th Ed. Mc Graw-Hill, 1989.

9. Overton, K. H. and Picken, D. J., *Progr. Chem. Org. Nat. Prod.* **34** (1977) 249.
10. Miller, S. L. *Science* **117** (1953) 528; Miller, S. L., Urey, H. C. and Oró, J. *J. Mol. Evol.* **9** (1976) 59.
11. Wolman, Y., Haverland, W. J. and Miller, S. L. *Proc. Natl. Acad Sci. USA* **69** (1972) 809.
12. Shevlin, P. B., Mc Pherson, D. W. and Melius, P. *J. Am. Chem. Soc.* **103** (1981) 7007.
13. Borowska, Z. and Mauzerall, D. *Proc. Natl. Acad. Sci. USA* **85** (1988) 6577.
14. Oró, J. *Biochem. Biophys. Res. Commun.* **2** (1960) 407; *Nature* **191** (1961) 1193.
15. Ferris, J. P. and Hagan, Jr., W. J. *Tetrahedron Report* **162, 40** (1984) 1093; *Chem. Eng. News.* **62** (35) (1984) 22.
16. Eschenmoser, A. and Loewenthal, E. *Chem. Soc. Revs.* **1992** 1.
17. Joyce, G. F., Schwartz, A. W., Miller, S. L. and Orgel. L. E. *Proc. Natl. Acad. Sci. USA* **84** (1987) 4398.
18. Cech, T. R. *Scientific Amer.* **1990** 76.
19. Sutherland, J. D. and Weaver, G. W. *Tetrahedron Lett.* **1994** 9105.

Chapter 2

Chemical Ecology

2.1 Introduction. General. Definitions

When C. Darwin in 1859 published his epochal "On the origin of species by means of natural selection" the time was not ripe for relating evolution and coevolution to metabolic processes although nearly a century earliar C. von Linné had voiced his belief that related plants contained related compounds. At that time organic chemistry was in its infancy.

From studies of the environment's effect on species, or conversely the adaptation of species to the environment, evolved a branch of biology defined as ecology. In particular, anatomic and morphological characteristics and changes of the species as a result of external pressure were observed. Life must submit to certain environmental factors of physical nature: climatic conditions (temperature, seasonal changes, altitude, light, wind, flooding and drought), conditions of soil (pH, salinity, fertility, toxic metals), pressure from herbivores and predators, marine and terrestrial environments etc. Bewildering discoveries have been made of life forms existing at a depth of several thousand meters in total darkness, close to hot fissures on the Atlantic floor. There are other examples of acido- and thermophilic bacteria, which are adapted to the conditions in hot springs at 80°C. Once the physical environmental conditions were defined, a large number of organisms adapted to these conditions and began a Darwinian struggle for space and food in competition with other species. The ecosystem is in a state of dynamic equilibrium with a great many interspecies relationships. Apart from early observations in the 1700s by French biologists that fragrances of flowers attracted insects for the purpose of pollination in exchange for nectar, the idea that organisms exert chemical control over one another originated much later.[1,2] This development started in the 1920s leading in 1959 to the discovery and structural determination of the first sex attractant, bombykol, **1** which was isolated from the female silk moth *Bombyx mori.*[3]

It was found that emission of scent in extraordinarily low concentrations had a profound effect on the sexual behaviour of animals and that production of toxic compounds could protect species from being consumed by others. In short, natural products play a fundamental role in the coexistence of species in the ecosystem. A new branch of biology was born—chemical ecology, ecological chemistry, Darwinian chemistry—whichever you prefer. Chemical ecology is thus defined as the discipline describing those relationships/interactions between species, that can be related to an effect of naturally produced compounds. The destructive effects of man-made compounds in nature, i.e. pollution, are not discussed in this context. Chemical ecology is an interdisciplinary science under steady development and is dependent on close cooperation between chemists, botanists and zoologists. The time is gone when scientists viewed secondary metabolites as waste products; we still know too little about them. It is, indeed, difficult to believe that organisms should allocate so large a proportion of metabolic resources for purposes devoid of sense. We will never know, however, what effect a certain metabolite could have had at one stage of evolution. Its presence in a species may not reflect an immediate need in the present situation. The environment has changed considerably over the geological time scale while secondary metabolism in part has remained more or less stationary. In other words, metabolites which once possessed an ecological function may have lost it because predators have vanished. This would be like supporting an army against a nonexistent enemy.

The story of the steroid ecdysone **2** illustrates this point. Ecdysone stimulates the metamorphosis of larvae, i.e. an effect opposite to juvenile hormone activity. It is required in extremely low concentrations for normal insect metamorphosis; 25 mg of ecdysone was obtained by extraction of 500 kg of silkworms.[5] Shortly after its structure was elucidated the amazing discovery was made that the common yew, *Taxus baccata*, contained large quantities of β-ecdysone,[6] 25 mg of ecdysone was isolated from 25 g of dried root. Plant surveys show that it is a constituent in many plants, especially in ferns and gymnosperms which are relatively primitive plants. It may not be purely coincidental that these plants evolved at a time when insects were the dominating species of the animal kingdom. It is therefore conceivable that plant ecdysones had a more distinct ecological function, as an insect repellent at an earlier phase of evolution.

The principles of chemical ecology have given a natural explanation to traditional medicine, experienced and practised for generations by medicine men. Since secondary metabolites produced by one species influence the conditions for other species they could likewise affect health conditions in man. Indeed, 25 % of our medicines come from the plant kingdom. Only about 10 % of the Earth's plants have been properly identified and of those only a fraction has been characterized biochemically leaving an enormous unexplored potential. Tragically, this diversity is constantly dwindling. Approximately 50 % of the species are found in the vulnerable tropical region covering only 7 % of the land surface. This area is now being subjected to the worst, most senseless destruction in the history of man. If nothing is done 50% of all species existing today are predicted to vanish forever by the year 2050.

There is a very practical and natural division between primary metabolism
—the life processes common to all organisms—and secondary metabolism,
specific to classes of organisms or families or even to individuals. The presence of secondary metabolites is not necessarily critical to the survival of species but their presence could be advantageous. There is a strong correlation
between secondary metabolism and the Linnean classification, indicating an
evolutionary connection. These relationships constitute another branch of biology, chemotaxonomy.[4] Groups of secondary metabolites are used with success as markers for botanical classification. Organisms evolve through mutations affecting genes, which in turn affect protein structures and enzyme systems. Mutations affecting basic primary metabolism are harmful whereas
changes in secondary metabolism make the organism more or less competitive, "fit for survival and reproduction".

Definitions. It is appropriate to define here a few frequently used terms. The
chemical interactions between individuals or populations of different species
in the widest sense are defined as allelopathy and the active substances, allelochemicals. Allelopathy is by some researchers restricted to higher plant—
higher plant interactions.[7] A further division of allelochemicals has been
made such that agents giving advantage to the organism producing them are
called allomones and agents of advantage to the organism receiving them are
called kairomones. This division may seem meaningless since many agents
function both as allomones and kairomones or act as hormones, pheromones
or food components. Compounds functioning as messengers between members of the same species are defined as pheromones. Sex pheromones and
other pheromones for aggregation, alarm, defense, territory marking, organization of communities, recognition and food localization have been identified, which chemically control or induce behaviour in animals.[2]

In this chapter examples demonstrating typical animal-animal, plant-plant,
plant-animal and microorganism-plant relationships are presented.[8]

1

2 R = H,   α-Ecdysone
  R = OH, β-Ecdysone

2.2 Adaptation to the environment

Temperature. The so-called C_3-plants of the temperate zone directly fix CO_2 into 3-phosphoglycerate via ribulose diphosphate, whereas the C_4-plants, largely from the warmer tropical zone have a more complex pathway. They first fix CO_2 by phosphoenolpyruvate, PEP, into oxaloacetic acid; this is then reduced to malate from which CO_2 is oxidatively regenerated in another compartment (malate dehydrogenase) and fed into the Calvin cycle; the pyruvate is then recycled. These compartmental structures correlate with anatomical characteristics of the C_4-plants. This modified photosynthetic pathway, the Hatch–Slack pathway, very effectively fixes CO_2 at a low partial pressure— but at the expense of extra ATP.[9] C_4-plants, e.g. sugar cane, synthesize glucose more effectively per unit leaf area and grow faster than C_3-plants. The ability to synthesize glucose via the Hatch–Slack pathway is not restricted to grasses and is present in several large families, including *Euphorbiaceae* and *Asteraceae*.

As already mentioned some thermophilic microorganisms are adapted to the extreme conditions of hot springs with temperatures as high as 80 °C. This must necessarily pose difficult problems to the organism, e.g. denaturation of proteins and enzymes, and integrity of lipid membranes and cell walls. Plants are generally sensitive even to moderate heat shocks up to 40–50 °C, but adjust to these temperatures after preincubation at slightly lower temperatures. One response of the cells is the rapid synthesis of novel heat-resistant proteins, with a M.W. of *ca.* 70 000–90 000 Dalton.[10]

At the other end of the temperature scale we have metabolic adjustments to freezing. It has been found that insects and plants tolerating low temperatures synthesize glycol as an antifreeze. They exploited this method long before we applied it to our automobile engines. Increased concentrations of carbohydrates such as glucose, fructose, sucrose and mannitol, also make the plants more frost-resistant.[11] Some macromolecules preventing the crystallization of water, glycoproteins of M.W. 2000–40 000 Dalton, have been isolated from the blood of fish. Changes in the protein content of frost-resistant plants have been noted.[12] It has also been suggested that the ratio of unsaturated to saturated acids is higher in arctic plants. This has not been verified.[12a]

Drought. Biochemical changes have been observed in drought-stressed plants. The amino acid proline is much more abundant in plants adapted to drought.[13] This could be attributed to a special osmotic effect of the amino acid. Proline also plays a biochemical role in adaptation of plants to salinity. Zwitterionic quaternary ammonium compounds have a similar effect, acting as osmoprotectives in *Plumbaginaceae* by accumulating in the cytoplasm.[13a]

The opening and closing of stomata are hormonally controlled by the level of abscisic acid **3** and a few related sesquiterpenes.[14] It was noted that the concentration of **3** increased considerably during wilting of wheat plants but returned to normal with sufficient water supply.

Flooding. The accessibility of oxygen is dramatically reduced during flooding, a circumstance reflected metabolically. Many plants grow in areas, which are flooded regularly and must therefore possess a respiratory machinery able to cope with shifting aerobic/anaerobic conditions. The biochemical background of adaptation is not clear. It has been hypothesized that in intolerant plants lack of oxygen leads to toxic levels of accumulated ethanol/acetaldehyde or of lactic acid derived from glycolysis.

Salinity. Plants respond to salinity by accumulating proline,[15] e.g. in *Armeria maritima*. An osmotic mechanism related to the response in drought-resistant plants has been proposed. Other plants, e.g. *Chloris gayana*, react on NaCl stress by accumulating glycinebetaine **4** or choline **5** as protective agents.[16] Organic sulphates are widespread in marine plants as exemplified by sulphoniumbetaine **6** isolated from the green alga *Ulva lactuca*[17] and the flavone **7** isolated from *Zostera marina*.[17a] Presumably their formation is beneficial for the solubilization and inactivation of phenolics.

Algae are reported to be rich in polyols, mannitol, sorbitol and glycerol. These compounds have both an antifreeze and osmotic function. The green alga *Dunaliella parva* is reported to accumulate as much as 70 % glycol by dry weight.[18]

Selenium and heavy metal toxicity.[19] Selenium is a trace element that in higher concentrations is fatal to animals but less so to plants. It occurs in high levels in the soil in certain areas in Australia, Asia and North America and is taken up in damaging doses by grazing animals via *Astralagus* spp. which accumulate it from the soil. Selenium can substitute sulphur in amino acids and when they are incorporated into proteins the selenium analogues inactivate essential enzyme or peptide systems. An interesting question arises why selenium is of no harm to the adapted plants themselves. It is conceivable that these plants are able to distinguish between sulphur and selenium and incorporate normal S-amino acids into enzymatic proteins.

$(CH_3)_3N^+CH_2COO^-$ $(CH_3)_3N^+ - CH_2CH_2OH$

4   **5**

$(CH_3)_2S^+ CH_2 CH_2 - COO^-$

6

3

7

Certain plants have the amazing ability to thrive in soil containing toxic heavy metals, such as zinc, lead, copper, tin, silver, nickel and gold. The deactivation mechanism is not clear, but it presumably involves several components: 1. complexing with peptides, called phytochelatins rich in cysteine,[20] 2. absorption at the cell wall, 3. compartmentation and exclusion. Certain plants exhibit such a preference for some heavy metal ions that they serve as indicators of the metal content of the soil. Thus *Eriogonum ovalifolium* indicates the presence of silver in the soil in Montana, U.S.A. and *Phacelia sericea* is found in the gold mining districts of South Africa.

2.3 Chemistry of pollination

Nectar and pollen. Pollination, i.e. the fertilization step in plants, is performed by insects and in some cases birds, bats and mice, quite unwittingly in their search for nectar and pollen, which beside its primary role as carrier of the male gamete serves as a nutrient. The main constituents of nectar are the carbohydrates glucose, fructose, and sucrose (10–70 %).[21] Minor amounts of raffinose (α-galactosyl-6^G-sucrose), trehalose, (α-glucosyl-α-glucose) and maltose are occasionally found. Common amino acids and lipids are present at approximately 0.05 %. Pollen is a source of protein (15–30 %), carbohydrates (10 %) and fats (5 %).[22] These values vary considerably according to species. Many species of butterflies rely primarily on nectar as their protein source and typical "butterfly flowers" produce nectar, consistently rich in amino acids.[23] These preferences demonstrate chemically based evolutionary mechanisms.

Many plants form nutritious exudates attracting insects. Certain *Acacia* spp. are associated with fierce ants protecting the host from harmful animal visitors. The beneficial animal-plant relationship was experimentally demonstrated by removal of the ant colony, which led to deterioration of the host plant.[24] From an ecological point of view it is most interesting to note that non-ant *Acacia* spp. instead produce toxic cyanoglycosides as herbivore repellents.

Fragrances. Flowering plants emit volatile low molecule weight compounds that attract pollinators from a distance. We can vividly imagine the chaos of chemical messages prevailing in a flourishing meadow on a bright summer day. Some insects are specialists and are stimulated by a few attractants, others are generalists visiting many flowers. The components of the fragrance emitted are absorbed at active sites on the antennae that fit the structures of the compounds like a lock and key, thereby exciting the insect. At closer range the insect is guided by a combination of olfactory and visual signals and finally also by tactile signals leading to the sexual organs and the rewarding nectar. Several flowering plants rudely deceive their pollinators by emitting scent without producing nectar. Insects and bats that are active at night rely entirely on olfactory stimulus. Plants pollinated by these animals are strong smelling and tend to intensify the emission of fragrance in the evening.

The correct absolute stereostructure of a compound and the relative ratio of the compounds in the blend are important for optimum activity of the fragrance. This means that with a limited number of volatiles the selective power of the fragrance is unlimited. An important factor for the activity of a blend is that one component may synergistically enhance the activity of another.

Our human perception of fragrances as pleasant or offensive appears to have little relevance among insects, which visit the most toxic, foul-smelling plants. Several plants have an aminoid smell which lures carrion and dung beetles by chemical mimicry. Major constituents are fishy smelling simple monoamines such as methyl-, ethyl-, propyl-, and butylamines in *Arum* spp., and 3-methylindole (skatole).

Biosynthetically, there are three types of pollination stimulants, 1. fatty acid derivatives which are products of the polyketide pathway. 2, aromatics from the shikimic acid pathway, and 3. mono- and sesquiterpenes.

8. R = CH$_2$OH	**12.** R = OCH$_3$	**15.** R = COOCH$_3$
9. R = CHO	**13.** R = NO$_2$	**16.** R = CHO
10. R = COOCH$_3$	**14.** R = NH$_2$	
11. R = CH$_2$CHO		

17. R = CH$_2$OH, Geraniol
18. R = CH2OCOCH$_3$, Geranyl acetate
19. R = CHO, Geranial
20. R = COOCH$_3$, Methyl geranate

21. 3-Carene

22. β-Ionone
23. CH$_3$(CH$_2$)$_{13}$CH$_3$
24. CH$_3$(CH$_2$)$_7$OOCCH$_3$

25. (–)-Cadinene

The aromatics **8–16** occur in *Hypericum* spp.[25] and the monoterpenes **17–21** in *Actaea* spp.[26] β-Ionone **22** occurs in violets and the fatty acid derivatives **23** and **24** occur in *Magnolia* and *Cypripedium* fragrance, respectively.

Fragrances sometimes contain compounds with dual effects in that they act as feeding attractants and as sex pheromones. Studies in orchid pollination have revealed remarkable bee/orchid coevolutionary relationships.[27,28] Fragrance from *Ophrys* flowers, which attract specialized *Andrena* male bees, contain compounds also identified in the odour glands of females. The flower's texture, shape and colour resemble female bees which induces male mating behaviour on the labellum, and during these movements pollination is accomplished. The fragrance contains several monoterpenes, benzenoids and both (+)- and (–)-γ-cadinene **25**, but interestingly only the (–)-form is active. It has sometimes been observed that individual components have a comparatively low activity, whereas the composite mixture is highly active, a phenomenon termed synergism. In the present case a combination of tactile, olfactory and visual stimuli induces efficient pollination. There are both advantages and disadvantages to high coevolutionary specialization. The plant will not be attacked by many herbivores and the specialist is alone in the pantry. On the other hand the plant is vulnerable in its dependence on one or a few insects, whose presence is vital for its reproduction.

Another remarkable chemically controlled parasitism has been observed in the *Andrena/Nomada* relationship. *Nomada* bees parasitize the nests of *Andrena* bees where the *Nomada* female lays her eggs in cells prepared by the host. Despite the harm caused by the *Nomada* larva an encounter between *Andrena* and *Nomada* females does not lead to hostilities, presumably because both smell of farnesyl hexanoate and geranyl octanoate. The *Andrena* female produces the compounds herself whereas the *Nomada* female receives them from the male during mating.[29]

Colours. At short distance the insects respond to the colours of flowers.[29a] Butterflies generally prefer vivid, bright colours and bees white, yellow and blue colours. Animals with poor colour sense such as bats, moths, flies, and beetles pollinate flowers with drab, green and whitish colours. Birds prefer vivid scarlet blossoms.

Three groups of compounds are responsible for the flower pigments, 1. flavonoids, **26–36**, 2. carotenes, and 3. quinones. The flavonoids represent the most important group. They are phenolic compounds and occur in nature mostly as glycosides. The highly conjugated anthocyanidin ions **31–35** are represented over the whole colour scale from orange (*Antirrhinum*), red (*Rosa*), blue (*Centaurea*) to purple black (*Tulipa*). Increasing number of hydroxyl groups, higher pH and complexation with metal ions (Mg, Al) give a red shift. *O*-Methylation causes a slight hypsochromic shift.

Flavanone
26. 7,3',4'-Trihydroxyflavanone,
    ca. 280 nm

Chalcone
27. Butein, 382 nm, (*Butea, Acacia*)

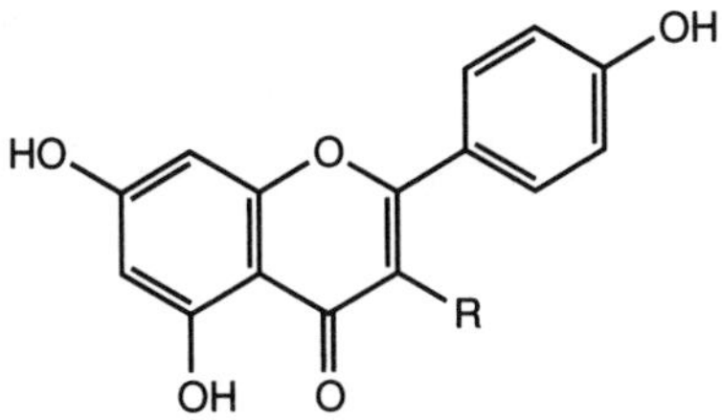

Flavone

28. R = H, Apigenin, 335 nm,
29. R = OH, Quercetin, 374 nm
   widespread

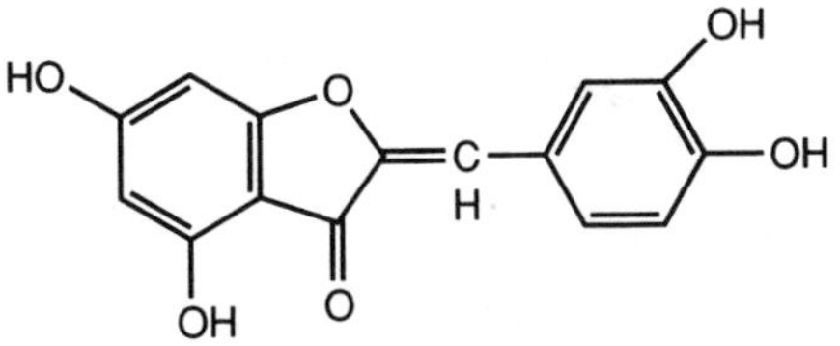

Aurone

30. Aureusidin, 399 nm (*Antirrhinum*)

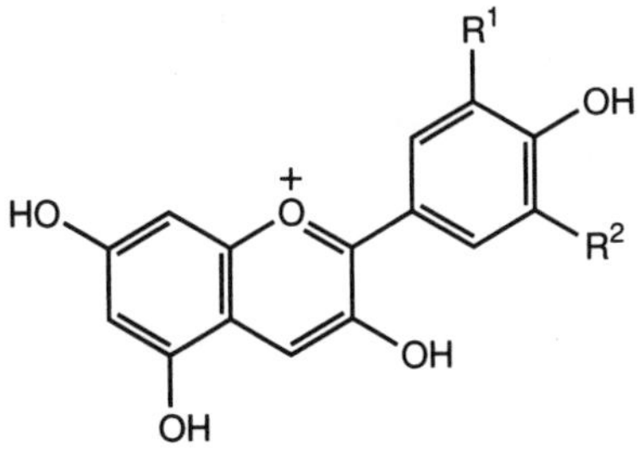

Anthocyanidin

31. R^1, R^2 = H, Pelargonidin, 510 nm
32. R^1 = H, R^2 = OH, Cyanidin, 525 nm
33. R^1, R^2 = OH, Delphinidin, 535 nm
34. R^1 = H, R^2 = OCH$_3$, Peonidin, 523 nm
35. R^1, R^2 = OCH$_3$, Malvidin, 532 nm.

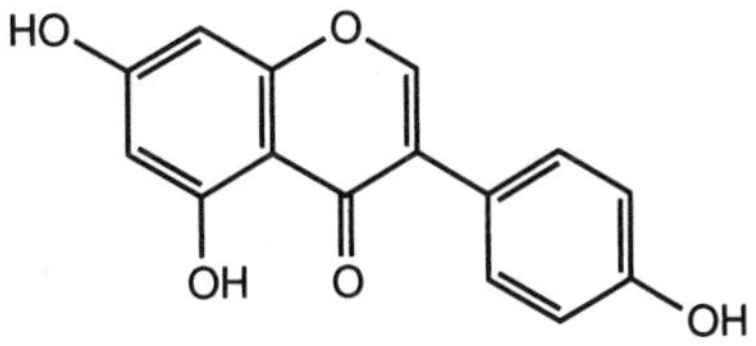

Isoflavone

36. Genistein (*Genista*), 328 nm

The polyconjugated ubiquitous carotenes **37–39** give yellow-orange colours to flowers such as *Tulipa, Narcissus* and *Calendula.*

In a few cases quinones have been identified as flower pigments,[30] **40–42.** Quinones are yellow to black coloured compounds, widespread in higher plants, lichens, insects and microorganisms. In higher plants they principally occur in the heartwood, bark and roots.

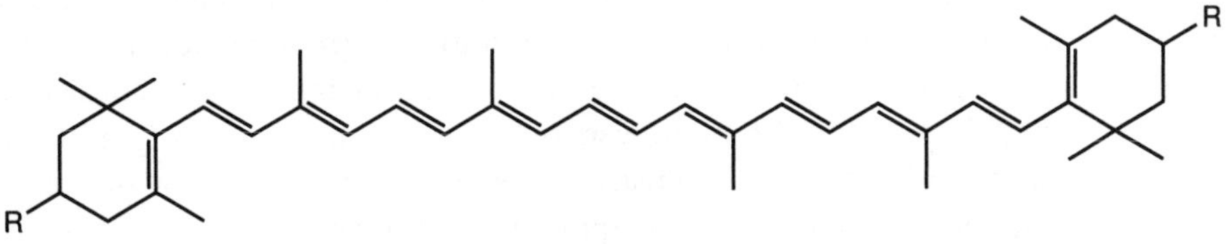

37. R = H, β-Carotene, 484 nm (*Narcissus*)
38. R = OH, Zeaxanthin, 483 nm (*Tulipa*)

39. Lycopene, 506 nm (*Calendula*)

40. Plumbagin (*Plumbago*)

41. Primin (*Primula sp.*)

42. Hypericin (*Hypericum*)

Most flavanones and flavones, e.g. **26, 28, 29, 36**, are colourless compounds absorbing light in the long wave UV region but are nevertheless important co-pigments giving different shades to the visible pigments.[31] Bees are sensitive to UV light, and the flavones have been found to serve as honey guides directing the insect to the nectar when it has lighted on the flower, as has been observed in *Rudbeckia hirta*.[32]

There is a very characteristic distribution of flavonoids in bird- and bee-pollinated blossoms of the same family (*Polemoniaceae*). It was found that bird- pollinated flowers of the *Polemoniaceae* family were consistingly richer in pelargonidin **31** than flowers pollinated by bees, which typically contained delphinidin **33**.[33] This illustrates evolutionary adaptation of colour to pollinators, which is not an exceptional phenomenon.

Flavonoids are favoured taxonomic markers for several reasons. They are structurally diverse, ubiquitous, metabolically adjustable and easily identified by chromatography in microscale.

2.4 Plant-animal relationships[34]

Insect feeding stimulants and repellents. The chemistry of plant defense. The general nutritional value of most plants varies comparatively little, so this cannot explain feeding preferences. Herbivores have a taste for and a dependence on certain plant constituents and are as fastidious as man, who likes peas, cabbage or salad but detests grass, which will be regurgitated. The panda feeds exclusively on *Bambusa* and the Australian koala bear on *Eucalyptus*. The cabbage fly larva *Pieris brassicae* has developed such a taste for cabbage that it prefers starvation to death to consuming any other food. The larva can be fooled to feed on other plants by infiltrating them with the particular feeding attractant sinigrin, a glycosinolate, **43**, that is a repellent and poison for most other insects. Sinigrin undergoes enzymatic hydrolysis and rearranges to acrid allyl isothiocyanate, **44.**

The *Pieris* butterfly must consequently find the right host for oviposition so that the larvae after hatching can develop normally.[35] It is of advantage for the *Pieris* spp. that they have their food source all to themselves and the cabbage plant is well protected against attack from other insects by the deterrent. A disadvantage for the specialized insect is its dependence on just one or a few plant species, a circumstance that makes reproduction risky in case host plants are not accessible. Actually, this state of things demonstrates the importance of crop rotation practised in the old days to fight pests. Crops as conspicuous plants have not evolved resistance to insect attacks and therefore artificial chemical defense must be applied to such new monocultures. Most conspicuously the flora has not succumbed to the fauna. There are several reasons for this. Abundance of green plants will certainly give rise to a wealth of herbivores that will cause overeating and gradually food deficiency, again leading to reduction of the herbivorous community. Before this hypothetical event will happen carnivorous predators will also increase in number thus reducing the number of herbivores. Furthermore, the plant chemical defense systems will intervene, making food unpalatable, distasteful or toxic. Mutations synthesizing the most efficient repellents will have the best chances to survive under the circumstances. Nature will consequently be in an ever-changing equilibrium. There is only one species, the human being, who on a global scale has been a serious threat to the natural balance and to mankind by unchecked overexploitations, destruction of rainforests in the Amazon and in Africa and by greed. Overexploitation of the mangrove swamps increases the risk of catastrophic floodings and causes serious destruction to the important coastal and offshore fauna and flora in these regions. In a slightly longer geological perspective it makes no difference how we treat nature because about 20000 years from now we will be in a new glacial period covering our cultural centers with a thick layer of ice.

Basically, plants defend themselves chemically according to two different strategies. Apparent plants such as oak or fir forming large forests easy to locate have been forced to arm themselves with a solid defense system consisting of large quantities of growth inhibitory tannins, polyphenols bound to glucose, **45**, or resins of terpenoid origin. Tannins form complexes with proteins making them less digestible. The tannin concentration of oak leaves increases with age. Opening buds and developing leaves are most vulnerable to attack by the winter moth *Operophtera brumata*, but already after a few days the leaves become tough and unpalatable to the first instar larvae which have to find other food sources. A synchronized bud opening and hatching of the eggs is imperative for the survival of the insect, circumstances advantageous to the oak tree in its most vulnerable period.[36] The metabolic cost for this type of heavy armoury is very high in comparison with the cost of the synthesis of highly active repellents. The alterations of the synthesis of antifeedants with plant development is an alternative strategy, practised to respond to seasonal variations of herbivory.

Unapparent, i.e. less recognizable plants, will probably use a less expensive synthetic strategy and produce small quantities of an active toxin interfering with the metabolism of the herbivores.

The monarch butterfly/milkweed (*Danaus plexippus/Asclepias curassavica*) relationship is another well investigated classical case demonstrating coevolutionary effects of plant toxins involving plant, insect and bird.[37] *A. curassavica* synthesizes physiologically active, bitter tasting cardiac glycosides, i.e. **46–48**, which are sequestered by *D. plexippus* larvae feeding on the plant. The glycosides serve as chemical defense against insect feeding and are toxic to higher animals. The adult butterfly uses the toxin as a protection against predators. If a blue jay, *Cyanocitta cristata*, should be foolish enough to eat a monarch butterfly it will vomit a few minutes later. The butterfly is associated with a characteristic colouration and, wise from experience, the bird recognizes it and refuses to eat another. The bird also avoids eating otherwise palatable species of butterflies with similar colouration. The butterflies thus practice mimicry for protection.[38] The production of latex by plants is assumed to be defensive. *Danaus plexippus* use a clever method to avoid the latex defence system by cutting the leaf veins to the feeding site before feeding on the milkweed.

Many toxic animals, e.g. tropical frogs, *Phyllobates* spp. and also the common ladybird, are vividly coloured to warn predators, and produce deterrents **49–51**. The Cholo Indians in Western Columbia use the skin secretions of *Phyllobates* spp. as arrow poison.[39] Recently it was found that the brightly coloured New Guinean bird, *Pitohui dicrous*, has a toxic plumage.[40] The structure of the toxin **50** was, rather sensationally, closely related to the *Phyllobates* toxin **49**. The poisonous frogs obtain the alkaloids from a dietory source and modifications of the structure occur in the organism.[40a]

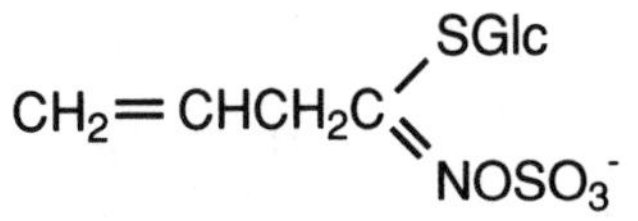

$$CH_2=CHCH_2C\begin{smallmatrix}SGlc\\ \\NOSO_3^-\end{smallmatrix}$$

43

$$CH_2=CHCH_2N=C=S$$

44

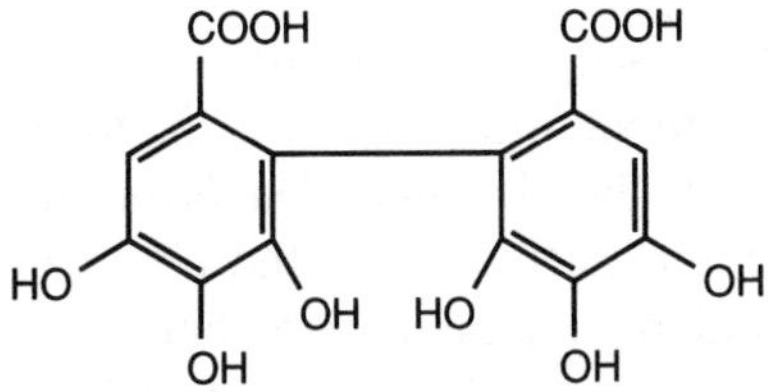

45. Hexahydroxydiphenic acid (occurs as glycoside)

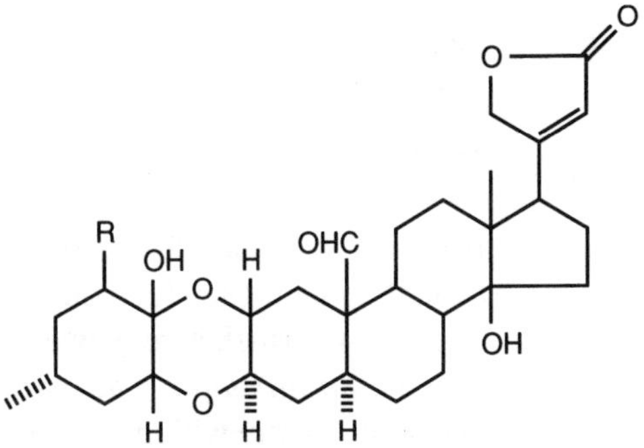

46, 47. R = ∿ OH, Calotropin, Calactin
48. R = O, Uscharidin

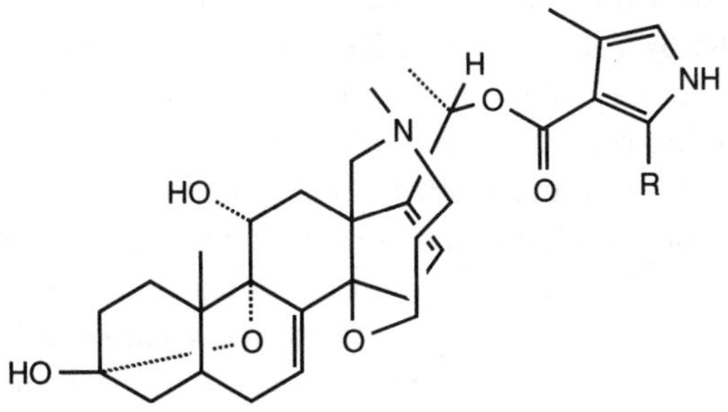

49. Batrachotoxin, R = CH₃
Phyllobates terribilis
50. Homobatrachotoxin, R = C₂H₅
Pitohui dicrous

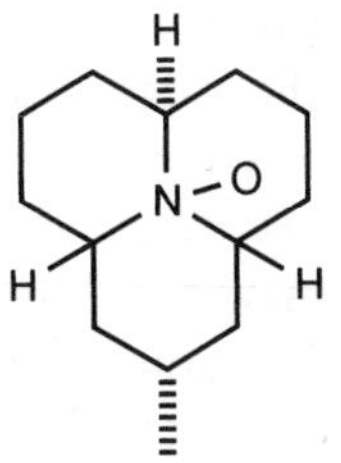

51. Coccinelline, ladybird

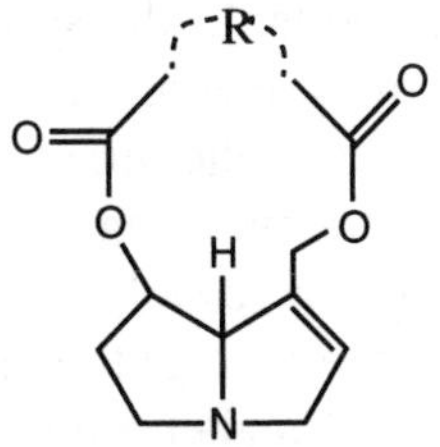

52. Pyrrolizidine alkaloid, R = dicarboxylic acid

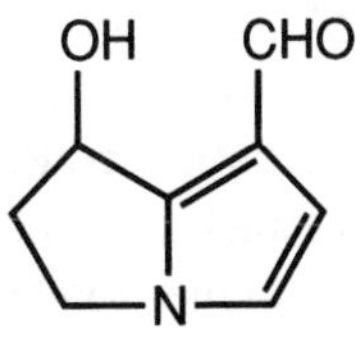

53. Hydroxydana-idal, male sex pheromone

54. Azadirachtin   **55.** Prunasin (R)

It is now well established that insects sequester and store toxic compounds from the plants they feed on and use them for defence, i.e. the toxic bianthra-quinone **42**, which is sequestered by the beetle *Chrysolina brunsvicencis* from *Hypericum hirsutum*. They store them in the cuticle thereby avoiding pro-blems of toxicity within the body and making them immediately accessible at predator attack. Occasionally the toxins are chemically modified into com-pounds with other properties. The dietary *Senecio* pyrrolizidine alkaloids have a threefold function in that they protect the plants from herbivores, and the adapted insect from predators and they are pheromone precursors.[41] The naturally occurring pyrrolizidine alkaloids **52** are oxidized by the male butter-fly of *Danaus* spp. to the pyrrole derivative **53,** used to attract the female and prepare her for mating. It was also discovered that in passing the food chain the pyrrolizidine alkaloids protected the aphid *Aphis jacobaea* feeding on *S. jacobaea* and its predator the ladybird, *Coccinella septempunctata*. She accumulated the toxin and supplemented her coccinelline **51** as deterrent.[41a] A widely known compound with antifeedant activities to numerous insects is azadirachtin from the Indian neem tree *Azadirachta indica*, **54**, effective against the desert locust *Schistocerca gregaria*.[42] Several low molecular weight phenolics, particularly in the form of glucosides, are part of the defence arse-nal protecting poplar, willow, aspen and birch from browsing.[42a, 45, 45a]

Approximately 1000 cyanogenic plants are known.[43] Toxic hydrocyanic acid is liberated by enzymatic cleavage of cyanohydrin glucosides, e.g. pruna-cin, **55** from *Prunus* spp. The lethal dose for human beings is *ca.* 0.1 g of HCN. Most herbivores avoid feeding on cyanogenic plants. In the spring the con-centration of cyanogenic glycosides in the young fronds of bracken fern *Pteridium aquilinum* is high, decreasing rapidly during the summer. As a com-pensation the tannin concentration goes up in the maturing fronds to meet herbivorous attacks. As a rule vital parts such as buds, developing leaves, and catkins have a higher load of toxins than the mature parts of the plant to deter grazing animals. Studies of feeding preferences of higher animals including man show, as was previously established for insects, that tannins and the ligni-fication products are significant chemical barriers. Less predictable (unap-parent) plants use metabolically less costly toxins as chemical defense mainly because of their short vegetative time. By smelling, licking and sampling the animal recognizes suitable parts of the plant.[44]

It should be kept in mind that all plants produce metabolites that to some extent have antifeeding properties. It is just the most spectacular cases we have been able to study in some detail. All the unpretentious interactions pass unnoticed. We know still very little about the mechanism of action. All classes of compounds from the main streams of biosynthesis are represented as allelochemicals. Many attractants and repellents, e.g. triterpene and flavone glycosides, are non-volatiles and the insects must therefore be guided to the host plant by other characteristic signal compounds, the structures of which still are unknown. For reviews of chemical interactions see ref. 45.

Hormone interactions. Several plants are capable of synthesizing steroids with structures identical to or resembling both human male and female sexual hormones. Their manufacture is presumably accidental but one can't completely exclude the possibility that they also have a function in the development of the plant. Many drugs used in traditional medicine by women for fertility control or as abortives could contain compounds of this kind, or mimics. Isoflavones are reported to have oestrogenic properties. The two isoflavones genistein **56** and formononetin **57** were responsible for the reduced lambing found among Australian sheep grazing on *Trifolium subterraneum*.[46]

56. $R^1, R^2 = OH$
57. $R^1 = H, R^2 = OCH_3$

58.

59.

Both the juvenile hormone (JH) **58** and analogues and the moulting hormone ecdysone **2**, see section 2.1, are required in a subtle balance for normal metamorphosis of the insect larva. The structure of **58** is closely related to the common sesquiterpene farnesol. These hormones or substances mimicking their endocrinological effects occur in many plants and there is evidence that their presence provides chemical defense against insects. Plant ecdysteroids and their interaction with insects have been reviewed.[46a]

The story of the "paper factor" illuminates the fashinating discovery of JH activity in plants.[47] The bug *Pyrrhocoris apterus* underwent normal metamorphosis when cultured in Prague by the biologist Dr. K. Sláma. However, when he moved his cultures to Harvard in 1964 they failed to transform into the winged adult after the fifth larval stage. The only experimental difference was that in Prague, Whatman filter paper was used in the Petri dishes, whereas the Harvard colonies were raised on American paper towels. Replacement of the towels with filter paper led again to normal development of the bugs. Apparently American paper pulp manufactured primarily from *Abies balsamea* contained some unknown factor hindering the metamorphosis, and indeed, by extraction of the paper, separation, biological testing and finally structural elucidation it was shown that the "paper factor" was a sesquiterpenoid metabolite, juvabione, **59** in *A. balsamea*, structurally related to the juvenile hormone **58**. Several active analogues have been isolated from various plants. It can be concluded that some plants practise incredibly intricate chemical defense methods against predation. There is nothing spectacular about the chemical structures of these highly active compounds and this is still more evident for most of the pheromones, section 2.5.2.

2.5. Animal-animal relationships

2.5.1. Chemical defense and warfare. Animals produce their defensive toxins either by *de novo* synthesis or by sequestering them from the diet as we have seen from the *Danaus* butterfly–blue jay case. All types of metabolites and animals are represented: molluscs, insects, fish, vertebrates and even birds. Confrontations and antagonism are perpetual among animals with fatal consequences for one party in a predator-prey relationship. Predators select those species of a genus that are most suitable to their taste, a selection not necessarily based entirely on nutritional judgement. A good deal of gustatory and olfactory sensing is involved. The principal response to predation is the development of chemical resistance. As a result toxic, unpalatable, less nutritious species evolve. These in turn, have a counter-effect on the predators, who progressively develop resistance to the active agents in their diet or look for other food sources. A mutual continual biochemical adaption takes place.

Various types of defense systems have evolved. The defense compounds are either contained in special glands or distributed in the blood, gut or elsewhere in the body. The secretions from glands either ooze out or are injected (wasp or scorpion sting, spider bite); this latter technique is often used to incapacitate prey. The conspicous, brightly coloured nudibranch, *Phyllidia varicosa*, a butterfly of the sea, lacks physical protection but is seldom eaten by fish. It secretes a strong smelling, volatile, ichthyotoxic, sesquiterpenoid, isocyano compound **60**, protecting the mollusc from predators. It was accidentally found that *P. varicosa* feeds on a sponge, *Hymeniacidon* sp. with the same smell, which produces substantial amounts of the compound.[48] Nudibranchs feed generally on sponges, coelenterates, bryozoa or algae, from which they accumulate chemical defensive agents, especially halogen compounds.[49,50] The sessile sponges are rather defenseless primitive animals, which produce toxins with antifeeding activity. Only adapted molluscs seem to be able to break the chemical barrier. Halogenated compounds are generally more toxic than corresponding hydrocarbons. All organisms are equipped with peroxidases, which in the saline, marine environment in principle have the capacity to halogenate (Cl, Br, I but not F) any metabolite. Today more than 1600 halogenated compounds are known to be produced in the marine biosphere, and the number is growing. It has been estimated that in the oceans *ca.* 6 million tons of methylhalides are formed biosynthetically per year![51] The terrestrial contribution of organohalogen compounds is comparably small.

It is reasonable to propose that organohalogen compounds play an important role in the marine chemical ecology, **61–63**. *Laurencia* spp. produce a wealth of halogenated hydrocarbons, fatty acid derivatives, terpenoids and aromatics. An isomer of **61**, which deters fish, is found in *Aplysia* spp., sea hares, whose diet includes *Laurencia.*[53] The prostaglandins are mammalian hormones and it was therefore interesting to discover a chlorinated analogue **62** in an octocoral.[54] *R. tigris* is a carnivorous nudibranch, which accumulates bipyrroles from its diet. When attacked, the small *Tambje* nudibranchs secrete large quantities of **63** deterring most predators except *R. tigris*, which can break the chemical barrier.[55] The pyrrole derivative was finally traced to a bryozoon.

Barnacles, *Balanus amphitrite amphitrite*, are not only a problem to marine vessels but also a nasty threat to many sessile marine animals. The octocoral *Renilla reniformis* secretes diterpenoides, **64a–c**, which inhibit the settlement of the *Balanus* larvae.[55a] The structure of these naturally occurring *Balanus* repellents gives leads to synthesis of simpler analogues of commercial interest.

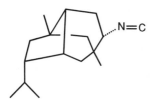

60. *Hymeniacidon* sp.
  sponge[48]

61. *Laurencia pinnatifida*,
  red alga[52]

62. *Telesto riisei*, octocoral[54]

63. *Roboastra tigris*, nudibranch[55]

64 a R = Me
64 b R = Et
64 c R = Pr

65. R = H, OH; n = 22 – 36

$CH_3(CH_2)_nCH_3$

66. n = 19 – 33

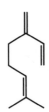

67. Myrcene

68. α-Pinene[59]

69. β-Pinene[59]

70.   **71.**   **72.**

R = alkene or alkene group

73. Trinervitane   **74.**

Ants and termites live in social colonies where reproduction, labour and defense are strictly specified. The sterile defense soldiers are equipped with a biting-injection organ and a large gland for defense chemicals amounting to 10–30 % of their dry weight. At least three different chemical strategies are used against intruders, e.g. raiding ants.[56] The first is biting and injection of a secretion containing lipophilic hydrocarbons, straight chain alkanes, mono-, sesqui- and diterpenes or macrocyclic lactones interfering with the healing of the punctured cuticle, **65–69.** However, it is conceivable that some of them serve as alarm pheromones. The second method involves daubing of an electrophilic and reactive contact poison,[57,58] **70–72.** The nitroalkene, β-ketoaldehyde and vinylketone functional groups react with nucleophiles in the cell and the long lipophilic R-group facilitates passage through membranes. The question arises how termites protect themselves against the poison. It has been suggested that they are able to detoxify the agents **70–72** by rapid reduction to unreactive products. Workers in the colony are less sensitive to the toxin than the raiding ants. The third method of chemical defense is ejection of an irritating glue entangling the predator. Insectivorous animals avoid the distasteful, sticky, diterpenoid secretion consisting of tri- or tetracyclic diterpenes **73** dissolved in a monoterpene solvent. It has been shown by labelling experiments that the defensive diterpenoids of type **73** are synthesized *de novo* by the termite soldiers. Benzoquinones have also been detected in the defensive secretion of termites. Use of glues as deterrents is by no means unique among insects. The glue, which hardens in the air, can be sprayed with precision at a distance or excreted thus disabling the predator. Soft-bodied earthworms and

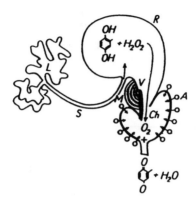

Fig. 1   The defensive glands of the bombardier beetle, *Brachynus* sp. 'after Schildt-
        knecht, H. and Holoubek, K. *Angew. Chem.* 73 (1961) 1. Used with the per-
        mission of VCH Verlagsgesellschaft.

slugs are chemically protected by a sticky slime, which most likely is proteina-
ceous in nature.

The bombardier beetle, *Brachynus* sp., is equipped with a most spectacular
defensive artillery. The jet chambers are shown in Fig. 1. The reservoir
contains a mixture of hydrogen peroxide and hydroquinone. Triggered by an
attack, the beetle opens the valve to the reaction chamber where catalase and
peroxidase enzymes convert H_2O_2 to O_2 and H_2O and oxidize hydroquinone
to benzoquinone in an exothermic reaction.[60] At nearly 100 °C the mixture is
discharged at high velocity by the oxygen pressure in the direction of the at-
tacker. A similarly constructed two-chambered reactor gland is used by the
millipede *Apheloria corregata* for the enzymatic production of hydrocyanic
acid and benzaldehyde from mandelonitrile. When attacked the millipede
opens the valve between the storage reservoir containing mandelonitrile and
the reaction chamber containing enzyme, and the poisonous repellent is excreted,[61] eqn. (1).

$$\text{PhCHCN} \xrightarrow{\text{Enz}} \text{PhCHO} + \text{HCN} \qquad (1)$$
$$\underset{\text{OH}}{|}$$

Several insects use quinones as repellents. The millipede *Narceus gordanus*
exudes toluquinone **74** from lateral glands.[62] Many insects practise defensive
reflex bleeding, regurgitation or defecation when attacked. The droplets of
distasteful haemolymph, exuded at the knee joint of ladybirds contain alkaloids, **51**, and the unusual diamine **75**.[63]

Cantharidin, **76** is the toxic principle in the haemolymph of the aphrodisiac Spanish fly and acts as a feeding deterrent. The secretions from the poison glands of ants show a diversified chemistry. *Acanthomyops claviger* synthesizes a series of simple straight chain hydrocarbons and monoterpenes, e.g. undecane, citral, citronellal, some of which also serve as alarm pheromones.[64] Iridodial **77** and isovaleric acid are major constituents of the anal gland defensive secretion of *Iridomyrmex nitidiceps*.[65] *Solenopsis* spp. typically manufacture 2,6-dialkylpiperidines **78** and 2,5-dialkylpyrrolidines **79** as defensive venoms. Piperazines **80** are synthesized by several ant spp.[66,67] Alkaloids from ants and other insects have been reviewed.[67a]

75.   **76.**

77.   **78.** R = alkyl   **79.** R¹, R² = alkyl

80. R = alkyl

As a response to danger some vertebrates emit strongly smelling secretions from anal glands or excrete urine and faeces. A classical case is the skunk *Mephitis mephitis* which releases evil smelling mercaptans and sulfides **81–83**, from its anal glands, effectively repelling most aggressors.[68] By muscular contraction the skunk can squirt the secretion several meters. Chemical investigations started more than 100 years ago by Wohler, Schwarz and O. Low. Students and colleagues at the university convinced Low to give up his project. The investigators concluded preliminarily that the secretion contained mercaptans. I accidentally run over a skunk with my car, which became lite-rally useless for any kind of human dating for one month. The secretion of mink, *Mustala sp.* contains thiethane, **84** and *Hyaena hyaena* produces the sul-fide **85.** These compounds are used for territory marking.

2.5.2. **Pheromones.** Chemical signal substances used for communication with-in the same species are defined pheromones. Best investigated are the insect pheromones but pheromonal interactions exist in all classes of animals from primitive single cellular species to mammals including man; even cases from the plant kingdom are known.

Sex pheromones. Both sexes emit attractants which consist of fairly simple, volatile, long chained unsaturated alcohols, acids, or esters, benzene deriva-tives, functionalized mono- or bicyclic aliphatic compounds, etc. with mole-cular masses in the 100–300 Dalton range. Low molecular weight facilitates evaporation of the lure thereby increasing its concentration in the air. The male sensory organ of *Bombyx mori* is exquisitely sensitive; a concentration of 10^{-16} M bombycol in air is detected. By turning upwind the male will be exposed to a positive concentration gradient which leads him to the female emitting the attractant. The antennae are thus able to record the intensity of the lure, but for this purpose one antenna should be enough.

   The two antennae of the recording system function as a coincidence meter, which informs the insect about the direction and also explains the zigzag flight course. If the male is flying at an angle to the female the two antennae will be differently exposed to the attractant; one antenna is to windward the other is to leeward. The male will shift his course until the aquisition of the two detectorantennae is the same, i.e. statistically towards the emitter. The phero-mone attracts the male at great distances, up to a few kilometres, and when the male has found his elected female he emits a short range attractant prepar-ing her for mating. Benzaldehyde is such an aphrodiasiac found in several species. Male butterflies of the *Euploea* genus produce a courtship pheromone having the terpenoid structure **88**.[69] The serpent's forked tongue has been a longstanding and mystisized problem.[69a] When the serpent follows the trail of a conspecific or prey its flicking tongue records the intensity of the pheromone signal. The forked structure is able to recognize a chemical gradient leading the snake along the trail. The principle is applied to sensory organs consisting of two symmetric parts, e.g. organs of hearing, sight and tactile sense.

Structures of some aliphatic insect pheromones are shown in **86–92**, cf. *Bombyx mori* and the *Ophrys* pollinators, section 2.2 and 2.3. The racemic **92** is not as active as the (+) form and the (–) form is inactive.[75] Addition of the (–) form to the (+) form lowers the response to the pheromone indicating that the inactive (–) form competes with the (+) form at the receptor site without stimulating the insect. Occasionally a mixture of enantiomers or a mixture of *E* and *Z* geometric isomers is more active than the pure isomers.[70] Often a blend of compounds is more active than one single component but only the main component is specific for the receptor. Otherwise the signals would create a chaotic situation. It has also been suggested that some male insects, e.g. the tsetse fly, produce anti-aphrodisiacs which are transferred to the femals during copulation to discourage other males.

(*R*)-3-hydroxybutyric acid and its dimer (*R*)-3-((*R*)-3-hydroxybutyryloxy)-butyric acid are sex pheromones emitted from the web of the female spider *Linyphia triangularis*. They elicit a peculiar web reduction behaviour in the male on webs of unmated females thereby reducing the chances of discovery by other males. Mated females construct new webs that no longer contain the lure. The (*S*)-form and the dehydrated derivative, crotonic acid, are inactive.[70a]

86. (–) Ipsdienol,
Ips pini[70]

87. (+) Grandisol,
Anthonomus grandis[71]

88. *Euploea* spp.[69]

89. exo-Brevicomin,
Dendroctonus brevicomis[72]

90. Methyl (*E*)-2,4,5-tetradecatrienoate,
Acanthoscelides obtectus[73]

$$CH_3(CH_2)_2CH = CHCH_2)_7OCOCH_3$$

91. (*Z*)-8-Dodecenyl acetate
　　Grapholitha molesta[74]

92. Disparlure, (7*R*, 8*S*-(+)-Epoxy-
　　2-methyloctadecane, gypsi moth,
　　Lymantria dispar

93. Fucoserratene　　　　　**94.** Ectocarpene　　　　　**95.** Multifidene

　　　　　96.　　　　　　　　　　　　　**97.**

Pheromones and synthetic mimics have been used in pest control. The size of the infestation can be determined by trapping insects and measures can then be taken for spraying the area with pesticides if necessary.

　　The situation is different for water-borne pheromones. The diffusion rate is extremely low in comparison to the rate in air but the water circulation increases the active space considerably, and most importantly, high molecular weight or low vapour pressure of the compound is no hindrance for activity. The emission rate is the limiting factor. Peptides are reported to be active as pheromones in several marine animals. Some kind of signal compounds must guide the sperm cells to aggregate around the egg. The female gametes of the brown algae *Fucus serratus, Ectocarpus siliculosus* and *Cutleria multifida* produce simple low molecular weight unsaturated hydrocarbons **93–95** that attract and effect a circular locomotive behaviour of the male gametes in their approach to the egg.[76] Other simple hydrocarbons have similar but weaker effects, a circumstance that raises serious problems concerning oil spills in sea water and their damaging effects on marine ecosystems. Compounds contained in oil could very well disturb sexual approach and hamper fertilization.

The defensive, evil smelling secretions of skunk, hyaena and mink play a role in sexual communication and recognition. Well documented is the effectiveness of the musky boar odour in arousing sows. The active compound is the steroid **96** which also is marketed as an artificial insemination aid. Boar meat is slightly tainted by this odour and this is one of the reasons why male pigs, reared for meat poduction, are castrated. Compound **96** has been detected in the sweat of men. It is not quite coincidental that one of the odourous macrocyclic ketones **97** of the scent glands of the civet *Viverra civetta* has a structure similar to **96**. The beaver, the musk ox, several species of deer and antelope, sheep, rats, rabbits, dogs, foxes and apes produce scents involved in sexual stimulation and recognition.[77] The Canadian garter snake *Thamnophis sirtalis parietalis* has a chemically controlled mating behaviour.[78] The female attracts the male by releasing a volatile pheromone, which the male catches on his tongue. After successful copulation the male plugs the cloaca of the female with an odorous repellent substance, mechanically obstructing further mating and making the female unattractive. There seems to be a contact pheromone stored in the skin of the female snake. Normally the male snakes show no homosexual behaviour, but they start courting other males which have been rubbed against sexually attractive females. The structures of the sex pheromones are still unknown.

Alarm and aggregation pheromones occur in most social insects and are often related to the defensive manoeuvres as a forewarning. Well known is the venomous formic acid, that simultaneously notifies the other members of the colony about the danger. Simple organic acids such as 2-methylbutyric, iso-butyric, methacrylic, tiglic, angelic, iso-valeric and senecioic acids have been detected in the defensive gland of the beetle *Pasimachus subsulcatus*.[79] Bee venom contains isopentenyl acetate **98** inducing other bees to sting at the same site. Alarm pheromones are low molecular weight compounds of simple structure such as citral, limonene, α-pinene, undecane and hexanal. (*E*)-β-farnesene **99** is an aphid (*Homoptera*) alarm pheromone causing them to spread, thereby helping others to escape. It is used in pest control, increasing the mobility of the aphids thereby increasing the efficiency of contact pesticides.[79a] The alarm pheromone of the leaf-cutting ant (*Atta texana*) and the ant *Pogonomyrmex barbatus* is (*S*)-(+)-4-methyl-3-heptanone, **100**, which is much more active than the (*R*)-(–)-isomer, indicating a chiral receptor.[80] The rice

weevil *Sitophilus oryzae* and the maize weevil *S. zeamais* cause serious economic losses of cereal grains. The males produce an aggregation pheromone,
sitophilure, identified as (4*S*,5*R*)-5-hydroxy-4-methyl-3-heptanone **102**.[81] The
pheromone and stereoisomers were synthesized for the purpose of monitor-
ing the population by baited traps. The attractancy of the synthetic stereoisomers was lower in tests using the weevils. The beetle *Gnathotrichus sulca-
tus* produces 6-methyl-5-heptene-2-ol, **101**,[82] and the weevil *Rhynchophorus
vulneratus*, a pest on coconut and oil palms, produces a similar compound,
(4*S*)-methyl-(5*S*)-nonanol as an aggregation pheromone.[82a]

$(CH_3)_2CH(CH_2)_2OCOCH_3$

98.

99.

100.

101. Sulcatol

102.

The shell of molluscs is no satisfactory protection against starfish, which
crawl on top of the mollusc and slowly pierce the shell. However, the mollusc
is forewarned by a steroidal secretion of the starfish, which gives it a chance
to escape. The mollusc can create a jet stream by rapid closure of its bivalve
via muscular contraction, thereby propelling itself out of reach of the starfish.

Trail pheromones. Foraging ants and termites leave scent marks showing the
trail from their nest to the food source, which other members of the colony
can follow, **103–106**. The termite *Reticulitermes flavipes* was found to be attracted by woods infected by the fungus *Lenzites trabea*. It was later found
that the metabolite produced by the fungus and the trail pheromone were
identical, **106**.[85] It has been demonstrated that there is colony specificity in the
trail pheromone **105** of the ant *Lasius neoniger*.[86] Ants belonging to different
colonies following crossing trails are thus able to recognize markers dropped
by their nest mates.

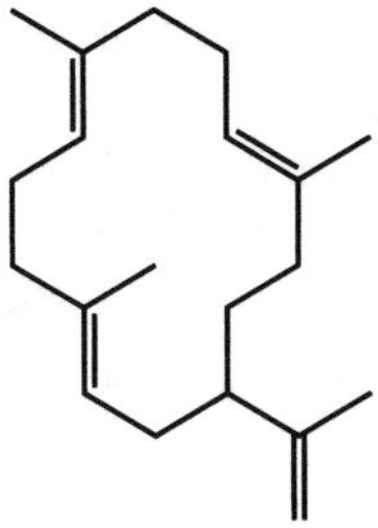

103. D,L-Neocembrane, Nasutitermes sp.[83]

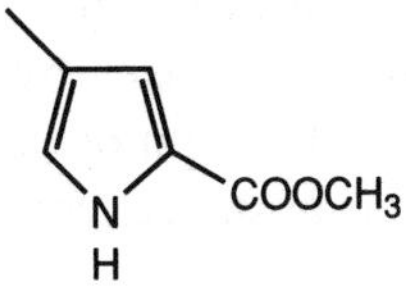

104. *Atta texana*[84]

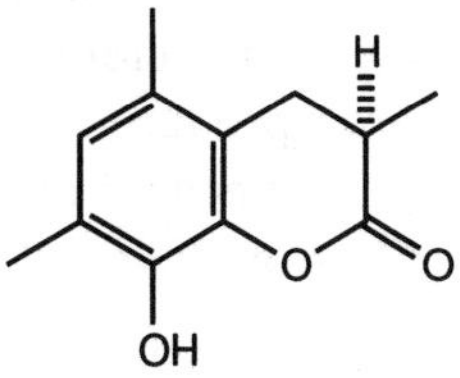

105. *Lasius* and *Formica* spp.[84a]

tr   cis   cis

$CH_3(CH_2)_2 CH = CHCH = CHCH_2CH = CHCH_2CH_2OH$

106. *Reticulitermes* sp.[85]

Oviposition pheromones. It is not a good idea that the female deposits her eggs on a plant or in a fruit which has been infested previously. This will inevitably lead to overcrowding and lack of food for the larvae.[87] The cherry fly *Rhagoletis cerasi* lays only one egg in the half-ripe fruit, and leaves a scent behind informing other females that the fruit already is occupied. The pheromone has been identified and synthesized, **107**,[87a,88] an unusual glycoside containing a sulphonic acid function. The apple maggot fly *R. pomenella* and *Ceratitis capitata* which attack hawthorn, *Crataegus* sp., also leave a chemical message to other presumptive ovipositing females. Related is the behaviour of the beetle *Tomiscus piniperda*, feeding on *Pinus sylvestris*, which recognizes the emission of verbenone, **108**, from an already infested tree. Verbenone is produced in increasing amount when the existing colony increases.[89] Parasitic female wasps ready to lay eggs must inject them into their plant-feeding caterpillar host. They recognize odours both from their prey and from the plant on which the caterpillars feed.[90] Studies of these special predator-prey relationships and oviposition deterrents could lead to development of natural biological weapons for pest control.

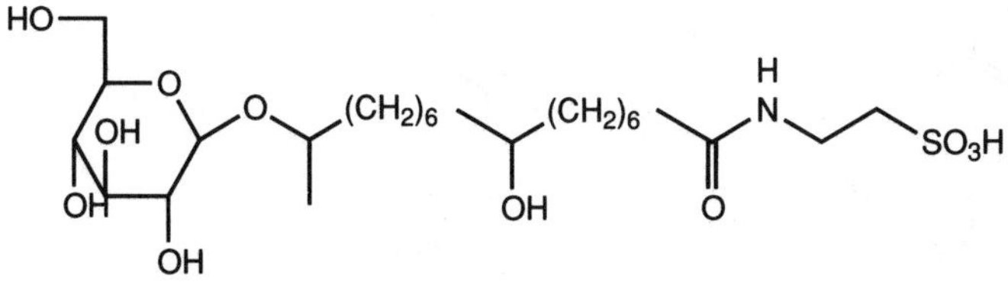

107.

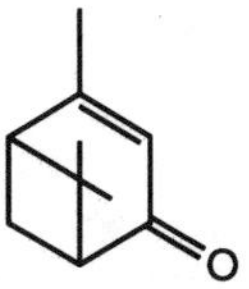

108.

Territorial pheromones. It is customary among many animals to make territorial claims by leaving scent marks as a warning signal. They either use the characteristic odour of their urine and droppings or they have glands which produce a smelly secretion, e.g. foxes, hyaenas and deer. A territorial defense mechanism exists in fish. The flight or fight reactions of certain fishes depend on their ability to smell a substance from the mucus of another fish.[91]

2.6. Plant-plant relationships

All plants produce chemicals, which to some degree affect the conditions of other species in the community. There are inhibitory as well as stimulatory effects. The idea that plants interact chemically was advanced in the early 1800s by de Candolle among others.[92] In most cases it is more a question of degree than a distinct effect, which in a long time perspective will be recognized as adaptation and evolution. One can suspect that plants, which successfully form monocultures, are potential manufacturers of phytotoxins.

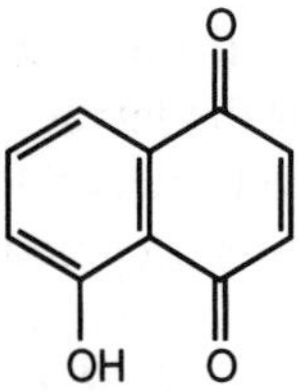

109.

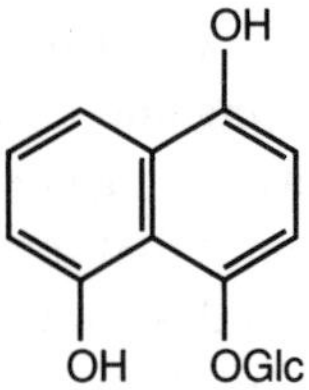

110.

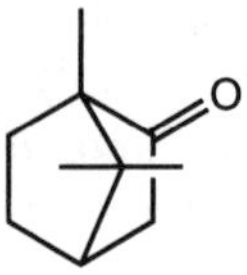

111.

112. Cineol

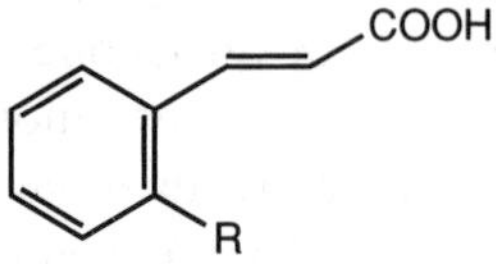

113. R = H, *tr*–Cinnamic acid
114. R = OH, *o*-Coumaric acid

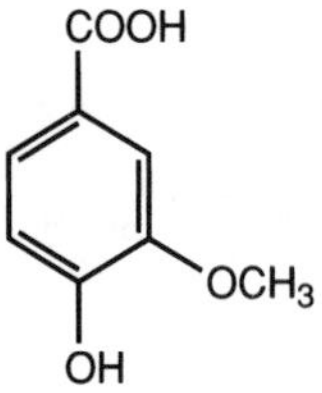

115. Vanillic acid

116. R = H, Hydroquinone
117. R = Glc, Arbutin

The undergrowth varies considerably in groves of different trees. It is established that the effect is not simply a matter of light, predation, competition for water and nutrients in the soil etc., but is an allelopathic effect. The undergrowth is sparse in spruce forests, *Picea* spp., as a result of the growth-inhibiting effects of the monoterpenes produced. The same effect is noticed in stands of *Eucalyptus*, endemic to Australia, especially when grown in the west, where adaptation to eucalypt allelopathy has not evolved. A well investigated case is the allelopathic effect of the walnut tree, *Juglans* sp. Herbs are virtually absent in a belt several meters wide surrounding the tree. This is due to juglone, **109**, which originally occurs in the non-toxic glycoside form **110** in the tree. It is exuded and washed by rain into the soil, where it becomes hydrolysed and oxidized into the toxic form **109**, which inhibits plant growth and seed germination.[93]

Another typical case is the effect of mint, *Salvia leucophylla*, and the sagebrush, *Artemisia californica*, dominating the dry Californian chaparral.[94] The shrubs produce a bare zone around the thickets. Eliminating all other factors, it was proved that volatile monoterpenes, notably camphor and cineol, **111, 112** inhibited the plant growth. At intervals fires destroy the canopies of the shrubs and as a result the grassland and annual herbs move into the bare patches. After *ca.* 6 years the shrubs have regenerated and start suppressing the surrounding vegetation.

It is a common belief among the Columbian Indians that *Duroia hirsuta*, a myrmecophilous tree of the Amazonas, produces a growth inhibiting compound suppressing the undergrowth.[94a] Recent work indicates that the iridoid plumericin could be responsible for the allelopathic effects.[94b] It is not clear whether the lack of vegetation under the trees also may be associated with the presence of ants. The bracken *Pteridium aquilinum* forms dense stands efficiently suppressing herbal growth.[95] The ground within the stands is only sparsely covered with grass. It was found that aqueous extracts of bracken were phytotoxic, containing cinnamic acid derivatives. The Mexican rubber plant *Parthenium argentatum* represents a case of self-inhibition. In the natural habitat it is evenly spaced in the desert but when grown closely as a crop it was noted that the plants developed best at the edges of the field. An exudate from the root containing *trans*-cinnamic acid **113** was found to inhibit the growth of plants including *P. argentatum*.[96] The Australian plant *Grevillea robusta* also produces a self-inhibiting exudate reducing its development in dense stands.[97] Phenolic compounds, e.g. **114–117**, widely produced by plants, are toxic to both plants (seed germination) and microorganisms. They usually occur as glucosides in the tissue. The remarkable formation of fairy rings by sunflowers or mushrooms is due to production of self-inhibiting compounds, which accumulate inside the stand. The plants preferably spread outwards into fresh soil. Allelopathic effects can cause problems in agriculture and horticulture if crop rotation is not observed. The phytotoxicity and allelopathy of terpenes have been reviewed.[98]

An indirect allelopathic effect is demonstrated by the heather, *Calluna vulgaris*, which inhibits the invasion of pine trees into the moor. It produces fungicidal phenolic compounds which inhibit mycorrhizal formation by *Boletus* spp. By removing the heather the growth of the trees is stimulated.[99]

2.7 Plant-microorganism relationships. Phytoalexins

Because of the enormous economic losses caused by microbial attacks on crops, the studies on plant-microorganism interactions have been focused on plant diseases. From an ecological point of view these interactions are mechanistically analogous to any other predator-prey relationship. In order to invade a plant the microorganism has to mechanically penetrate a surface barrier, which could contain repelling chemicals, before entering the tissue where the invader meets growth inhibiting compounds. The successful microorganism, the pathogen, which generally is host specific, multiplies in the plant tissue and produces toxic metabolites leading to disease symptoms, mutilation and eventually destruction of the host plant. The plant counteracts by changing its metabolic activities and by making attempts to detoxify the foreign toxins by oxidation or by glycosylation. There is a constant battle going on until one of the combatants succumbs. There are occasions when symbiosis of fungus and plant is of advantage to both organisms, e.g. mycorrhiza formation, and presumably there are many interactions representing peaceful coexistence without noticeable symptoms.

It is characteristic for the various interactions described so far that the defensive agents are present in the organisms before the attack. This is not always the case for the microbial attack on plants. The plant-microorganism allelopathy can grossly be classified by the presence of 1. pre-infectional allomones and 2. post-infectional allomones.[100] The first case corresponds to the situation described earlier. Certain wood is exceptionally resistant to decay, e.g. juniper *Juniperus communis*, rich in terpenoids, and therefore often used as fence stakes. In the heartwood of trees a variety of secondary metabolites is accumulated, many of which render it resistant to decay, e.g. pinosylvin **118**, widely distributed in *Pinus* spp.

118.   **119.**

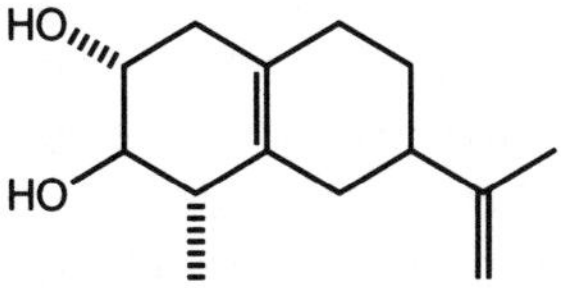

120. Rishitin
Solanum tuberosum

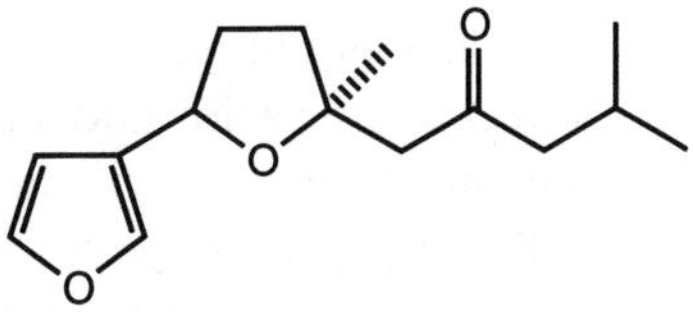

121. Ipomoeamarone
Ipomoea batatas

$$CH_3CH = CH(C \equiv C)_3 \; CH = CHCH(OH)CH_2OH$$

122. Safynol

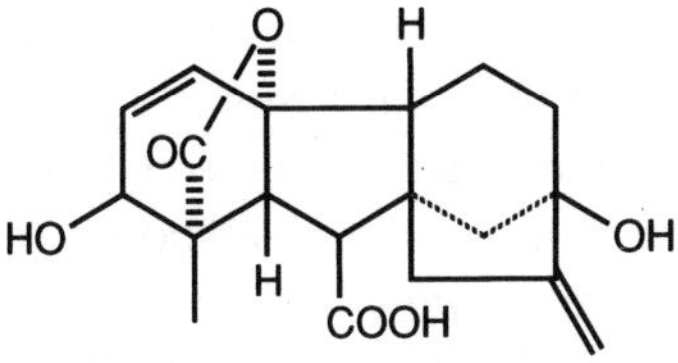

123. Gibberellic acid

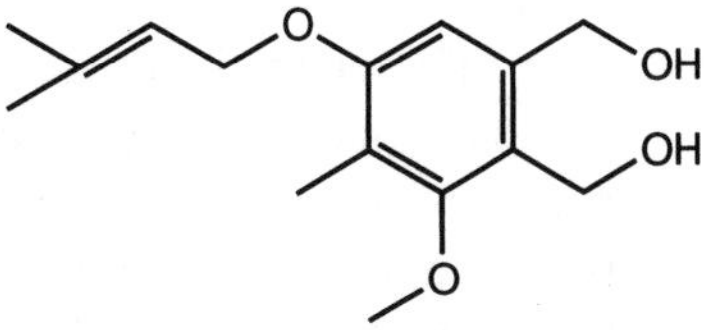

124. Zinniol

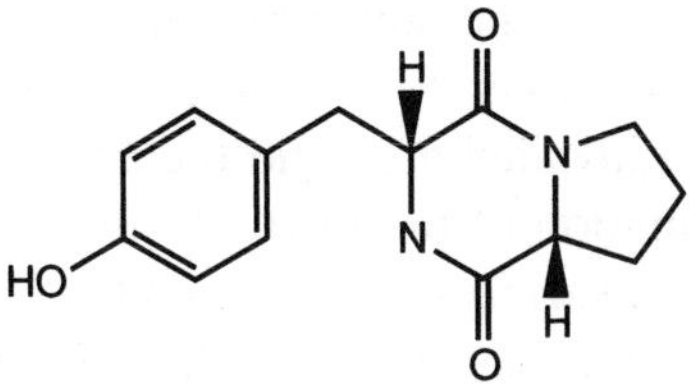

125. Maculosine

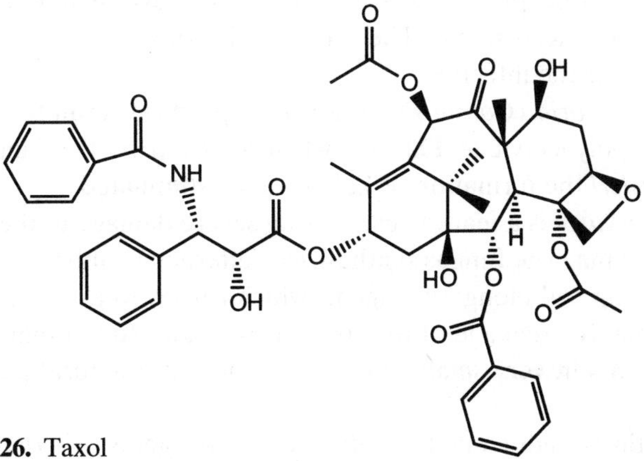

126. Taxol

Benzoic acid, which occurs in whortleberries and cranberries, *Vaccinium* spp., acts as a preservative. Commonly occurring hydroxy-substituted benzoic and cinnamic acids and a large number of hydroxy- and methoxy-substituted flavonoids have fungicidic properties. Other compounds have herbivore-repelling properties, such as cyanogenic glycosides, which release hydrocyanic acid and glycosinolates, which release mustard oils.

Formation of post-infectional allomones is a novel concept advanced by Müller and Börger[101] as a result of their studies on *Phytophthora infestans* infected potato tubers. They termed the compounds phytoalexins and defined them as low molecular weight antibiotics, synthesized *de novo* and accumulated in the cell after exposure to the microorganism. This induced defense has the advantage of diverting metabolic resources only when an insect or pathogen attack occurs.

The concept found experimental support by the isolation and structural determination of pisatin, **119**, an isoflavane derivate formed in peas, *Pisum sativum*, infected by the fungus *Monilina fructicola*.[102] Since then, several hundred phytoalexins of widely different structures have been isolated. *Solanum tuberosum* produces the sesquiterpene rishitin, **120** in response to fungal infection,[103] and sweet potatoes, *Ipomoea batatas*, the furanosesquiterpene **121** together with several related compounds when infected with spores of *Ceratocystis fimbriata*. Phenanthrene derivatives are formed as induced defence in the *Orchidaceae*, acetylenes in *Compositae* and anthranilic acid derivatives[104] in *Caryophyllaceae* and *Graminae*. The same phytoalexins are formed when the tissue is treated with mercuric ions.[104a] The defence is confined to the immediate neighbourhood of the invaded area and the efficiency of the defence is related to the rate of formation of the phytoalexin. Its toxicity is generally non-specific towards fungi. A variety of biotic and abiotic agents including viruses, polypeptides, polysaccharides, heavy metal ions, Hg^{2+}, Cu^{2+} and UV-light have been found to elicit formation of phytoalexins. These elicitors are thus able to channel normal metabolism into other products. The spectacular antibiotic phytoalexin is just one of several new compounds formed in the damaged tissue. The phytoalexin formation reaches a maximum *ca.* 48 h after the infection.

Carthamus tinctoria (*Compositae*) normally produces minute amounts of the fungicidal polyacetylene **122**, but when attacked by the fungus *Phytophthora drechsleri* the formation of **122** is greatly stimulated.

Giberella fujikuroi is a feared pest causing severe damage to the rice crops in Japan and China. The fungus synthesizes a diterpene, gibberellic acid, **123**, acting as a strong cell elongating agent, which causes weakening of the cell wall and ultimately breakage of the straw. It was later found that **123** is synthesized by plants in very small amounts and acts as a natural plant growth hormone.

Fungi infecting weeds could be useful biocontrol agents. *Alternaria cichorii* causes foliar blight of Russian knapweed, *Acroptilon repens*. Extracts of the fungus containing zinniol **124** as one of the phytotoxins produced caused the same type of necrotic lesions as the intact fungus.[105] A relative to *A. repens* is the spotted knapweed, *Centaurea maculosa*, plaguing pasture land in western North America. It was shown that *Alternaria alternata* produced a knapweed-specific phytotoxin, the piperazine derivative maculosin, **125**, which showed promising herbicidal potential.[106] This kind of observation could give leads to the preparation of synthetic analogues.

It is an interesting thought that microbes thriving on a plant could be adapted to produce the same compounds as their host. Taxol, **126**, is a promising antitumour agent formed in the inner bark of yew, *Taxus brevifolia*. By scanning several associated fungi it was found that *Taxomyces andreanae* produced small quantities of **126**, an observation of interest in connection with future production of larger quantities of the drug.[107] Several comprehensive reviews covering phytoalexins have appeared.[108]

Bibliography

1. Ehrlich, P. R. and Raven, P. H., *Evolution* **18** (1965) 586.
2. Whittaker, R. H. and Feeny, P. *Science* **171** (1971) 757.
3. Butenandt, A., Beckmann, R. and Hecker, H., *Hoppe-Seyler's Z. physiol. Chem.* **324** (1961) 71.
4. Hegnauer, R. *Chemotaxonomie der Pflanzen*. I–IX (1962–1990) Birkhäuser, Basel.
5. Butenandt, A. and Karlson, P. *Z. Naturforsch.* **9b** (1954) 389; Karlson, P., Hofmeister, H., Hummel, H., Hocks, P. and Spiteller, G. *Chem. Ber.* **98** (1965) 2394.
6. Takemoto, T., Ogawa, S., Nishimoto, N., Arihari, S. and Bue, K. *Yakugaku Zasshi* **87** (1967) 1414.
7. Muller, C. H. *Recent Adv. Phytochem.* **3** (1970) 106.
8. Harborne, J. B. *Introduction to Ecological Biochemistry*, Academic Press, 1989; *Adv. Chem. Ecol.* in *Nat. Prod. Rep.* **10** (1993) 327; Eisner, T. and Meinwald, J. (Eds.) *Chem. Colloquium, Proc. Natl. Acad. Sci. USA* **92** (1995)1. are recommended for extended reading.
9. Leninger, A. L. *Biochemistry* 2nd Ed. Worth Publs. 1978; Stryer, L. *Biochemistry* 3rd. Edn. W.H. Freeman, 1989.
10. Schoeffl, F., Lin, C. Y. and Key, J. L. *Ann. Proc. Phytochem. Soc. Eur.* **23** (1984) 129.
11. Levitt, J. *Responses of Plants to Environmental Stresses*. 2nd Ed. Academic Press, 1980.
12. Alberdi, M. and Corcuera, L. J. *Phytochemistry* **30** (1991) 3177.
12a. Dorne, A. J., Cagel, G. and Douce, R. *Phytochemistry* **25** (1986) 65.
13. Singh, T. N., Aspinall, D. and Paleg, L. G. *Nature, New Biol.* **236** (1972) 188.
13a. Hanson, A. D., Rathinasabapathi, B., Rivoal, J., Burnet, M., Dillon, M. O. and Gage, D. A. *Proc. Natl. Acad. Sci. USA* **91** (1994) 306.
14. Wright, S. T. C. and Hiron, R. W. P. *Nature* **224** (1969) 719.
15. Stewart, G. R. and Lee, J. A. *Planta* **120** (1974) 279.
16. Storey, R. and Wyn Jones, R. G. *Phytochemistry* **16** (1977) 447.
17. Dickson, D. M. J., Wyn Jones, R. G. and Davenport, J. *Planta* **150** (1980) 158.

17a. Barron, D., Varin, L., Ibrahim, R. K., Harborne, J. B. and Williams, C. A. *Phyto-chemistry* **27** (1988) 2375.
18. Ben Amotz, A. and Avron, M. *Plant Physiol.* **51** (1973) 875.
19. Robb, D. A. and Pierpoint, W. S. (Eds.) *Metals and Micronutrients, Uptake and Utilization by Plants.* Academic Press, 1983.
20. Grill, E., Winnacker, E. L. and Zenk, M. H. *Proc. Nal. Acad. Sci.* **84** (1987) 439.
21. Percival, M. S. *New Phytol.* **60** (1961) 235.
22. Stanley, G. and Linskens, H. F. *Pollen: Biology, Biochemistry and Management.* Springer-Verlag, Berlin, 1974.
23. Baker, H. G. and Baker, I. *Nature,* **241** (1973) 543.
24. Janzen, D. H. *Studies in Biology* No. 58. E. Arnold, London, 1975.
25. Dahl, A., Wassgren, A.-B. and Bergström, G. *Biochem. Syst. Ecol.* **18** (1990) 157.
26. Pellmyr, O., Bergström, G. and Groth, I. *Phytochemistry* **26** (1987) 1603.
27. Kullenberg, B. *Zool. Bidr. Uppsala* **34** (1961) 1.
28. Kullenberg, B. and Bergström, G. *Endevour* **34** (1975) 59; Borg-Karlson, A.-K. *Phytochemistry,* **29** (1990) 1359.
29. Tengö, J. and Bergström, G. *Science* **196** (1977) 1117.
29a. Lunan, K. *Experientia,* **49** (1993) 1002.
30. Thomson, R. H. *Naturally Occurring Quinones,* Academic Press, 1971. *Naturally Occurring Quinones, III. Recent Advances,* Chapman and Hall, 1987.
31. Goto, T. *Prog. Chem. Org. Nat. Prod.* **52** (1987) 114.
32. Thompson, W. R., Meinwald, J., Aneshansley, D. and Eisner, T. *Science* **177** (1972) 528.
33. Harborne, J. B. and Smith, D. M. *Biochem. Syst. Ecol.* **6** (1978) 127.
34. For general discussions of various aspects of plant-animal relationships such as plant apparency, chemical defence, feeding preferences etc. see chapters in *Recent Adv. Phytochem.* **10** (1975), Wallace, J. W. and Mansell, R. L. (Eds.) and *Herbivores. Their Interaction with Secondary Plant Metabolites.* Rosenthal, G. A. and Janzen, D. H. (Eds.) Academic Press, 1979.
35. Erickson, J. M. and Feeny, P. *Ecology,* 55 (1974) 41.
36. Feeny, P. *Ecology,* **51** (1970) 565; Rhoades, D. F. and Cates, R. G. *Recent Adv. Phytochem.* **10** (1975) 168.
37. Roeske, C. N., Seiber, J. N., Brower, L. P. and Moffit, C. M. in *Recent Adv. Phytochem.* **10** (1975) 93.
38. Rothschild, M. in *Insect-Plant Interactions.* van Emden, H. (Ed.). Oxford Univ. Press. **1973**, 59.
39. Witkop, B. and Gössinger, E. *The Alkaloids,* **21** (1983) 139.
40. Dumbacher, J. P., Beehler, B. M., Spande, T. F., Garaffo, H. M. and Daly, J. W. *Science,* **258** (1992) 799.
40a. Daly, J. W., Secunda, S. I. and Cover, Jr. J. F. *Toxicon* **32** (1994) 657.
41. Edgar, J. A. and Culvenor, C. C. J. *Nature,* **248** (1974) 614; **250** (1974) 646; Pliske, T. E. and Eisner, T. *Science* **164** (1969) 1170.
41a. Witte, L., Emke, A. and Hartmann, T. *Naturwissenschaften* **77** (1990) 540.
42. Morgan, E. D. and Thornton, M. D. *Phytochemistry,* **12** (1973) 391.
42a. Sunnerheim, K., Palo, R. T., Theander, O. and Knutson, P. G. *J. Chem. Ecol.* **14** (1988) 549.
43. Conn, E. E. in *Herbivores, Their Interaction with Secondary Plant Metabolites.* Rosenthal G. A. and Janzen, D. H. (Eds.) Academic Press, 1979, 387.
44. Rhoades, D. F. and Cates, R. G. *Recent Adv. Phytochem.* **10** (1975) 168.
45. Harborne, J. B. *Nat. Prod. Rep.* **3** (1986) 323; **6** (1989) 85; **10** (1993) 327.
45a. Haslam, E. *Plant Polyphenols,* Cambridge Univ. Press, Cambridge, 1989.

46. Bradbury, R. B. and White, D. E. *Vitamins and Hormones.* **12** (1954) 207.

46a. Camps, F. in *Ecological Chemistry and Biochemistry of Plant Terpenoids*, Harborne, J. B. and Tomas-Barberan, F. A. (Eds.) Clarendon Press, Oxford, 1991, p. 331.

47. Williams, C. M. in *Chemical Ecology*, Sondheimer, E. and Simeone, J. B. (Eds.) Academic Press, New York, 1970, p. 103.

48. Burreson, B. J., Scheuer, P. J., Finer, J. and Clardy, J. *J. Am. Chem. Soc.* **92** (1975) 4764.

49. Krebs, H. C. *Prog. Chem. Org. Nat. Prod.* **49** (1986) 151.

50. Gribble, G. W. *J. Nat. Prod.* **55** (1992) 1353.

51. Isidorov, V. A. *Organic chemistry of the Earth's Atmosphere.* Springer, Berlin, 1990.

52. Gonzáles, A. G., Martin, J. D., Martin, V. S., Norte, M., Pérez, R., Ruano, J. Z., Drexler, S. A. and Clardy, J. *Tetrahedron* **38** (1982) 1009.

53. Kinnel, R. B., Dieter, R. K., Meinwald, J., Van Engen, D., Clardy, J., Eisner, T., Stallard, M. O. and Fenical, W. *Proc. Natl. Acad. Sci. USA* **76** (1979) 3576.

54. Baker, B. J., Okuda, R. K., Yu, P. T. K. and Scheuer, P. J. *J. Am. Chem. Soc.* **107** (1985) 2976.

55. Carté, B. and Faulkner, D. J. *J. Chem. Ecol.* **12** (1986) 795.

55a. Keifer, P. A. Rinehart, Jr. K. L. and Hooper, I. R. *J. Org. Chem.* **51** (1986) 4450.

56. Prestwich, G. D. *Tetrahedron* **38** (1982) 1911; in *Natural Product Chemistry*, Atta-ur-Rahman (Ed.) Springer, Berlin, 1986, 318; *Scientific American*, 1983, August, 68.

57. Quennedey A. *Recherche* **6** (1975) 274.

58. Spanton, S. G. and Prestwich, G. D. *Tetrahedron* **38** (1982) 1921.

59. Baker, R. and Walmsley, S. *Tetrahedron* **38** (1982) 1899.

60. Schildknecht, H. and Holoubek, K. *Angew. Chem.* **73** (1961) 1.

61. Eisner, H. E., Eisner, T. and Hurst, J. J. *Chem. Ind. (London)* **1963** 124.

62. Monro, A., Chadha, M. S., Meinwald, J. and Eisner, T. *Ann. Entomol. Soc. Am.* **55** (1962) 261.

63. Tursch, B., Braekman, J. C., and Daloze, D. *Experientia* **32** (1976) 401.

64. Wilson, E. O. in *Chemical Ecology*, Sondheimer, E. and Simeone, J. B. (Eds.) Academic Press, 1970, p. 133.

65. Cavill, G. W. K., Robertson, P. L., Brophy, J. J., Clark, D. V., Duke, R., Orton, C. J. and Plant, W. D. *Tetrahedron* **38** (1982) 1931.

66. Wheeler, J. W., Avery, J., Olubajo, O., Shamim, M. T., Storm, C. B. and Duffield, R. M. *Tetrahedron* **38** (1982) 1939.

67. Jones, T. H., Blum, M. S. and Fales, H. M. *Tetrahedron* **38** (1982) 1949.

67a. Numata, A. and Ibuka, T. *The Alkaloids* **31** (1987) 193.

68. Andersen, K. K., Bernstein, D. T., Caret, R. L. and Romanczyk Jr., L. J. *Tetrahedron* **38** (1982) 1965.

69. Francke, W., Schulz, S., Sinnwell, V., König, W. A. and Roisin, Y. *Liebigs Ann.* **1989** 1195.

69a. Schwenk, K. *Science* **263** (1994) 1573.

70. Brand J. M., Young, J. C. and Silverstein, R. M. *Prog. Chem. Org. Nat. Prod.* **37** (1979) 1.

70a. Schulz, S. and Toft, S. *Science* **1993** 1635.

71. Tumlinson, J. H., Hardee, D. D., Gueldner, R. C., Thomson, A. C., Hedin, P. A. and Minyard, J. P. *Science* **166** (1969) 1010.

72. Silverstein, R. M:, Brownlee, R. G., Bellas, T. E., Wood, D. L. and Browne, L. E. *Science* **159** (1968) 889.

73. Horler, D. F. *J. Chem. Soc.* (C) **1970**, 859; Pircle, W. H. and Boeder, C. W. *J. Org.*

Chem. **43** (1978) 2091.

74. Roelofs, W. L., Comeau, A. and Selle, R. *Nature* **224** (1969) 723.
75. Iwaki, S., Marumo, S., Saito, T., Yamada, M., and Katagiri, K. *J. Am. Chem. Soc.* **96** (1974) 7842.
76. Müller, D. G. in *Marine Natural Products Chemistry*, Falkner, D. J. and Fenical, W. H. (Eds.), Plenum Press, New York, 1977, p. 351.
77. Albone, E. *Chem. Britain*, **13** (1977) 92.
78. Crews, D. and Garstka, W. R. *Scientific American* **247** Nov. (1982) 136.
79. Davidson, B. S., Eisner, T., Witz, B. and Meinwald, J. *J. Chem. Ecol.* **15** (1989) 1689.
79a. Picket, J. A. in *Ecological Chemistry and Biochemistry of Plant Terpenoids*. Harborne, J. B. and Tomas-Barberan, F. A. (Eds.) Clarendon Press, Oxford, 1991, p. 297.
80. Riley, R. G., Silverstein, R. M. and Moser, J. C. *Science* **183** (1974) 760; McGurk, D. J., Frost, J., Eisenbrown, E. J., Vick, K., Drew, W. A. and Young, J. *J. Insect Physiol.* **12** (1966) 1435.
81. Walgenbach, C. A., Phillips, J. K., Burkholder, W. E., King, G. G. S., Slessor, K. N. and Mori, K. *J. Chem. Ecol.* **1987** 2159; Schmuff, N. R., Phillips, J. K., Burkholder, W. E., Fales, H. M., Chen, C.-W., Roller, P. P. and Ma, M. *Tetrahedron Lett.* **1984** 1533; Mori, K. and Ebata, T. *Tetrahedron* **42** (1986) 4421.
82. Byrne, K. W., Swigar, R. M., Silverstein, R. M., Borden, J. H. and Stokkink, E. *J. Insect. Physiol.* **20** (1974) 1895.
82a. Mori, K., Kiyota, H., Malosse, C. and Rochat, D. *Liebigs Ann.* **1993** 1201.
83. Kitokara, Y., Kato, T., Kobayashi, T. and Moore, B. P. *Chem. Lett.* **1976** 219.
84. Tumlinson, J. H., Silverstein, R. M., Moser, J. C., Brownlee, R. G. and Ruth, J. M. *Nature* **234** (1971) 348.
84a. Kern, F. and Bestmann, H. J. *Z. Naturforsch.* **49C** (1994) 865.
85. Matsumura, F., Coppel, H. C. and Tai, A. *Nature* **219** (1968) 963.
86. Trainello, J. F. A. *Naturwissenschaften* **67** (1980) 361.
87. Schoonhoven, L. M. *J. Chem. Ecol.* **16** (1991) 157; Thiery, D. and Gabel, B. *Experientia* **49** (1993) 998.
87a. Hurter, J., Boller, E. F., Städler, E., Blattmann, B., Buser, H. R., Bosshard, N. U., Damm, L., Koslowski, M. W., Schöni, R., Raschdorf, F., Dahinden, R., Schlumpf, E., Fritz, H., Richter, W. J. and Schreiber, J. *Experientia*, **43** (1987) 157.
88. Ernst, B. and Wagner, B. *Helv. Chim. Acta*, **72** (1989) 165; Küchler, B., Voss, G. and Gerlach, H. *Liebigs Ann.* **1991** 545.
89. Byers, J. A., Lanne, B. S. and Löfquist, J. *Experientia*, **45** (1989) 489.
90. Tumlinson, H. H., Lewis, W. J. and Vet, L E. M. *Scientific American*, March **1993** 46.
91. von Frisch, K. *Z. Vergleich. Physiol.*, **29** (1941) 46; Pfeiffer, W. *Experientia*, **19** (1963) 113; Todd, J. H., Atema, J. and Bardach, J. E. *Science*, **158** (1967) 260.
92. de Candolle, A. P. *Theorie Elémentaire de la Botanique*, Paris, 1813.
93. Bode, H. B. *Planta* **51** (1958) 440.
94. Muller, C. H. *Rec. Adv. Phytochem.* **3** (1970) 106.
94a. Schultes, R. E. and Raffauf, R. F. *The Healing Forest*, Dioscorides Press, Portland, Oregon, 1990, p. 380.
94b. Page, J. E., Madriñan, S. and Towers, G. H. N. *Experientia* **50** (1994) 840; Jensen H. M. *Thesis*, Chem. Inst. Univ. Aarhus, Denmark. 1993.
95. Gliessman, S. R. and Muller, C. H. *J. Chem. Ecol.* **4** (1978) 337.
96. Bonner, J. and Galston, A. W. *Botan. Gazz.* **106** (1944) 185.

97. Webb, L. J., Tracey, J. G. and Haydock, K. P. J. *Appl. Ecol.* **4** (1967) 13.
98. Fischer, N. H. in *Ecological Chemistry and Biochemistry of Plant Terpenoids*, Harborne, J. B. and Tomas-Barberan, F. A. (Eds.) Clarendon Press, Oxford, 1991, p. 377.
99. Harley, J. L. *Ann. Rev. Microbiol.* **6** (1952) 367.
100. Ingham, J. L. *Phytopath. Z.* **78** (1973) 314.
101. Müller, K. O. and Börger, H. *Arb. biol. Abt. Reichsanst. Land Forst.* **23** (1940) 189.
102. Perrin, D. R. and Bottomley, W. *J. Am. Chem. Soc.* **84** (1962) 1919.
103. Tomiyama, K., Sakuma, T., Ishizaka, N., Sato, N., Katsui, N., Takasugi, M. and Masamune, T. *Phytopathology* **58** (1968) 115.
104. Niemann, G. J. *Phytochemistry* **34** (1993) 319.
104a. Schneider, J. A., Lee, I., Naya, Y., Nakanishi, K., Oba, K. and Uritani, I. *Phytochemistry* **23** (1984) 759.
105. Stierle, A., Herschenhorn, J. and Strobel, G. A. *Phytochemistry* **32** (1993) 1145.
106. Stierle, A., Cardellina II, J. H. and Strobel, G. A. *Proc. Natl. Acad. Sci. USA* **85** (1988) 8008; Strobel, G. A. *Scientific American* **1991,** July, 50.
107. Stierle, A., Strobel, G. A. and Stierle, D. *Science* **260** (1993) 214; Stierle, A., Strobel, G., Stierle, D., Grothaus, P. and Bignani, G. *J. Nat. Prod.* **58** (1995) 1315.
108. Brooks, C. J. W. and Watson, D. G. *Nat. Prod. Rep.* **2** (1985) 427; **8** (1991) 367; Threlfall, D. R. and Whitehead, I. M. in Harborne, J. B. and Tomas-Barberan, F.A. (Eds.) *Ecological Chemistry and Biochemistry of Plant Terpenoids.* Clarendon Press, Oxford, 1991, p. 159; Smith, D. A. and Banks, S. W. *Phytochemistry* **25** (1986) 979; Kemp, M. S. and Burden, R. S. *Phytochemistry,* **25** (1986) 1261.

Chapter 3

Carbohydrates and primary metabolites

3.1 Classification. Structure of glucose

Carbohydrates (sugars) is the general term for polyhydroxy compounds usually containing carbonyl functions. They are classified according to the numbers of carbons, C_7 heptoses, C_6 hexoses, C_5 pentoses, etc., or according to the number of units in the molecule, monosaccharides, disaccharides and polysaccharides. Sugars containing an aldehyde function are termed aldoses and if they contain a keto function, ketoses. They are present in all living material where they play a central role in primary biosynthesis, production and storage of energy and matter. Hexoses are by far the commonest type of sugars and glucose is one of the most widely distributed and abundantly

$$
\begin{array}{ccccc}
\text{CO}_2\text{H} & & \overset{1}{\text{CHO}} & & \\
| & & |_2 & & \\
\text{CHOH} & & \text{CHOH} & & \\
| & & |_3 & & \\
\text{CHOH} & \xleftarrow{\text{HNO}_3} & \text{CHOH} & \xrightarrow{\text{PhNHNH}_2} & \\
| & & |_4 & & \\
\text{CHOH} & & \text{CHOH} & & \\
| & & |_5 & & \\
\text{H COH} & & \text{H COH} & & \\
| & & |_6 & & \\
\text{CO}_2\text{H} & & \text{CH}_2\text{OH} & & \\
\end{array}
$$

Glucaric acid	(+) Glucose	(1)
Mannaric acid	(+) Mannose	

$$
\begin{array}{ccc}
\text{CH=NNHPh} & & \text{CH}_2\text{OH} \\
| & & | \\
\text{C=NNHPh} & & \text{CO} \\
| & & | \\
\text{CHOH} & \xleftarrow{\text{PhNHNH}_2} & \text{CHOH} \\
| & & | \\
\text{CHOH} & & \text{CHOH} \\
| & & | \\
\text{H COH} & & \text{H COH} \\
| & & | \\
\text{CH}_2\text{OH} & & \text{CH}_2\text{OH} \\
\end{array}
$$

Glucosazone   (-) Fructose

occurring compounds. The structural elucidation of glucose by Fischer in 1896 belongs to the classical era of stringent structural reasoning long before modern spectroscopy was available. No absolute configuration was known at the time and Fischer suggested that the configuration of C^5 was as written in (1) and by chance he proved to be right. The C^5 carbon of (+)glucose was shown to have the same configuration as C^2 of (+)glyceraldehyde and compounds related to (+)glyceraldehyde were said to belong to the D-series because (+)glucose was dextrorotatory. The enantiomers belong to the L-series (levorotatory). The small capital D and L thus refer to the absolute configuration and are not connected with the sign of rotation, which could change within a series of related compounds. For historical and practical reasons these notations continue to be used in certain fields, e.g. in carbohydrate and amino acid chemistry along with the more general Cahn-Ingold-Prelog convention based on the priority rules.

$$CHO$$
$$H \underset{}{\overset{}{|}} OH$$
$$CH_2OH$$

Fischer projection
D-Glyceraldehyde
R-Glyceraldehyde

The gross structure of glucose, mannose and fructose was known, and it was found that they gave the same osazone, i.e. they have identical configuration at C^3–C^5 and fructose must have the carbonyl group at C^2. When glucose and mannose were oxidized by nitric acid, different optically active glucaric acids were obtained (Fig. 1). The symmetric structures **1** and **2** can therefore be eliminated and also **3** because the configuration of glucose and mannose differs only at C^2. A change of the configuration at C^2 of **3** leads to either **1** and **2**. Thus, the structure of mannaric and glucaric acids must be represented by **4–6**.

When (–)arabinose was reacted with hydrocyanic acid and hydrolysed, a mixture of gluconic and mannonic acids was obtained, and oxidation with nitric acid gave an optically active dibasic acid, **7** (2). The configuration at C^2–C^4 in arabinose is thus identical to the configuration at C^3–C^5 in glucose and mannose and since the only optically active dicarboxylic acid from arabinose is represented by **7**, it follows that the structure of (+)glucose and (+)mannose is either **8** or **9**, the structure of (–)fructose is **10** (Fig. 2), and the structures of arabinose, gluconic and mannonic acids are as depicted in (2).

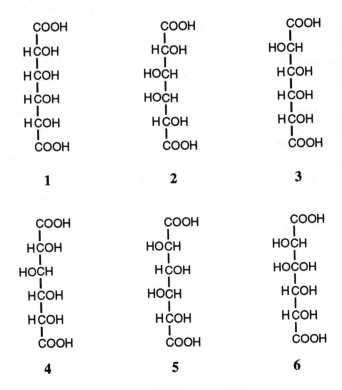

Fig. 1   Structure of glucaric acids

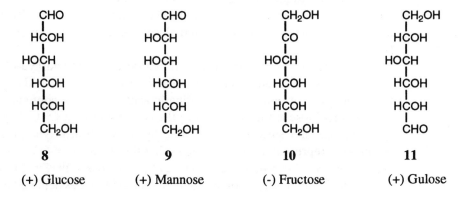

Fig. 2   Structure of hexoses

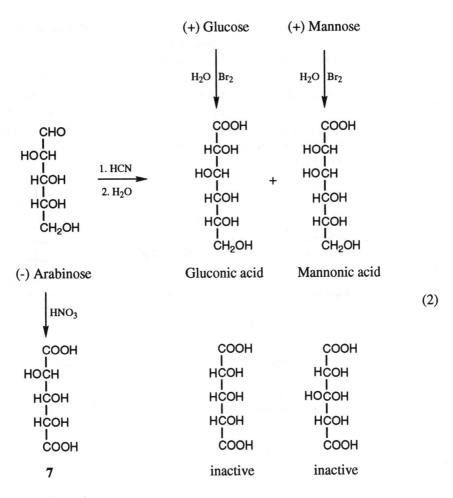

(+) Glucose   (+) Mannose

H_2O | Br_2 H_2O | Br_2

CHO
|
HOCH
|
HCOH   1. HCN
| ————————→
HCOH   2. H_2O
|
CH_2OH

COOH   COOH
| |
HCOH   HOCH
| |
HOCH   +   HOCH
| |
HCOH   HCOH
| |
HCOH   HCOH
| |
CH_2OH CH_2OH

(-) Arabinose   Gluconic acid   Mannonic acid

(2)

| HNO₃

COOH   COOH   COOH
| | |
HOCH   HCOH   HCOH
| | |
HCOH   HCOH   HOCOH
| | |
HCOH   HCOH   HCOH
| | |
COOH   COOH   COOH

7   inactive   inactive

optically active

A fourth sugar was available to Fischer, (+)gulose, which on oxidation with nitric acid gave the same glucaric acid as (+)glucose. This observation is only compatible with structure **11** for (+)gulose, **8** for (+)glucose and **9** for (+)mannose. An interchange of the C^1 and C^6 functions in **9** leads to the same compound.

3.2 Conformation and stereoisomerism

The conformations of a molecule refer to those spatial arrangements of chain and substituents which can be achieved by rotations about single bonds. Stereoisomerism refers to isomeric pairs of molecules which are non-superimposable mirror images. The open form of hexoses exists in equilibrium with their hemiacetals or hemiketals (3a,b). Glucose is predominantly present in the pyranose form and both the α- and β-anomers have been isolated. By definition, α denotes the isomer which has C^1-OH and C^5-C^6 in the *trans* position. Cyclization creates a new asymmetric centre. The anomeric carbon atom is distinguished from the other carbon atoms by the fact that it is linked to two oxygen atoms. The anomers which are stereoisomers differ in physical properties and in optical rotation. In solution the two forms approach an equilibrium, a reaction that can be followed by measuring the optical rotation. This change is called *mutarotation*. The six-membered rings exist in chair and boat forms (Fig. 3).

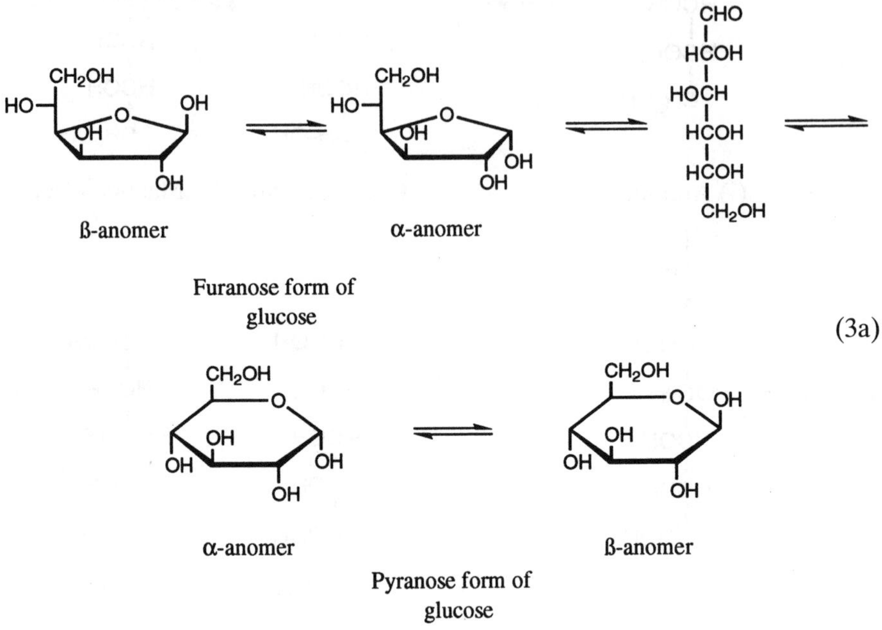

β-anomer   α-anomer

Furanose form of
glucose

(3a)

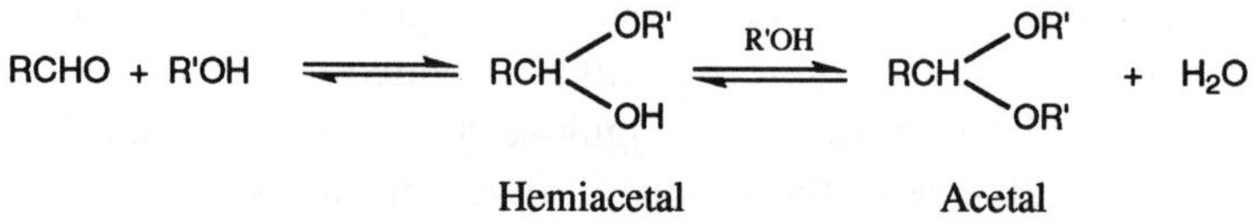

α-anomer   β-anomer

Pyranose form of
glucose

$$RCHO + R'OH \rightleftharpoons RCH \begin{smallmatrix} OR' \\ \\ OH \end{smallmatrix} \xrightarrow{R'OH} RCH \begin{smallmatrix} OR' \\ \\ OR' \end{smallmatrix} + H_2O \qquad (3b)$$

Hemiacetal   Acetal

A   B

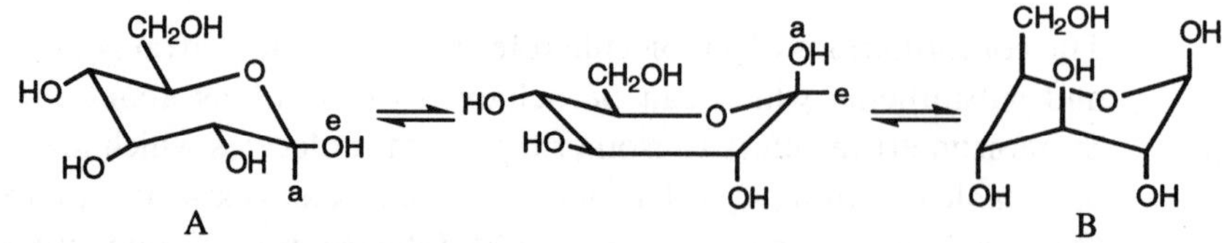

Fig. 3   Chair and boat forms of β-glucose. e denotes equatorial bond, a axial bond

(4)

Because of reduced steric interaction, the chair form is normally more stable than the boat form and of the two chair forms A is more stable because of 1,3-diaxial interactions and destabilizing polar effects in B. 1,3-Diaxial repulsion is a controlling factor for conformation in six-membered rings but quite unexpectedly the pyranoid tribenzoate in (4) occurs preferably in the axial form[1]. The anomeric factor is supposedly to blame, i.e. the repulsion between the halogen and the lone pairs of the ring oxygen seems to govern the conformation. ^{1}H NMR spectroscopy has been used effectively for the determination of conformations. Not only are the shifts of axial and equatorial protons different, but the coupling constants, $^3J_{aa}$, of 1,2-diaxial protons are larger (*ca.* 9 Hz) than coupling constants of J_{ee} and J_{ae} of 1,2-diequatorial and 1,2-axial-equatorial protons (*ca.* 2–4 Hz). The coupling constants, $^1J_{13_{C^1H^1}}$, are also consistently different for axial and equatorial anomeric protons and they have been used for structural determination.[2]

3.3 Photosynthesis

Photosynthesis, which entails carbon dioxide fixation and conversion of light energy into chemical energy, is the reversed process of respiration, i.e. combustion. It is the central pathway of carbohydrate biosynthesis in green plants. Energy is required for the process and is provided by light from the sun, which is absorbed by the chlorophylls and accessory carotenoids of photosystems I and II in green plants. The details of the overall reaction (5) are extraordinarily complicated, but are now known to a great extent. Life on earth depends on this reaction and it is of utmost importance that man economizes his use of the large remaining areas of green vegetation in the temperate and tropical zones, e.g. in the Amazon region. They are part of our renewable

$$n\,CO_2 + n\,H_2O \underset{\text{Respiration}}{\overset{\substack{\text{Assimilation} \\ h\nu,\ \text{ATP, NADPH}}}{\rightleftharpoons}} (CH_2O)n + nO_2 \qquad (5)$$

energy sources. Animals use the energy in their respiration process and carbon dioxide is returned, thereby contributing to the oxygen: carbon dioxide balance. The exceedingly large combustion of fossil fuels in the last decades seems to have slowly increased the carbon dioxide content in the atmosphere, which in the long run could bring about climatic changes with unforeseen consequences.

Photosynthesis has two phases: first, the *light reaction*, in which visible light energy ($\lambda < 690$ nm) is used for the production of the reducing reagent NADPH and energy rich ATP. Oxygen is produced during this phase, at the separate oxygen evolving center (OEC) (6). In the second phase, the *dark reaction*, NADPH and ATP are used for the carbon dioxide reduction to give carbohydrates. The terms "energy rich" and "energy releasing" commonly used in the literature in conneccction with ATP actually mean "reactive". It is not a question of transferring a quantum of energy but simply of using ATP as a phosphorylating agent, thus making functions in other molecules (substrates) more reactive, e.g. $-COOH$, $-CONH_2$, $-OH$ and SO_4^{2-} are transformed to $-COOP$, $-C(=NH)OP$, OP and $HOSO_2OP$. The nucleophile reacts with one of the P-O-P bonds of ATP and one or two phosphate units are transferred to the nucleophile.

$$H_2O + NADP^+ + P + ADP \xrightarrow{\ h\nu\ } {}^1\!/_2O_2 + NADPH + ATP + H^+ \qquad (6)$$

At photosystems I and II electrons are boosted by the light energy to a higher level of energy and transported by the central electron transport chain to their final destination, $NADP^+$, which forms NADPH.

It was found by using oxygen-labelled water that the oxygen produced did come from water and not from carbon dioxide. Some sulphur bacteria use hydrogen sulphide as a hydrogen donor, while other bacteria use certain organic substrates such as 2-propanol.

Water is catalytically split or oxidized at the OEC thereby providing the electrons required for the reduction of $NADP^+$ (6). The mechanism of water oxidation is not well understood despite the formal simplicity of the reaction as rewritten in (7).

$$2H_2O \xrightarrow{\ h\nu\ } 4H^+ + 4e^- + O_2 \qquad (7)$$

One problem is that single electron transfer from water to give $HO^\bullet$ is thermodynamically very unfavourable because of the high oxidation potential, $E^\circ_{HO^\bullet/H_2O} = +2.4$ V. The two-electron transfer reaction giving intermediate formation of hydrogen peroxide has $E^\circ_{H_2O_2/H_2O} = +1.36$ V at pH 7, whereas a concerted four-electron transfer has $E°_{O_2/H_2O} = +0.81$ V at pH 7. This value can now be compared more favourably with the oxidation potential of the natural photosynthetic one-electron oxidant formed by loss of one electron from photosystem II, $E°_{PII+/PII}=+1.1V$.[3] The OEC should therefore have an oxidation potential close to that value. The role of the catalyst is to avoid thermodynamically unfavourable electron transfer steps requiring high oxidation potentials and be able to store four oxidation equivalents to be released in one step.

The structure of the catalyst core is not fully known. It is a multinuclear complex containing four oxygen bridged Mn ions, stabilized by chloride and calcium ions, embedded together with essential polypeptides in the membrane of the chloroplast. The low potential S_0 state probably consists of a Mn^{4+}, Mn^{3+}_3 complex, (or Mn^{2+}, Mn^{3+}_3), which is charged by successive light flashes to the unstable high potential S_4 state, Mn^{5+}, Mn^{4+}_3, (Mn^{3+}, Mn^{4+}_3) capable of releasing its four oxidation equivalents in a concerted reaction[4]. The Mn^{5+} ion of the S_4 state is formally equivalent to the electrophilic $Mn^{5+}=O$ intermediate proposed to be the reactive species in catalytic olefin epoxidation and hydroxylation of saturated hydrocarbons.

$$S_4 \qquad\qquad S_2$$

$$Mn^{5+}, Mn_3^{4+} \text{ complex} \qquad Mn^{3+}, Mn_3^{4+}$$

$$(8)$$

$$O_2 + Mn^{4+}, Mn_3^{3+}$$

$$S_0$$

The mechanism of the O-O bond formation, which is still open to speculation, can be understood in the following way. The strong positive charge of the central S_4 Mn-core is expected to withdraw electrons from the ligated oxygen atoms, which makes at least one of them prone to react with the electrophilic hydroxyl ion, giving rise to the peroxidic bond. This reaction is the reversal of the well established manganese catalysed fission of hydrogen peroxide and is favoured because of the remaining strong positive potential of the S_2 state, which is capable of concertedly oxidizing the peroxide (considered to be a transition state) to elementary oxygen. This step does not require the same high oxidation potential as the initial O-O bond formation. Reaction (8) shows a section of the reactive site of the Mn complex. Manganese complexes containing the $(Mn_4O_{2-4})^{n+}$ core have been synthesized, which mimic the essential structural features of the native OEC, such as the metal nuclearity of four, the oxygen bridges between the metal centers, the oxidation levels and the atomic distances.[5] Conformational changes of the enzyme complex at the final oxidation step to give the S_4 state could conceivably bring the core from its originally hydrophilic environment into a more lipophilic environment, which increases its oxidation potential and facilitates the O-O-coupling. After the release of oxygen the complex returns to its S_0 conformation and the cycle can be repeated. A rare example of O-O bond formation, mechanistically analogous to the water oxidation as formulated in (8), has been reported for a reaction involving iron complexes.[5a]

Great expectations are tied to research on artificial photosynthesis by visible light[5b]. Mastering photochemical water cleavage could solve the Earth's energy problem in an environmentally acceptable and inexpensive way. The formidable task is to find the most suitable light-harvesting unit and redox catalyst and to achieve charge separation, i.e. separation of the hydrogen and oxygen formation.

In the second phase, ATP and NADPH are used for the carbon dioxide reduction. Calvin and coworkers found by using radioactive CO_2 that 3-phosphoglyceric acid was one of the first intermediates and the label was predominantly localized in the carbonyl group. They proposed the following sequence of reactions (9–11). Ribulose-1,5-diphosphate enolizes and the anion is carboxylated at the 2-position giving an unstable intermediate which undergoes a cleavage to two moles of 3-phosphoglycerate (9). 3-Phosphoglycerate now enters the reverse glycolytic pathway (10,11). ATP transfers 3-phosphoglycerate to the more reactive 3-phosphoglyceroyl phosphate. It is reduced by NADPH to 3-phosphoglyceraldehyde, one molecule of which rearranges to dihydroxyacetone phosphate and undergoes an aldol condensation with another molecule of 3-phosphoglyceraldehyde to form fructose-1,6-disphosphate that ultimately isomerizes to glucose. Each of these steps is catalysed by its specific enzyme. Assimilation of radioactive carbon dioxide for a very short period should thus give glucose labelled on carbon 3 and 4.

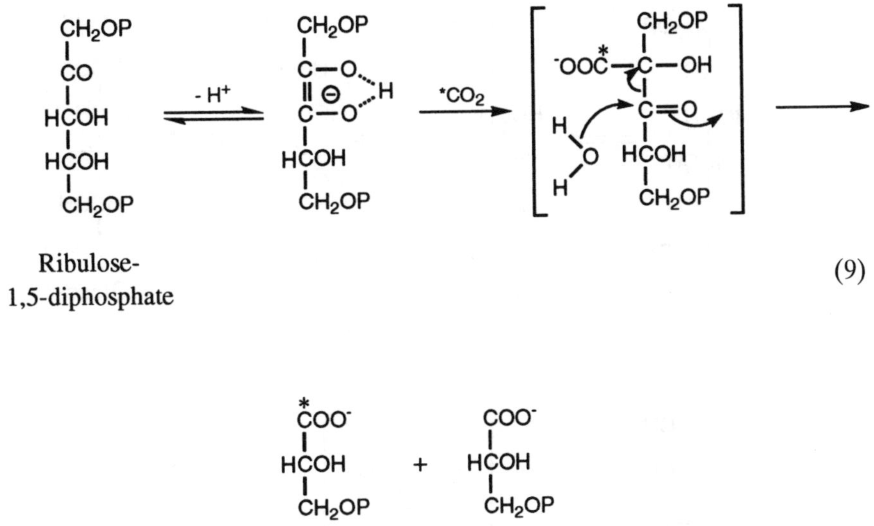

Ribulose-
1,5-diphosphate

(9)

3-Phosphoglycerate

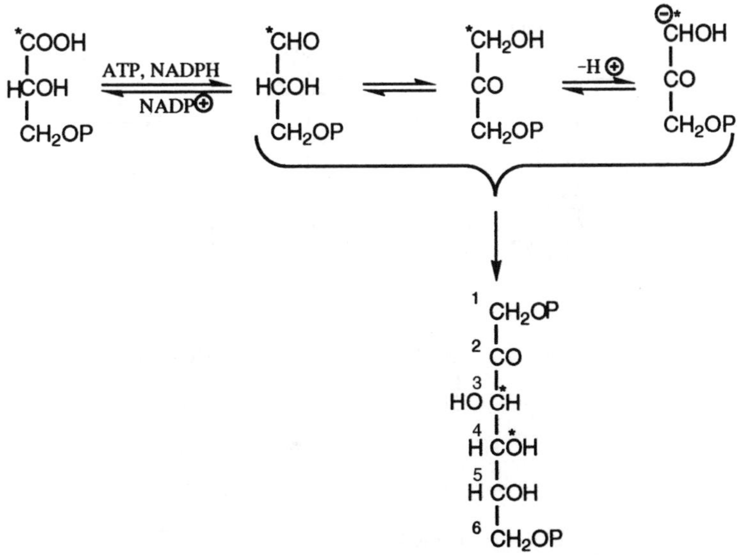

(10)

Fructose-1,6-
diphosphate

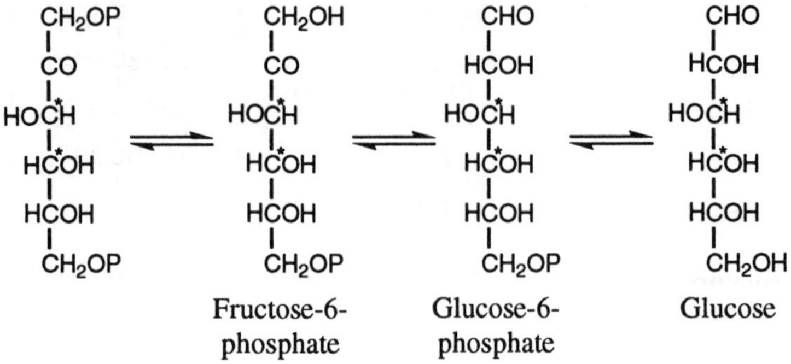

$$
\begin{array}{ccccccc}
\text{CH}_2\text{OP} & & \text{CH}_2\text{OH} & & \text{CHO} & & \text{CHO} \\
| & & | & & | & & | \\
\text{CO} & & \text{CO} & & \text{HCOH} & & \text{HCOH} \\
| & & | & & | & & | \\
\text{HOCH} & & \text{HOCH} & & \text{HOCH} & & \text{HOCH} \\
| & \rightleftharpoons & | & \rightleftharpoons & | & \rightleftharpoons & | \\
\text{HCOH} & & \text{HCOH} & & \text{HCOH} & & \text{HCOH} \\
| & & | & & | & & | \\
\text{HCOH} & & \text{HCOH} & & \text{HCOH} & & \text{HCOH} \\
| & & | & & | & & | \\
\text{CH}_2\text{OP} & & \text{CH}_2\text{OP} & & \text{CH}_2\text{OP} & & \text{CH}_2\text{OH}
\end{array}
\qquad (11)
$$

Fructose-6-
phosphate   Glucose-6-
                   phosphate   Glucose

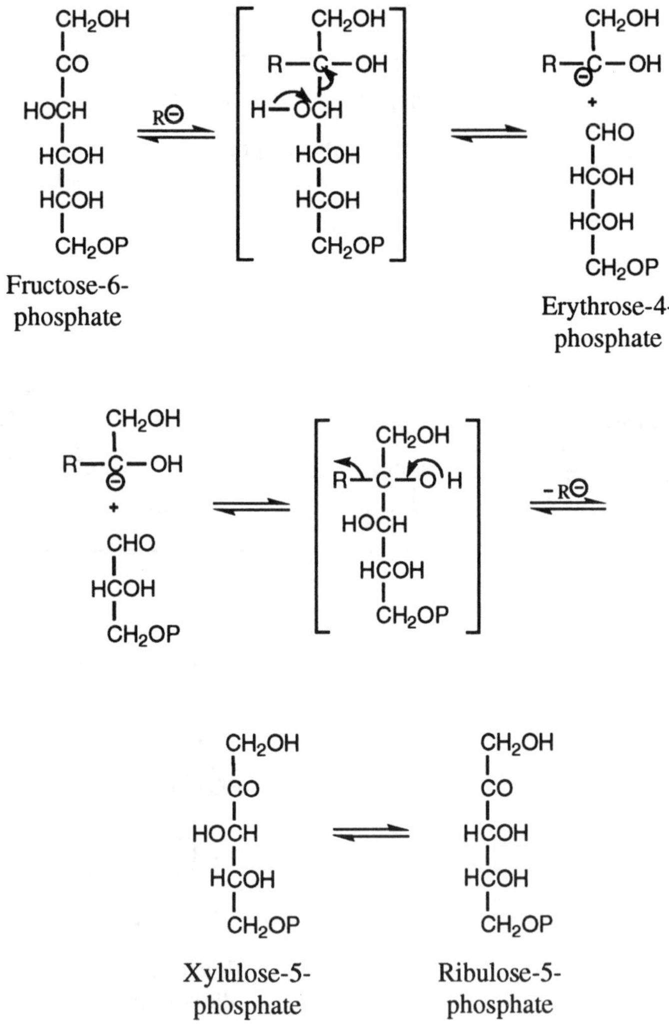

Fructose-6-
phosphate

Erythrose-4-
phosphate

$$ (12) $$

Xylulose-5-
phosphate   Ribulose-5-
                   phosphate

Equations (9–11) account for the production of one glucose molecule from carbon dioxide and ribulose-1,5-diphosphate, but one ribulose-1,5-diphosphate molecule is regenerated for each molecule of carbon dioxide reduced. In order to make the photosynthesis of carbohydrates self-consistent Calvin proposed the cycle (12,13). The Calvin cycle means that of the six fructose-6-phosphate molecules formed from six ribulose-1,5-diphosphate molecules and six molecules of carbon dioxide, one fructose-6-phosphate molecule is diverted into one glucose molecule, whereas the other five fructose-6-phosphate molecules are recycled into six new ribulose-1,5-diphosphate molecules. By the action of the coenzyme thiaminpyrophosphate (RH, vitamin B_1) fructose-6-phosphate is cleaved in a retrobenzoin condensation fashion and recondensed with 3-phosphoglyceraldehyde to ribulose-5-phosphate via xylulose-5-phosphate as an intermediate, a reaction catalysed by the enzyme transketolase (12).

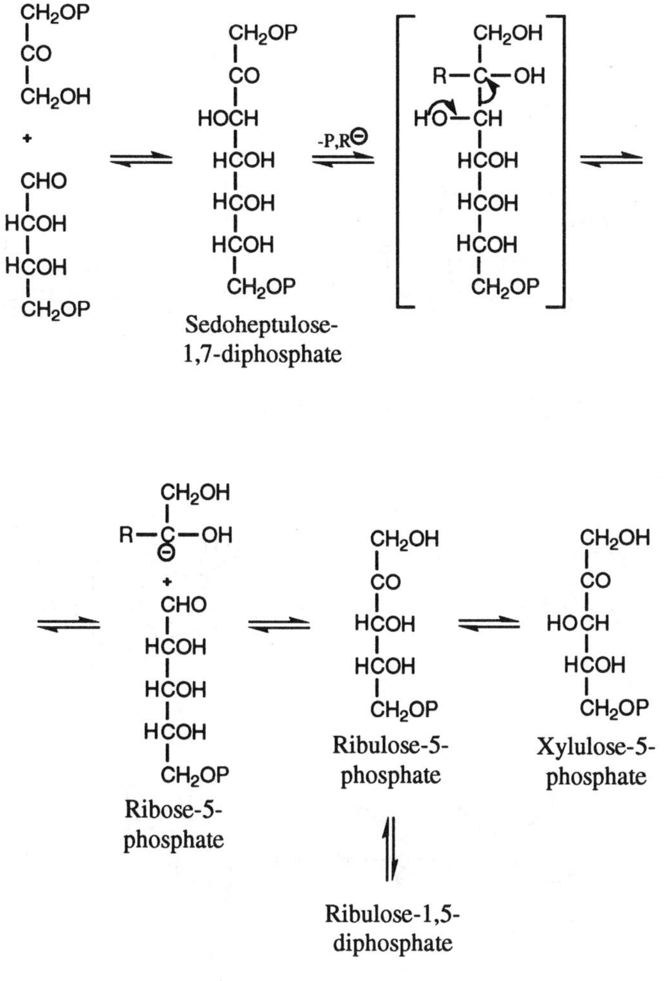

Sedoheptulose-1,7-diphosphate

(13)

Ribose-5-phosphate

Ribulose-5-phosphate

Xylulose-5-phosphate

Ribulose-1,5-diphosphate

The other cleavage product, erythrose-4-phosphate, condenses with dihy-droxy-acetone phosphate to sedoheptulose-1,7-diphosphate which undergoes cleavage by transketolase. Ribose-5-phosphate is formed (13) and the C_2-fragment gives xylulose-5-phosphate with 3-phosphoglyceraldehyde. Both pentoses rearrange finally to ribulose-5-phosphate which is phosphorylated by ATP to ribulose-1,5-diphosphate.

The mechanism of transketolization is formulated in (14) and (15). The de-protonated form of thiamine pyrophosphate attacks the carbonyl group and the anion formed in the cleavage is stabilized. In this respect the coenzyme re-sembles the action of the cyanide group in the benzoin condensation or the thiazolium ion catalyst in the Stetter reaction. Thiamine pyrophosphate plays the same role in the decarboxylation of pyruvic acid (16).

(14)

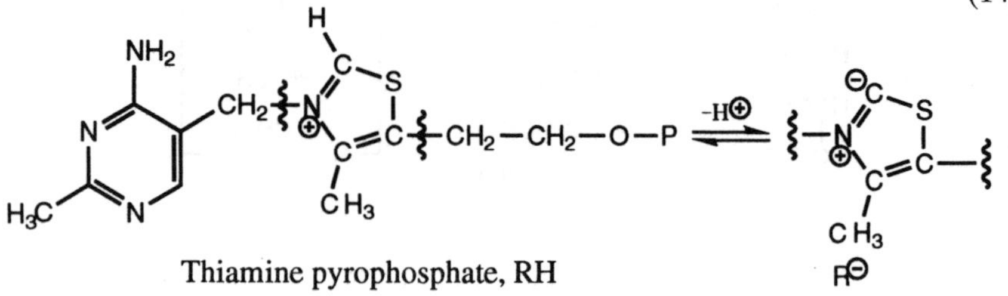

Thiamine pyrophosphate, RH

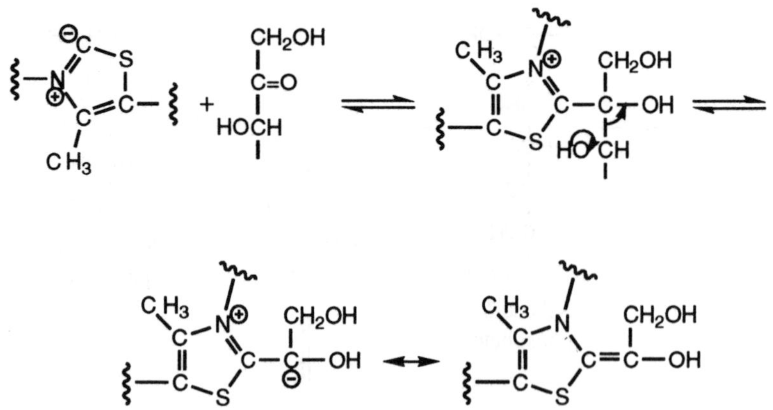

(15)

(16)

The reactions of the Calvin cycle can be summed up according to (17). Ribulose-1,5-diphosphate appears on both sides indicating that it is a true component of the photosynthesis and that another molecule is regenerated in the process.

$$6 \text{ Ribulose-1,5-diphosphate} + 6 \text{ CO}_2 + 18 \text{ ATP} +$$
$$12 \text{ NADPH} + 12 \text{ H}^{\oplus} + 12 \text{ H}_2\text{O} \rightarrow 6 \text{ ribulose-1,5-} \qquad (17)$$
$$\text{diphosphate} + \text{glucose} + 18 \text{ P} + 18 \text{ ADP} + 12 \text{ NADP}^{\oplus}$$

3.4 Breakdown of glucose. Glycolysis. The citric acid cycle

Glucose is both a reservoir of energy in the cell and the starting material for the biosynthesis of a vast number of compounds. In the muscle, glucose is broken down anaerobically to lactate as the end-product and energy is released. This process is called glycolysis, and we have seen in the previous section that the process is reversible. Lactate is brought by the blood to the liver where glucose is resynthesized. In the closely related alcoholic fermentation, pyruvic acid is decarboxylated (16) and reduced (18).

$$\text{CH}_3\text{CHO} \xrightarrow{\text{NADH}} \text{CH}_3\text{CH}_2\text{OH} \qquad (18)$$

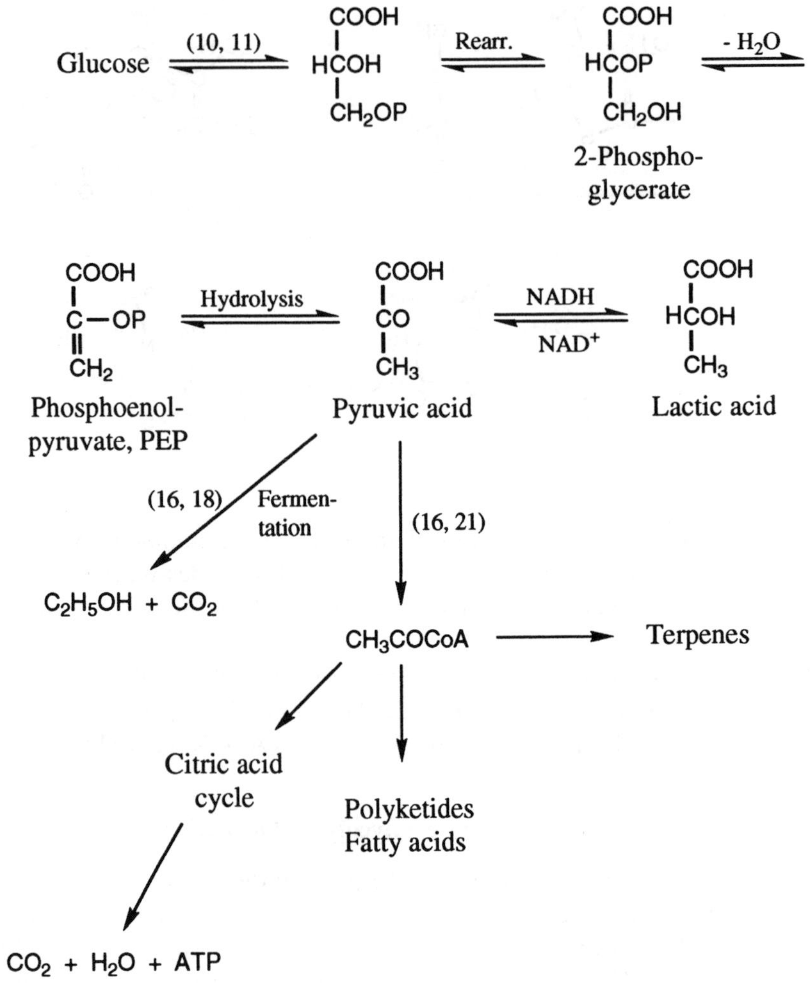

Fig. 4   Breakdown of glucose

By oxidative decarboxylation of pyruvic acid, acetyl coenzyme A is formed. This is the key compound in three vital processes (Fig. 4):

1. the citric acid cycle which, coupled with oxidative phosphorylation, in essence is respiration;
2. fatty acid synthesis; and
3. terpene synthesis.

In this section we will examine in some detail the formation of acetyl co-enzyme A and its combustion to carbon dioxide (19) in the citric acid cycle, a reaction that releases much more energy than glycolysis (20). In both processes about 50 per cent of free energy is conserved in ATP.

$$\text{Glucose} + 6\ O_2 + 36\ P + 36\ ADP \rightarrow 6\ CO_2 + 36\ ATP + 42\ H_2O \qquad (19)$$
$$\Delta G^{O'} = -423 \text{ kcal/mol}$$

$$\text{Glucose} + 2\ P + 2\ ADP \rightarrow 2\ \text{lactate} + 2\ ATP + 2\ H_2O \qquad (20)$$
$$\Delta G^{O'} = -32.4 \text{ kcal/mol}$$

The first step, decarboxylation, of the transformation of pyruvic acid to acetyl CoA is mediated by the coenzyme thiamine pyrophosphate according to (16). The enamine is oxidized by the coenzyme lipoic acid of the enzyme complex and the acetyl group is then transferred to coenzyme A which leaves the complex (21). The dihydrolipoic acid is reoxidized by FAD.

Lipoate   S-Acetyldihydrolipoate   (21)

CH₃COCoA

Acetyl CoA enters the citric acid cycle and condenses with oxaloacetic acid, 9, to form citric acid, 1. (Fig. 5) Elimination and addition of water give isocitric acid and oxidation of the hydroxyl group followed by decarboxylation give α-ketoglutaric acid, 4.

In much the same way as oxidative decarboxylation of pyruvic acid (16, 21), α-ketoglutaric acid is decaboxylated to succinyl CoA, 5, which is dehydrogenated to fumaric acid, 7. Addition of water gives malic acid which on oxidation regenerates oxaloacetic acid and the circuit is completed (Fig. 5). For every turn of the cycle one molecule of acetyl CoA is consumed, and two moles of carbon dioxide are evolved together with eight hydrogens or eight

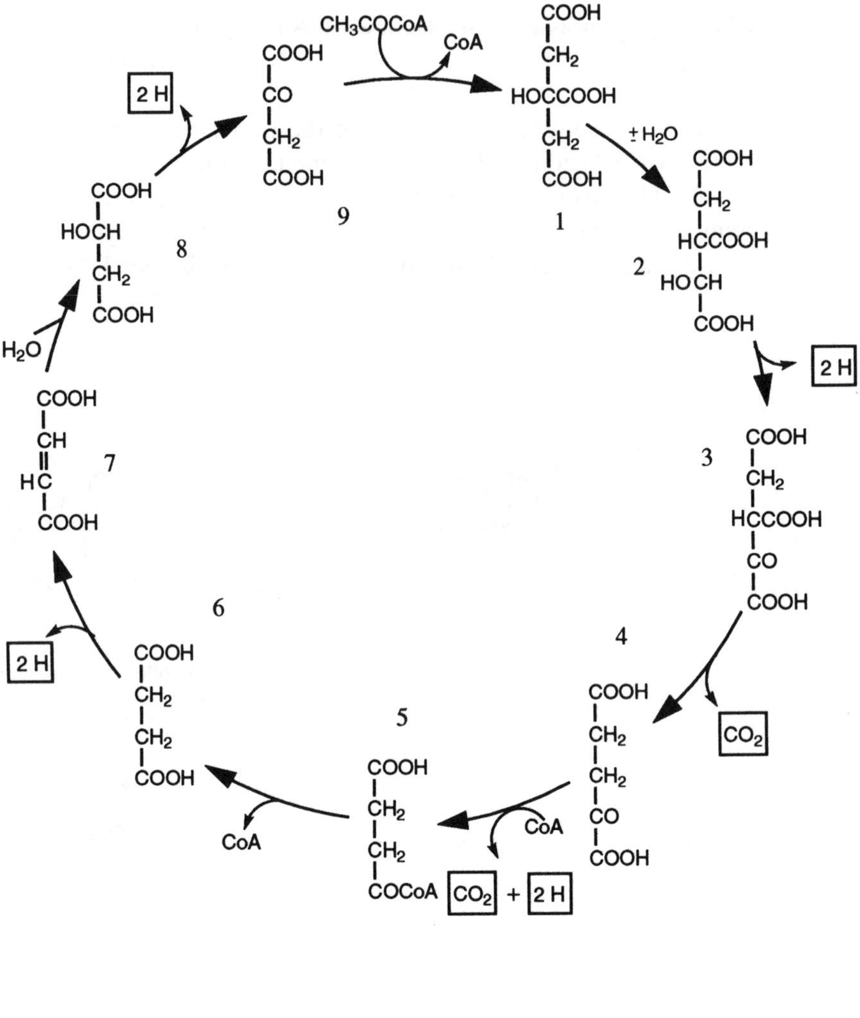

Fig. 5   The citric acid cycle. Two CO_2 and 8H are produced. 1, citric acid; 2, isocitric acid; 3, oxalosuccinic acid; 4, α-ketoglutaric acid; 5, succinyl CoA; 6, succinic acid; 7, fumaric acid; 8, malic acid; 9, oxaloacetic acid. The 8H are oxidized to water with concomitant production of free energy and ATP

electrons and eight protons. The hydrogens are oxidized to water with the production of large amounts of energy (19), part of which is conserved as ATP, a process called oxidative phosphorylation. The citric acid cycle was formulated by Krebs as the result of an ingenious piece of experimentation and reasoning, hence it is sometimes called the Krebs cycle.

3.5 Monosaccharides

Glucose is the precursor of other hexoses, disaccharides, polysaccharides and glycosides. Glucose-6-phosphate, fructose-6-phosphate and mannose-6-phosphate are directly interconvertible but transformation of glucose to other hexoses requires the assistance of a nucleoside triphosphate, usually uridine triphosphate, UTP, which is attacked at the α-P by the phosphate group of glucose-1-phosphate. Diphosphate is split off and uridine diphosphoglucose is formed (22). UDP-glucose can be transformed to most other sugars by epimerizations, oxidations and reductions, or directly reacted with other alcohols or sugars to glycosides.

Glucose-1-
phosphate   Uridine triphosphate, UTP

(22)

Uridine diphospho-
glucose, UDP-glucose

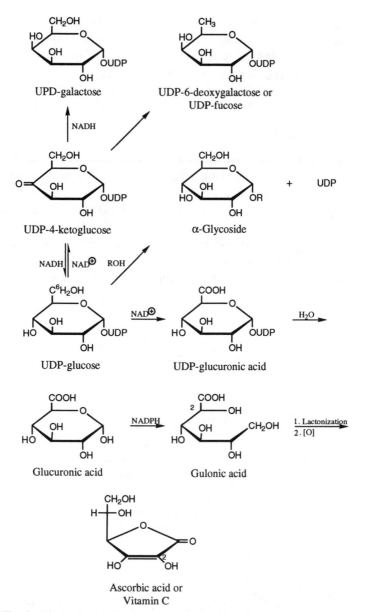

Fig. 6   Biosynthesis of galactose and ascorbic acid

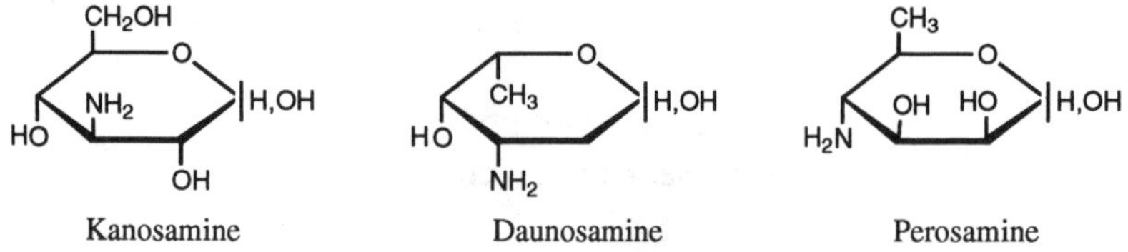

Kanosamine   Daunosamine   Perosamine

Fig. 7   Amino sugars from glycosidic antibiotics

These reactions have interesting mechanistic implications. Studies of the glucose-galactose interconversion demonstrate that no exchange of C^4–H or C^4–O takes place with the solvent, i.e. the ordinary S_N2 substitution with inversion of configuration at C^4 or a dehydration–hydration sequence are excluded. A redox process[6] involving $NAD^\oplus$ as oxidant is attractive for two reasons. It is in accordance with the lack of C^4–H exchange with the solvent and furthermore feeding 4-keto-D-glucose to the enzyme system containing 4–^{3}H–NADH results in the formation of 4–^{3}H–D-galactose but, strangely enough, 4–^{3}H–D-glucose was not detected. The redox process requires that the intermediate 4-keto sugar must rotate 180° before being reduced.[7] Fig. 6 shows the biosynthesis of ascorbic acid and fucose, a common constituent of cell walls in bacteria and algae, and the biosynthesis of galactose, an important sugar occurring in milk and a common constituent of many polysaccharides, e.g. agar and gum arabic. Hydrolysis of milk sugar, lactose, gives one mole of glucose and one mole of galactose. The mechanism of C^6-deoxygenation is shown in Fig. 9.

Oxidation of C^6 of glucose to a carboxyl group gives glucuronic acid, the precursor of ascorbic acid, or vitamin C. Reduction of the aldehyde group gives gulonic acid which lactonizes to the five-membered gulonolactone. Oxidation of C^2 and enolization gives ascorbic acid. Ascorbic acid is a vitamin, an essential nutrient for human beings, who require *ca.* 70 mg a day, but are unable to synthesize it themselves. Deficiency leads to the symptoms of

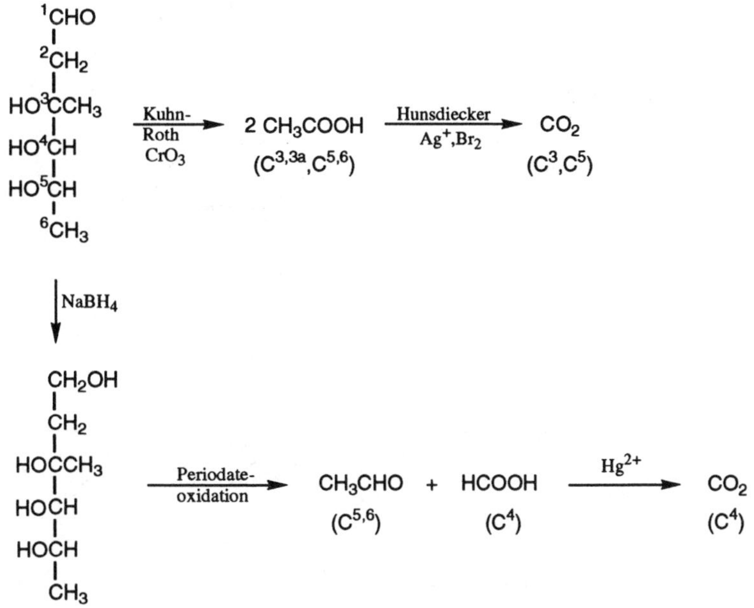

Fig. 8 Degradation of mycarose

scurvy. The physiological processes in which ascorbic acid partakes are still unknown. It is excreted in human urine as the 2-O-sulphate. Ascorbic acid is produced in large quantities industrially from glucose.

Fig. 7 shows some rare sugar constituents of glycosidic antibiotics containing the amino function. It is introduced by oxidation of a hydroxyl group with NADP$^\oplus$ to a carbonyl group followed by transamination (section 7.4) and reduction.

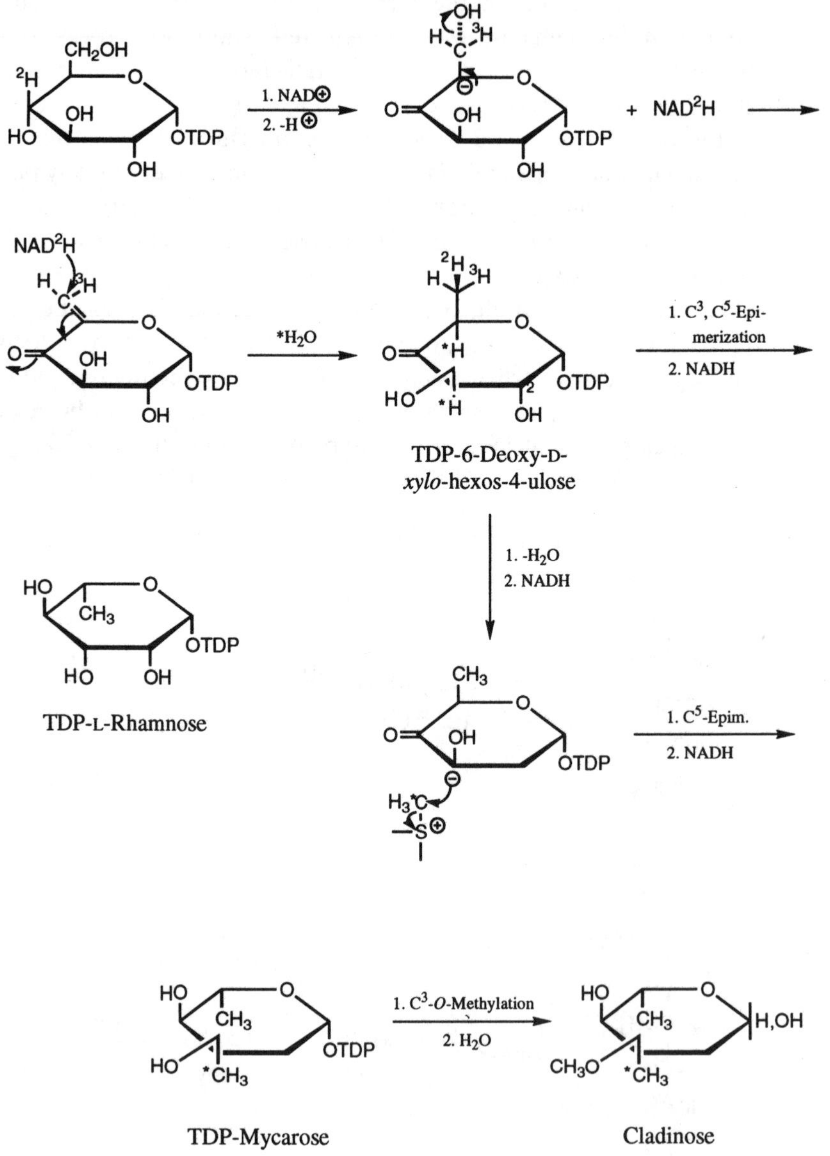

TDP-6-Deoxy-D-
xylo-hexos-4-ulose

TDP-L-Rhamnose

TDP-Mycarose   Cladinose

Fig. 9   Biosynthesis of rhamnose, mycarose and cladinose

The first branched sugars, apiose[8] (Fig. 11) and hamamelose[9] were isolated several years ago from parsley *Petroselium crispum*, and *Hamamelis virginiana*, respectively, and they were for a long period regarded as curiosities until in recent times branched sugars were detected in microorganisms as components of antibiotics,[10-12] i.e. mycarose, cladinose, streptose, and aldgarose.

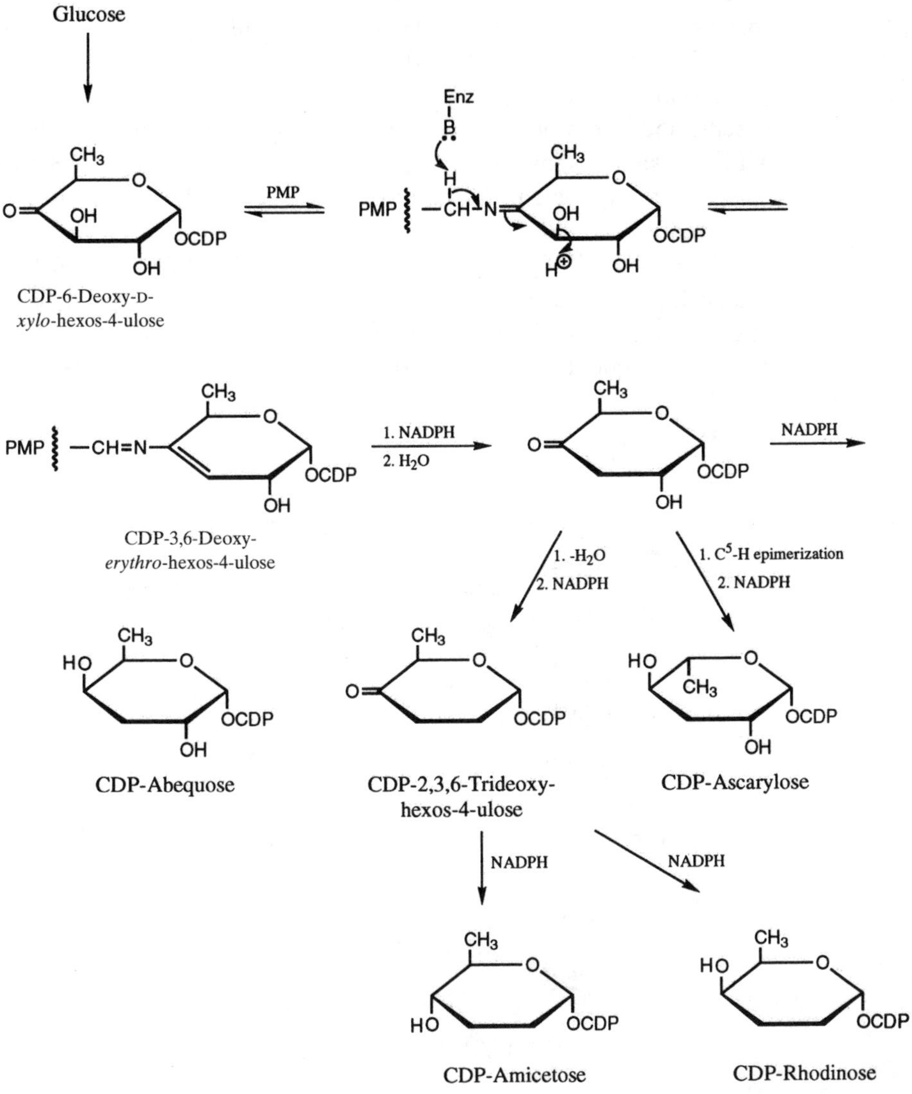

Fig. 10   Biosynthesis of some deoxy sugars

Mycarose is biosynthesized from intact glucose in cell-free extracts from *Streptomyces rimosus*, methionine acting as a methyl donor.[12] The early biosynthetic studies were carried out in the 'prespectroscopic' period by using Kuhn–Roth oxidation, borohydride reduction, periodate cleavage, and Hunsdiecker decarboxylation (Fig. 8). The fact that TDP-6-deoxy-D-*xylo*-hexos-4-ulose (TDP = thymidine diphosphate, section 9.2) acts as an intermediate precursor for rhamnose[13] and mycarose, gives the key to the mechanisms of deoxygenation and C-methylation (Fig. 9). Oxidaton of C^4–OH of TDP-glucose by $NAD^{\oplus}$ to a carbonyl group activates C^3–H and C^5–H. A carbanionic species is generated and C^6–OH is lost to yield an α, β-unsaturated ketone which is then reduced by NADH. The label of C^4–^{2}H of the substrate is recovered at C^6. The steric course of the hydride transfer is completely stereospecific and intramolecular and involves displacement of the hydroxyl group with inversion as shown by ^{2}H and ^{3}H labelling.[14] C^3–H and C^5–H undergo exchange with the solvent. Consequently C^2 is reduced before C^3 is methylated.

Formation of C^3-deoxy sugars can be formulated similarly by generation of a carbanion at C^2. The enzymatic formation of 2-keto-3-deoxy-6-P-gluconate from 6-P-gluconic acid has been carefully studied.[15] A carbanion is generated at C^2, facilitating the elimination of the C^3–OH. The intermediate enol rearranges subsequently to the 2-keto derivative (23).

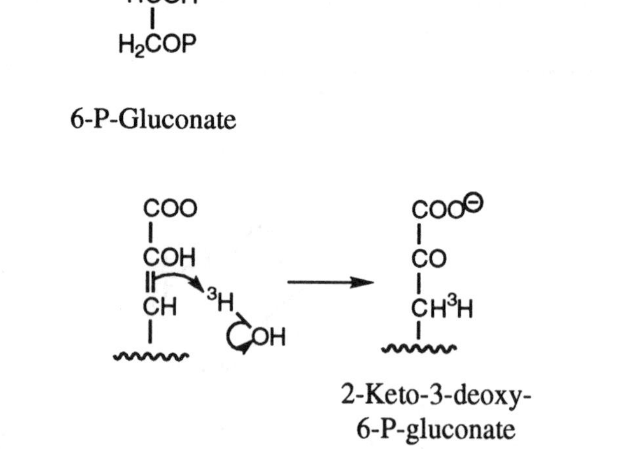

6-P-Gluconate

(23)

2-Keto-3-deoxy-
6-P-gluconate

In consistency with this mechanism one atom of ^{3}H was incorporated at C^3 when the reaction was carried out in ^3H_2O; this rules out a hydride reduction. It is also demonstrated that the 2-keto-3-deoxy sugar formed does not undergo proton exchange at C^3. Unreacted 6-P-gluconic acid was found to be radioactive, indicating a fast anionic pre-equilibration at C^2. It is observed spectroscopically that the formation of the 2-keto product is slower than the disappearence of the substrate which suggests that the intermediate enol form has a comparatively long lifetime. It is actually stabilized by the conjugated carboxyl function. The ketonization proceeds non-enzymatically as shown by the formation of equal amounts of 3S and 3R stereoisomers when the reaction was carried out in D_2O.

CDP-6-deoxy-D-*xylo*-hexos-4-ulose (CDP = cytidine diphosphate) serves as precursor for 3-deoxy sugars[16] in an enzymatic process mediated by pyridoxamine phosphate (PMP) (Fig. 10). This cofactor forms an imine with the

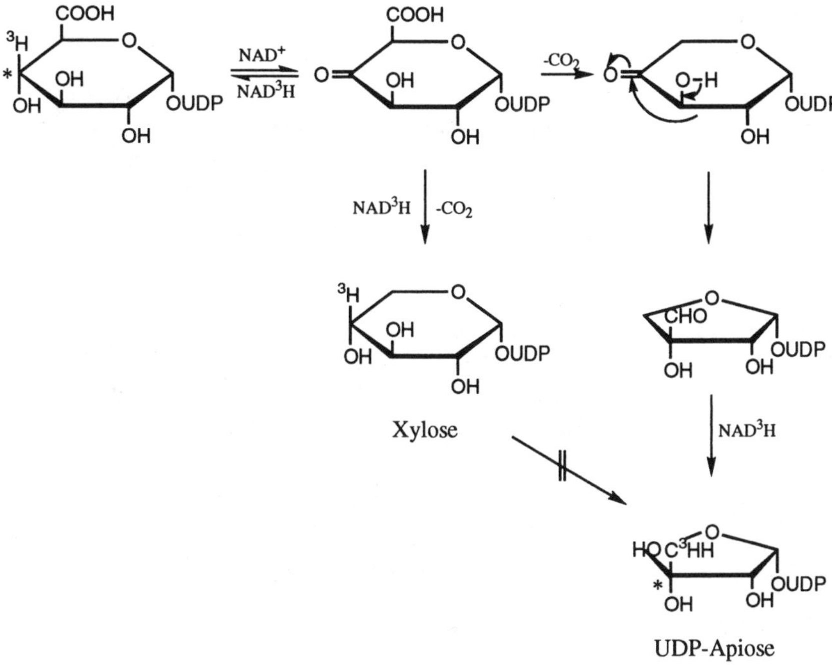

Fig. 11   Biosynthesis of UDP-apiose and xylose

4-keto sugar which transforms to an enimine by expulsion of the C^3–OH. The findings that both C^4=O and C^3–OH undergo ^{18}O exchange in H_2^{18}O, and that methylene labelled ^{3}H-pyridoxamine undergoes proton exchange, suggest reversibility of the reactions and lend strength to the postulated mechanism. Hydrolysis of the enimine could lead directly to 3,6-dideoxy-*erythro*-hexos-4-ulose but since NADPH is required as cofactor, a reductive step must precede the hydrolysis. Rather surprisingly incubation of the enzyme with (4*R,S*)–^{3}H–NADPH does not lead to incorporation of ^{3}H at the methylenes of pyridoxamine phosphate and the sugar residue. A stereospecific ^{3}H washout by the solvent has been offered as an explanation. The 4-keto function is eventually reduced by NADPH to the bacterial 3,6-dideoxy sugars abequose and ascarylose. Elimination of the C^2–OH from 3,6-dideoxy-*erythro*-hexos-4-ulose followed by reduction give rise to other bacterial sugars, 2,3,6-trideoxyhexoses amicetose and rhodinose (Fig. 10).

An early hypothesis that apiose is of isoprenoid origin is not substantiated. It was based on structural similarities and occurrence of apiose in the rubber tree, *Hevea brasiliensis*. If we disregard the isotopic scrambling caused by the

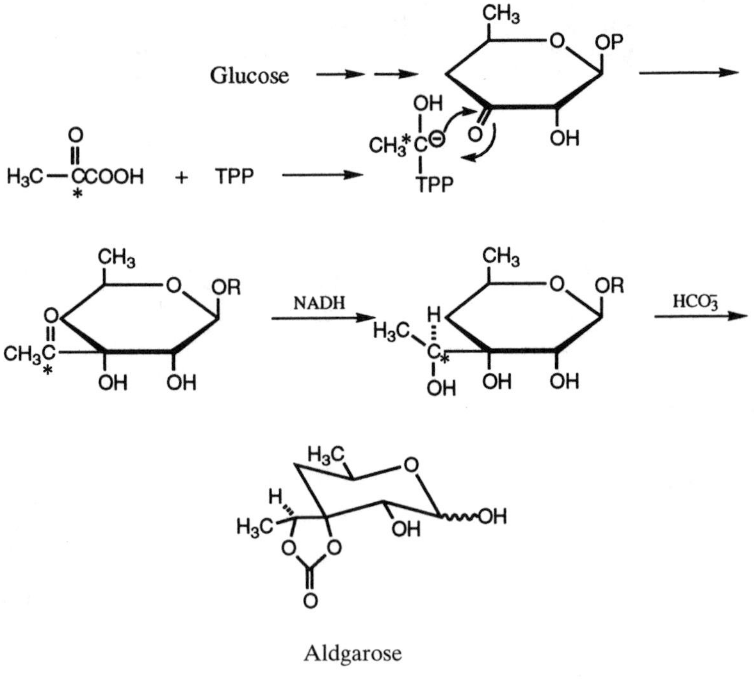

Aldgarose

Fig. 12   Proposed biosynthesis of aldgarose

enzymes during incubation, tracer studies reveal that glucose is the precursor[17, 18] (Fig. 11). It loses C^6 via decarboxylation of UDP-glucuronic acid and the hydroxymethyl group is formed by ring contraction. $C^4–^3H$ labelled UDP-glucuronic acid gives UDP-apiose with ^3H in the hydroxymethyl group, a reaction which is mediated by $NAD^\oplus$. C^3 and C^4 of glucose appear principally in $C^{3'}$ and C^3, respectively, of apiose which excludes the possibility that the ring contraction occurs via the 3-keto derivative.

Aldgarose, a constituent of the macrolide antibiotic aldgamycin E from cultures of *Streptomyces lavendulae*[19], has a two carbon side chain in which labelled pyruvate, but not acetate, is incorporated; methionine and ethionine do not function as precursors. Glucose is exclusively incorporated in the hexose portion. The unique cyclic carbonate group derives from bicarbonate.[11] It is suggested that the two carbon fragment is introduced via a thiamine phosphate mediated decarboxylation of pyruvate. The enamine generated (16) condenses with a hexos-3-ulose (Fig. 12). This pathway is still not fully substantiated, since known thiamine deactivators have no influence on the incorporation of pyruvate in this enzymatic system. A pyridoxamine phosphate mediated decarboxylation of pyruvate is an alternative route to the two carbon fragment and alanine could conceivably be a precursor.

We have earlier met some pentoses as intermediates in the photosynthetic process. They appear also as products from the degradation of glucose via the phosphogluconate pathway. In this glucose-6-phosphate is dehydrogenated to 6-phosphogluconate and further oxidized and decarboxylated to ribulose-5-phosphate. ß-Keto acids are known to decarboxylate with greatest ease and enzymes dexterously use this route to shorten the chain by one carbon atom. Isomerization gives the whole family of aldo- and ketopentoses (Fig. 13). Fig. 11 shows an alternative way to xylose from glucuronic acid.

Reduction of the carbonyl function of sugars in the open chain form leads to sugar alcohols. The most common sugar alcohols are sorbitol, mannitol, and galactitol obtained from glucose, mannose, and galactose, respectively (Fig. 14). Sorbitol was first isolated from the berries of mountain ash, *Sorbus aucuparia*. It occurs also in the red alga, *Bostrychia scorpioides* (14 per cent). Mannitol is widely distributed in nature, e.g. in many brown algae (*Fucus, Laminaria, Halidrys*) and in manna ash, *Fraxinus ornus*. Galactitol occurs in many plants and exudates. It has a plane of symmetry and is therefore optically inactive.

Related to the sugar alcohols are the cyclitols which are derived from glucose-6-phosphate, as shown by ^{14}C labelling.[20] A likely route is depicted in (24). C^5 is oxidized and C^6 condenses with C^1; epimerization leads to the other cyclitols. A synthesis of *myo*-inositol inspired by the proposed biosynthetic cyclization step has been carried out in the laboratory.[21]

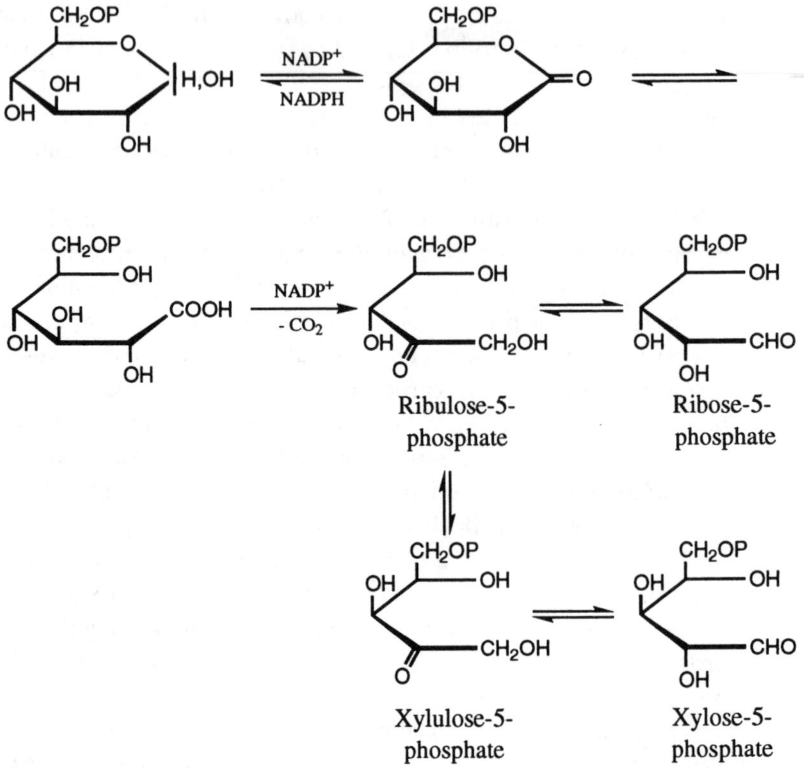

Fig. 13   Degradation of glucose to pentoses via the phosphogluconate pathway

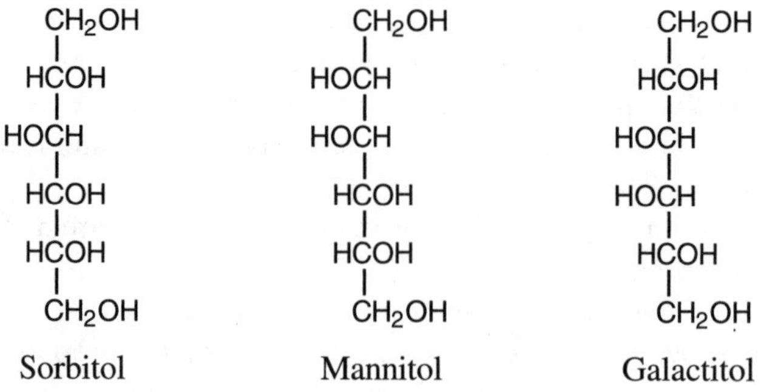

Fig. 14   Structure of common sugar alcohols

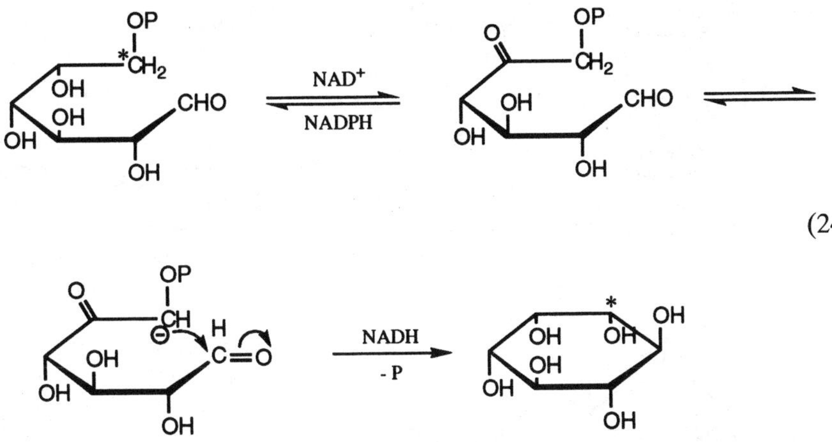

(24)

myo-Inositol

3.6 Disaccharides and glycosides

Disaccharides are formed from two monosaccharides joined by an acetal or ketal link. They belong to a group of glycosides where the alcohol is another sugar. They are classified according to their reducing power. Non-reducing are those with both carbonyl functions blocked as acetals, e.g. sucrose and trehalose. Examples of reducing sugars are lactose (4-*O*-β-D-galactopyranosyl-D-glucopyranose or in shorthand β-D-Gal*p*-1-4-D-Glc*p*), maltose, cellobiose, melibiose, etc. (Fig. 15). Sucrose is manufactured in large amounts from beet *Beta vulgaris* and cane *Saccharum officinarum*. The world production in 1970 was *ca.* 70 million tons. It gives one mole of glucose and one mole of fructose on hydrolysis. Trehalose, like maltose and cellobiose, gives two moles of glucose on hydrolysis. It is found in lower plants and insects. Maltose is formed by enzymatic degradation of starch and cellobiose by controlled hydrolysis of cellulose. They differ in the configuration of the C^1-O-C^4 linkage, maltose being an α-glucoside and cellobiose a β-glucoside. Lactose and melibiose give one mole of galactose and one mole of glucose on hydrolysis. They are both galactose glycosides; lactose forms a β-glycosidic linkage to the C^4 of glucose and melibiose an α-glycosidic linkage to the C^6 of glucose.

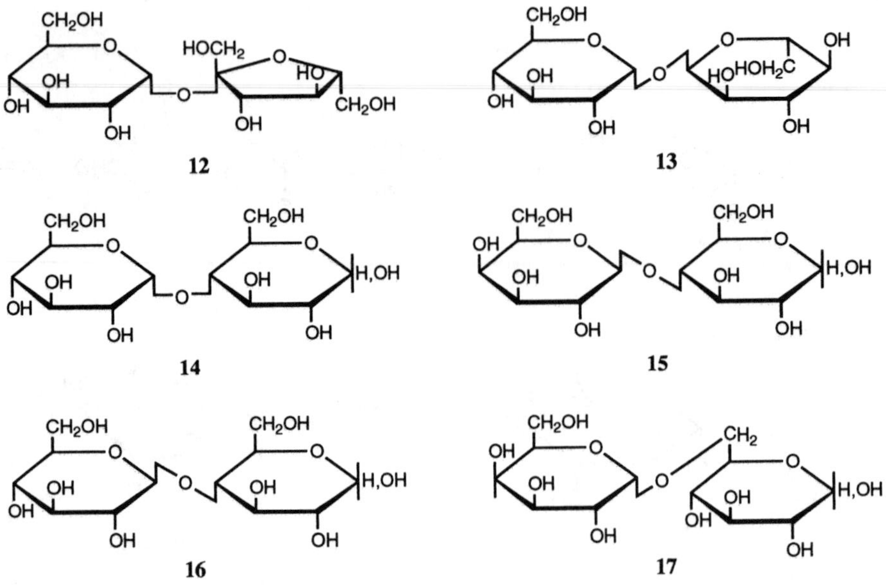

Fig. 15   Structures of disaccharides **12**, sucrose; **13**, trehalose; **14**, maltose; **15**, lactose; **16**, cellobiose; **17**, melibiose

Sucrose is biosynthesized from glucose activated as UDP-glucose and fructose-6-phosphate (25). UDP functions as an effective leaving group. The configuration at C^1 of UDP-glucose and at C^1 in the glucose moiety of sucrose formed is α, i.e. the displacement proceeds with retention of configuration. A one-step S_N2 backside substitution at C^1 by fructose is therefore excluded. Two explanations have been advanced for the outcome of the reaction. UDP-glucose is absorbed on the enzyme surface and thus shielded from a backside attack. A carbonium ion at C^1, stabilized by the ring oxygen, is formed when UDP dissociates and fructose-6-phosphate enters from the same side. Alternatively the enzyme participates actively with, for example, an amine function by forming a covalent bond with C^1 with inversion of configuration and the disaccharide is formed in a second S_N2 displacement by attack of fructose (25). The exact nature of this glycosidic formation is still unknown.

On the other hand, the lactose synthesis proceeds by backside attack of glucose at C^1 of α-UDP-galactose thus forming the β-1,4-linkage (26).

The structures of the disaccharides were established by complete methylation followed by hydrolysis and identification of the methylated products. The structure of lactose was deduced from the following findings:

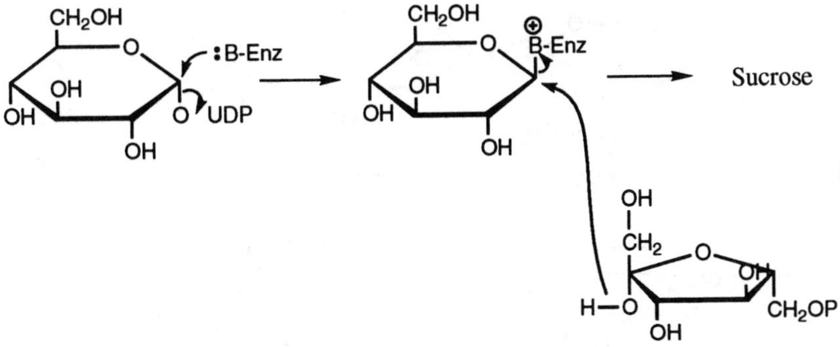

(25)

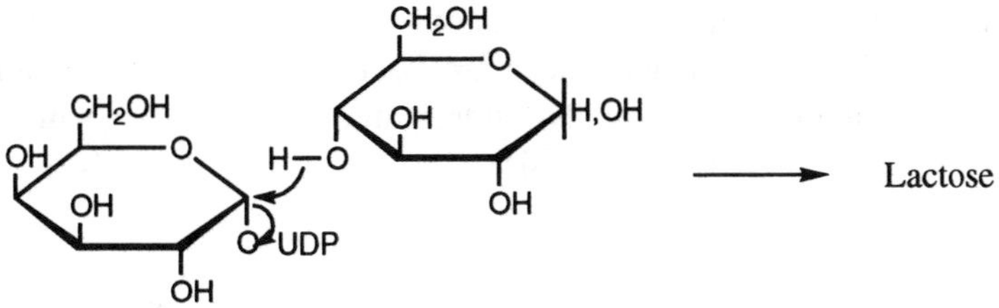

Lactose (26)

1. hydrolysis with dilute mineral acid gave the monosaccharides glucose and galactose;
2. lactose reduced Fehling's solution and it could be oxidized by bromine water to an acid which gave galactose and gluconic acid, i.e. lactose is a galactoside;
3. complete methylation of lactose gave octamethyllactose which on hydrolysis gave 2,3,4,6-tetra-*O*-methylgalactose and 2,3,6-tri-*O*-methylglucose (27); the acid gave 2,3,4,6-tetra-*O*-methylgalactose and 2,3,5,6-tetra-*O*-methylgluconic acid;
4. lactose was cleaved by β-glycosidase which is specific for β-glycosidic bonds.

These results show that **15** (Fig. 15) represents the structure of lactose.

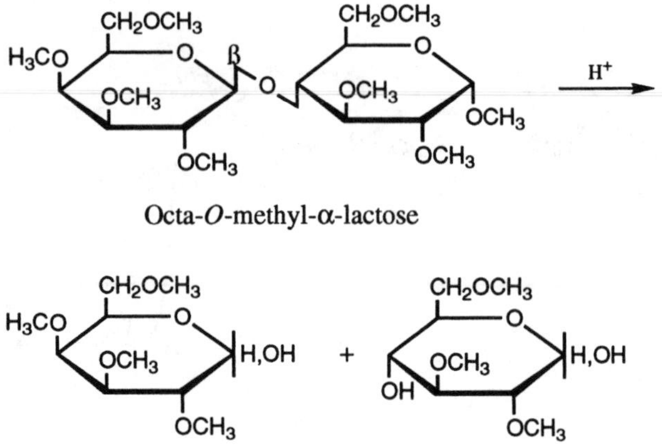

Octa-*O*-methyl-α-lactose

(27)

2,3,4,6-Tetra-*O*-   2,3,6-Tri-*O*-
methylgalactose   methylglucose

Glycosides of phenols or other alcohols are widely distributed in nature. Flavonoids and anthocyanins, pigments of flowers and berries, have their hydroxyls linked to sugars. The sharp flavour of mustard and horseradish is caused by an interesting thioglycoside, sinigrin, **20** (Fig. 16), which rearranges to an isothiocyanate on hydrolysis (28):

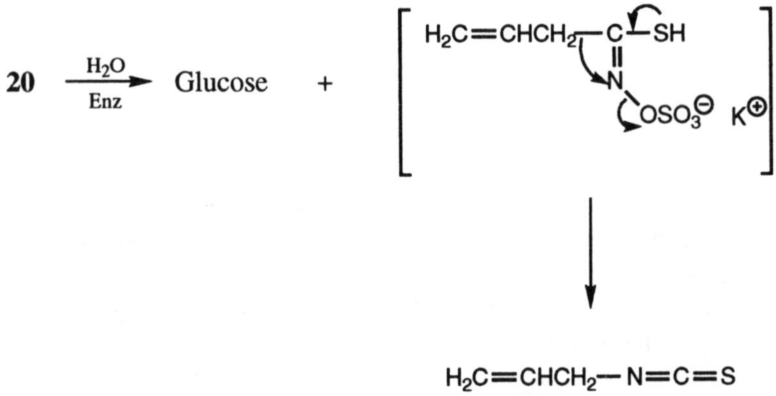

(28)

$$H_2C{=}CHCH_2{-}N{=}C{=}S$$

Amygdalin belongs to the cyanogenetic group of glycosides. It occurs in bitter almonds, *Prunus amygdalus*, and is hydrolysed by the enzyme emulsin to benzaldehyde, hydrocyanic acid, and two D-glucose molecules. Further structural work revealed that it is the gentiobioside of benzaldehyde cyano-hydrin, **21** (Fig. 16).

The cardiac glycosides belong to the steroid glycosides present in the *Strophanthus* genus, the foxglove *Digitalis purpurea* and in the lily of the valley *Convallaria majalis*. Preparations from *Strophanthus* are used by African tribes as arrow poisons for hunting, and *Digitalis* found use in medieval ordeals as an test of innocence, often with a fatal outcome. The active agents have a powerful action on the heart muscle; *ca.* 0.1 mg of strophanthin injected into the blood stream stops the heart of a mouse after a couple of minutes. In very small dosages the compounds have a beneficial effect in the treatment of heart ailments.

The *N*-glycosides are of the utmost importance as they are structural units in coenzymes, nucleic acids, nucleotides, etc. This group of compounds is discussed in Chapter 9.

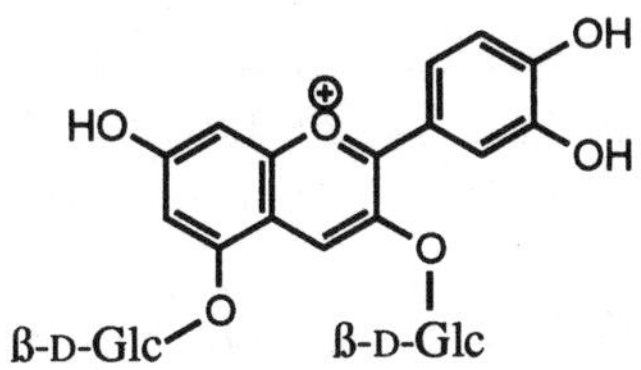

18

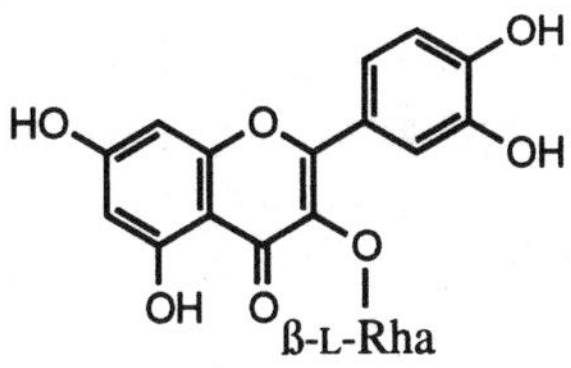

19

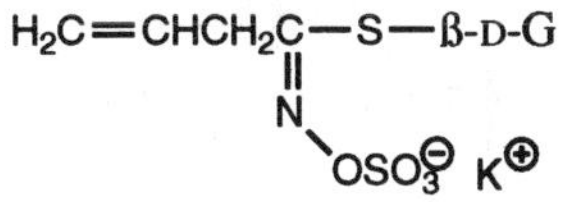

20

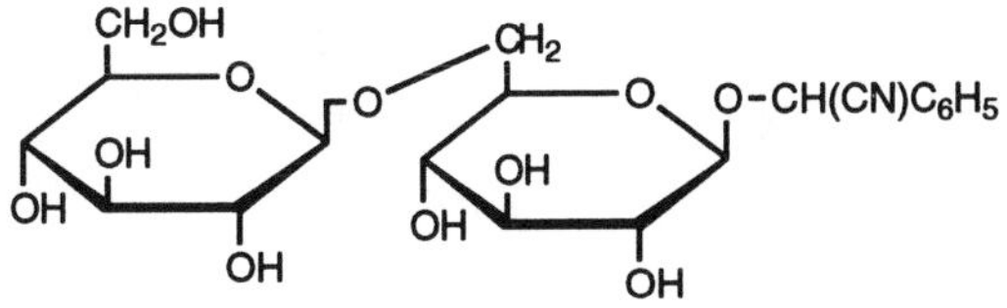

21

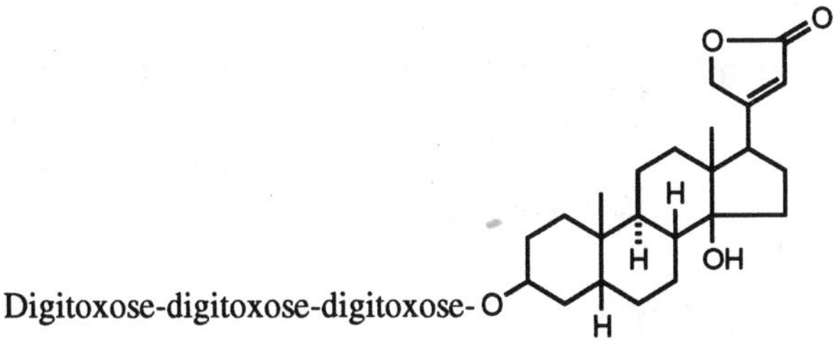

22

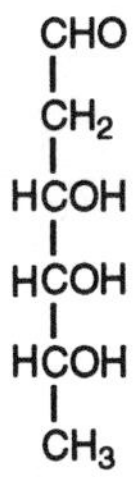

Digitoxose

Fig. 16 Structures of some glycosides occurring in nature: **18**, pyrylium salt from red roses; **19**, quercitrin, an L-rhamnoside; **20**, sinigrin; **21**, amygdalin; **22**, digitoxin

3.7 Polysaccharides

Polysaccharides are divided into homopolysaccharides containing one type of monomer (glucose), e.g. starch and cellulose, and heteropolysaccharides, containing two or more different monomers, e.g. hyaluronic acid (D-glucuronic acid and N-acetylglucosamine). They function either as storage (starch, glycogen, dextran, inulin) or as stuctural polysaccharides, i.e. cell wall constituents (cellulose, hemicellulose, chitin, pectin, alginic acids, hyaluronic acid). Starch is the most abundant storage polysaccharide. It can be separated into α-amylose, a linear polymer of α-1,4-linked-D-glucose units with a molecular weight of 100 000–500 000, and amylopectin, a branched polymer with a backbone of α-1,4-linked-D-glucose units branched with α-1,6-linkages, molecular weight 10–100 million (Fig. 17). Amylose is cleaved to α-maltose by β-amylase, an enzyme occurring in malt. The term β-amylase does not refer to the sugar bond cleaved, but–unfortunately–to a group of enzymes. Degradation of amylopectin by ß-amylase stops at the branching points and gives a product of rather high molecular weight, known as limit dextrin. Complete degradation can be brought about by a special 1,6-α-glucosidase. The amylose chains are wound like a helix and can enclose large amounts of iodine forming an intense blue addition compound.

Glycogen is the reserve polysaccharide of animal cells and is especially abundant in the liver. It has the same general structure as amylopectin but is more branched. Glycogen or starch is synthesized in the cell according to (29).

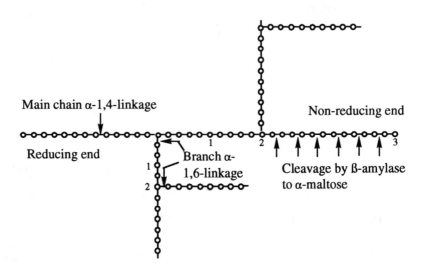

Fig. 17   Amylopectin model. Complete methylation and hydrolysis give 2,3,6-tri-*O*-methylglucose from 1 as main product, and small amounts of 2,3-di-*O*-methyl-glucose from 2 and 2,3,4,6-tetra-*O*-methylglucose from the non-reducing end unit 3

$$\text{UDP-D-Glucose} + (\text{Glucose})_n \xrightarrow{\text{Enz}} (\text{Glucose})_{n+1} + \text{UDP} \tag{29}$$

The reaction needs a primer, a polysaccharide having at least four glucose units, which is attacked at the non-reducing end.

Dextrans are highly branched polysaccharides of D-glucose produced by certain bacteria, e.g. *Leuconostoc mesenteroides* and *Betacoccus arabinosaceus*. The linkages in dextrans of different strains may vary: 1–2, 1–3, 1–4, and 1–6. Dextran is used as a blood plasma substitute. Inulin is a starch-like polysaccharide occurring particularly in the *Compositae* family. It is built up from β-1,2-linked D-fructose residues, and has a molecular weight of *ca.* 6000.

Cellulose is the most abundant cell wall constituent in the plant kingdom. It is made up linearly by β-1,4-linked D-glucose units and has a molecular weight of *ca.* one million. Complete methylation and hydrolysis give only 2,3,6-tri-*O*-methylglucose and minute amounts of 2,3,4,6-tetra-*O*-methylglucose from the end group, proving the pyran structure and no branching. α- or β-amylase does not attack cellulose and most mammals are unable to digest it, but some ruminants which have bacteria-produced cellulase in their intestinal tract, can do so. Cellulose is synthesized in plants from UDP- or GDP-

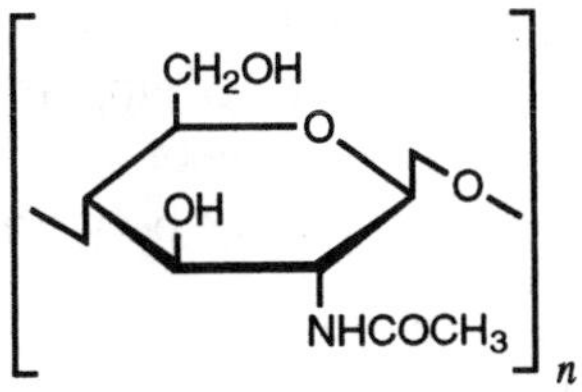

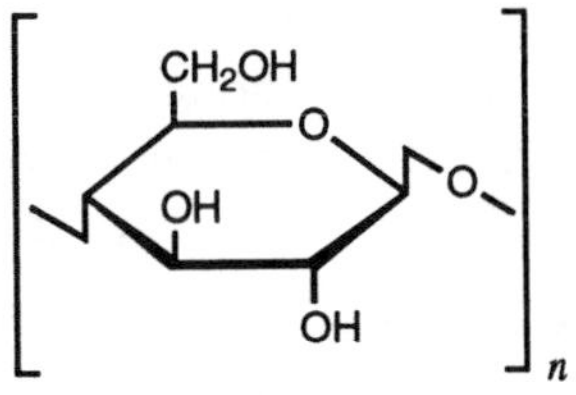

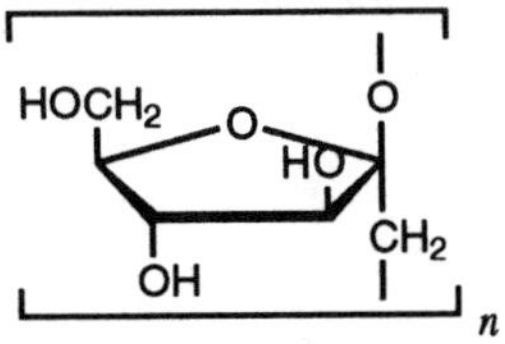

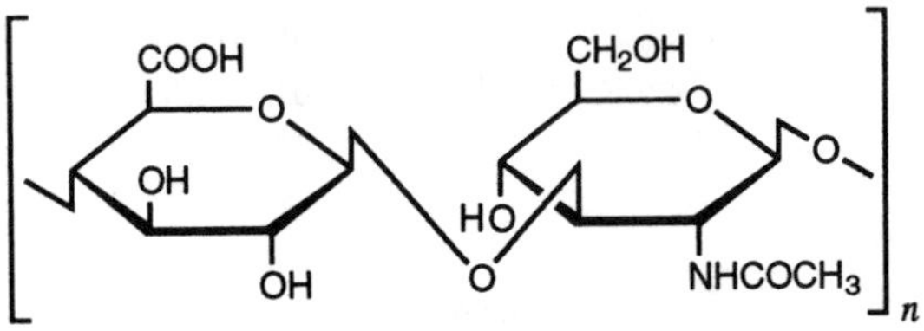

Fig. 18 Building units of polysaccharides: **23**, chitin; **24**, cellulose; **25**, inulin; **26**, hyaluronic acid

glucose and a glucose polymer in the presence of cellulose synthase, see (29). Hemicelluloses are heteropolysaccharides containing pentoses (xylose, arabinose) and occur together with cellulose in wood and straw. Alginic acid is a gelatinous material present in the cell wall of most brown algae. It is extracted commercially from giant kelp *Macrocystis pyrifera, Laminaria digitata*, and *Ascophyllum nodosum* and used as an emulsifier in foodstuffs, pharmaceuticals, and cosmetics. Hydrolysis gives *ca.* two moles of D-mannuronic acid and one mole of L-guluronic acid which in the polymers are β-linked in 1,4-position. The shell of crustaceans (lobster) and the exoskeleton of insects contain a β-1,4-homopolymer, chitin, structurally similar to cellulose with *N*-acetyl-2-amino-2-deoxyglucose as the building block. Hyaluronic acid is a mucopolysaccharide occurring in the vitreous humour of the eye and in the synovial fluid in joints. Hydrolysis gives equal amounts of D-glucuronic acid and *N*-acetyl-D-glucosamine linked in β-1,3-position. This disaccharide is then linked in the β-1,4-position (Fig. 18).

3.8 Problems

3.1 A green plant is illuminated for a very short period in an atmosphere containing $^{14}CO_2$. In which positions will ribose-5-phosphate and sedoheptulose-7-phosphate be labelled?

3.2 A polysaccharide is exhaustively methylated and hydrolysed. Equal parts of 2,3,4-tri-*O*-methylgalactose and 2,3,6-tri-*O*-methylglucose are obtained. The polysaccharide has a reducing end group, and it is enzymatically degraded by an α-glycosidase. Suggest a stucture. The molecular weight is 4.2×10^6. How many hexose residues does the polymer contain?

3.3 It is suggested that the epimerization of glucose to galactose proceeds via 3-keto-UDP-1-glucose. Discuss this reaction sequence assuming that the C^4–H is retained.

3.4 Propose a biosynthetic pathway for garosamine from TDP-glucose. The methyl groups are supposed to originate from methionine. (Grisebach, H. *Adv. Carbohydr. Chem. Biochem.* **35** (1978) 81.)

TDP-Garosamine

3.5 Trifluoroethanol reacts with methyl vinyl ketone in the presence of yeast, *Saccharomyces cereviciae*, to give the diketone I. Discuss the reaction mechanisms and suggest enzymes and cofactors for the reactions involved.

$$CF_3CH_2OH + = \overset{O}{\diagup}\diagdown \quad \overset{Yeast}{\longrightarrow} \quad CF_3 \overset{O}{\diagup}\diagdown\overset{}{\diagup}\diagdown\overset{O}{\diagup}$$

3.6 Discuss a plausible biosynthetic route to the aminocyclitol antibiotic streptomycin, a fermentation product of *Streptomyces griseus*. All three units are biosynthesized from D-glucose-6-phosphate. When 6-^{14}C-D-glucose-6-phosphate was fed to the microorganism, the label was recovered at the starred positions. Streptose was labelled in the formyl and methyl groups when the microorganism was incubated with 3,6-di-^{14}C-D-glucose. The ^{13}C NMR spectrum of streptomycin incubated with 6-^{13}C-glucose showed three enhanced peaks at 13.4, 61.2, and 72.4 PPM. (Grisebach, H., *Adv. Carbohydr. Chem. Biochem.* **35** (1978) 81.)

Streptidine

Streptose

2-Deoxy-2-methylamino-L-glucose

Streptomycin

Bibliography

1. Durette, P. L. and Horton, D. *Carbohydr. Res.* **18** (1971) 57.
2. Perlin, A. S. in *Int. Rev. Sci. Org. Chem. Ser. II*, Hey, D. H. (Ed.) **7** (1976) 1. Butterworth, London; Bock, K. and Thögersen, H. *Annu. Rev. NMR Spectros.* **13** (1982) 1.
3. Luneva, N. P., Knerelman, E. I., Shafirovich, V. Ya. and Shilov, A. E. *New. J. Chem.* **13** (1989) 107.
4. Brudvig. G. W., Beck, W. F. and de Paula, J. C. *Ann. Rev. Biophys. Chem.* **18** (1989) 25; Pecoraro, V. L. *Photochem. and Photobiol.* **48** (1988) 249; Wieghardt, K. *Angew. Chem.* **101** (1989) 1179; Joliot, P. and Kok, B. in *Bioenerg. Photosynth.* Govindgee (Ed.) Academic Press, New York, 1975, p. 387. Yachandra, V. K., DeRose, V. J., Latimer, M. J., Mukerji, I., Sauer, K. and Klein, M. P. *Science* **260** (1993) 675; *J. Am. Chem. Soc.* **116** (1994) 5239.

5. Vincent, J. B., Christmas, C., Chang, H-R., Li, Q., Boyd P. D. W., Huffman, J. C., Hendrickson, D. N. and Christou, G. *J. Am. Chem. Soc.* **111** (1989) 2086.

5a. Guajardo, R. J., Hudson, S. E., Brown, S. J. and Mascharak, P. K. *J. Am. Chem. Soc.* **115** (1993) 7971.

5b. Bard, A. J. and Fox, M. A. *Acc. Chem. Res.* **28** (1995) 141.

6. Glaser, L, in *The Enzymes*, 3rd Edn. Boyer, P. (Ed.) **6** (1972) 355. Academic Press, New York.

7. Walsh, C. *Enzymatic Reaction Mechanisms*. Freeman, San Fransisco, 1979, p. 347.

8. Vongerichten, E. *Liebigs Ann.* **318** (1901) 121.

9. Fischer, E. and Freudenberg, K. *Chem. Ber.* **52** (1919) 177.

10. Umezawa, S. *Int. Rev. Sci. Org. Chem. Ser. II*, **7** (1976) 149. Butterworth, London.

11. Grisebach, H. *Adv. Carbohydr. Chem. Biochem.* **35** (1978) 81.

12. Grisebach, H. *Biosynthetic Patterns in Microorganisms and Higher Plants*, J. Wiley, New York, 1976, p. 66.

13. Glaser, L. and Zarkowsky, H. in *The Enzymes*, 3rd Edn., Boyer, P. (Ed.) **5** (1971) 465. Academic Press, New York.

14. Snipes, C. E., Brillinger, G. U., Sellers, L., Mascaro, L. and Floss, H. G. *J. Biol. Chem.* **252** (1977) 8113.

15. Meloch, H. P. and Wood, W. A. *J. Biol. Chem.* **239** (1964) 3505, 3517.

16. Rubenstein, P. and Strominger, J. L. *J. Biol. Chem.* **249** (1974) 3776, 3782, 3789.

17. Watson, R. R. and Orenstein, N. S. *Adv. Carbohydr. Chem. Biochem.* **31** (1975) 135.

18. Grisebach, H. and Dobereiner, U. *Biochem. Biophys. Res. Commun.* **17** (1964) 737.

19. Aschenbach, H. and Karl, W. *Chem. Ber.* **108** (1975) 759, 780.

20. Eisenberg, F. and Bolden, A. H. *Biochem. Biophys. Res. Commun.* **12** (1963) 72.

21. Kiely, D. E. and Sherman, W. R. *J. Am. Chem. Soc.* **97** (1975) 6810.

Chapter 4

The shikimic acid pathway

4.1 Biosynthesis of shikimic acid

A very large number of compounds exhibit a characteristic C_6-aromatic–C_3-side chain structure, e.g. aromatic amino acids, cinnamic acids, coumarins, flavonoids, lignin constituents, etc., and it was recognized early on that they must have some common origin. The biosynthesis of these componds was elucidated by mutant studies of *Escherichia coli* and tracer studies particularly by Davis and Sprinson. Shikimic acid (2) was isolated as early as 1885 by Eykman from the Japanese plant *Illicium anisatum* long before we were aware of its biosynthetic significance. Its name derives from the Japanese name of this plant, but the compound was later found to be widespread. The role of shikimic acid was revealed from the observation that it could replace the essential amino acids phenylalanine, tyrosine, and tryptophan in auxotrophic *E. coli* mutants, i.e. it must be an intermediate in the biosynthetic sequence.

Erythrose-4-phosphate is a compound of far reaching importance in biosynthesis. We have seen that it is an intermediate in the regenerative process of ribulose-1,5-diphosphate in photosynthesis (section 3.3) and it appears in the pentose cycle. An analysis of the distribution of a ^{14}C label in shikimic acid, biosynthesized from specifically labelled ^{14}C-glucose in *E. coli* led to the proposition[1,2] that erythrose-4-phosphate starts the biosynthetic sequence leading to shikimic acid by condensation with phosphoenol pyruvic acid (PEP) to 3-deoxy-D-*arabino*-heptulosonic acid-7-phosphate (DAHP) (1). Elimination of phosphoric acid gives the ketone, formally in its enol form, that cyclizes to 3-dehydroquinic acid. Further elimination of water and reduction then gives shikimic acid. This is a sound, straightforward mechanism that was suggested at an early stage without detailed experimental proof. More recent studies revealed that the reaction is not quite so simple and they illustrate clearly that all facets of a reaction step have to be considered before acceptance can be given to a mechanistic interpretation.

^{18}O labelling shows that all the label is recovered in the phosphate liberated, i.e. it is the C-O bond that is broken rather then the P-O bond in the PEP condensation with erythrose-4-phosphate. By using specifically tagged Z and E 3-^3H PEP it was shown that the condensation with erythrose-4-phos-

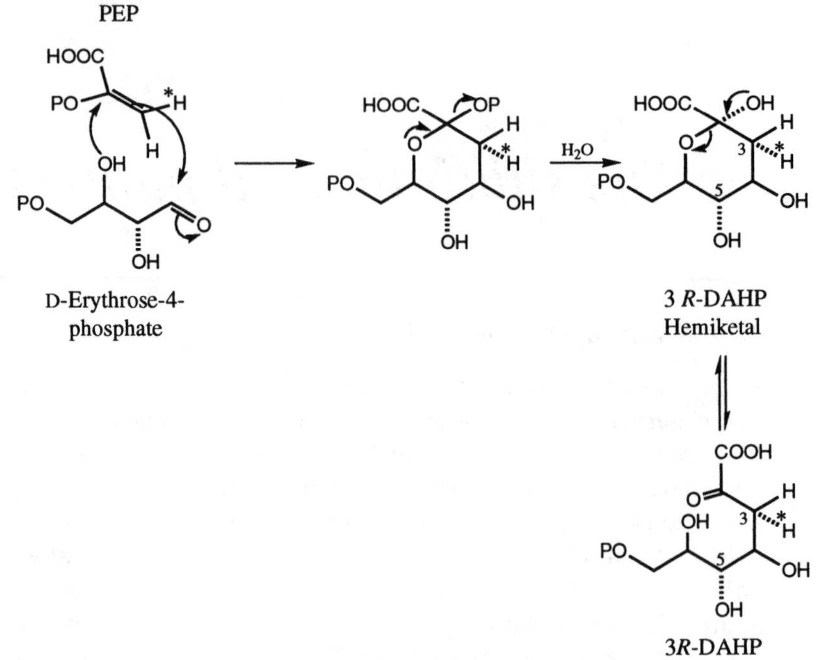

PEP

D-Erythrose-4-
phosphate

3 *R*-DAHP
Hemiketal

(1)

3*R*-DAHP

DAHP $\xrightarrow[\text{2. --POH}]{\text{1. NAD}^+}$ $\xrightarrow{\text{NADH}}$

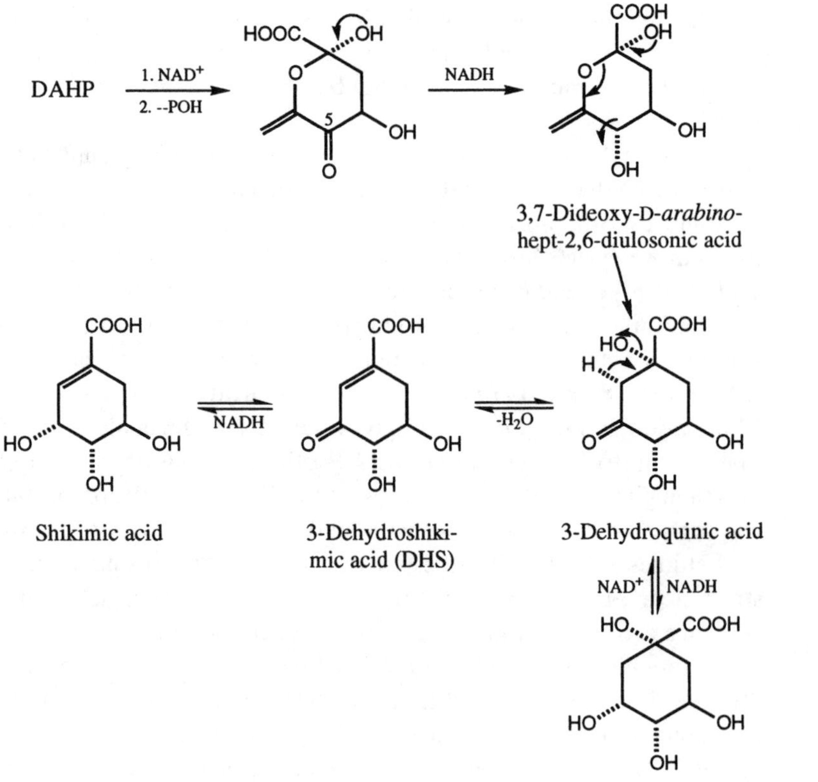

3,7-Dideoxy-D-*arabino*-
hept-2,6-diulosonic acid

(2)

Shikimic acid

3-Dehydroshiki-
mic acid (DHS)

3-Dehydroquinic acid

$\text{NAD}^+ \| \text{NADH}$

phate is stereospecific (1). (*Z*)-PEP gives rise to (3*S*)-DAHP and (*E*)-PEP to (3*R*)-DAHP, i.e. the *si* face of PEP adds to the *re* face of erythrose-4-phosphate.[3] Reaction (1) seems to account for most of the facts but it does not fully allow for the observation of the early release of phosphate and the idea of formation of an intermediate covalent bond between PEP and the enzyme.

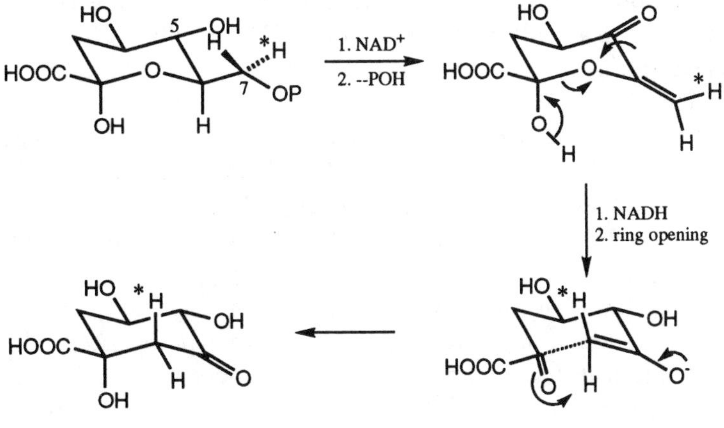

(3)

3-DHQ

The formation of 3-dehydroquinic acid exhibits some unexpected mechanistic features. The enzymatic reaction requires the NAD⁺-NADH couple and 5-³H labelled DAHP shows a small isotope effect. Apparently the hemiketal form is oxidized at C^5 which facilitates α–β–elimination of the C^7-phosphate.[4,5] Stereospecific labelling with ^2H at C^7 evidences that this enzymatic elimination proceeds in a *cis*-fashion in contrast to many acid or base catalysed *trans*-eliminations (2,3).[6,7] The C^5-keto group is reduced and ring opening followed by recyclization through a chair-like transition state gives dehydroshikimic acid. There are indications that the dehydration of dehydroquinic acid occurs in a *cis*-fashion via Schiff's base formation between a lysine residue of the enzyme and the keto function of 3-DHQ as demonstrated by inactivation of the enzyme by treament with sodium borohydride.[8]

Quinic acid (2), widely found in nature, is an offshoot of the pathway; once formed it is normally not easily metabolized again. However, several microorganisms are able to convert it into 3-dehydroquinic acid.[9]

4.2 Aromatic amino acids

The further transformation of shikimic acid involves some remarkable reactions in biosynthesis. Shikimic acid is phosphorylated regioselectively at C^3 and then it reacts with PEP at C^5. The phosphorylation serves the purpose of forming a more efficient leaving group for the following elimination. The tetrahedral PEP adduct has been isolated and characterized[10] and it has been demonstrated by labelling experiments that the addition and the subsequent elimination reactions proceed with opposite stereochemistry, i.e. either in a syn/anti or in an anti/syn fashion,[11] Fig. 1. The elimination of phosphoric acid to give chorismic acid proceeds in an overall 1,4-*trans* fashion[12] as proved by specific labelling of C^6.

This reaction is enzymatically controlled since most 1,4-conjugate eliminations in cyclohexene systems proceed predominantly in a *syn* fashion *in vitro*. Chorismic acid is an unstable intermediate, located at the branching point of the metabolic path (Fig. 2). One branch goes to anthranilic acid and indole derivatives, and the other via Claisen rearrangement to prephenic acid, or properly *pre*-phenylalanine, and phenyl- and *p*-hydroxyphenylpyruvic acid. The transamination (section 7.4) can in some species occur already at the prephenic acid stage.[13] Biosynthesis of *p*-aminobenzoic acid (Fig. 3) is an off-shoot from the stem. An outline of the shikimic acid pathway is given in Fig. 2.

There is a preference for a chair transition state when chorismic acid rearranges to prephenic acid as shown by configurational analysis of the side chain methylene group, C^3, of phenylalanine and tyrosine. The side chain is attacked on the *si* face by C^1; a boat conformation would lead to an attack on the *re* face and give the opposite configuration of C^3, Fig 1.[11] It ought to be noted that the present route to tyrosine does not require molecular oxygen in contrast to the oxidative route presented in section 4.3, Fig. 9, in conjunction with biological hydroxylation of phenylalanine.

Prephenic acid rearranges rapidly *in vitro* by the action of acids to phenylpyruvic acid but more slowly by alkali in an intramolecular redox reaction to β-*p*-hydroxyphenyl lactate[14] (4). Support for the unusual 9,4-hydride shift was produced by an investigation with a substrate labelled at the methine carbon.[15]

The transformation of chorismic acid to anthranilic acid is formulated in Fig. 3. Support for the elimination-addition reaction is found in the isolation of the related *trans*-2-amino-3-hydroxy-2,3-dihydrobenzoic acid[16] and *trans*-2,3-dihydroxy-2,3-dihydrobenzoic acid[17] from natural sources. The hydroxybenzoic acids are formed from isochorismic acid by acid catalysed elimination of water and pyruvic acid. The *p*-amino derivatives presumably originate directly from chorismic acid without intervention of isochorismic acid.[18] However, later work suggests that another isochorismic acid with the OH-group in 6-position could be on the pathway to 4-amino-4-deoxychorismic acid, (see ref. 23).

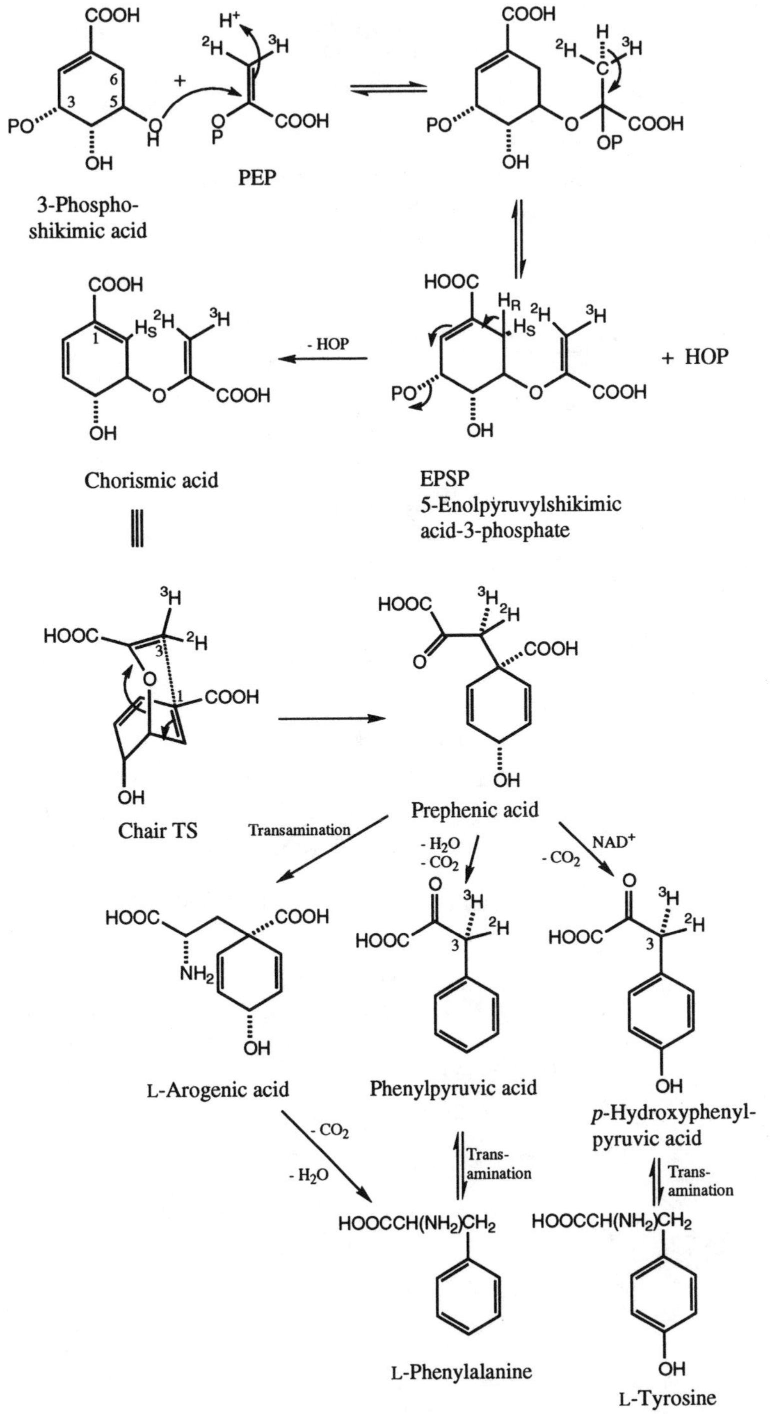

Fig. 1   Biosynthesis of phenylalanine and tyrosine

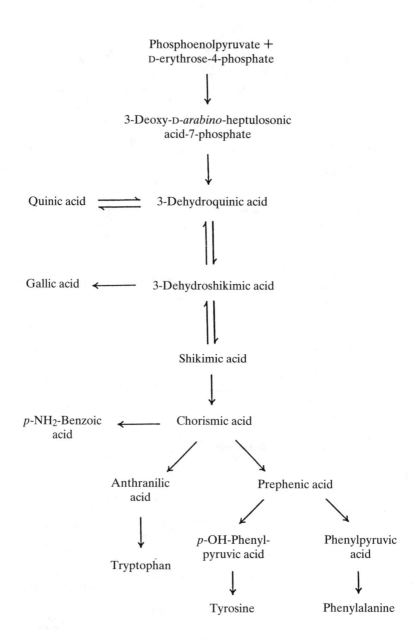

Fig. 2   Outline of shikimic acid pathway

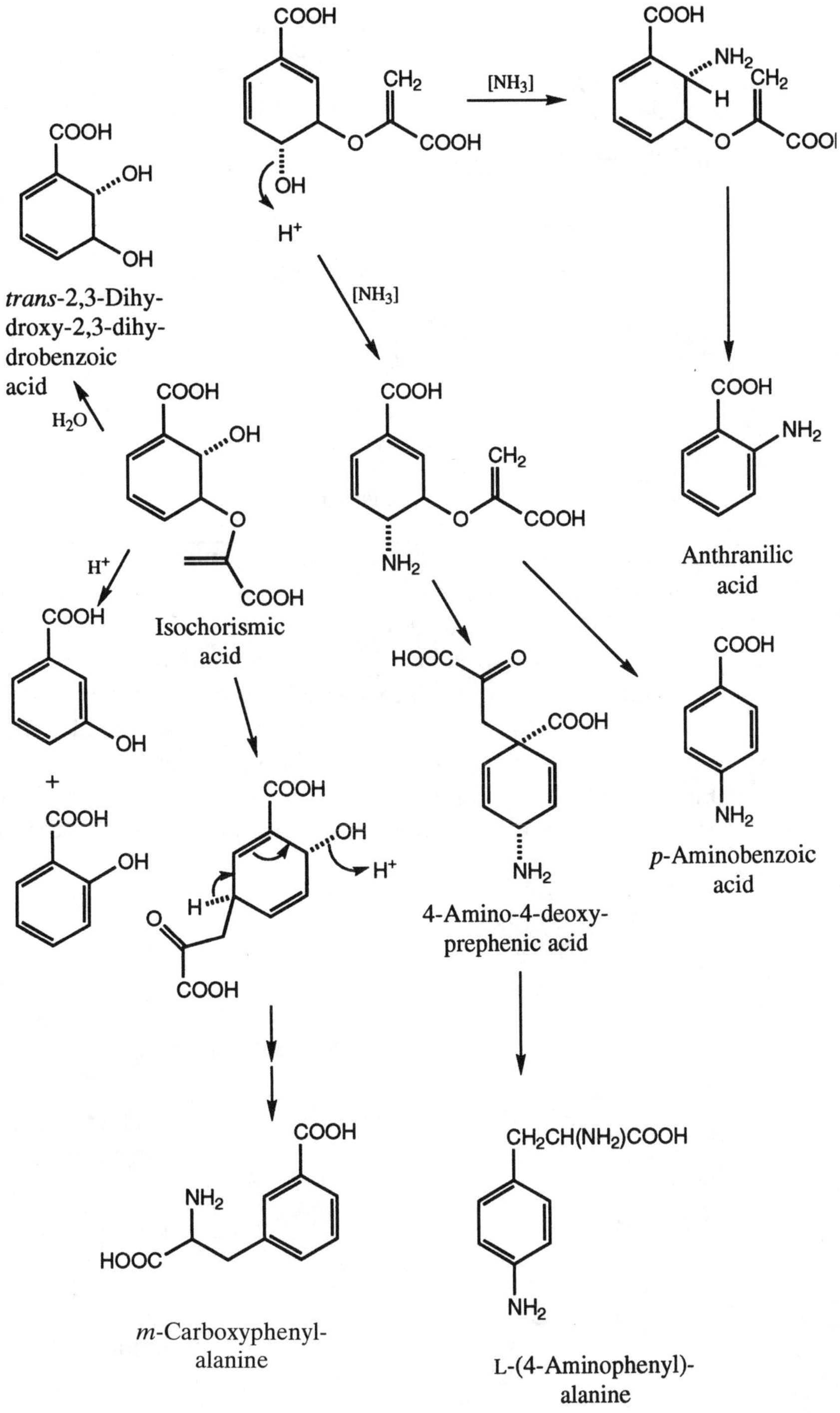

Fig. 3 Biosynthesis of amino- and hydroxybenzoic acids and substituted phenyl-alanines

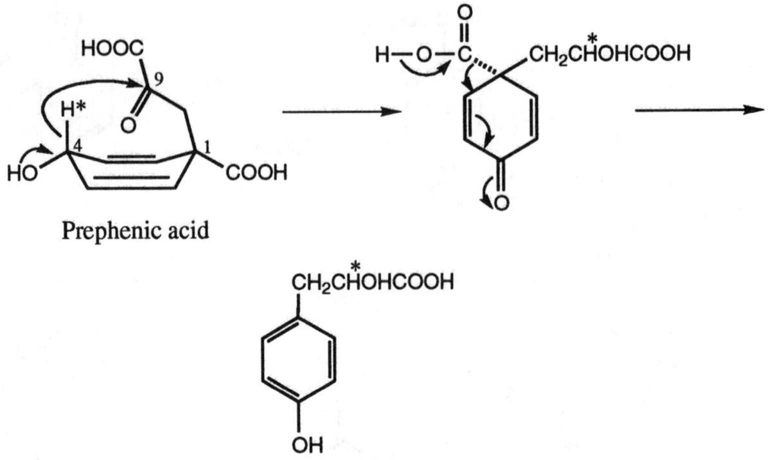

(4)

Prephenic acid

β-*p*-Hydroxyphenyllactic acid

The further reactions of anthranilic acid are full of unexpected events before the indole nucleus is completed (Fig. 4). An enzyme complex promotes the reaction with 5-phosphoribosyl-1-pyrophosphate. The nucleotide, *N*-(5′-phosphoribosyl)-anthranilic acid, undergoes an Amadori rearrangement via the Schiff's base and cyclizes to indole-3-glycerylphosphate.[19] Alkylation with serine and expulsion of glyceraldehyde-3-phosphate complete the sequence.[20]

This last reaction requires pyridoxal which forms a Schiff's base with serine thus facilitating the elimination of water and the indole derivative adds to the aminoacrylate-pyridoxal complex formed. See section 7.3 for further mech-

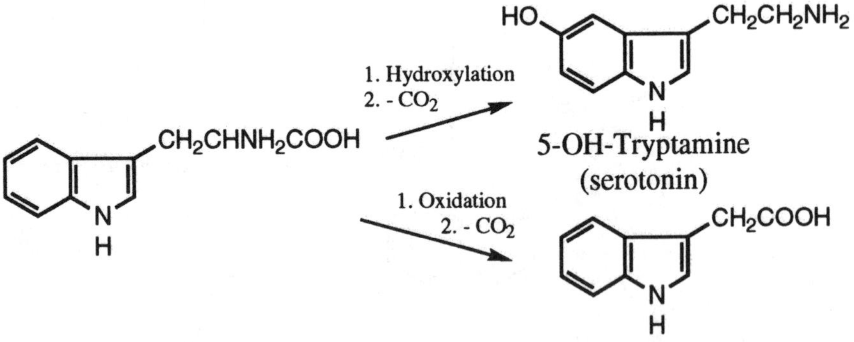

1. Hydroxylation
2. - CO_2

5-OH-Tryptamine
(serotonin)

1. Oxidation
2. - CO_2

Indoleacetic acid
(auxin)   (5)

anistic details of pyridoxal promoted reactions. Tryptophan is the precursor of indoleacetic acid, a plant growth hormone controlling cell elongation, and of the neurotransmitter 5-hydroxytryptamine (serotonin) (5). In a reaction, which essentially is a reversal of the serine alkylation, tryptophan is degraded to indole, pyruvic acid, and ammonia[21] (see Problem 4.1).

By feeding experiments in *Reseda lutea* it has been shown that labelled shikimate is also built into *m*-carboxyphenylalanine.[22] This can be explained by a Claisen rearrangement of isochorismic acid, a reaction analogous to the chorismic-prephenic acid rearrangement, Fig. 3. However, it has also been proposed that the *m*-carboxy derivative actually derives from the isomeric isochorismic acid with the hydroxy group in 6-position. This compound is very labile and rearranges to the *m*-carboxyprephenic acid derivative (6).[23] Conceivably this isochorismic acid could be on the pathway to 4-amino-4-deoxy-chorismic acid.

(6)

m-Carboxyphenyl-
pyruvic acid

4.3 Biological hydroxylation. Redox reactions

Redox reactions are common reactions in biosynthesis, constituting key steps in most biosynthetic schemes. Astonishingly few types of redox reactions, representing a few basic mechanistic principles, adequately cover the redox requirements of the cell. The H_2O-O_2-couple is the ultimate source and sink of electrons in these reactions, see photosynthesis. We have already met them in connection with oxidation and reduction of carbonyl groups in the sequence $-CH_2OH \rightleftarrows -CHO \rightleftarrows -COOH$ and tacitly accepted hydroxylation, i.e. insertion of oxygen into unactivated aliphatic and aromatic C–H bonds. The redox

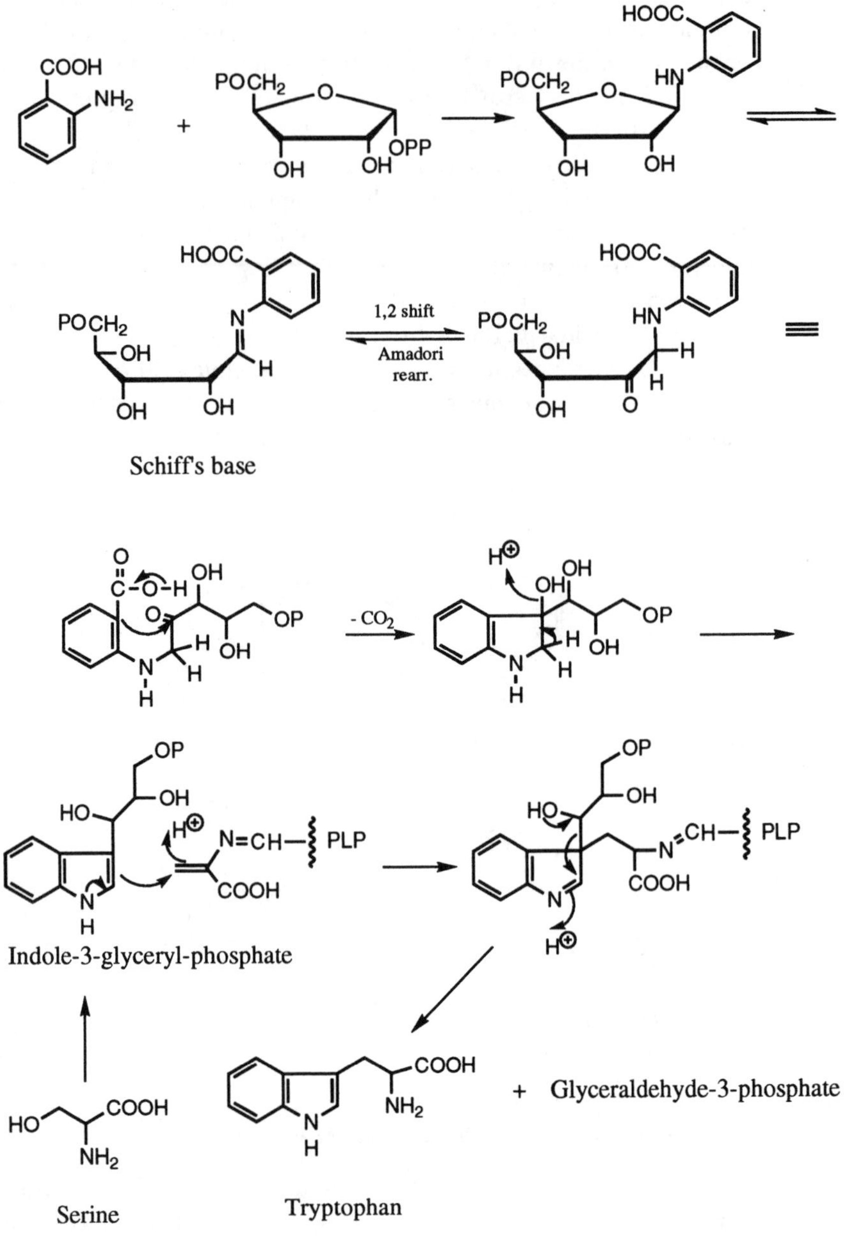

Fig. 4   Biosynthesis of tryptophan from anthranilic acid

reactons of carbonyl functions with the $NAD^{\oplus}$/NADH or FAD/$FADH_2$ couples have been fairly easy to accommodate with mechanistic organic chemistry. These reactions are promoted by enzyme systems classified as dehydrogenases or oxidases and they do not involve the immediate participation of molecular oxygen.

Dioxygenases catalyse the insertion of both oxygen atoms and monooxygenases (hydroxylases) catalyse the insertion of one oxygen, the other oxygen atom forms water. The reactions are symbolically formulated (7–9).

$$RH_2 + NAD^{\oplus} \rightleftharpoons R + NADH + H^{\oplus} \qquad (7)$$

$$RH_2 + O_2 \rightarrow R(OH)_2 \qquad (8)$$

$$RH_2 + O_2 + NADPH + H^{\oplus} \rightarrow RHOH + H_2O + NADP^{\oplus} \qquad (9)$$

The monooxygenases are of special importance. They are responsible for hydroxylation of aliphatic compounds, e.g. steroids and the α-oxidation of fatty acids; hydroxylations of aromatics, e.g. conversion of phenylalanine to tyrosine; dealkylation of amines, ethers, and thioethers; oxidation of amino groups to nitro; sulphide to sulphone, etc. They operate in defence mechanisms by oxidizing foreign substances, such as drugs, making them more water soluble and therefore apt to be excreted. Compounds like DDT, which are difficult to oxidize, accumulate in fat cells with fatal consequenses. It is characteristic of algae, where diffusion of metabolites into the surroundings is facilitated to such an extent, that they have a poorly developed oxygenase enzyme system. Hardly any biochemical reaction has been more extensively investigated than oxidation or in a wider sense the chemistry of respiration. X-Ray crystallographers have determined the structure of several oxidizing enzymes, biochemists have given a good account of the function of the respiratory chain and the mechanism of electron transport, and mechanistic studies have revealed many secrets of substrate behaviour and product formation but, paradoxically, we still cannot satisfactorily formulate the act of bond breaking and bond making of oxygen insertions, and the precise nature of the attacking oxygen species is still not clear.

We have to start out from the chemistry of oxygen and organic peroxides, and eventually contemplate on the unknown steps by analogy. Ground state oxygen, $^3O_2$, is a triplet with an uncoupled pair of electrons in different orbitals and with their spins parallel; 22.5 kcal above ground state lies singlet oxygen, $^1O_2$, with paired electrons. It is a short-lived and more reactive species with a lifetime of *ca.* 2 μs in water. It has been inferred that oxygen, activated to the singlet state, could be responsible for certain oxidations in the living cell. It causes demethylation of amines, oxidizes sulphides to sulphoxides,

forms allylic peroxides by an ene reaction with olefins, and adds in a Diels–
Alder reacton to 1,3 dienes.[24] This last reaction explains in a simple way the
formation of the peroxidic anthelmintic ascaridol[25] in *Chenopodium* spp.
(10a), a reaction atypical for $^3O_2$. But the general reaction pattern of $^1O_2$ is
different from that of $^3O_2$ and is more difficult to accommodate with the oxidation products formed in nature.[25a] $^3O_2$ is therefore the most plausible oxidative agent. The formation of ascaridol can conventionally be explained by
the iron catalysed reaction (10b).

(10)

Ascaridol

α-Terpinene

Oxygen is stepwise reduced by four electrons to water and all the reactive
intermediates are known. The reduction starts by formation of the superoxide
radical ion which gives well characterized salts with alkali metals. The ion is
easily generated polarographically at a low reduction potential, E_o (O_2/O_2^-)
= –0.33 V,[26] and it can be kept for several days in aprotic solvents as a tetra-
alkylammonium salt (11). It decomposes rapidly in contact with water in a
redox reaction. As a base it adds a proton and forms the reactive hydrogen
peroxide radical (12), pK_s = 4.9,[27] i.e. at physiological pH only a few per cent
are present in the acidic form. By protonation, which proceeds extremely
rapidly, k_{-12} = 4.7 × 10^{10} l mol^{-1}s^{-1},[28] the chemical character of the superoxide
ion is drastically changed. From being a good nucleophile and a mild reducing
agent it is turned into a strong oxidant and this change in oxidation potential
is of importance for our understanding of its reactions.

$$O_2 \underset{-e^{\ominus}}{\overset{e^{\ominus}}{\rightleftharpoons}} O_2^{\ominus \cdot} \tag{11}$$

$$HO_2^{\cdot} \overset{fast}{\rightleftharpoons} H^{\oplus} + O_2^{\ominus \cdot} \tag{12}$$

$$O_2^{\ominus \cdot} + O_2^{\ominus \cdot} \overset{slow}{\rightleftharpoons} O_2^{2\ominus} + O_2 \tag{13}$$

$$HO_2^{\cdot} + O_2^{\ominus \cdot} \overset{fast}{\longrightarrow} HO_2^{\ominus} + O_2 \tag{14}$$

$$O_2^{\ominus \cdot} + Cu^{2\oplus} \rightleftharpoons Cu^{\oplus} + O_2 \tag{15}$$

$$Cu^{\oplus} + HO_2^{\cdot} \rightleftharpoons Cu^{2\oplus} + HO_2^{\ominus} \tag{16}$$

The superoxide ion undergoes a very slow disproportionation to peroxide and oxygen (13), predominantly caused by impurities and moisture. This electron transfer is accelerated enormously by protonation, $k_{14} = 8.5 \times 10^7$ l mole^{-1} s^{-1}.[27] Actually, (14) represents a reaction that leads to singlet oxygen. The electron transfer is facilitated by transition metal catalysis, e.g. by Cu$^{\oplus}$ in dismutase (15,16), and these reactions are nearly diffusion controlled, $k = 2.3 \times 10^9$ l mole^{-1}s^{-1}.[29,30,31] The superoxide ion is often described in the literature as an oxidant. This is a misunderstanding; it is basically a reductant but acts as an oxidant in consequence of protonation in protic solvents. It is unreactive against Por Fe$^{2\oplus}$ but reduces Por Fe$^{3\oplus}$. The peroxide radical oxidizes Por Fe$^{2\oplus}$ but is unreactive against Por Fe$^{3\oplus}$.[32]

The superoxide ion is an efficient nucleophile (17) and reacts rapidly with alkyl bromides in dimethylsulphoxide with formation of peroxides and alcohols. The hydrogen peroxide radical carries all the characteristics of a short-lived reactive radical. It is a hydrogen abstractor (18). The O–H bond energy for homopolar fission of alkyl peroxides or hydrogen peroxide (19,20) is 89–90 kcal mole^{-1} which means that many H abstractions are approximately thermally neutral.[33] The C–H bond energies of the methylene groups in propane and 1,4-pentadiene are 95 and 80 kcal mole^{-1}, respectively. The hindrance for reaction (18) is the activation energy which has been measured in a few cases at *ca.* 10 kcal; k_{18} is consequently comparatively small, *ca.* 0.1–10 l mole^{-1} s^{-1}.[34]

$$R\,Br + O_2^{\ominus \cdot} \rightarrow ROO^{\cdot} + Br^{\ominus} \tag{17}$$

$$HOO^{\cdot} + RH \rightarrow HOOH + R^{\cdot} \tag{18}$$

$$t\text{-BuOOH} \rightarrow t\text{-BuOO}^{\cdot} + H^{\cdot} \quad \Delta H = 90 \text{ kcal} \tag{19}$$

$$HOOH \rightarrow HOO^{\cdot} + H^{\cdot} \quad \Delta H = 89 \text{ kcal} \tag{20}$$

$$R^{\cdot} + O_2 \overset{fast}{\longrightarrow} ROO^{\cdot} \underset{H+}{\overset{e-}{\longrightarrow}} ROOH \underset{2H+}{\overset{2e-}{\longrightarrow}} ROH + H_2O \tag{21}$$

This is partly compensated for by close orientation of reagent and substrate on the enzyme and slow diffusion rate of the intermediates from the active site. For thermodynamic reasons the peroxide radical will not react, or at best reacts slowly, with non-activated C–H bonds of saturated hydrocarbons (CH_3, CH_2, CH groups) but it is a likely candidate for the post as "activated oxygen" for radical reactions with activated C–H bonds, such as allylic methylene groups. The term non-activated hydrocarbons in this context means in general hydrocarbons of low acidity with pK values > 20–25. Of importance for the following discussion is also the diffusion controlled reaction of $^3O_2$ with carbon centred radicals. This reaction is one of the propagation steps in autooxidation (21).[35] The suggestion of the involvement of $O_2^-\cdot$ in biological hydroxylation was put forward in the late fifties.[36]

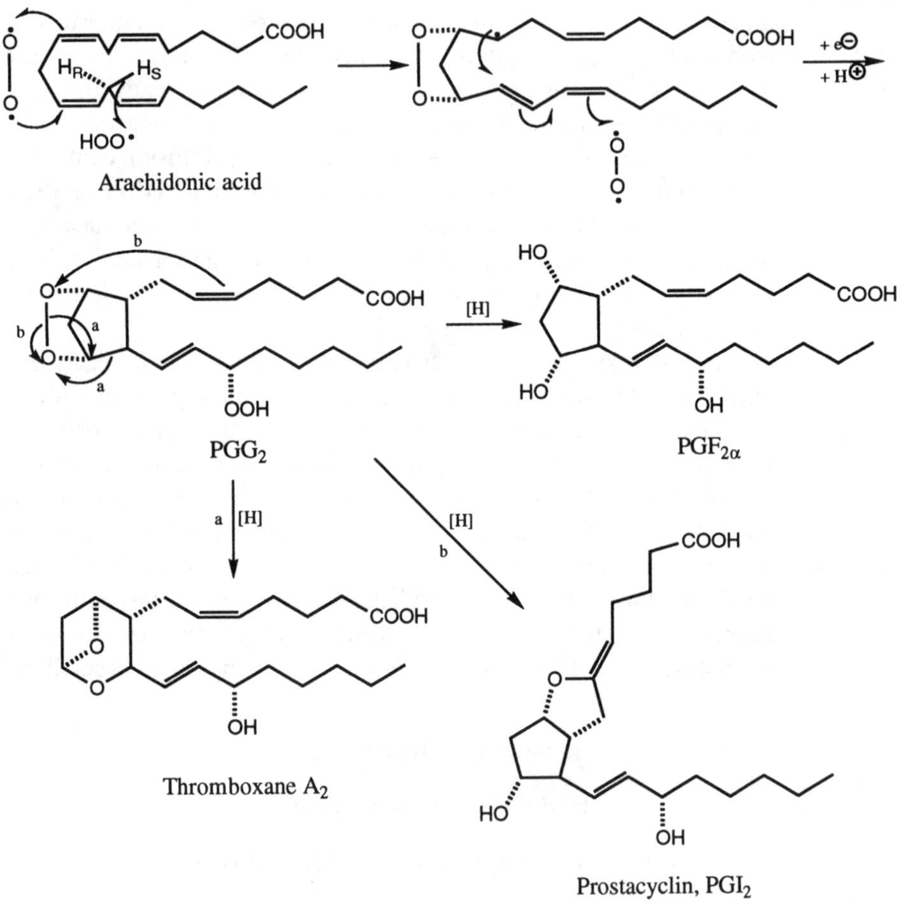

Fig. 5 Biosynthesis of prostaglandins and thromboxanes from arachidonic acid

There is evidence that the first intermediate is a peroxide in many aliphatic hydroxylations. Since these intermediates are unstable, they decompose in the work up procedure of natural products, but are now isolated to an increasing extent from natural sources.[37] A mechanism based on reactions (11, 12, 18, 21) explains the biosynthesis of the prostaglandins[38] (Fig. 5), see section 5.4.

In a lipoxygenase reaction C^{13}–H_S is removed from arachidonic acid; the radical formed reacts with oxygen and cyclizes. Another molecule of oxygen is added and a one-electron reduction gives the peroxide. Cleavage of the O–O bond is eventually accomplished by a peroxidase or by a bifunctional oxygenase. It ought to be noted that in the formation of the endoperoxide both oxygen atoms are incorporated, formally corresponding to the activity of a dioxygenase. The intermediacy of radicals is confirmed by adding 2-methyl-2-nitrosopropane as scavenger to a mixture of linoleic acid, oxygen and lipoxygenase.[39] 2-Methyl-2-nitrosopropane reacts rapidly with carbon centred radicals and competes with oxygen for the radical giving a stable nitroxide which can be recorded by ESR (22). The spectrum gives a characteristic triplet from

Linolenic acid   (22)

nitrogen and a smaller doublet from the proton at C^9 or C^{13} which vanishes on introducing deuterium at these positions. Deuteration of C^{11} did not change the spectrum, indicating that the more stable conjugated radical was trapped in accordance with the biosynthetic scheme for prostaglandins. The proposed pathway is mimicked in the laboratory in the serial cyclization of peroxy olefins (23).[40] *O*- or *N*-demethylation is explained by H abstraction from the methyl, actually the preferred point of attack in radical reactions, and the intermediate α-hydroxyamine or α-hydroxyether is easily hydrolysed to the free amine or alcohol.

(23)

Haber and Weiss[41] pointed out several years ago the intermediacy of hydroxyl and hydroperoxide radicals in the Fe^{2+} catalysed decomposition of hydrogen peroxide, the Fenton reaction (24–27). It is closely related to the reactions taking place at the cytochromes or other enzymes catalysing oxygenation.

$$Fe^{2+} + H_2O_2 \rightarrow Fe^{3+} + OH^{\cdot} + OH^{-} \qquad\qquad (24)$$
$$OH^{\cdot} + H_2O_2 \rightarrow HOO^{\cdot} + H_2O \qquad\qquad (25)$$
$$HOO^{\cdot} \rightleftharpoons H^{+} + O_2^{-\cdot} \qquad\qquad (26)$$
$$Fe^{3+} + O_2^{-\cdot} \rightleftharpoons Fe^{2+} + O_2 \qquad\qquad (27)$$

The peroxide bond is weak with a homopolar dissociation energy of 51.0 kcal $mole^{-1}$. In Fenton's reaction it is cleaved by addition of an electron and the very reactive hydroxyl radical is formed. It is a strong oxidant (28) with weak acidic properties, $pK_s = 11.8$,[42] (29) and it abstracts hydrogen atoms from any aliphatic molecule in a nearly diffusion controlled reaction,[43] $K_{30} = 10^9$ l $mole^{-1}$ s^{-1}, as a result of the high energy of the O–H bond; $\Delta H_{31} = 119$ kcal $mole^{-1}$, i.e. (30) is always exothermic.

$$OH^{\cdot} + e^{-} \rightarrow OH^{-} \qquad\qquad (28)$$
$$OH^{\cdot} \rightleftharpoons O^{-\cdot} + H^{+} \qquad\qquad (29)$$
$$OH^{\cdot} + RH \rightarrow H_2O + R^{\cdot} \qquad\qquad (30)$$
$$H_2O \rightarrow OH^{\cdot} + H^{\cdot} \qquad \Delta H = 119 \text{ kcal} \qquad\qquad (31)$$
$$H_2O_2 + 2e^{-} \rightarrow 2OH^{-} \qquad\qquad (32)$$

In view of the high chemical reactivity and oxidation potential of the hydroxyl radical, it is improbable that it is formed or has any function in the cell. Its formation is circumvented by cleavage of hydrogen peroxide in a two-electron transfer process (32) or by decomposition by catalase.

It is thus necessary to explore other routes, which give a better mechanistic explanation of the hydroxylation of non-activated aliphatics, e.g. of methane. The dissociation energy of its C–H bond is 102 kcal $mole^{-1}$.

With the relevant chemistry of oxygen in mind, we now turn to enzymatic hydroxylation as it is enacted at the P450 cytochromes (P for pigment and 450 for the wavelength of its UV absorption band). The prosthetic group is a high spin Fe^{3+}-porphyrin complex (the resting oxidation state) that adds the substrate, one electron and one molecule of oxygen, which is reduced to the nucleophilic superoxide radical (33). The iron atom in the cytochromes changes reversibly between the Fe^{2+} and Fe^{3+} forms, in contrast to the situation in the oxygen carrier haemoglobin, where the iron atom keeps its Fe^{2+} form. Oxyhaemoglobin is inactive as a hydroxylation catalyst.

The reaction of the superoxide with an electrophile, E^+, i.e. H^+ or strategically located effector group, a carboxylate, leads to the peroxide radical which in principle is capable of abstracting a hydrogen atom from the substrate as discussed previously (33, 12, 18). However, from here the reaction takes another direction. The iron-superoxide complex is reduced by a second electron to give the unstable Por $Fe^{3+}OOH$ species (a) which gives the active oxidant, the Fe^{5+}–oxo-derivative (33b–g), by splitting off a hydroxyl ion. The Por Fe^{3+} OOH species can also be prepared in a shorter way from hydrogen peroxide and Por Fe^{3+}.

The electrons used for the reduction, i.e. the activation of oxygen, are delivered by NADPH via FAD, which serves as a switch between the two-electron donor (the hydride donor) NADPH and the one-electron acceptor, the porphyrin-iron. Most characteristically, NADPH is inert to oxygen, whereas $FADH_2$ is highly reactive undergoing both one- and two-electron transfer.

Several resonance structures are possible for the oxo derivative, (33 b–g). Physico-chemical investigations favour the radical structure (d) and (f), which have unpaired electron densities on the oxygen and in the π-orbital of the porphyrin ring, respectively, and especially structure (d) endowing oxygen radical properties is useful for discussing the hydroxylation process[44]. The weight of the different electronic structures of the oxo derivative is modulated by the fifth porphyrin-iron ligand. The structures (e) and (f) have been identified at $-90°$ C both giving epoxides with olefins. Structure (f) was approximately 10 times more reactive than (e).[45]

Still another isomeric structure has been suggested, the N-oxide, where the oxygen is inserted into the iron-nitrogen bond. However, this species has different spectroscopic properties compared with (b–f) and does not react with olefins[46]. We will return to the slipped geometry[47] later in our discussion.

The monooxygen structure of the reactive oxygenating intermediate is supported by the fact that monooxygen donors such as ClO^-, C_6H_5IO, $NaIO_4$,

33 d

N-oxide

Por Fe^{III}

$\Updownarrow e\ominus$

Por Fe^{II} + O$_2$ $\rightleftharpoons$ Por Fe^{III} O$-$O$^{\bullet}$ $\rightleftharpoons$ Por Fe^{III} + O$-$O$\ominus^{\bullet}$

$\Updownarrow$ H$^+$, $e\ominus$ $\Updownarrow$ H$\oplus$ (33)

Por Fe^{III} + HOO$\ominus$ $\rightleftharpoons$ Por Fe^{III}$-$O$-$OH   HOO$^{\bullet}$

$\downarrow$ Cata-   (a)   Dismu- $\Big|$ (16, 17)
lase $\Updownarrow$ tation

H$_2$O + O$_2$ Por Fe^{III}$-$O$\oplus$+ OH$\ominus$ H$_2$O$_2$ + O$_2$
(b)

$\downarrow$

Por Fe^{III} + $^{\bullet}$O$^{-}_{\bullet}$ $\rightleftharpoons$ Por Fe^{V}O$\ominus$ $\longleftrightarrow$ Por Fe^{IV}O$^{\bullet}$ $\longleftrightarrow$ Por Fe^{V}= O $\longleftrightarrow$ $^{\oplus}$Por Fe^{IV}= O

(g)   (c)   (d)   (e)   (f)

amine-N-oxides and oxaziridines in combination with synthetic transition metal (Fe, Mn, Co, Cr) porphyrine or salene complexes mimic effectively the reactions of the native P450/O$_2$/NADPH system. The reaction is used synthetically, e.g. for the preparation of epoxides (34).[48]

R $\diagup$= $\xrightarrow[\text{C}_6\text{H}_5\text{IO}]{\text{catalyst}}$ R$\diagup\triangle$O   (34)

catalysts

M = Fe, Mn, Cr, Co   Salene complex

The dissociation of (d) according to equation (35), corresponds formally to the acid-base equilibrium (29) where protonation turns O^{-} $^{\bullet}$ into the strong oxidant and hydrogen abstracting agent HO$^{\bullet}$. As a result of spin delocalization the (d) form is presumably less reactive than HO$^{\bullet}$, ΔH_{31} = 119 kcal mol^{-1}. Hydrogen peroxide reacts rapidly with (d) to give HOO$^{\bullet}$ (36) which implies that the bond strength of $Fe^{IV}O–H$ is 119 (HO–H) $> Fe^{IV}O–H > 90$ (HOO–H) kcal mol^{-1}, i.e. sufficient to break homolytically non-activated carbon–hydrogen bonds, Fig. 6. The oxygen atom slips slightly to make space for a metal-substrate interaction which weakens the C–H bond.[48a] An Fe-substrate interaction is presumably necessary to oxidize methane by non-haem methanotrophic bacteria, [49] (cf. the analogous reaction: $Pt^I + CH_4 \rightarrow Pt = CH_2 + H_2$). The first step is the formation of the carbon radical which in the second step reacts with the Fe^{IV}-hydroxyl group to give the hydroxy compound. The homolytic pathway is supported first by the observation of a considerable isotope effect, i.e. hydrogen abstraction is the rate determining step. Second, the relative reaction rates of the aliphatic hydrogens are *tert*-H $>$ *sec*-H $>$ *prim*-H, in agreement with expected homolytic reactivities. The selectivity of cytochrome P450 is close to that of the HOO$^•$ radical. The two-step reaction is supported by investigations of several cases of non-stereospecific bio-hydroxylations. The concerted reaction, i.e. insertion of the oxygen atom, represents the highly stereospecific reaction with retention of configuration, but high stereospecificity can in principle also be attained by steric constraint at the active site hindering rotation around the carbon-carbon bond. At the other end of the reaction spectrum we have non-stereospecific H-abstraction and subsequent non-stereospecific addition of the OH-group. In between, there are cases representing stereospecific H-abstraction, while the addition of the OH-group leads to mixtures of stereoisomers. A well investigated case is the cytochrome P450$_{cam}$ catalysed hydroxylation of camphor, where the hydrogen abstraction is non-stereospecific but the OH-addition leads to a single stereoisomer as proved by the D-labelling technique.[50]

$$PorFe^{IV}–O^• \rightleftharpoons PorFe^{IV} + O^{-•} \qquad (35)$$

$$PorFe^{IV} – O^• + H_2O_2 \rightarrow PorFe^{IV}OH + HOO^• \qquad (36)$$

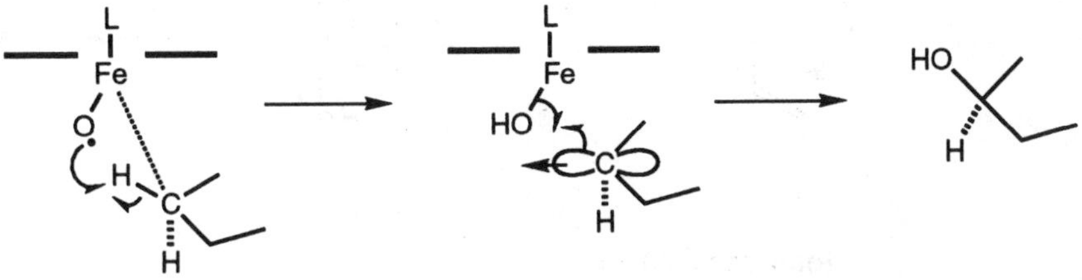

Fig. 6   Hydroxylation of an aliphatic compound showing retention of configuration. The Fe atom is located slightly above the porphyrin plane

An objection has been raised against the two-step mechanism with an intermediate free radical reacting with the Fe^{IV}-OH oxygen in a second step, also called the oxygen rebound step. Kinetic investigatons have shown that no correlations exist between rates of rearrangements and product ratios of the putative intermediate radical, which is contrary to expectations.[50a] Furthermore, the calculated rates of the rebound step turned out to be unreasonably high. To take this into account we assume that there is a considerable interaction between the transition metal catalyst and the radical (dotted line in Fig. 6) which retards the rearrangement of certain radicals to the more stable species. That means, provided space allows, that a skew four-centred transition state is formed and the oxygen transfer is characterized as an inner space reaction. This situation is especially favourable for hydrogen abstraction from methyl groups.

The principle of the oxygen rebound reaction has been applied to several oxygenations in the following sections and can be extended to formation of sulphur-carbon bonds in, for example, the biosynthesis of penicillins and formation of nitrogen-carbon bonds in alkaloids.

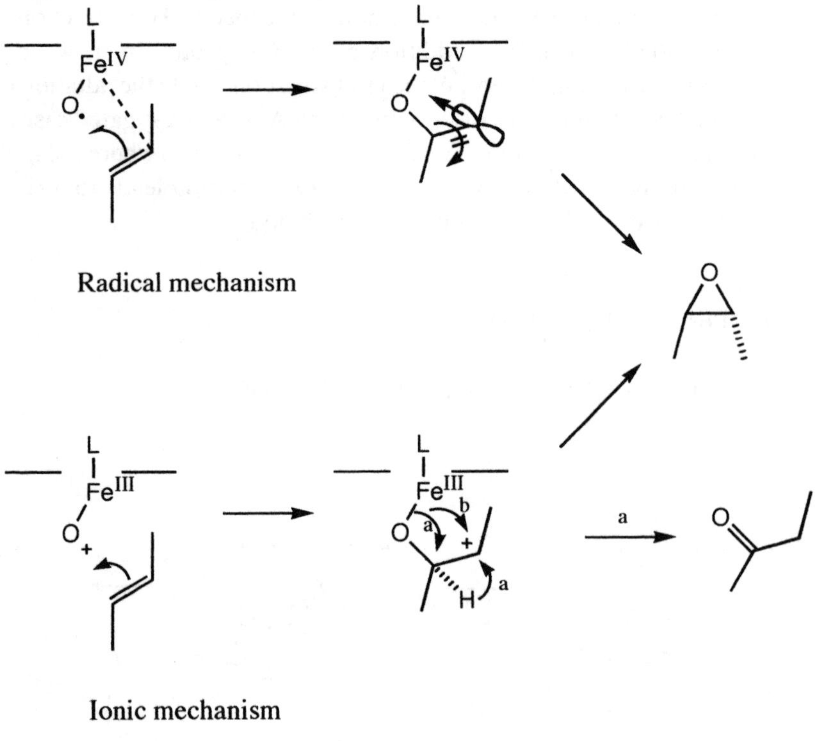

Radical mechanism

Ionic mechanism

Fig. 7   Epoxidation of olefins. L = Ligand

The iron oxo porphyrins efficiently oxidize olefins and aromatics into epoxides and phenolics as part of their metabolism. Both radical and cationic mechanisms have been proposed based on experimental and theoretical studies.[44,51]

The olefinic oxidation is rationalized in the following way, Fig. 7. The 33d form adds in a radical fashion to the double bond giving rise to the intermediary C-centred radical. It is tacitly accepted that this radical reacts with the PorFe ^{IV}–O bond in a second step to give the epoxide with retention of configuration. This implies that the enzyme imposes rotational constraints and that the intermediate has a short lifetime. The epoxides are occasionally accompanied by small amounts of carbonyl products, the formation of which is best explained by the ionic mechanism involving the 33b form and a 1,2-hydrogen shift, route a. As expected, electron rich olefins react faster than electron poor ones.

The oxidation of aromatics is more complex,[52] Fig. 8. The radical pathway leading to the arene epoxides is identical to olefinic epoxidation. Acid catalysed opening of the epoxides gives phenols either via the NIH-shift involving formation of the ketone and subsequent enolization, route a, or via direct aromatization, route b, which means loss of label. Electrophilic substitution

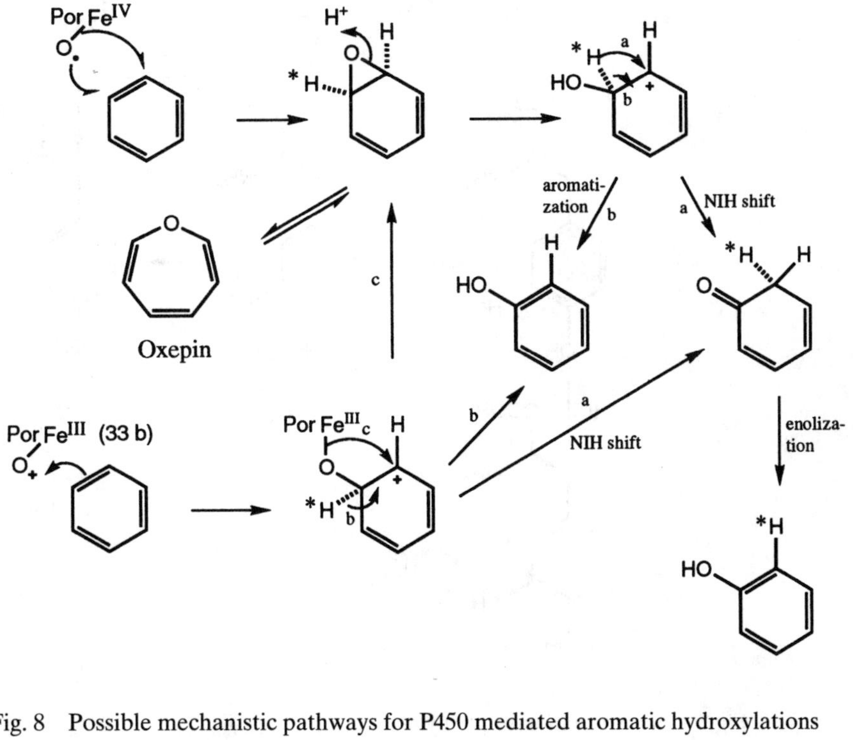

Fig. 8   Possible mechanistic pathways for P450 mediated aromatic hydroxylations

could operate by invoking the 33b form. Model reactions show that arene epoxides indeed rearrange into phenols and exhibit the same isotope effects as the native system. The NIH shift, named after a research group at the National Institute of Health, USA, is characteristic of aromatic hydroxylation. Phenylalanine labelled at C^4 with tritium or deuterium retained *ca.* 90 per cent of the label at C–3.[53] The NIH shift is not restricted to *para*-hydroxylation, but functions for *ortho*-hydroxylation as well. However, hydroxylation *ortho* or *para* to an existing hydroxyl group leads to loss of label.[54] The mechanism is depicted in Fig. 9. Two paths lead to loss of label. In path a the label is directly eliminated by aromatization and in path b in the enolization step. The high retention indicates high migratory aptitude of the proton and an unusually high isotope effect. It has been argued that the enolization therefore is enzymatically controlled.

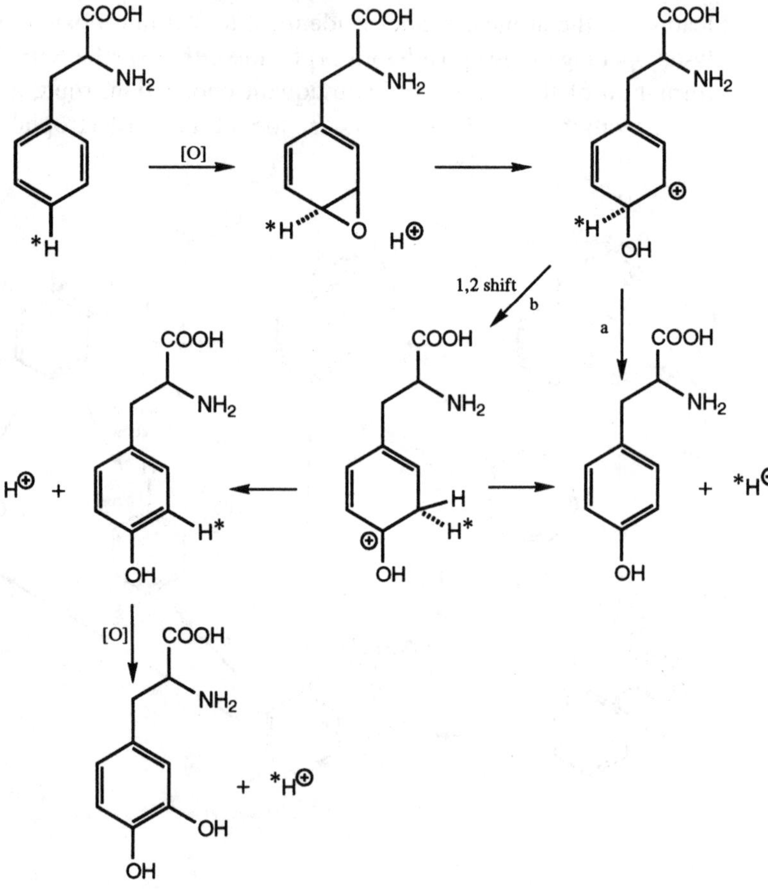

Fig. 9   Hydroxylation of phenylalanine

In the benzene series these epoxides are unstable, but in higher condensed systems, such as naphthalene, or pyrene, epoxides have been isolated.[55] Epoxides of polycyclic aromatics display carcinogenic properties. That means, in fact, that the detoxifying mechanism leads to still more harmful derivatives. It is suggested that opening of the epoxide by nucleophilic amino groups in nucleic acids initiates chemical carcinogenesis. Benzene epoxide is shown to be in equilibrium with the oxepin isomer and the equilibrium is markedly shifted towards the seven-membered ring system (Fig. 8). Related naturally occurring metabolites, which contain either the hexadiene structure or the oxepin structure, have been isolated. The substituents of the cyclohexadiene structure are always *trans*-located (Fig. 10), which is typical for opening of an oxiran ring by a nucleophile.

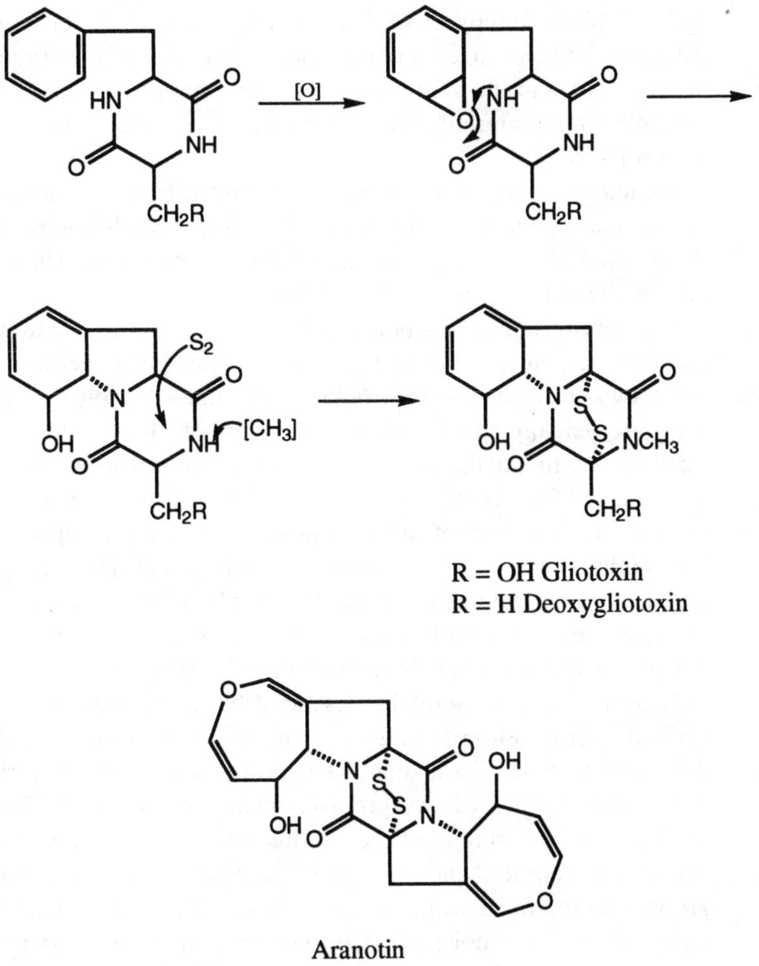

R = OH Gliotoxin
R = H Deoxygliotoxin

Aranotin

Fig. 10   Proposed biosynthesis of gliotoxin

The biosynthesis of gliotoxin[56] is illustrative of the process of aromatic epoxidation. The precursor is believed to be the cyclodipeptide formed from phenylalanine and serine. The arene oxide is opened in a *trans* fashion and a disulphide bridge is introduced. It has been established that all nine carbons of phenylalanine are incorporated in gliotoxin and it is shown that ^{14}C-labelled cyclo-(L-alanyl-L-phenylalanyl) is used by *Trichoderma viridis* to produce the 3-deoxy analogue of gliotoxin. Furthermore, a structurally related oxepin derivative, aranotin, has been isolated.

Further support for the intermediacy of epoxides in aromatic hydroxylation has been gained from studies of the mechanism of aromatization of some substituted benzene-1,2-oxides.[57] 1-Methoxycarbonylbenzene-1,2-oxide rearranges quantitatively to methyl salicylate at pH 7; 83 per cent of the product is formed by migration of the methoxycarbonyl group, i.e. C^2–O oxiran cleavage, and 17 per cent by C^1–O cleavage. Depending upon the nature of the substituent C^2–O cleavage can also lead to loss of the substituent. Thus, 1-hydroxymethylbenzene-1,2-oxide gives 8 per cent phenol and 92 per cent salicylalcohol without migration of the substituent, i.e. entirely via C^1–O cleavage 1; see Problem 3.7.

Metabolic studies with synthetic 1–^{2}H and 2-^{2}H naphthalene-1,2-oxide give approximately the same retention of label in 1-naphthol as microsomal hydroxylation of 1-^{2}H- and 2-^{2}H-naphthalene,[58] again supporting the epoxide as intermediate in aromatic hydroxylation.

It is appropriate at this point to discuss some of the reactions of flavin dependent oxygenases. In the respiratory chain flavins act as converters between two-electron oxidants, NAD$^\oplus$, and one-electron oxidants, Por Fe$^{3\oplus}$. They are oxidants in their own capacity as cofactors of dehydrogenases and oxidases and finally they can serve as oxygen activators in oxygenase systems. The reduced form is stepwise oxidized to the quinoid form by oxygen with intermediate formation of the semiquinone radical and superoxide ion which can combine to a flavin-4a-hydroperoxide (Fig. 11). The semiquinone radical is rather stable and can easily be detected by ESR spectroscopy. The 4a-hydroperoxide or its homolytic fission product is thought to be the active species in hydroxylation of phenols and indoles (37,38).[59]

Aromatics are sequentially degraded by monohydroxylation to phenols, *ortho*-dihydroxylation to catechols and finally ring opening by iron-dependent dioxygenases via a dienone hydroperoxide (39). The ring is cleaved either between the hydroxyl groups or next to one of them. The mechanism is not known with certainty. We know that both oxygen atoms of molecular oxygen are incorporated and spectral data indicates coordination of the phenolic group with the ferric centre of the cofactor. The ferric ion abstracts an electron and the catechol radical formed combines with oxygen to a hydroperoxide or a highly strained dioxetane which rearranges in an obscure way. The intermediacy of the cyclodienone peroxide is supported by isolation of

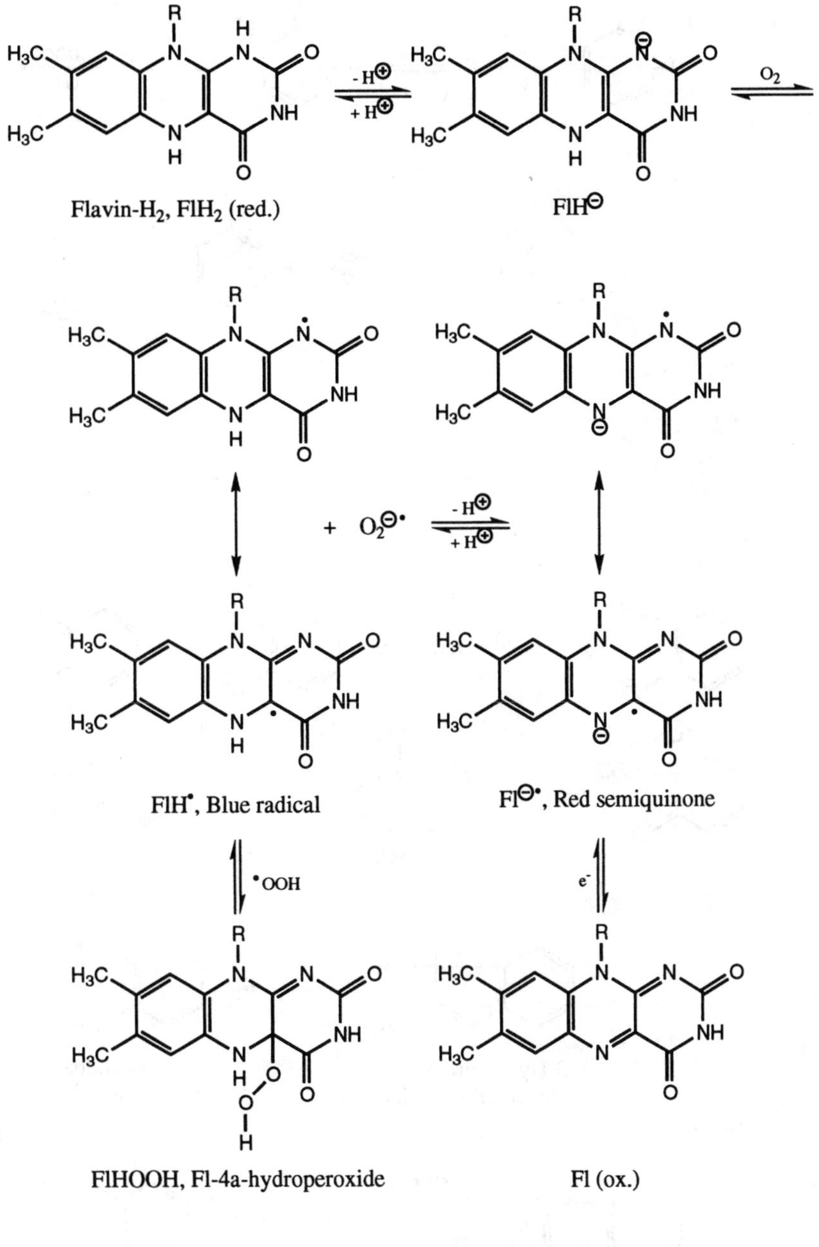

Fig. 11 Stepwise one-electron transfers; oxidation of flavin-H₂ by oxygen

$$FlHOOH \rightleftharpoons FlH^{\bullet} + {}^{\bullet}OOH \rightleftharpoons FlH^{\bullet} + O_2^{\ominus\bullet} + H^{\oplus} \tag{37}$$

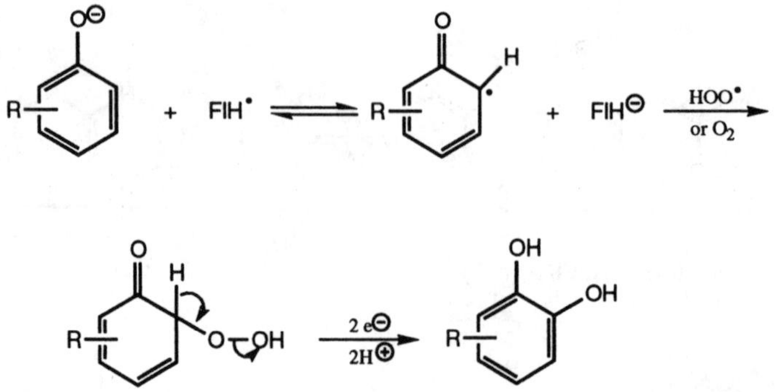

(38)

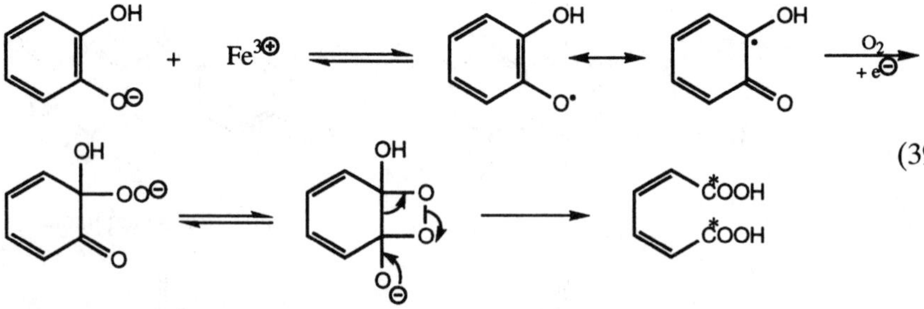

(39)

3-Hydroperoxy-2,3-
dimethylindolenine

2-Acetamidoaceto-
phenone

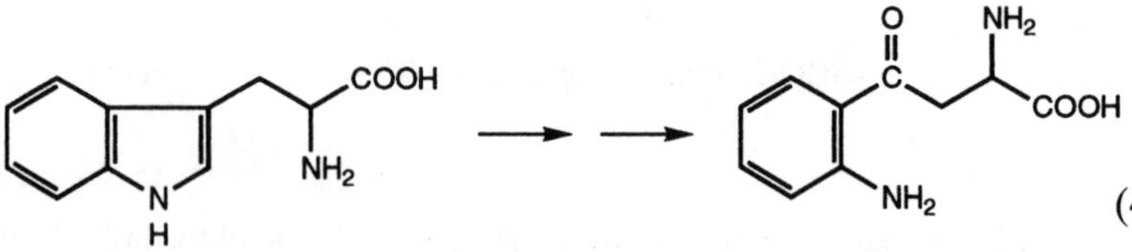

(40)

Tryptophan

Kynurenine

(41)

such derivatives and the finding that they indeed are cleaved to dicarbonyl compounds. The heteroaromatic 3-hydroperoxy-2,3-dimethylindolenine rearranges to 2-acetamidoacetophenone on treatment with base[59] (40), a reaction that is analogous to the one occurring at one stage of the biosynthesis of kynurenine a degradation product of tryptophan (41).

Related are the non-haem iron oxygenases which require α-oxoglutarate, molecular oxygen and ascorbate. They are involved in the hydroxylation of amino acids such as proline and lysine. The mechanism is not well understood but can be modelled on the P450 oxygenases. One of the problems is to show how α-oxoglutarate functions as a reducing agent. Molecular oxygen is incorporated into both the substrate and the succinic acid formed. One mechanism favours intermediary formation of persuccinic acid. Ligated ferrous ions reduce oxygen to the superoxide radical ion which adds to the carbonyl group to yield persuccinic acid and carbon dioxide. The peracid then reacts with the substrate in a non-haem iron catalysed reaction similar to the conventional cytochrome P450 reaction to give the hydroxylated product (42). The ascorbate is presumably required to control the Fe^{2+}/Fe^{3+} equilibrium.

The conversion of tyrosine to homogentisic acid via *p*-hydroxyphenylpyruvate, where the keto function serves as effector group, follows the same scheme (43).

$$L_5Fe^{2+} + {}^*O_2 \longrightarrow L_5Fe^{3+} + {}^*O_2^{-\bullet} \quad \xrightarrow{\underset{\displaystyle RC-COOH}{\overset{\displaystyle O \atop \displaystyle \|}{}}}$$

(42)

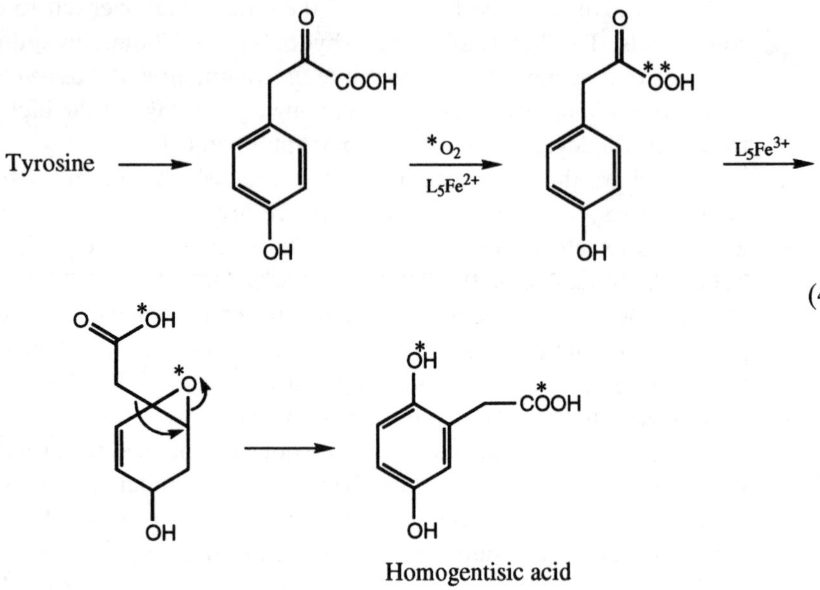

(43)

Homogentisic acid

4.4 Cinnamic and benzoic acids

The ubiquitously distributed cinnamic acids arise from phenylalanine by enzymatic elimination of ammonia followed by aromatic hydroxylation and methylation. It was earlier believed that the biosynthesis took the route via phenylpyruvic acid which was reduced and dehydrated, reactions that *per se* have ample precedents, but when it was discovered that enzymes were able to eliminate ammonia directly from the amino acid, this route was considered to represent the major pathway. This reaction is categorized as a concerted α,β-elimination assisted by a basic centre on the enzyme abstracting the β-proton (44). The elimination takes place in the stereoelectronically most favourable *trans*-planar conformation by exclusive loss of the pro-$(3S)$ hydrogen leading directly to *trans*-cinnamic acid.[60] The reaction is analogous to the classical

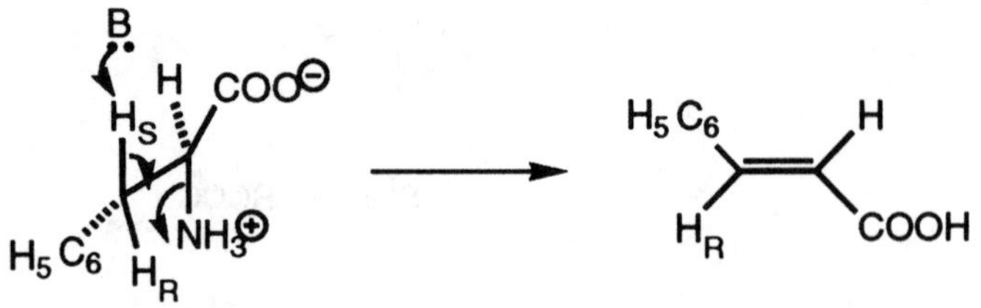

(44)

Hofmann elimination. A closer investigation of the phenylalanine ammonia lyase, PAL, revealed that the reaction is reversible which means that the re-addition of ammonia must take place opposite to the polarity of the double bond. Chemical inhibition of the enzyme is achieved by carbonyl reagents such as sodium cyanide and sodium borohydride. Reduction of the enzyme

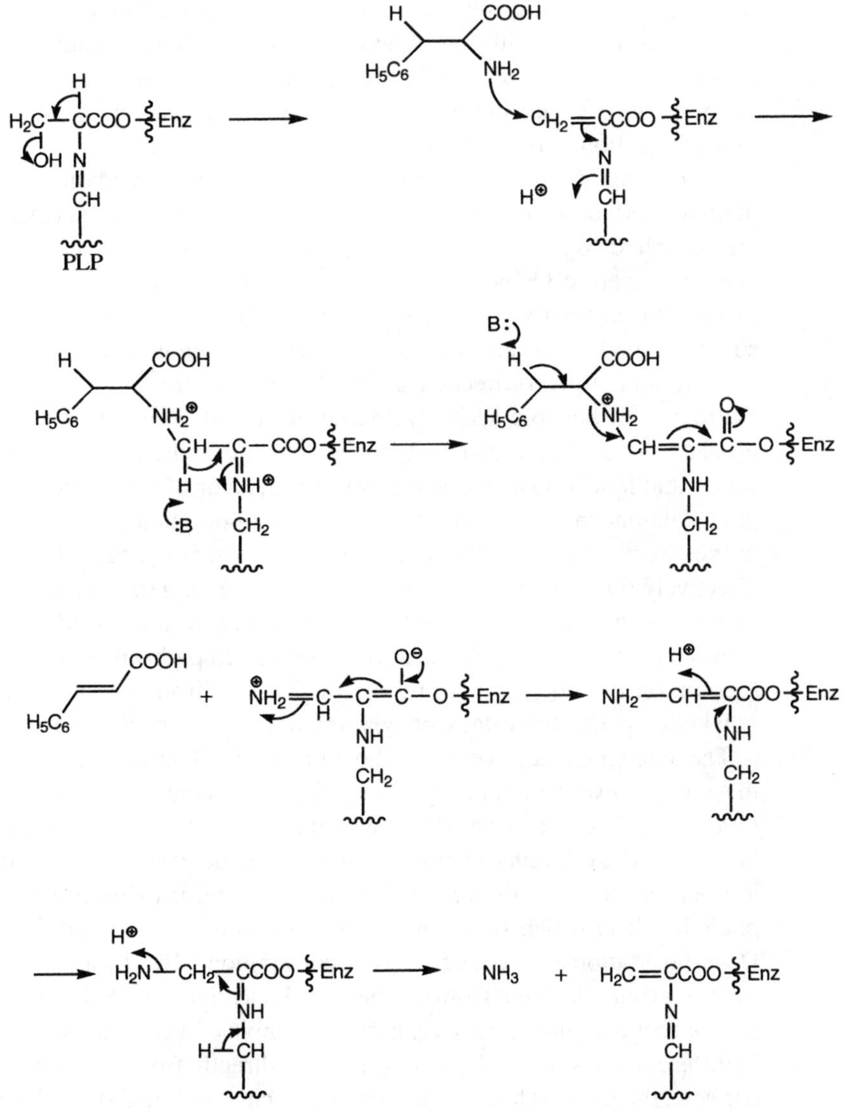

Fig. 12   Hypothetical mechanism for PAL activity

with tritiated sodium borohydride gives on subsequent hydrolysis alanine with tritium predominantly confined to the methyl group.[61] These observations suggest that the active site of the enzyme is a dehydroalanine residue probably generated from serine (Fig. 12). The amino group of serine forms a Schiff's base with another subunit of the enzyme activating the α-proton and resulting in elimination of water. The β-carbon of the dehydroalanine residue reacts in a Michael-type reaction with nucleophiles, such as hydride ions from sodium borohydride or cyanide ions, blocking its activity. The amino group of the substrate binds in the same way to the dehydroalanine residue and this could improve its ability as a leaving group as demonstrated by shuffling around charges and protons in the prosthetic group (Fig. 12). Expulsion of ammonia in the final step regenerates the dehydroalanine residue making it ready for action in the next cycle.

Shortening of the side chain of cinnamic acids by β-oxidation is one of the main routes leading to benzoic acids (*cf.* section 5.9).[62] The exact timing of the aromatic hydroxylation varies from plant to plant but in general the hydroxylation is more effective at the C_6C_3 level that at the C_6C_1 level in higher plants. We can arrive, in principle, at vanillic acid, either along the route coumaric acid → caffeic acid → ferulic acid, or via coumaric acid → *p*-OH-benzoic acid → protocatechuic acid (Fig. 13).[63] There is also a shortcut to the hydroxylated benzoic acids. Dehydration and dehydrogenation of 3-dehydroshikimic acid give protocatechuic acid and gallic acid directly.[64] It was found that labelled glucose was a better precursor of gallic acid than labelled phenylalanine in *Geranium pyrenaicum*, an observation that speaks in favour of the direct route.[65] On the other hand phenylalanine was metabolized more effectively than glucose to gallic acid in *Rhus typhina* suggesting that phenylalanine → cinnamic acid → *p*-OH-cinnamic acid → caffeic acid → 3,4,5-trihydroxycinnamic acid → gallic acid is the preferred pathway here.[66] *Epicoccum nigrum*, on the other hand, produces gallic acid from orsellinic acid via the polyketide pathway by decarboxylation and oxidation (Fig. 14).

The side chain can eventually be completely eliminated as in the biosynthesis of the hydroquinone glycoside arbutin, a metabolite of *Bergenia crassifolia*. *p*-Hydroxybenzoic acid was shown to be an efficient precursor for arbutin. The hydroxylation of benzoic acid follows the general scheme developed in the precedeing section. *p*-Hydroxybenzoic acid is oxidized by $Fe^{3\oplus}$ to the phenol radical which then couples with oxygen or a peroxide radical (45). Decarboxylation and reduction give hydroquinone. Fig. 15 shows the caffeoyl ester of arbutin isolated from cowberries, *Vaccinium* spp. Salicylic acid, also a constituent of many berries, is another example of a compound that can arise in different ways. It can be biosynthesized directly from benzoic acid. Alternatively, the hydroxylation can occur at the cinnamic acid stage. It can also, at a much earlier stage, be derived from isochorismic acid by elimination of pyruvate (see Fig. 3) and, finally it can be made via the polyketide pathway.

Berries, rich in salicylic or benzoic acids can be preserved without addition of preservatives (cowberries, cloudberries, *Vaccinium vitis idea, Rubus chamae-morus*) because of the bacteriostatic properties of these compounds.

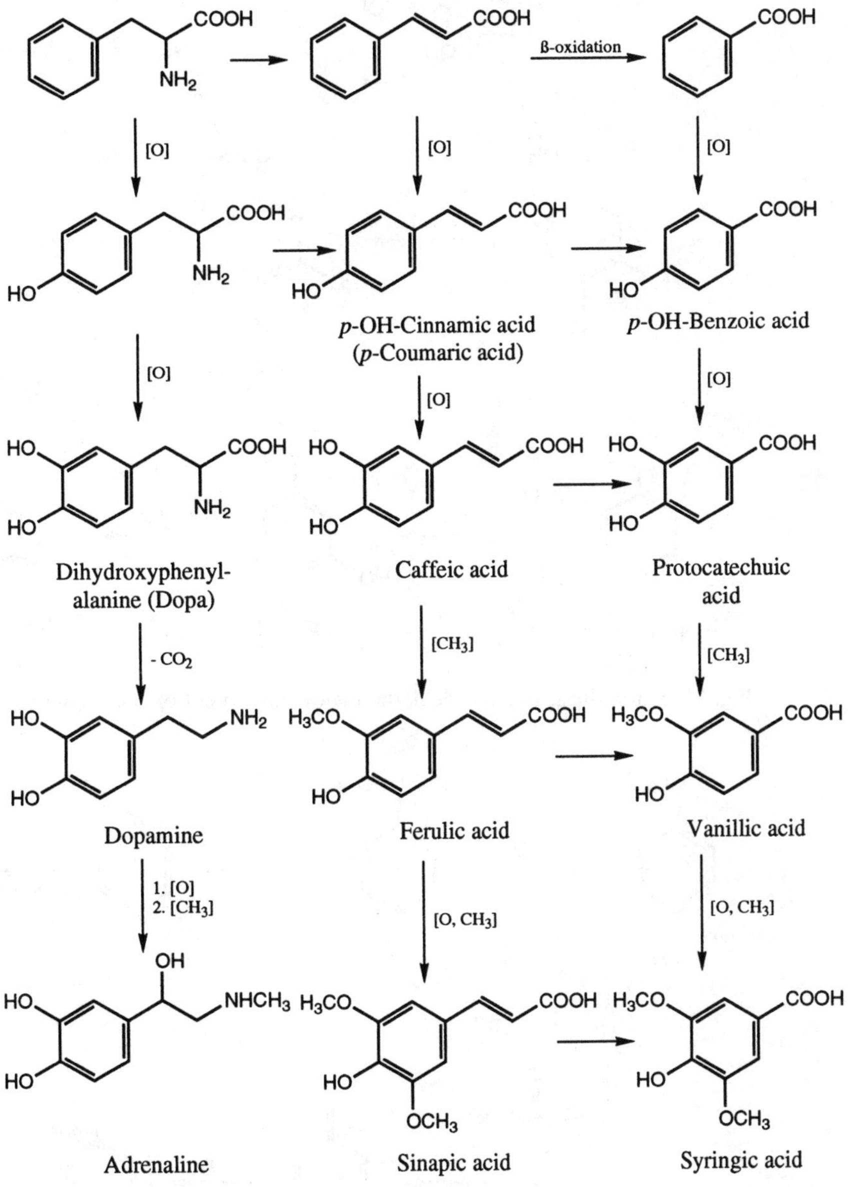

Fig. 13   Biosynthetic network of cinnamic and benzoic acids and biosynthesis of adrenaline

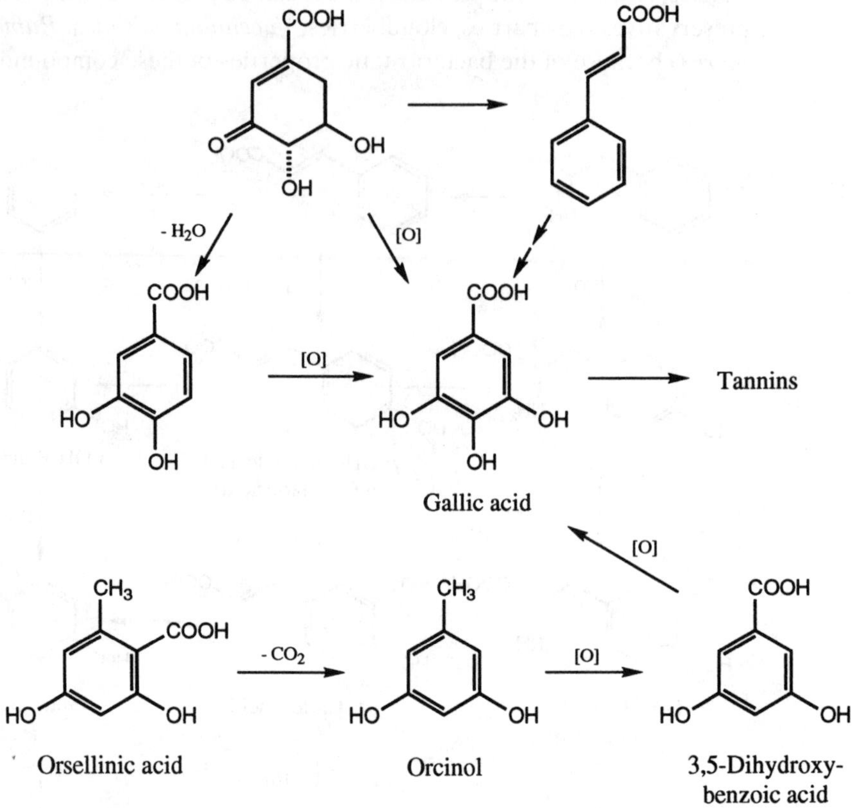

Fig. 14 Biosynthesis of gallic acid, the major constituent of gallotannins

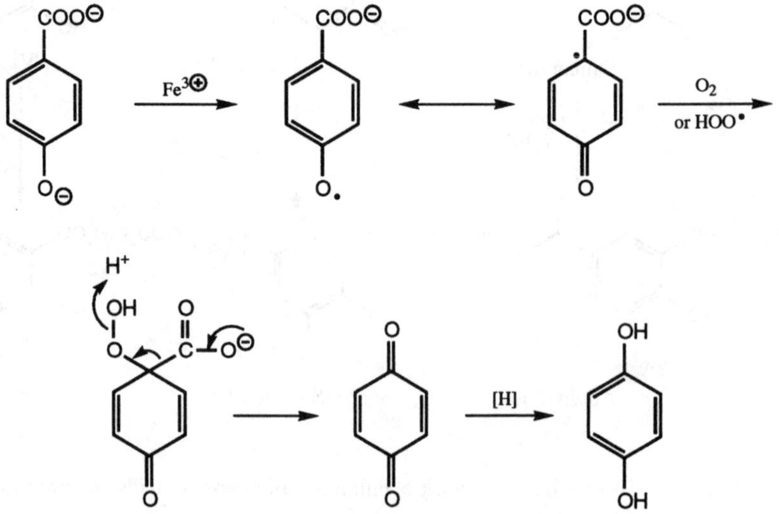

(45)

The cinnamic and benzoic acids occur mostly as esters and glycosides, Fig. 15. 3-Caffeoylquinic acid (chlorogenic acid) was isolated in crystalline form by Payen in 1846 from coffee and was later found to be a common metabolite in plants. The esterification takes place via the activated CoA ester of the acid.

The order of events leading to salicin (46), widely distributed in *Salicaceae*, has been looked at in some detail. ^{14}C-Labelled phenylalanine, cinnamic acid and *o*-coumaric acid were incorporated into salicyl alcohol in *Salix purpurea*

(46)

Helicin   Salicin

with increasing order of efficiency, whereas salicylic acid was poorly incorporated.[67] In another series of investigations with *Vanilla planifolia* the synthesis of vanillin was studied. Phenylalanine, cinnamic acid, and ferulic acid were incorporated with increasing ease, whereas labelled vanillic acid was poorly converted which means that in higher plants free benzoic acids are poorly reduced. It was also found that glycoside formation occurs at a higher oxidation level than salicylic alcohol since administration of this compound to the plant led to glycosidation of the $-CH_2OH$ group.[68] β-Oxidation of cinnamic acids according to the classical fatty acid degradation scheme should give rise to acetyl CoA and benzoyl CoA, representing a more energy-rich carboxyl moiety that consequently could be reduced by NADH more easily. It was, in fact, possible to trap labelled acetyl/CoA originating from cinnamic acid. Glycosidation occurs probably at the salicylaldehyde stage. These findings are summarized in (46) for the biosynthesis of salicin.

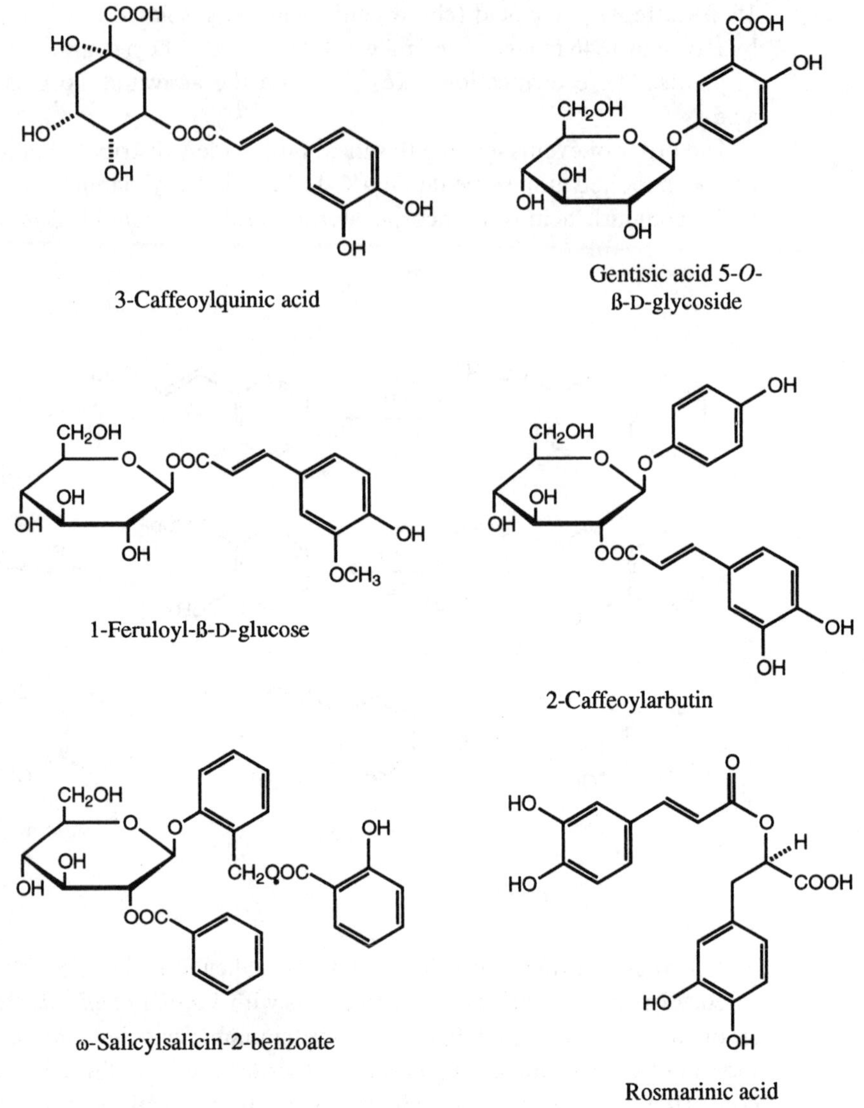

3-Caffeoylquinic acid

Gentisic acid 5-*O*-
ß-D-glycoside

1-Feruloyl-ß-D-glucose

2-Caffeoylarbutin

ω-Salicylsalicin-2-benzoate

Rosmarinic acid

Fig. 15   Naturally occurring esters and glycosides of cinnamic and benzoic acids

Rosmarinic acid (*Hyptis* spp.), reported to possess anti-inflammatory properties, is an ester of caffeic acid and 3,4-dihydroxyphenyllactic acid. The former part is derived from phenylalanine, whilst the lactic acid part is derived
from tyrosine via transamination, reduction of 4-hydroxyphenylpyruvic acid
by NADPH and *meta*-hydroxylation.[69]

The C_6C_2 unit is comparatively rare in natural products except in alkaloids, which frequently contain the arylethylamine unit. It is formed by:

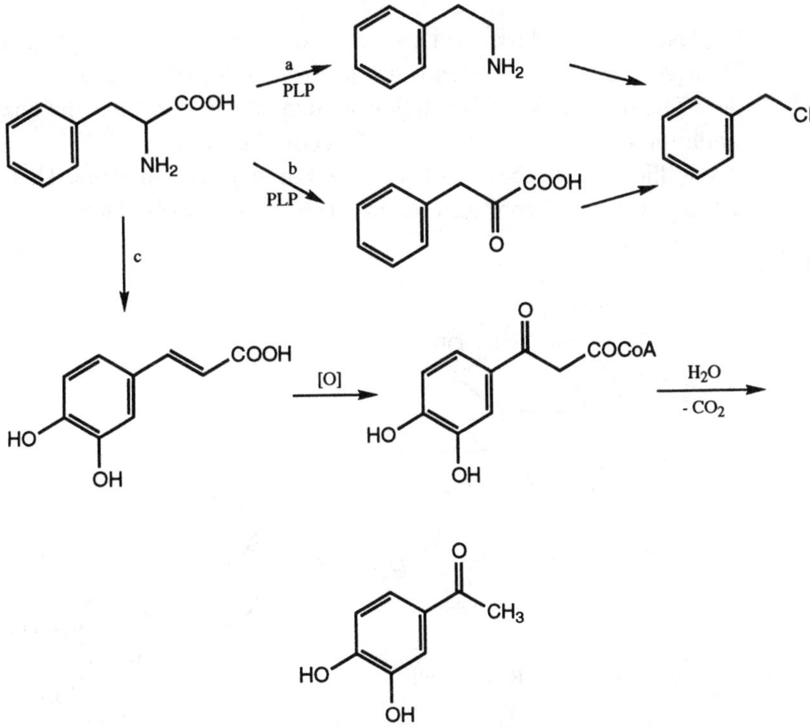

(47)

Pungenin

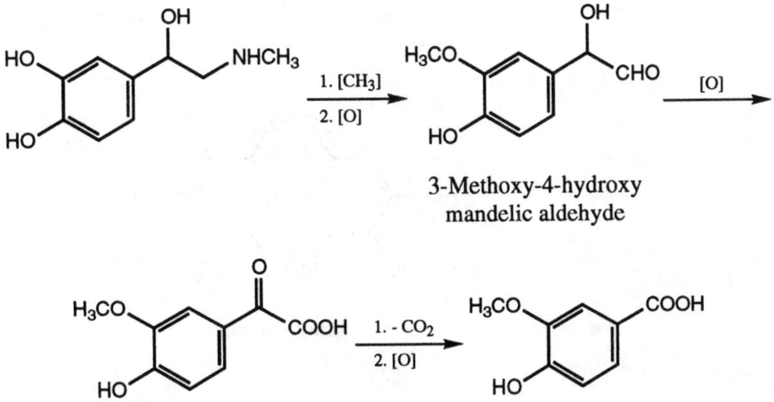

3-Methoxy-4-hydroxy
mandelic aldehyde

(48)

Vanillic acid

1. decarboxylation of phenylalanine;
2. decarboxylation of phenylpyruvic acid;
3. decarboxylation of benzoylacetic acid (47).

The last mentioned reaction gives rise to an acetyl group as in pungenin from *Picea pungens*.[70] β-Keto acids are known to decarboxylate with exceptional ease. The one-carbon degradation sequence (47b) parallels the enzymatic degradation of pyruvic acid (section 3.3, equation (16)).

Vanillic acid is occasionally detected in mammalian urine. This compound is biosynthesized from adrenaline according to (48). The *meta*-hydroxyl is

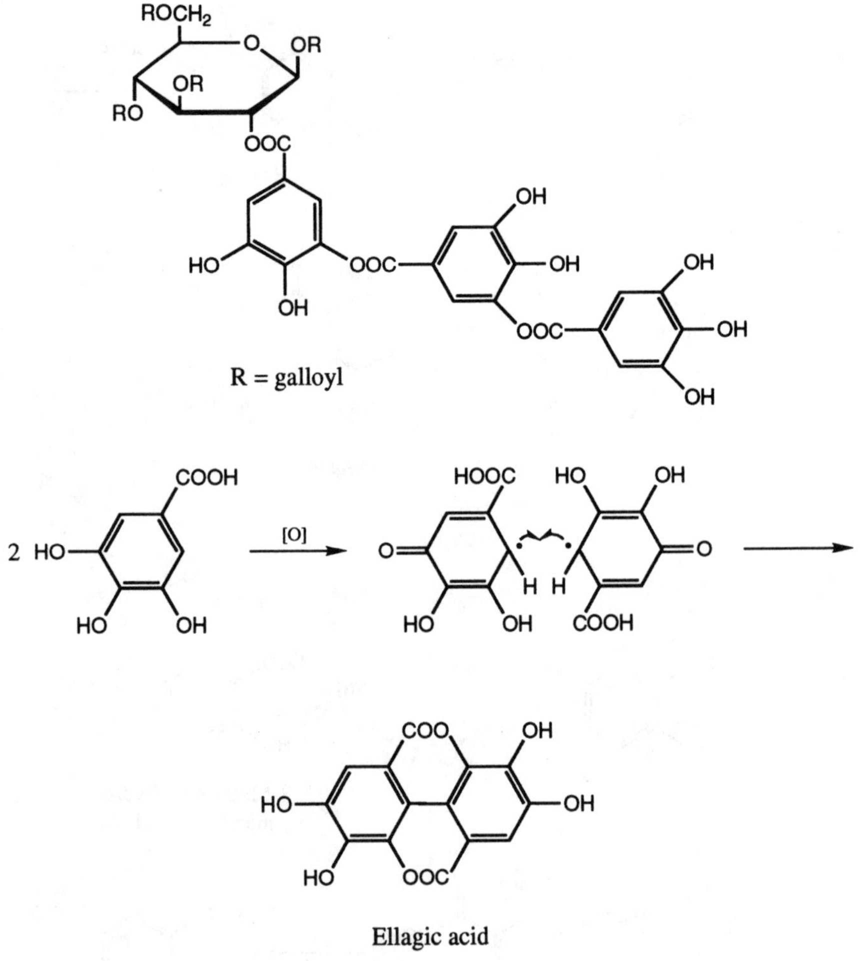

Fig. 16   Structure of a gallotannin and biosynthesis of ellagic acid by phenol coupling

methylated, oxidation gives mandelic aldehyde, and further oxidation and de-carboxylation eventually lead to vanillic acid. In plants vanillic acid is biosyn-thesized from ferulic acid by β-oxidation. The biosynthesis of adrenaline is shown in Fig. 13.

Tannins are considered to be excretion products of many plants, but are ac-tually involved in the defence mechanism against parasites and grazing ani-mals. Bark and heartwood of trees (e.g. *Quercus*) are rich in these compounds and so are gall which develop as the result of attack by certain insects. The structures of tannins are complex. The water soluble, hydrolysable tannins are polygalloyl or ellagoyl glycosides. There are often as many as nine gallic acids per glucose unit, i.e. at least one of the gallic acid residues has its hydroxyls esterified by other gallic acid molecules. This link is called a depsi-de bond. Ellagic acid is a dimer of gallic acid formed by phenol oxidation (Fig. 16). Gallotannins give glucose and gallic acid on hydrolysis with acids or en-zymes. Scheele isolated gallic acid in 1787 by this latter method from galls. Large amounts of oak and quebracho wood were formerly used in the vege-table tanning procedure but it is now to a great extent replaced by mineral tanning. Tannins form relatively insoluble complexes with the proteins of hides which render them smooth and resistant thereby converting the hides into leather.

4.5 Coumarins

Coumarins are widely distributed in plants,[71] particularly in the *Umbelliferae* and *Rutaceae* families. They are lactones which open on treatment with base to *cis-o*-hydroxycinnamic acids and spontaneously cyclize again on acidification. Irradiation of the *cis*-cinnamate causes *cis–trans* isomerizaton (49). It is evident

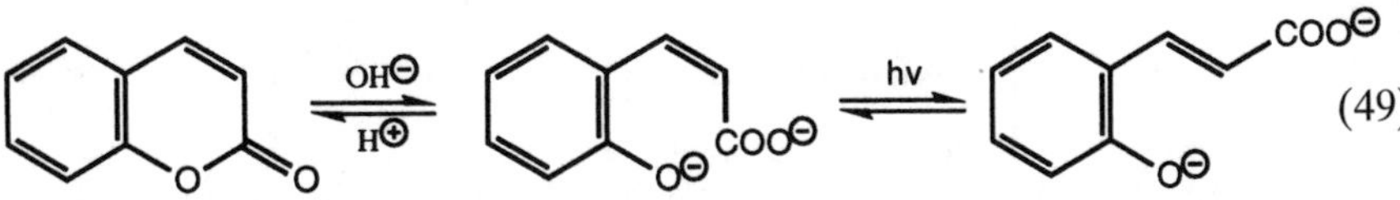

Coumarin   (49)

o-Coumarate

R = H Umbelliferone
R = CH₃ Herniarin

R = H Asculetin
R = CH₃ Scopoletin

Fig. 17   Structures of some coumarin derivatives

that coumarins derive from shikimic acid via cinnamic acids and organisms possessing an enzyme system capable of *o*-hydroxylating cinnamate are thus able to biosynthesize coumarins. Most coumarins are hydroxylated at C⁷. Fig. 17 shows some representative derivatives. It is found by tracer experiments in lavender, *Lavandula officinalis*, that glucose, phenylalanine, and cinnamic acid are efficiently incorporated into both coumarin itself and herniarin. *o*-Coumaric acid and *o*-coumarylglycoside are selectively converted into coumarin, whereas *p*-coumaric acid and 2-glucosyloxy-4-methoxycinnamic acid are selectively converted into herniarin. That places cinnamic acid at a

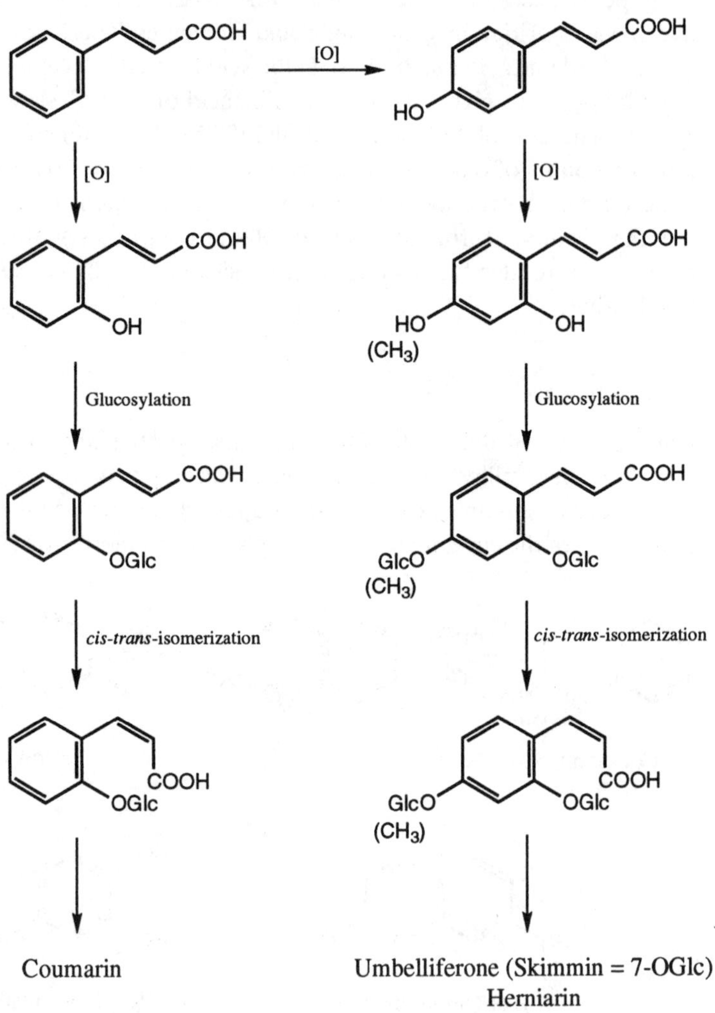

Fig. 18   Biosynthesis of coumarin, umbelliferone, and herniarin. Glc = glucose

branching point of the biosynthetic sequence. The first step to herniarin is a *p*-hydroxylation of cinnamic acid and the first step to coumarin is an *o*-hydroxylation. It was also found that in the intact cell herniarin is predominantly present as a glycoside, presumably the *cis*-2-β-glycoside. The stage at which methylation occurs has not been determined with certainty. The results are summarized in Fig. 18.[72] Studies of the biosynthesis of umbelliferone gave analogous results.

The mechanism of the *o*-hydroxylation was the subject of some dispute when it was demonstrated that the ring oxygen of the coumarin residue in the antibiotic novobiocin (section 1.6, Fig. 7) from *Streptomyces niveus* originated from the carboxyl group[73] as schematically formulated in (50) for oxidation of *cis*-cinnamic acid. When the *trans*-form is oxidized the 2-hydroxy group presumably comes from molecular oxygen. By the use of ^{14}C it was shown that the coumarin moiety of novobiosin originated from tyrosine, i.e. the 7-OH is introduced at a much earlier stage. When the spirolactone was fed to *Lavandula officinalis*, poor incorporation was observed.[74]

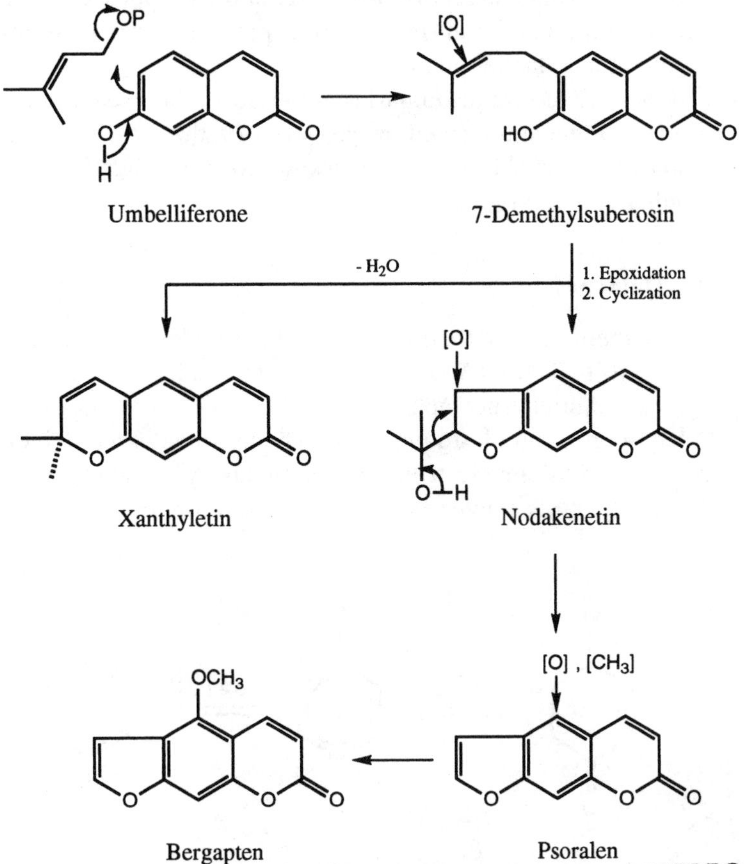

Fig. 19   Biosynthesis of furanocoumarins

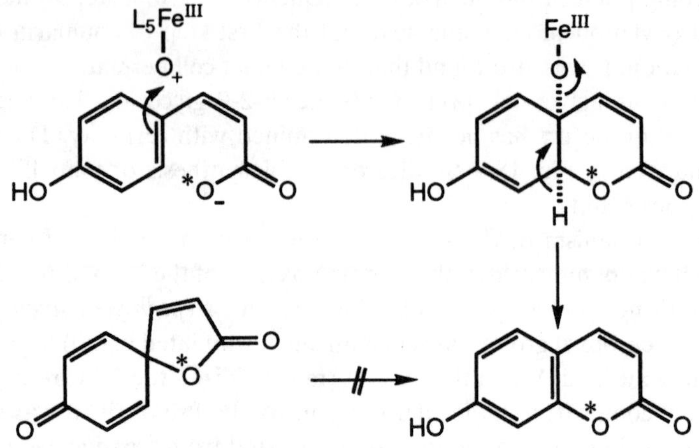

(50)

The fact that the coumarins predominantly occur as glycosides in the intact plant cell favours direct *ortho*-hydroxylation as opposed to oxidative cyclization in the microorganism. Different pathways are apparently followed in moulds and in higher plants.

A variety of furanocoumarins are known. The two-carbon chain is of isoprenoid origin and formed by cleavage of a three-carbon fragment. Feeding studies have established the pathway shown in Fig. 19, involving P450 dependent monooxygenase.[75]

4.6 Quinones

The quinones derive their name from the simplest member, *p*-benzoquinone, obtained in 1838 by Woskresensky as an oxidation product of quinic acid. They structurally encompass pigments, antibiotics, vitamin K, and coenzymes (ubiquinones and plastoquinones). The last mentioned quinones function in metabolism as one-electron transfer agents by virtue of their ability to form reversibly rather stable semiquinone radicals on reduction (51).

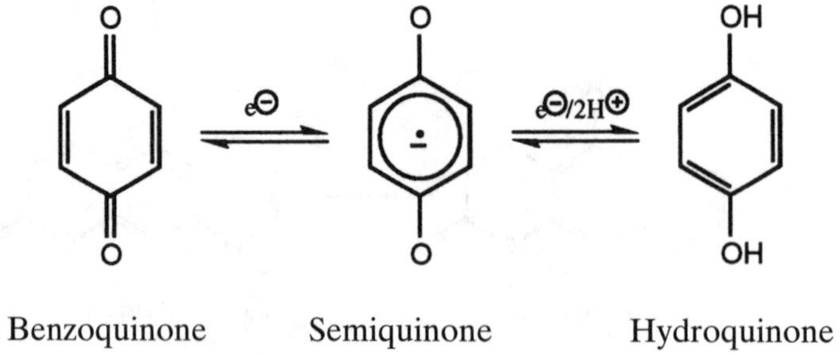

(51)

Benzoquinone   Semiquinone   Hydroquinone

Quinone biosynthesis shows a very diversified picture. It often differs in moulds and higher plants and the gross structure of a compound often gives poor guidance as to the origin of the compound. Plumbagin and 7-methyl-juglone in *Plumbago europaea* come from the polyketide pathway, but juglone in *Juglans regia* is derived from skikimic acid (Fig. 20). Most of the higher

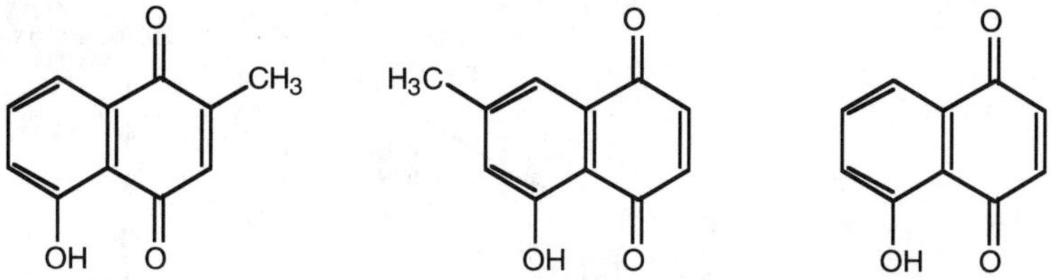

Plumbagin   7-Methyljuglone   Juglone

Fig. 20   Compounds of similar structure but of different biogenesis

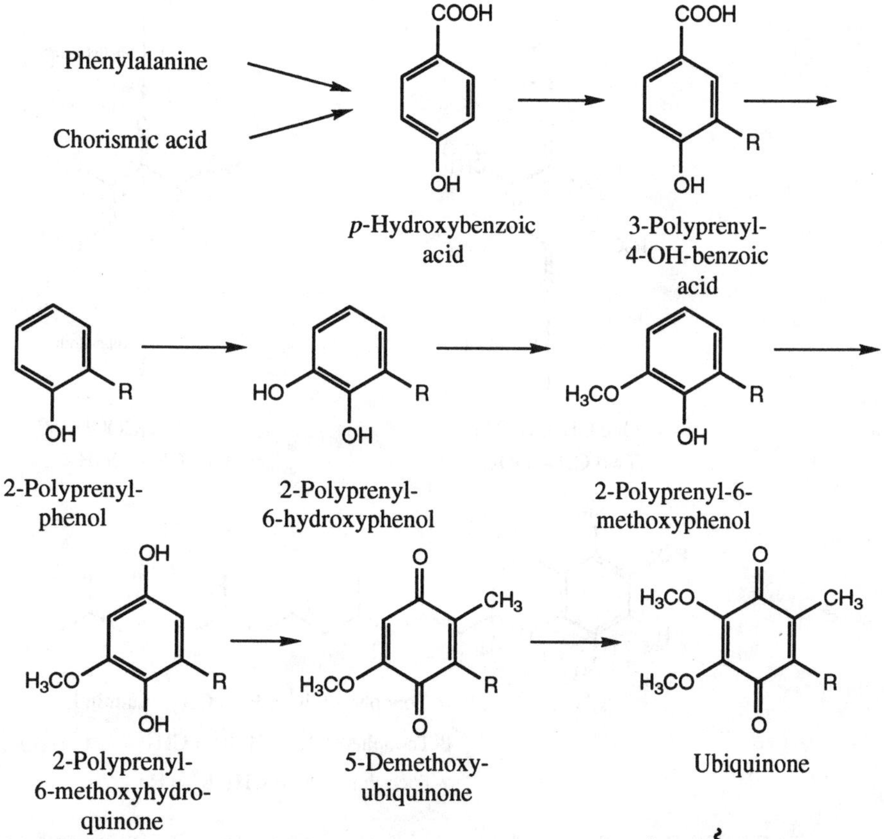

Phenylalanine
Chorismic acid

p-Hydroxybenzoic
acid

3-Polyprenyl-
4-OH-benzoic
acid

2-Polyprenyl-
phenol

2-Polyprenyl-
6-hydroxyphenol

2-Polyprenyl-6-
methoxyphenol

2-Polyprenyl-
6-methoxyhydro-
quinone

5-Demethoxy-
ubiquinone

Ubiquinone

Fig. 21 Biosynthesis of ubiquinones. R = polyprenyl chain, $n = 6 - 10$

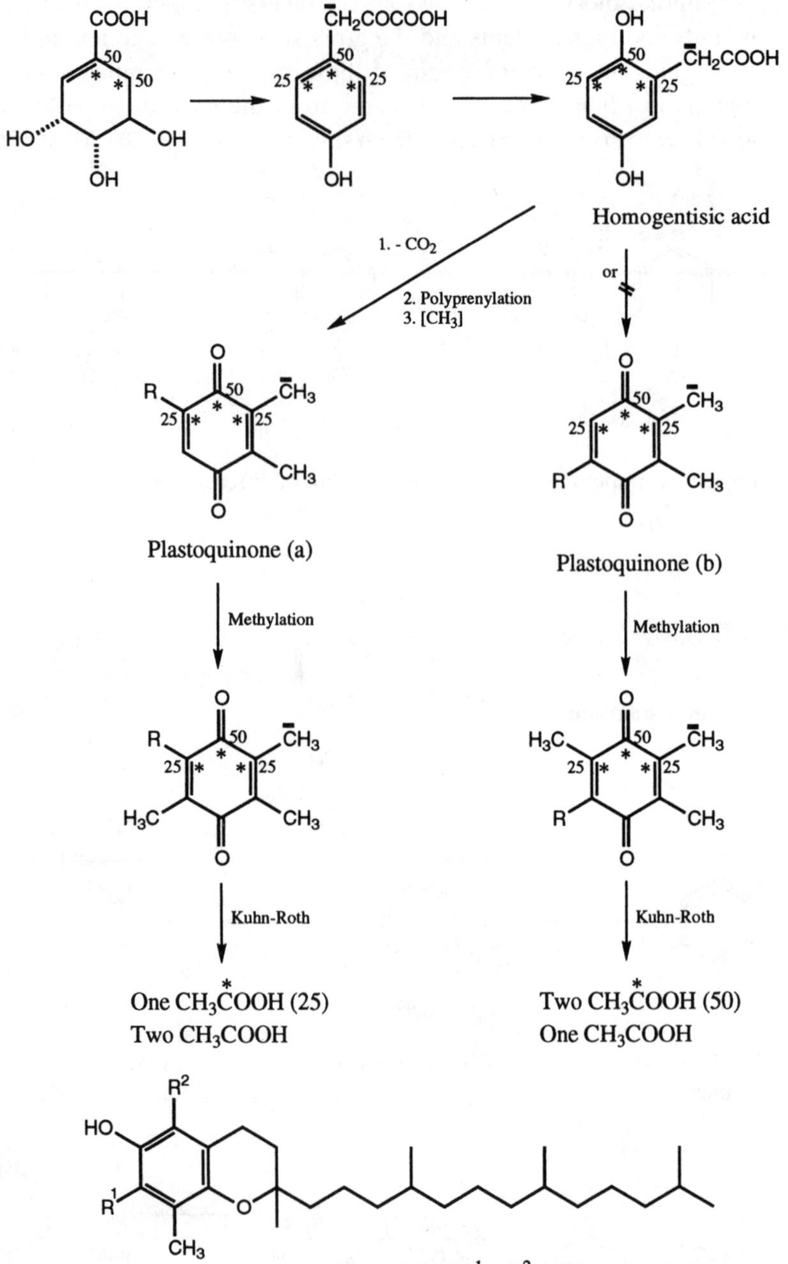

α-Tocopherol, $R^1 = R^2 = CH_3$, vitamin E

ß-Tocopherol, $R^1 = H$, $R^2 = CH_3$

γ-Tocopherol, $R^1 = CH_3$, $R^2 = H$

Fig. 22 Biosynthesis and degradation of plastoquinones. R = polyprenyl. Structure of tocopherols quinones arise by the polyketide pathway or by mixed pathways. Both ubiquiones and plastoquinones have mixed biogenesis with a polyprenoid side chain, but since the benzoquinone nucleus is derived from shikimic acid, elaborated in a intriguing way, they will be treated separately in this chapter. Quinone rings formed in other ways are referred to under the corresponding headings.

The biosynthesis of ubiquinone, involved in the electron transport chain, was elucidated particularly by isolation of metabolites in the photosynthetic bacterium *Rhodospirillum rubrum*[76] and by the use of mutants of *Escherichia coli*.[77] The intermediates accumulated by the different mutants were isolated and identified by mass spectrometry and NMR spectrometry. *p*-Hydroxybenzoic acid, formed by elimination of pyruvic acid from chorismic acid in bacteria or by degradation of phenylalanine in plants and mammals, is alkylated by polyprenyl phosphate (Fig. 21).[78] The phenolic acid is decarboxylated to polyprenylphenol, hydroxylated and *O*-methylated in the 6-position. At this stage *p*-hydroxylation occurs; *cf*. the biosynthesis of arbutin in *Salix*, where oxidation to hydroquinone either is coincident with decarboxylation or immediately follows decarboxylation of *p*-hydroxybenzoic acid. The hydroxylation involves molecular oxygen as shown by use of $^{18}O_2$.[79] the *C*- and *O*-methyls are derived from methionine.[80] The structurally related plastoquinones have a somewhat different history. They derive from tyrosine or *p*-hydroxypyruvate which oxidatively rearrange to homogentisic acid followed by polyprenylation, methylation, and decarboxylation (Fig. 22). The mechanism of this rather unusual rearrangement is discussed in section 4.3. The order of events is not known with certainty, but toluquinol and gentisyl alcohol are not intermediates. The stereochemistry of the decarboxylation of chirally deuterated homogentisic acid was studied by the malate synthase-fumarase method (section 6.2). It was established that the reaction proceeded with retention of configuration.[81] The exact orientation of the two methyl groups with respect to the polyprenyl chain was solved by the use of 1,6-^{14}C-shikimic acid.[82] The plastoquinone was completely methylated and degraded according to Kuhn-Roth (chromic acid-sulphuric acid oxidation). For substitution pattern (a) 25 per cent of the activity should be recovered as acetic acid and for substutution pattern (b) 50 per cent. Finally, if both routes (a) and (b) are followed, we expect 37.5 per cent recovery. The measured activity was 23 per cent indicating prenylation in the *meta* position to the α-carbon of homogentisic acid. It was shown by use af C^2H_3 labelling that one of the methyls originated from methionine.

The tocopherols, vitamin E, essential to normal reproduction, are structurally related to the plastoquinones. The structure is that of a chroman composed of a phytyl group and in the hydroquinone ring 1-3 methyl groups, Fig. 22.

In vitamin K (menaquinones), which affects the clotting of blood, all seven carbon atoms of shikimic acid are retained (Fig. 23). The remaining three

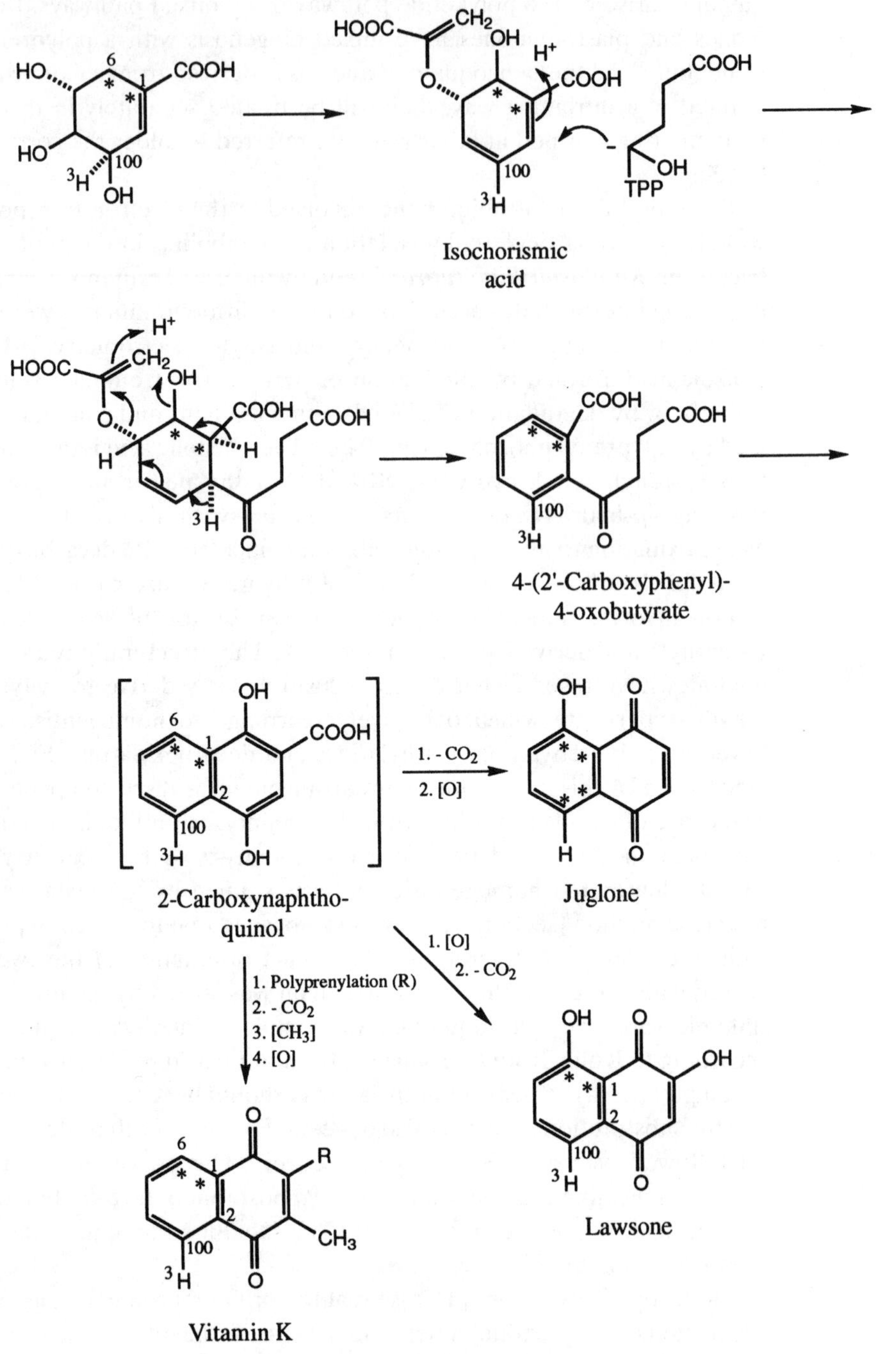

Fig. 23   Biosynthesis of vitamin K, lawsone and juglone

carbon atoms originate from the three central carbons in glutamic acid or 2-ketoglutaric acid which reacts with isochorismate rather than chorismate, Fig. 23.[83] The reaction is mediated by thiamine pyrophosphate. It was also observed that 4-(2´-carboxyphenyl)-4-oxobutyrate was efficiently incorporated, but the likely intermediate 2-carboxynaphthoquinol was poorly incorporated. Experiments with 1,6-[14]D or 3-^{3}H labelled shikimic acid have shown that the ring junction occurs at $C^{1,2}$ in vitamin K, juglone, and lawsone in both plants and bacteria.[84] The observed activities obtained by degradation show that the biosynthesis of juglone involves a symmetrical intermediate, probably naphthoquinone, whereas no such intermediate appears on the route to lawsone and vitamin K. Hence the hydroxylation and prenylation takes place at the position of the carboxyl group. However, the related 2-carbomethoxy-3-prenyl-1,4-naphthoquinone has been isolated from *Galium mollugo* showing that *ortho*-prenylation takes place here[85] (*cf.* biosynthesis of alizarin, section 5.10).

In a related investigation[86] with maize seedlings, *Zea mays*, the labels of 2-^{14}C-4-(2´-carboxyphenyl)-4-oxobutyrate and 3-^{14}C-4-(2´-carboxyphenyl)-4-oxobutyrate were incorporated at C^3 and C^2, respectively, of phylloquinone showing asymmetric alkylations (Fig. 24). The phytyl group is introduced at the position of the carboxyl group.

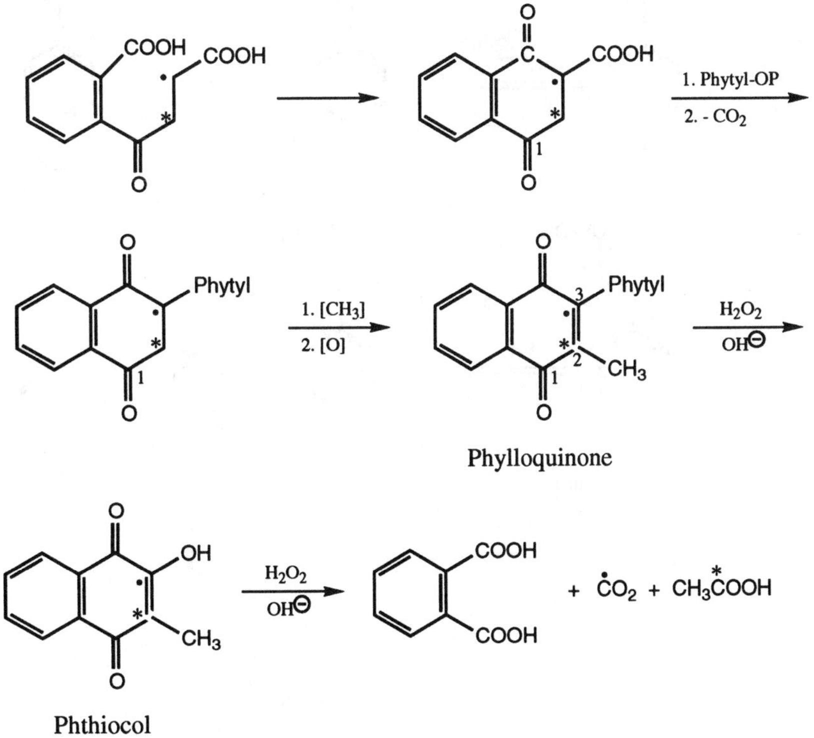

Fig. 24   Biosynthesis and degradation of phylloquinone

4.7 Lignin constituents

Lignin is universally distributed primarily in all woody tissues where it acts as a matrix for the cellulose fibres and adds to the strength and stability of the cell wall, without which growth of any tall plant would be impossible. It turned out that the structure of this extremely complex, polymeric, and insoluble network of aromatic building blocks could be rationalized in a comparatively simple manner. It was suggested in 1933 by Erdtman that lignin is formed chiefly by oxidative, radical polymerization of coniferyl alcohol. The natural polymerization was mimicked by treatment of coniferyl alcohol with a mushroom enzyme extract, whereby a lignin-like product was obtained.[87]

Cell-free enzyme preparations from *Forsythia suspensa* reduce ferulic acid to coniferyl alcohol (52).[88] Ferulic acid and *p*-coumaric acid are the best substrates. Sinapic acid reacts slowly and completely, methylated acids not at all. The enzyme requires magnesium ions for full activity. The reduction can proceed further via the phosphate to the formation of an allyl or propenyl side

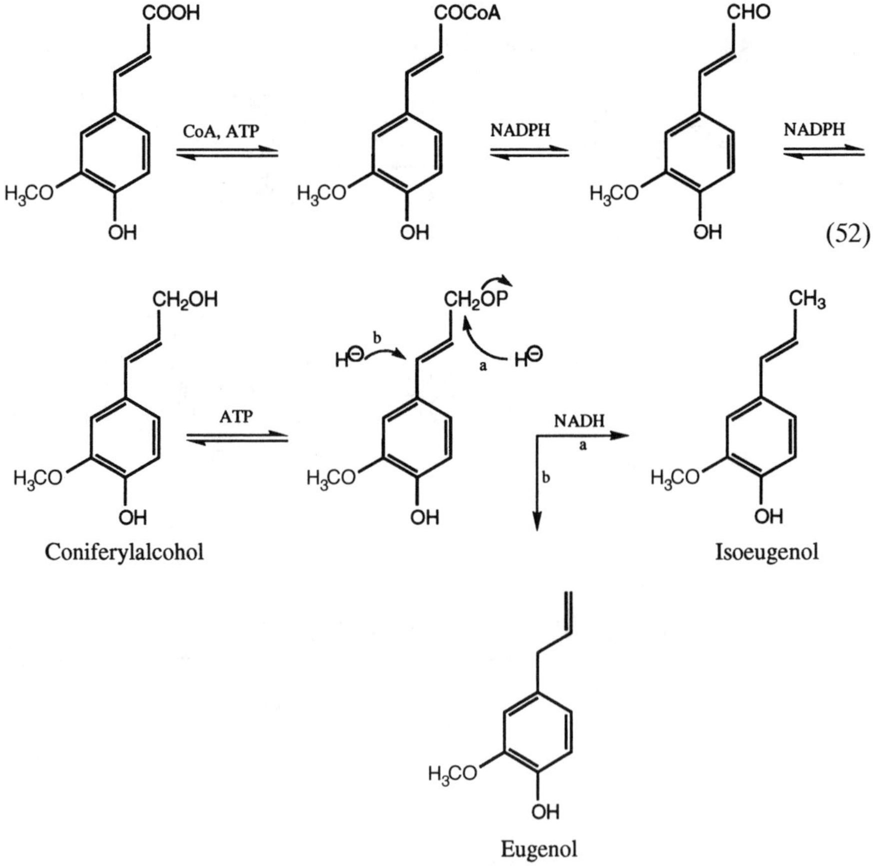

(52)

Coniferylalcohol

Isoeugenol

Eugenol

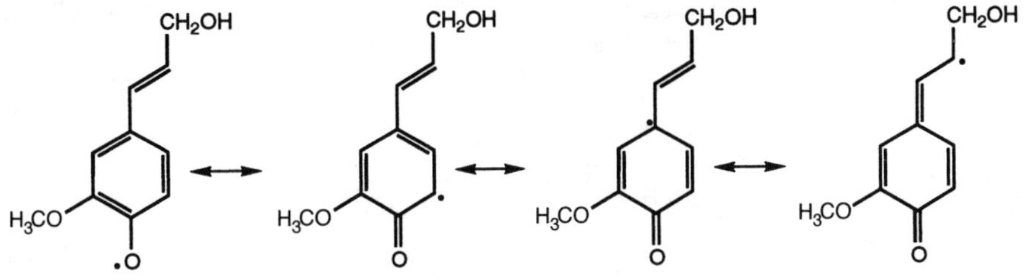

Fig. 25   Resonance structures of the coniferyl radical

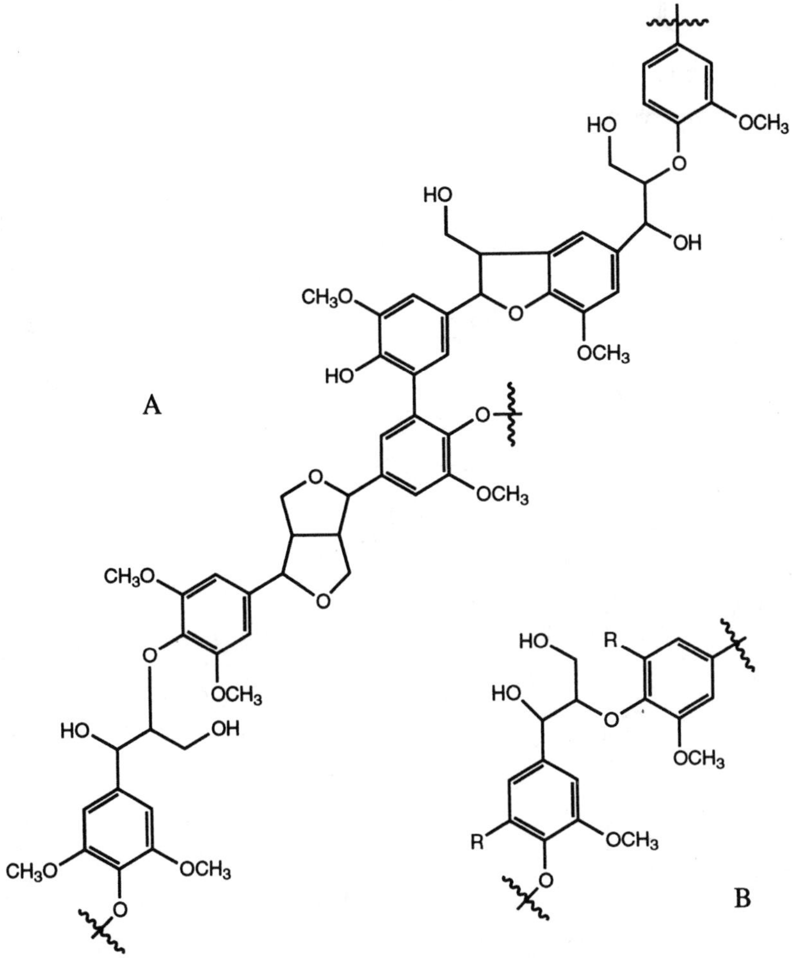

Fig. 26   A. Hypothetical lignin model. B. The most frequently occurring substructure.
R = H, OCH3

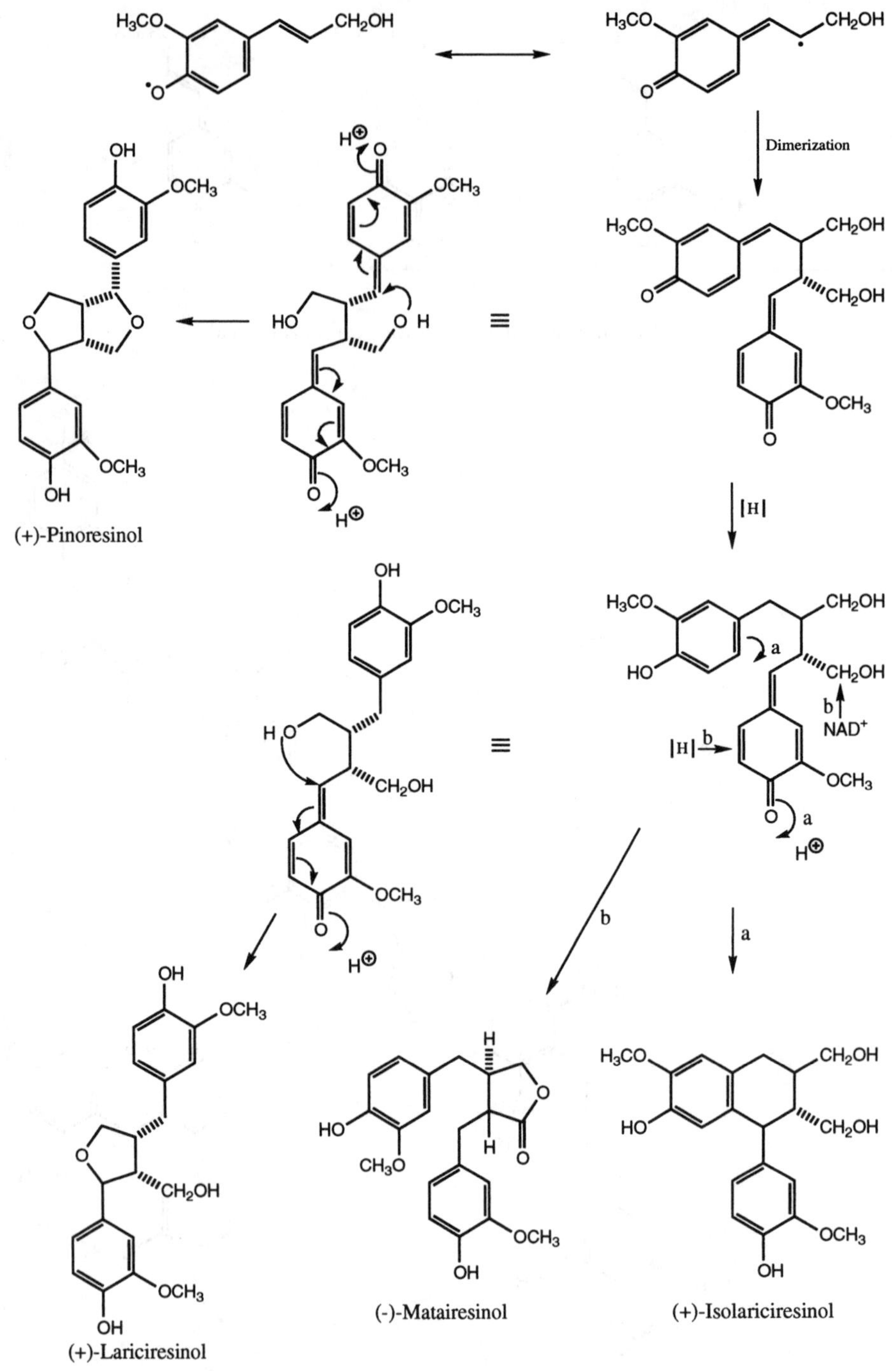

Fig. 27   Biosynthesis of some lignans from coniferyl alcohol

chain. The next step is a phenol oxidase mediated oxidation of the phenolate to phenoxy radicals, the electron spin density of which determines the site of dimerization or polymerization. Using the resonance formalism, one can predict the active sites (Fig. 25). This is an example of a general type of reaction called phenol oxidation which we shall encounter on several occasions. The radicals can pair at several positions and if we first consider the presence of several precursors, second the different possibilities for radical coupling of the building blocks in the course of polymerization and third the formation of stereoisomers, we arrive at an unlimited number of different lignins. The lignification has been assumed to occur exclusively by polymerization of the *E*-form of the various cinnamyl alcohols but later work shows that also the *Z*-form is efficiently incorporated.[89] The most frequently occurring substructure in wood lignin is the 1-arylglycerol 2-arylethers, 30–50 % of the building

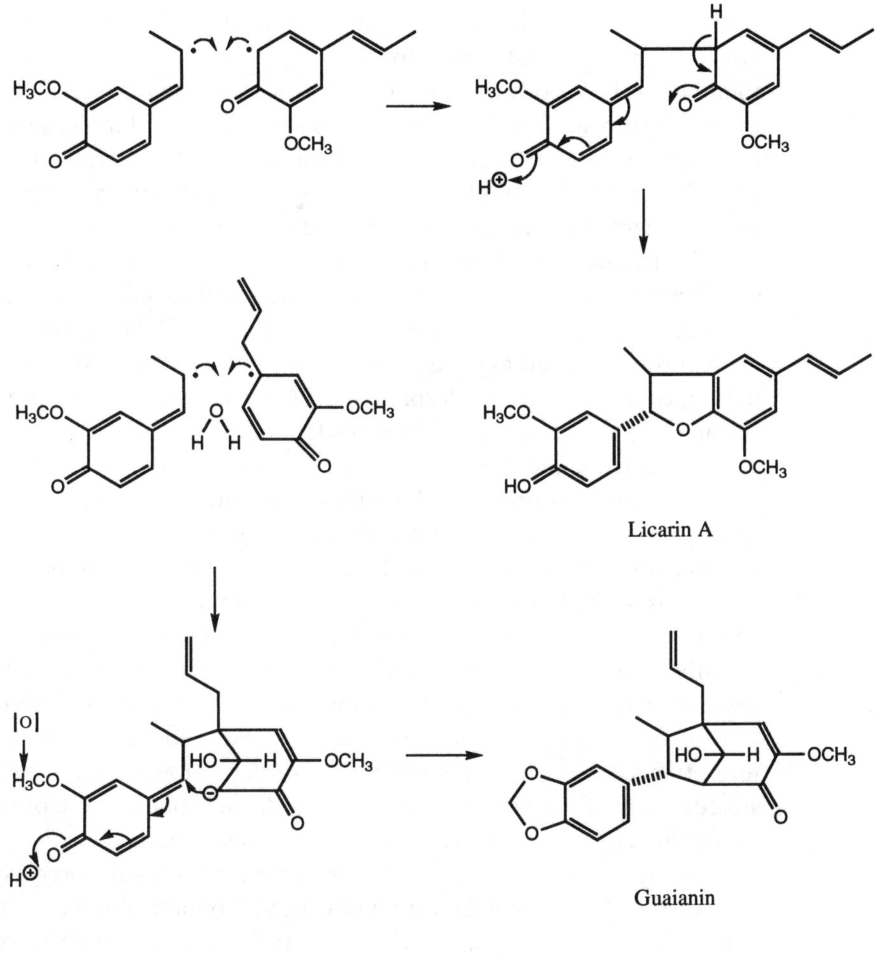

Licarin A

Guaianin

Fig. 28   Synthesis of lignans from eugenol and isoeugenol

units,[90] Fig 26. The plants often dramatically increase the lignification rate in tissues attacked by microorganisms as a defence mechanism. A great number of dimers have been isolated and identified from nature. Radical combination is probably the preferred reaction and not addition of one radical to the double bond of another cinnamyl alcohol. In Figs 27 and 28 the general principle of dimer formation is illustrated. By the term lignans we understand dimers formed by reactions involving the side chains. They exhibit a number of physiological effects, such as antiviral, antitumor, cytotoxic, insecticidal and ichthyotoxic effects.

4.8 Total synthesis

Synthesis is no longer practised for the purpose of proving or confirming structures. This is a cumbersome procedure in our days of advanced spectroscopy. Nevertheless syntheses of all kinds of naturally occurring compounds are constantly executed more for the purpose of exercising the noble art of total synthesis. Woodward has with brilliance been able to solve the immense problems surrounding the synthesis of such complicated molecules as strychnine, reserpine, sterols, chlorophyll, vitamin B_{12}, etc., thereby widening our perspectives of the potential of organic synthetic chemistry. A large number of novel elegant syntheses have been performed by workers in the field for the benefit of organic synthesis in general. A remarkable example of the impact of this kind of work is given by the many new and useful methods developed in conjunction with the synthesis of prostaglandins.[91] These compounds are distinguished by their high physiological activity and their extreme scarcity in nature, circumstances which sparked off the synthetic activity. Some syntheses are of purely theoretical and academic interest, whereas others are of great practical value. Natural products are usually optically active, i.e. only one enantiomer is produced by nature. Traditionally most laboratory syntheses start from inactive materials and consequently they give a racemic product which has to be resolved, i.e. in the first instance 50 per cent of the product is wasted, if we want to compete synthetically with nature. Only in very few instances has it been possible to convert efficiently one enantiomer into the other and usually only one is physiologically active. Therefore, chemists have increasingly turned their attention toward inexpensive optically active starting materials, e.g. sugars, suitable for ready transformations into other classes of compounds.[92] One or several asymmetric centres then stereospecifically control the formation of the new ones. These principles are exemplified by the instructive synthesis of (–)shikimic acid and (–)quinic acid from easily available D-arabinose.[93] Shikimic acid contains three asymmetric centres, the absolute configuration of which is contained in the configuration of C^2, C^3 and C^4 of D-arabinose (Fig. 29). Arabinose is catalytically reduced to arabitol and selectively converted into the 1,5-ditrityl derivative. The bulky

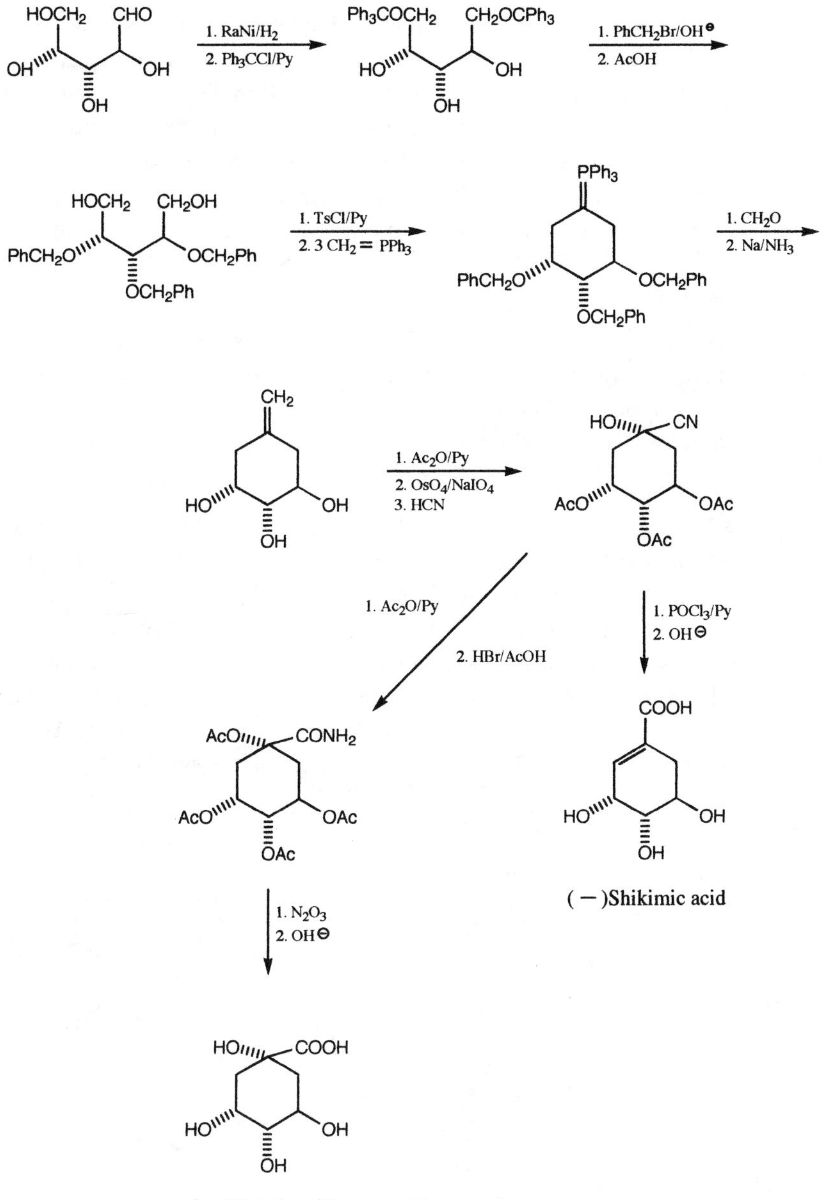

Fig. 29   Total synthesis of (–)quinic acid and (–)shikimic acid

trityl groups react much faster with the primary alcohols than with the secondary ones. The C^2, C^3 and C^4 hydroxyls are protected by benzyl groups which at a later stage can be reductively eliminated. The two primary 1,5-hydroxyls are now set free by detritylation with acetic acid and activated by tosylation for the following Wittig-type reactions. Treatment of the ditosyl derivative with excess of methylenetriphenylphosphorane gives a new cyclic ylide. Exo-methylenation with formaldehyde followed by debenzylation, acetylation, and cleavage of the olefin with the osmium oxide/periodate couple gives a cyclohexanone derivative. Reaction with hydrocyanic acid gives the cyano-hydrin which by hydrolysis can be transformed either into (–)shikimic acid or (–)quinic acid, identical in all respects with the natural compounds. The in-coming cyano group is directed *trans* to the 3,4-*cis* acetyl groups.

4.9 Problems

4.1 Tyrosine is degraded by the enzyme tyrosine phenol lyase to phenol, pyru-vic acid, and ammonia. Formulate a mechanism for this fragmentation under the supposition that pyridoxal serves as cofactor. The reactions are actually shown to be reversible. (Yamada, H., Kumagai, H., Kushima, N., Torrii, H., Enei, H. and Okumura, S. *Biochem. Biophys. Res. Commun.* **46** (1972) 370.)

4.2 Formulate some intermediate steps for the biosynthesis of betanidin, the red pigment in beetroot, from two molecules of tyrosine. Dopa is readily in-corporated and by use of ^{15}N-tyrosine it is demonstrated that the intact amino acid is incorporated in both fragments. The labels of 1-^{14}C-$3,5$-$^{3}H_2$-^{15}N-tyro-sine show up at the marked positions. Monodecarboxylation of betanidin le-ads to loss of labelled C^{19} and migration of $\Delta^{17,18}$ to $\Delta^{14,15}$. (Fischer, N. and Dreiding, A. S. *Helv. Chim. Acta* **55** (1972) 649; Liebisch, H. W., Matscheiner, B. and Schütte, H. R. *Z. Pflanzenphysiol.* **61** (1969) 269.)

Betanidin

4.3 Several lignans possess the general structure I with different configurations. Two other lignans II and III originate from a common intermediate formed by dimerization of eugenol. Rationalize the biosynthetic formation of these compounds (Gottlieb, O. R. *Phytochemistry* **28** (1989) 2545).

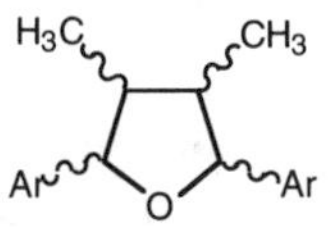

I 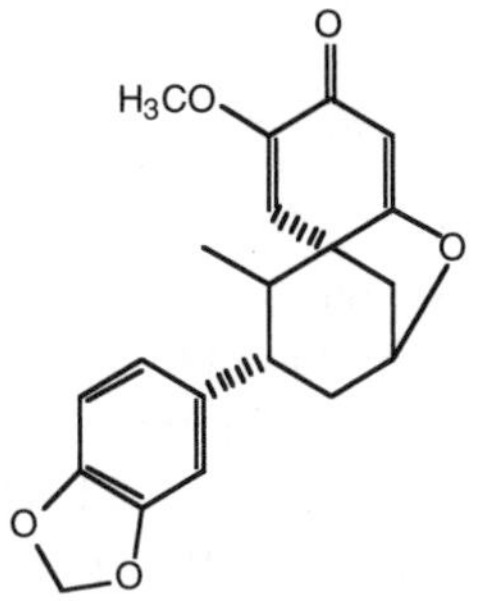   II   III

Futoenone   *epi*-Burcellin

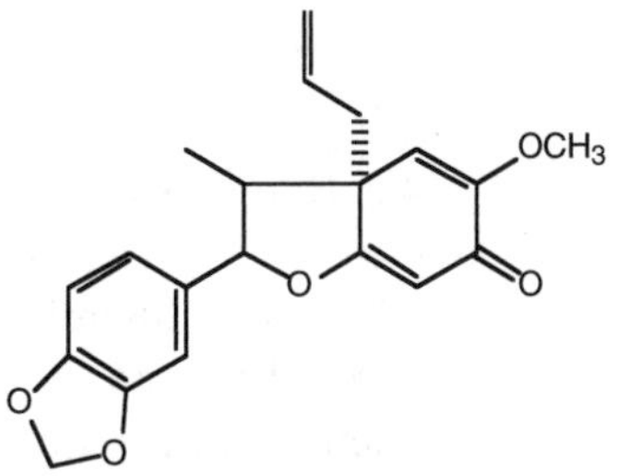

4.4 Only the C_6C_1 unit of the alkaloid ephedrine originates from phenylalanine. The remaining carbon atoms of the skeleton are furnished by pyruvic acid in a reacton presumably with benzoyl CoA and catalysed by pyruvate decarboxylase from bakers yeast. Propose a reaction mechanism for this transformation and discuss the further reactions leading to ephedrine. (Grue-Sörensen, G. and Spenser, I. D. *J. Am. Chem. Soc.* **116** (1994) 6195.)

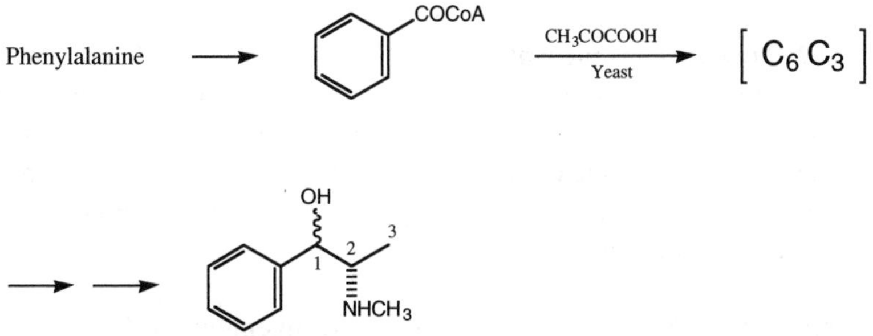

Phenylalanine

Ephedrine

4.5 Suggest biosynthetic pathways for the plant naphthoquinones alkannin and cordiachrome C. Geranyl phosphate is a precursor. (Schmid, H. V. and Zenk, M. H. *Tetrahedron Lett.* **1971**, 4151. Moir, M. and Thomson, R. H. *J. Chem. Soc. Perkin I*, **1973**, 1352, 1556; Inouye, H., Ueda, S., Inoue, K. and Matsumura, H. *Phytochemistry* **18** (1979) 1301.)

Alkannin   Cordiachrome C

4.6 Rather surprisingly it is suggested (and disputed) that biosynthesis of eugenol in *Ocimum basilicum* involves loss of the terminal side chain carbon which subsequently is replaced by a C_1 unit from methionine. How would you account for the unexpected events in the side chain, (a) if the loss and introduction of the C_1 unit occur already at the amino acid stage; (b) if the loss and introduction of the C_1 unit occur at the stage of coniferyl alcohol? (Canonica, L., Manitto, P., Monti, D. and Sanchez, A. M. *J. Chem. Soc. Chem. Commun.* **1979**, 1073.)

Eugenol

4.7 Adrenaline is biosynthesized from tyrosine (Fig. 13, section 4.4). Oxygen insertion into the side chain occurs at one stage. Many of these insertions are known to involve molecular oxygen but let us suppose that an investigation gives the result that the hydroxy group originates instead from water. How would you mechanistically account for this finding? Search the literature for the correct answer.

4.8 Formulate mechanistically the rearrangements of the 1-substituted benzene-1,2-oxides described on page 140 (Ref. 57). Suggest a suitable experiment for the determination of the direction of cleavage of the oxiran ring in 1-methoxycarbonylbenzene-1,2-oxide.

Bibliography

1. Sprinson, D. B. *Adv. Carbohydr. Chem.* **15** (1961) 235.
2. For authoritative reviews see (a). Haslam, E. *Shikimic Acid. Metabolites and Metabolism.* J. Wiley, Chichester, 1993. (b). Dewick, P. M. *Nat. Prod. Reports* **10** (1993) 233 and previous articles in the series.
3. Floss, H. G., Onderka, D. K. and Carroll, M, *J. Biol. Chem.* **247** (1972) 736.
4. Sprinson, D. B., Rothschild, J. and Sprecher, M. *J. Biol. Chem.* **238** (1963) 3170.
5. Widlanski, T. S., Bender, S. L. and Knowles, J. R. *J. Am. Chem. Soc.* **109** (1987) 1873; *Biochemistry* **28** (1989) 7572.
6. Turner, M. J., Smith, B. W. and Haslam, E. *J. Chem. Soc. Perkin I* **1975** 52.
7. Bartlett, P. A. and Satake, K. *J. Am. Chem. Soc.* **110** (1988) 1628.
8. Butler, J. R., Alworth, W. L. and Nugent, M. J. *J. Am. Chem. Soc.* **96** (1974) 1617.
9. Cain, R. B., Bilton, B. J. and Darrah, J. A. *Biochem. J.* **108** (1968) 797.
10. Anderson, K. S. and Johnson, K. A. *Chem. Rev.* **90** (1990) 1131.
11. (a) Asano, Y., Lee, L. L., Shieh, T. L., Spreafico, F., Kowal, C. and Floss, H. G. *J. Am. Chem. Soc.* **107** (1984) 4314. (b) Grimshaw, C. E., Sogo, S. G. Copley, S. D. and Knowles, J. R. *J. Am. Chem. Soc.* **106** (1984) 2699. (c) Hoare, J. H. and Berchtold, G. A. *J. Am. Chem. Soc.* **106** (1984) 2700. (d) Sogo, S. G., Widlanski, T. S., Hoare, J. H., Grimshaw, C. E., Berchtold, G. A. and Knowles, J. R. *J. Am. Chem. Soc.* **106** (1984) 2701.
12. Hill, R. K. and Newcome, G. R. *J. Am. Chem. Soc.* **91** (1969) 5893.
13. Patel, N., Pierson, D. L. and Jensen, R. A. *J. Biol. Chem.* **252** (1977) 5839; Lingens, F. and Keller, E. *Naturwissenschaften* **70** (1983) 115.
14. Plieninger, H. *Angew. Chem. Int. Ed.* **1** (1962) 3711.
15. Danishewsky, S. and Hirama, M. *Tetrahedron Lett.* **1977** 4565.
16. McCormick, J. R. D., Reichenthal, J., Kirsch, U. and Sjolander, N. O. *J. Am. Chem. Soc.* **84** (1962) 3711.
17. Young I. G., Jackman, L. M. and Gibson, F. *Biochem. Biophys. Acta* **177** (1969) 381.
18. Teng, C-Y. P., Ganem, B., Doctor, S. Z., Nichols, B. P., Bhatnagar, R. K. and Vining, L. C. *J. Am. Chem. Soc.* **107** (1985) 5008.
19. Lingens, F. *Angew. Chem. Int. Ed.* **7** (1968) 350.
20. Tatum, E. L. and Bonner, D. *Proc. Natl. Acad. Sci. USA* **30** (1944) 30.
21. Metzler, D. A., Ikawa, M. and Snell, E. E. *J. Am. Chem. Soc.* **76** (1954) 648.
22. Larsen, P. O., Onderka, D. K. and Floss, H. G. *Biochem. Biophys. Acta* **381** (1975) 397.
23. Mattia, K. M. and Ganem, B. *J. Org. Chem.* **59** (1994) 720.
24. Wasserman, H. H. and Murray, R. W. *Singlet Oxygen.* Academic Press, New York, 1979.
25. Schenck, G. O. and Ziegler, K. *Naturwissenschaften* **32** (1945) 157.
25a. For a review see Know, J. P. and Dodge, A. D. *Singlet Oxygen in Plants, Phytochemistry* **24** (1985) 889.
26. Wood, P. M. *FEBS Letters* **22** (1974) 44.
27. Behar, D., Czapski, G., Rabani, J., Dorfman, L. M. and Schwarc, H. A. *J. Phys. Chem.* **74** (1970) 3209; Afanas'ev, I. B. Kuprianova, N. S. *J. Chem. Soc. P II* **1985** 1361.
28. Divisek, J. and Kastening, B. *J. Electroanal. Chem.* **65** (1975) 603.
29. Rigo, A., Tomat, R. and Rotilio, G. *J. Electroanal. Chem.* **57** (1974) 291.
30. Ross, F. and Ross, A. B. *Selected Specific Rates of Reactions of Transients from Water in Aqueous Solution*, Nat. Bur. Stands., Washington, June 1977.

31. Afanas'ev, J. B. *Russian Chem. Revs.* **48** (1979) 527. Eng. Transl.
32. Butler, J., Jayson, C. G. and Swallow, A. J. *Biochem. Biophys. Acta* **408** (1975) 215.
33. Golden, D. M. and Benson, S. W. *Chem. Revs.* **69** (1969) 125.
34. Howard, J. A. in *Free Radicals II*, Kochi, J. (Ed.), John Wiley, New York, 1973, p. 3.
35. Pryor, W. A., *Free Radicals*, McGraw-Hill, New York, 1966, p. 287.
36. Fridovich, I. and Handler, P. *J. Biol. Chem.* **233** (1958) 1578.
37. Castel, D. A. *Nat. Prod. Rep.* **9** (1992) 289.
38. Hamberg, M., Svensson, J., Wakabayashi, T. and Samuelson. B. *Proc. Natl. Acad. Sci. USA* **71** (1974) 345.
39. deGroot, J. J. M. C., Garssen, G. J., Vliegenhart, J. F. G. and Boldingh, J. *Biochem. Biophys. Acta* **326** (1973) 279.
40. Porter, N. A., Roe, A. N. and McPhail, A. T. *J. Am. Chem. Soc* **102** (1980) 7574.
41. Haber, F. and Weiss, *J. Proc. Roy. Soc.* **A 412** (1934) 332.
42. Weeks, J. L. and Rabani, J. *J. Phys. Chem.* **70** (1966) 2100.
43. Dorfman, L. M. and Adams, G. E. *Reactivity of the Hydroxyl Radical in Aqueous Solution*, Nat. Bur. Stands., Washington, June 1973.
44. Akhtar, M. and Wright, J. N. *Nat. Prod. Rep.* **8** (1991) 527; Meunier, B. *Chem. Revs.* **92** (1992) 289.
45. Yamaguchi, K., Watanabe, Y. and Morishima, I. *J. Chem. Soc. Chem. Commun.* **1992** 1721.
46. Groves, J. T. and Watanabe, Y. *J. Am. Chem. Soc.* **108** (1986) 7836.
47. Tatsumi, K. and Hoffman, R. *Inorg. Chem.* **20** (1981) 3771; Sevin, A. and Fontecave, M. *J. Am. Chem. Soc.* **108** (1986) 3266.
48. Meunier, B., Guilmet, E., De Carvalho, M.-E. and Poilblanc, R. *J. Am. Chem. Soc.* **106** (1984) 6668.
48a. Holthausen, M. C., Fiedler, A., Schwarz, H. and Koch, W. *Angew. Chem. Int. Ed.* **34** (1995) 2282.
49. Rosenzweig, A. C. and Lippard, S. J. *Acc. Chem. Res.* **27** (1994) 229; Arndtsen, B. A., Bergman, R. G., Mobley, T. A. and Peterson, T. H. *Acc. Chem. Res.* **28** (1995) 154.
50. Gelb, M. H., Heimbrock, D. C. Malkonen, P. and Sligar, S. G. *Biochemistry*, **21** (1982) 370; Fourneron, J. B., Archelas, A. and Furstoss, R. *J. Org. Chem.* **54** (1989) 2478.
50a. Newcomb, M., LeTadic, M.-H., Putt, D. A. and Hollenberg, P. F. *J. Am. Chem. Soc.* **117** (1995) 3312.
51. *Cytochrome P450: Structure, Mechanism and Biochemistry.* Ortiz de Montellano, P. R. (Ed.), Plenum Press, 1986.
52. Korzekwa, K., Trager, W., Gouterman, M., Spangler, D. and Loew, G. H. *J. Am. Chem. Soc.* **107** (1985) 4273.
53. Guroff, G., Reifsnyder, A. and Daly, J. W. *Biochem. Biophys. Res. Commun.* **24** (1966) 720.
54. Nagatsu, T., Lewitt, M. and Udenfriend, S. *J. Biol. Chem.* **239** (1963) 2910.
55. Jerina, D. M., Daly, J. W., Witkop, B., Zalzman-Nirenberg, P. and Udenfriend, S. *J. Am. Chem. Soc.* **90** (1968) 6525.
56. Bu'Lock, J. D. and Ryles, A. P. *Chem. Commun.* **1970** 1404.
57. Chao, H. S. J. and Berchtold, C. H. *J. Am. Chem. Soc.* **103** (1981) 898.
58. Boyd, D. R., Daly, J. W. and Jerina, D. M. *Biochemistry*, **11** (1972) 1961.
59. Muto, S. and Bruice, T. C. *J. Am. Chem. Soc.* **102** (1980) 7559.
60. Hanson, K. R., Wightman, R. H., Staunton, J. and Battersby, A. R. *Chem. Commun* **1971**, 185.
61. Hanson, K. R. and Havir, E. A. *Arch. Biochem. Biophys.* **141** (1970) 1.

62. Geissman, T. A. and Hinreiner, E. *Bot. Rev.* **18** (1952) 165.
63. Zenk, M. H. and Müller, G. *Z. Naturforsch.* **19B** (1964) 398.
64. Gross, S. R. *J. Biol. Chem.* **233** (1958) 1146.
65. Conn, E. E. and Swain, T. *Chem. Ind.* **1961**, 592.
66. Zenk, M. H. *Z. Naturforsch.* **19B** (1964) 83.
67. Zenk, M. H. *Phytochemistry* **6** (1967) 245.
68. Pridham, J. B. and Saltmarsh, M. J. *Biochem. J.* **87** (1963) 218.
69. Häusler, E., Peterson, M. and Alfermann, A. W. *Z. Naturforsch.* C **46** (1991) 371.
70. Neish, A. C. *Can. J. Bot.* **37** (1959) 1085.
71. Murray, R. D. H., *Prog. Chem. Org. Nat. Prod.* **35** (1978) 199.
72. Brown, S. A. in *Biosynthesis of Aromatic Compounds*, G. Billek (Ed.), Pergamon Press, Oxford, 1969, p. 15.
73. Bunton, C. A., Kenner, G. W., Robinson, M. J. T. and Webster, B. R. *Tetrahedron* **19** (1963) 1001.
74. Austin, D. J. and Meyers, M. B. *Phytochemistry* **4** (1965) 245.
75. Hamerski, D. and Matern, U. *Eur. J. Biochem.* **171** (1988) 369; Hamerski, D., Beier, R. C., Kneusel, R. E., Matern, U. and Himmelspach, K. *Phytochemistry* **29** (1990) 1137.
76. Friis, P., Daves, Jr., G. D. and Folkers, K. *J. Am. Chem. Soc.* **88** (1966) 4754.
77. Gibson, F. *Biochem. Soc. Trans.* **1** (1973) 317.
78. Sippel, C. J., Goevert, R. R., Slachmann, F. N. and Olson, R. E. *J. Biol. Chem.* **258** (1983) 1057; Kang, D., Takeshige, K., Isobe, R. and Minakami, S. *Eur. J. Biochem.* **198** (1991) 599.
79. Uchida, K. and Aida, K. *Biochem. Biophys. Res. Commun.* **46** (1972) 130.
80. Jackman, L. M., O'Brien, I. G., Cox, G. B. and Gibson, F. *Biochem. Biophys. Acta* **141** (1967) 1.
81. Krügel, R., Grumbach, K.-H., Lichenthaler, H. and Rétey, J. *Biorg. Chem.* **13** (1985) 187.
82. Whistance, G. R. and Threlfall, D. R. *Phytochemistry* **10** (1971) 1533.
83. Campbell, I. M. *Tetrahedron Lett.* **1969** 4777; Simantiras, M. and Leistner, E. *Phytochemistry* **28** (1989) 1381.
84. Leduc, M. M., Dansette, P. M. and Azerad, R. G. *Eur. J. Biochem.* **15** (1970) 428; Chung, D.-O., Maier, U. H., Inouye, H. and Zenk, M. H. *Z. Naturforsch.* **C49**(1994) 885.
85. Heide, L. and Leistner, E. *J. Chem. Soc. Chem. Commun.* **1981** 334.
86. Hutson. K. G. and Threlfall, D. R. *Phytochemistry* **19** (1980) 535.
87. Freudenberg, K. and Richtzenhain, H. *Chem. Ber.* **76** (1943) 997.
88. Mansell, R. L., Gross, G. G. Stöckigt, J., Franke, H. and Zenk, M. H. *Phytochemistry* **13** (1974) 2427.
89. Lewis, N. G., Dubelsten, P., Eberhardt, T. L. Yamamto, E. and Towers, G. H. N. *Phytochemistry* **26** (1987) 2229.
90. Wallis, A. F. A., Lundquist, K. and Stromberg, R. *Acta Chem. Scand.* **45** (1991) 508.
91. Caton, M. P. L. *Tetrahedron Reports* 68. *Tetrahedron* **35** (1979) 2705.
92. Fraser-Reid, B. and Anderson, R. C. *Prog. Chem. Org. Nat. Prod.* **39** (1980) 1.
93. Bestman, H. J. and Heid, H. A. *Angew. Chem. Int. Ed.* **10** (1971) 336.

Chapter 5

The polyketide pathway

5.1 Introduction

Acetic acid or its biosynthetic equivalent, acetyl CoA, occupies a central position in the synthesis of natural compounds (see Fig. 5, Chapter 1). Linear Claisen condensation leads to β-keto esters (1a), which either by reduction and repeated condensation give fatty acids or by further direct condensation give polyketides. They, in turn, can cyclize to a vast number of aromatics. Acetyl CoA is also the starting material for terpenoids. In a branched condensation the keto function of acetoacetyl CoA reacts with another acetyl CoA molecule to form β-hydroxy-β-methylglutaryl CoA which is transformed into the 'active' isoprene unit (1b) and ultimately to the terpenoids. This pathway is treated separately in Chapter 6. From the intermediate formation of the β-keto ester, both in the biosynthesis and in the β-oxidative degradation of fatty acids, it may be concluded that one process is simply the reversal of the other. However, more recent research has shown that they are significantly different. They proceed in different compartments of the cell with different sets of enzyme systems: biosynthesis in the cytosol, β-oxidation in the mitochondria. The presence of citrate was found to stimulate synthesis, but was not required for breakdown. Unexpectedly, carbon dioxide was

$$CH_3COCH_2COCoA + CH_3COCoA \xrightarrow[\text{branched}]{b} \quad \begin{array}{c} HO \quad CH_2COOH \\ C \\ H_3C \quad CH_2COCoA \end{array}$$

β-Hydroxy-β-methylglutaryl CoA

a | linear

$$CH_3COCH_2COCH_2COCoA$$

Triketide

Cyclization / \ Reduction

Aromatics   Fatty acids

Isoprene unit

(1)

shown to be essential for synthesis although by isotopic labelling it proved not to be incorporated. Biosynthesis employs NADPH for reduction, whereas oxidation employs $NAD^{\oplus}$, and in biosynthesis (3R)-3-hydroxy acids occur as intermediates, whereas (3S)-3-hydroxy acids are formed in the β-oxidation sequences. The simple condensation step (1a, which does occur, but is not the common pathway) has thus to be modified. In the subsequent sections a detailed account is given of the metabolism of fatty acids, olefins, acetylenes, some branched derivatives, aromatics and compounds with mixed biogenesis.

5.2 Fatty acids, fats[1]

Acetyl CoA is formed in the cell either by degradation of fatty acids, decarboxylation of pyruvic acid obtained via glycolysis or degradation of certain amino acids. The thio group has a twofold effect. It activates the carbonyl function as an electrophile and the methyl group as a nucleophile, i.e. acetyl CoA is considerably more active than an ordinary ester in a Claisen condensation, but still not sufficiently reactive to be utilized by the fatty acid synthetase complex for that purpose. A further activation of the α-carbon is necessary and this is brought about by carboxylation of acetyl CoA to malonyl CoA. The reaction is mediated by ATP and biotin as a carbon dioxide carrier (2,3). The cofactor biotin is anchored by its carboxyl group to the ε-amino group of a lysine residue of the enzyme. In a reversible reaction with ATP, bicarbonate forms a reactive mixed anhydride of carbonic and phosphoric acids which carboxylates biotin to N^1-carboxybiotin. In agreement with (2) the label of $HC^{18}O_3^{\ominus}$ ends up in the inorganic phosphate formed.[2] N^1-carboxybiotin is an unstable intermediate, the existence of which was established by transformation into its methyl ester by diazomethane, identical to authentic N^1-methoxycarbonylbiotin.[3,4] Stereochemical investigations carried out on 2-^{3}H labelled propionyl CoA show that the pro-R hydrogen is lost and is replaced by carbon dioxide with retention of configuration.[5] No hydrogen exchange occurs since the rate of ^{3}H release is equal to the rate of carboxylation. It was also demonstrated by use of chiral acetyl CoA that the carboxylation occurs with retention of configuration. A concerted cyclic mechanism has been suggested (3). Intuitively one could now be led to believe that the ready formation of a carbanion from malonate is the immediate cause of the condensation. However, no isotopic scrambling with the solvent was observed in the methylene group of dideuteriomalonyl CoA, nor was any isotope effect on the rate of condensation observed which eliminates this mechanism.[6] One can conclude that the condensation occurs with concomitant exothermic decarboxylation generating the anion, so giving an extra push to the reaction (6). The condensation takes place on a multifunctional enzyme complex, the fatty acid synthetase, FAS. Malonyl CoA is attached to a thiol group of the acyl carrier protein, ACP, and is condensed with the acetyl group, as starter, stra-

$$HOCO_2^{\ominus} \quad + \quad ATP$$

$$\Updownarrow$$

Biotin + $HO-\overset{O}{\underset{}{C}}-O-\overset{O}{\underset{O^{\ominus}}{P}}-O^{\ominus}$ $\longrightarrow$ (2)

N^1-Carboxybiotin

$\rightleftharpoons$ $^{\ominus}OOCCH_2COCoA$ + biotin-Enz   (3)

$$CH_3COCoA \quad + \quad HSEnz \rightleftharpoons CH_3COSEnz \quad + \quad CoA \tag{4}$$

$$^{\ominus}OOCCH_2COCoA \quad + \quad HSEnz \rightleftharpoons {}^{\ominus}OOCCH_2\,COSEnz \quad + \quad CoA \tag{5}$$

$\rightleftharpoons$ $CH_3COCH_2COSEnz$ + $HSEnz$ + CO_2 (6)

$$CH_3COCH_2COSEnz \xrightleftharpoons{NADPH} CH_3\overset{OH}{\underset{|}{C}}HCH_2COSEnz \tag{7}$$

$$CH_3\overset{OH}{\underset{|}{C}}HCH_2\,COSEnz \xrightleftharpoons{-\,H_2O} CH_3CH = CHCOSEnz \tag{8}$$

$$CH_3CH = CHCOSEnz \xrightleftharpoons{NADPH} CH_3CH_2CH_2COSEnz \tag{9}$$

etc.

$$n\overset{\Delta}{C}\overset{*}{H}_3COOH \longrightarrow \overset{\Delta}{C}H_3\overset{*}{C}H_2\overset{\Delta}{C}H_2\overset{*}{C}H_2\overset{\Delta}{C}H_2\overset{*}{C}H_2- \tag{10}$$

tegically bound to a vicinal thiol group (4–6). The β-keto ester formed is reduced with NADPH (7), dehydrated (8), and finally reduced with NADPH to
the homologue containing two more carbons (9). We know that the hydroxy
acyl intermediate of (7), has (3*R*)-configuration, that ^{3}H is preferentially retained from (2*S*)-2-^{3}H-malonate and that *trans*-enoate is exclusively obtained
by dehydration (8). Enzymatic dehydration of stereospecifically synthesized
(2*R*, 3*R*)-3-hydroxy-2-^{3}H-butanoate was furthermore found to give *trans*-
butenoate with retention of ^{3}H indicating that the elimination proceeds in a
syn-fashion. Consequently, the condensation (6) must proceed with inversion
of configuration to fit the overall stereo outcome of the process (Fig. 1).[7] The
reduction occurs on the *si* face at C^2 of the *trans*-enoate. *Syn*-elimination is
stereoelectronically somewhat less favourable than *trans*-elimination. Nevertheless, it occurs occasionally in conjunction with enzyme mediated reactions,
cf. the *syn*-elimination of water from 5-dehydroquinic acid to 5-dehydroshiki-
mic acid (section 4.1).

The butyryl SEnz can now act as starter and be processed in reactions
(4–10). This scheme has been verified by feeding the organism with labelled
acetic acid (10). For the synthesis of palmitic acid, the most frequently occurring fatty acid, one acetyl CoA as a starter, seven malonyl CoA units, and
fourteen NADPH, are needed. It is rather remarkable that in most organisms
the chain elongation practically stops at C_{16}. The enzyme readily accepts a
C_{14}CoA ester as starter but reluctantly a C_{16} CoA. The successive addition of
two carbon units to acetyl CoA explains why even numbered acids are com-

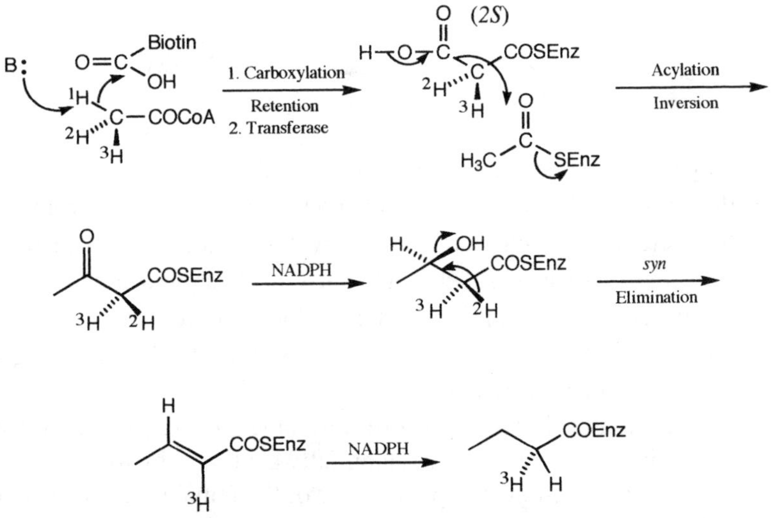

Fig. 1   Stereochemistry of the fatty acid biosynthesis

monest. However, small amounts of odd numbered acids do occur. They normally originate from propionic acid as starter. We shall later meet several cases where other acids are used as starters. Table 1 shows the structure of some naturally occurring fatty acids. The first number refers to the number of carbon atoms in the acid, the second to the number of double bonds and the bracketed number to the position and structure of the double bond.

Table 1 Naturally occurring fatty acids

Structure	Name
$CH_3(CH_2)_8COOH$ 10 : 0	Capric acid
$CH_3(CH_2)_{10}COOH$ 12 : 0	Lauric acid
$CH_3(CH_2)_{12}COOH$ 14 : 0	Myristic acid
$CH_3(CH_2)_{14}COOH$ 16 : 0	Palmitic acid
$CH_3(CH_2)_{16}COOH$ 18 : 0	Stearic acid
$CH_3(CH_2)_7CH = CH(CH_2)_7COOH$ 18 : 1 (9c)	Oleic acid
$CH_3(CH_2)_7CH = CH(CH_2)_7COOH$ 18 : 1 (9tr)	Elaidic acid
$CH_3(CH_2)_4(CH = CHCH_2)_2(CH_2)_6$ COOH 18 : 2 (9c, 12c)	Linoleic acid
$CH_3CH_2(CH = CHCH_2)_3(CH_2)_6COOH$ 18 : 3 (9c, 12c, 15c)	Linolenic acid
$CH_3(CH_2)_4(CH = CHCH_2)_4(CH_2)_2COOH$ 20 : 4 (5c, 8c, 11c, 14c)	Arachidonic acid

Recent investigations of the early steps of the polyketide synthesis deserve comment. Previously, the enzyme-bound acetyl group, $CH_3COSEnz$ (= acetyl ACP), was regarded as the primary starting unit. A general condensing enzyme system has now been discovered which directly uses acetyl CoA as a starting unit without prior transacylation (4) onto ACP (11).[8]

$$CH_3COCoA + HO_2C\diagup\diagdown COSEnz \longrightarrow \text{(structure)} COSEnz + CoA + CO_2 \qquad (11)$$

Dogma states that the condensations, i.e. the Claisen reaction, subsequent reductions and dehydrations to give polyketides proceed to completion before the metabolite is released from the synthetase complex. This implies that intermediates such as short-length fatty acids should not be exchanged nor should conceivable intermediates be incorporated when added to the medium. The thioesterases present are consequently discriminating and act preferably on the final product.

Branched-chain fatty esters of normal length (C_{16}–C_{20}) have been identified as intermediates in long-chain fatty acids and hydrocarbons. 3,11-Dimethylnonacosan-2-one, the pheromone of the female German cockroach, *Blattella germanica*, is biosynthesized from 13 acetates and 2 propionates according to (12) via the hydrocarbon which in the final step is oxidized to the ketone.

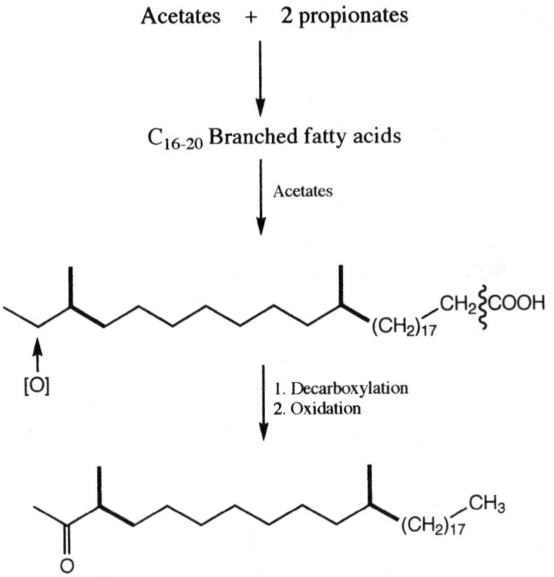
Acetates + 2 propionates
C16-20 Branched fatty acids
Acetates
CH2 COOH
(CH2)17
[O]
1. Decarboxylation
2. Oxidation
CH3
(CH2)17
O

3,11-Dimethylnonacosan-2-one

(12)

Long-chain unsaturated fatty acids ($C_{24–34}$), e.g. 26:3 (5c, 9c, 19c), isolated from the lipid fraction of marine sponges are biosynthesized by chain extension with dietary palmitic acid as the starter unit.[10] The order of desaturation is not rigorous because both 26:1(5c) and 26:1(9c) are incorporated into 26:2(5c, 9c).

A variation of the stereo outcome of the hydration and reduction of the α,β-unsaturated thioester has occasionally been observed depending on the source of the enzyme.[11] It was discovered that in the same organism, *Cladosporium cladosporioides*,[12] the fatty acid enoyl reductase for oleic acid operated with opposite stereochemistry compared with the polyketide enoyl reductase for cladosporin at C-11 but with the same stereochemistry at C-9.

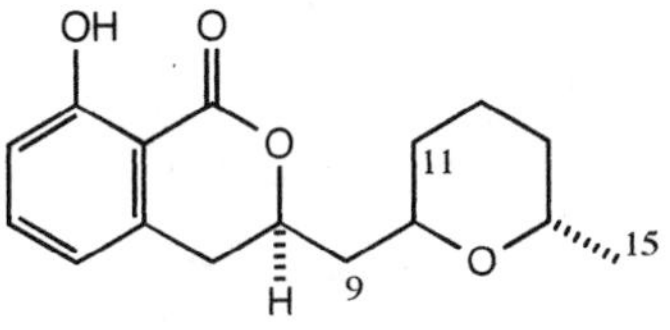
OH
O
O
11
O
15
H
9

Cladosporin

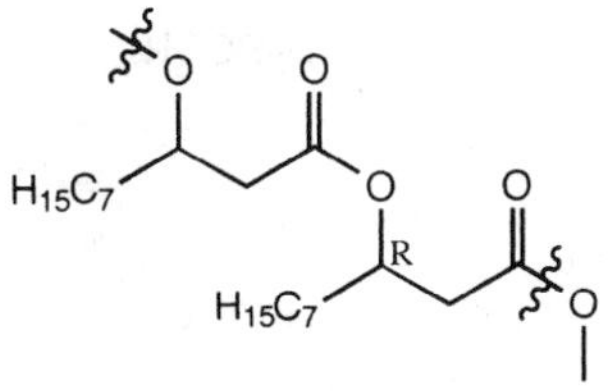
O
O
H15C7
O
O
R
H15C7
O

Poly-(*R*)-3-hydroxydecanoate

Pseudomonas pudida produces poly-(*R*)-3-hydroxydecanoate, the stereo-structure of which indicates that the monomer is formed directly via biosynthesis (*R*-form) and not by β-oxidation (*S*-form). However, feeding experiments show that decanoic acid is incorporated which suggests that β-oxidation indeed is operating with concomitant epimerization.[13] This illustrates unexpected complexities in biosynthetic research.

There is increasing evidence for incorporation of partially assembled polyketide fragments particularly in macrolides such as dehydrocurvularin,[14] methymycin,[15] mycinamicin,[16] and aspyrone,[17] furthermore in condensed aromatics such as tetracyclines[18] from *Streptomyces* spp. and in polyethers such as monensin A[19] from *S. cinnamonensis*. These studies will be discussed in the following sections.

Lipids is the collective name for compounds soluble in hydrocarbons. They include fats, waxes, and phosphoglycerides as well as hydrocarbons of quite different biogenesis such as steroids. The term is reserved by most biochemists for compounds yielding fatty acids and glycerol on hydrolysis, i.e. fats, and they function as an energy depot. They are biosynthesized in steps from fatty acid CoA and glycerol-3-phosphate (13). The most common acids are palmitic, stearic, and oleic acids. In some lipids one or several hydroxyl groups are *O*-alkylated by a long chain alcohol. The phospholipids (lecithins) are important constituents of cell membranes. They contain a betain head group, e.g. phosphatidyl choline, in which one of the OH groups of phosphatidic acid (13) is esterified by choline, $(CH_3)_3N^{\oplus}CH_2CH_2OH$ (Fig. 2).

Sphingolipids, e.g. galactocerebroside, are closely related compounds containing sphingosine, with the amino function acylated by a fatty acid. The

$$
\begin{array}{ccc}
CH_2OCOR^1 & & CH_2OCOR^1 \\
| & \xrightarrow{\;R^2COCoA\;} & | \\
HCOH & & HCOCOR^2 \\
| & & | \\
CH_2OPO(OH)_2 & & CH_2OPO(OH)_2
\end{array}
\xrightarrow{\;H_2O\;}
$$

Phosphatidic
acid

(13)

$$
\begin{array}{ccc}
CH_2OCOR^1 & & CH_2OCOR^1 \\
| & \xrightarrow{\;R^3COCoA\;} & | \\
HCOCOR^2 & & HCOCOR^2 \\
| & & | \\
CH_2OH & & CH_2OCOR^3
\end{array}
$$

Diacylglycerol   Triacylglycerol

hydroxyl group is functionalized by phosphoric acid or a sugar. They are found in cerebral membranes and nerve endings where they function as impulse transmitters.

Waxes are defined as esters of fatty acids with long chain alcohols. They function as water repellant coatings on skin, feathers, fruits, etc.

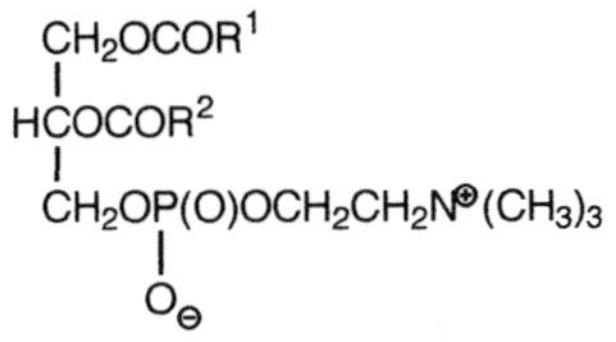

CH$_2$OCOR^1
|
HCOCOR^2
|
CH$_2$OP(O)OCH$_2$CH$_2$N$^\oplus$(CH$_3$)$_3$
|
O$_\ominus$

Phosphatidylcholine (lecithin)
(R^1 and R^2 = fatty acids)

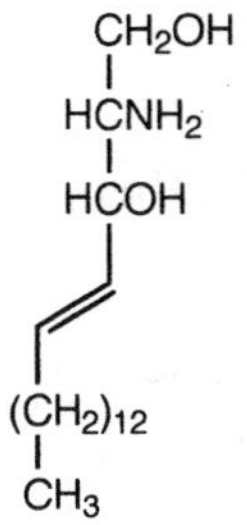

CH$_2$OH
|
HCNH$_2$
|
HCOH

(CH$_2$)$_{12}$
|
CH$_3$

Sphingosine

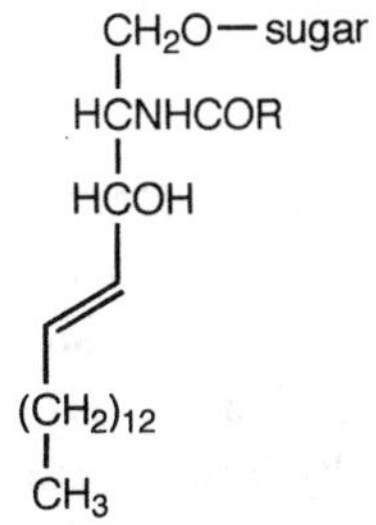

CH$_2$O—sugar
|
HCNHCOR
|
HCOH

(CH$_2$)$_{12}$
|
CH$_3$

Galactocerebroside
(Sugar = galactose; R = alkyl)

Fig. 2   Structures of a phosphoglyceride and a sphingolipid

5.3 Branched fatty acids

Branched fatty acids are common in a variety of organisms. They are formed, either by priming the reaction with a branched starter, e.g. isobutyryl CoA, or α-methylbutyryl CoA (14), or by condensation with a alkylated malonyl CoA which arises by carboxylation of homologues of acetic acids (15,16). Occasionally, methylation of the linear chain occurs with adenosyl methionine as the methylating agent.

$$CH_3CH_2\overset{|}{\underset{CH_3}{C}}HCOSEnz \ + \ n\,CH_3COCoA \ \longrightarrow \ CH_3CH_2\overset{|}{\underset{CH_3}{C}}H(CH_2)_{2n}COOH \qquad (14)$$

$$CH_3CH_2COCoA \ + \ {}^{\ominus}OOC\!\!-\!\!\text{biotin}\!\!-\!\!Enz \ \longrightarrow \ CH_3\overset{\nearrow COO^{\ominus}}{\underset{\searrow COCoA}{C}}H \qquad (15)$$

Methylmalonyl CoA

$$\text{(16)} \qquad CH_3COCH\!\!-\!\!COSEnz \qquad \underset{CH_3}{|}$$

In this manner, an unusual fatty acid **1** is synthesized in the uropygial gland of the goose (Fig. 3). It is derived from one acetyl CoA and four methylmalonyl residues.[20] When ^{14}C-labelled 2-cyclopentenylcarboxylic acid, derived from cyclopentenylglycine, was administered to plants belonging to the *Flacourtiaceae*, labelled cyclopentenyl fatty acids such as chaulmoogric acid **2**, used in the treatment of leprosy, were synthesized.[21] The cyclohexyl moiety of 11-cyclohexylundecanoic acid,[22] **3**, from *Alicyclobacillus acidocaldarius* derives from shikimic acid via dehydration and hydrogenation to cyclohexane carboxylic acid which serves as starter unit.

Tuberculostearic acid or 10-methylstearic acid **4** and sterculic acid **5** are synthesized in a different manner in that the methyl group comes from methionine.[23] Labelling experiments[24] in cultures of *Lactobacillus plantarum* with 9,10-$^2H_2$-oleate and 8,8,11,11-$^2H_4$-oleate showed no scrambling or loss of label, ruling out rapid equilibration of edge or corner protonated cyclopropyl species and also any mechanism involving allylic activation of the double bond. Incubation with ^2H_3C-methionine produced *ca.* 10 per cent of 19-$^2H_1$-dihydrosterculic acid which indicated that exchange in the ^2H_3C species occurred at some stage of the biosynthesis. It was established that the exchange did not take place in the methyl group of methionine prior to alkylation. The findings, which are incompatible with intermediate exomethylene formation, might be explained by reversible cyclopropane group formation as shown in (17). The intermediate carbonium ion is either reduced to **4**, path a, or rearranges to the cyclopropane. This intramolecular electrophilic aliphatic substitution has ample analogues in terpene chemistry. Dehydrogenation of the cyclopropane ultimately gives the cyclopropene **5**.

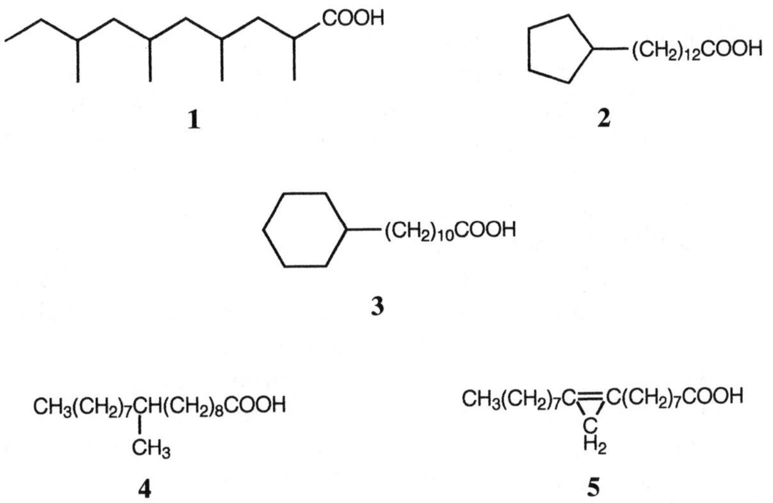

Fig. 3 Branched fatty acids: **1**, metabolite in the uropygial gland of the goose; **2**, chaulmoogric acid; **3**, 11-cyclohexylundecanoic acid; **4**, tuberculostearic acid; **5**, sterculic acid

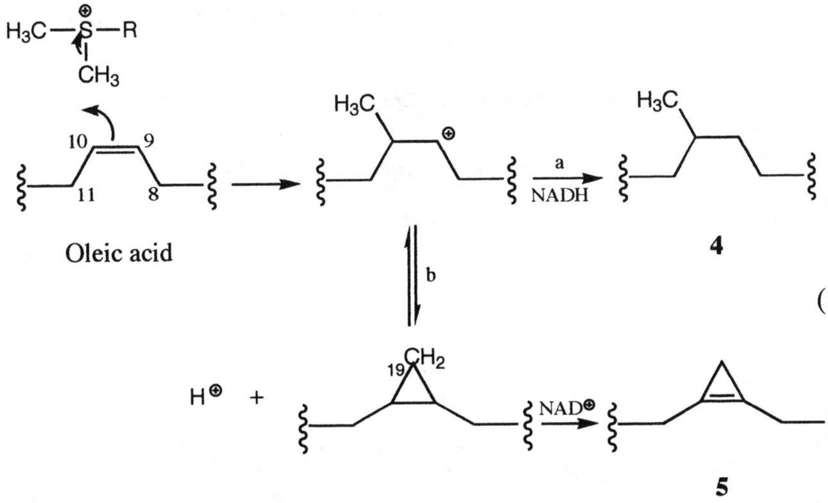

(17)

5.4 Olefinic acids. Prostaglandins

Unsaturated fatty acids are common and constitute a large portion of the lipids. In contrast to saturated acids, which throughout follow an essentially invariant biosynthetic route, the formation of unsaturated fatty acids follows fundamentally different biosynthetic routes depending upon the surrounding aerobic or anaerobic conditions. Under aerobic conditions unsaturation is introduced into the preformed saturated acids, while anaerobes economize by preserving the double bond obtained by dehydration of the intermediary 3-hydroxyalkanoylthioesters. It ought to be noted that the elimination of water here gives the 3,4-unconjugated *cis*-double bond. The (Z)-3-alkanoylthioester directly undergoes chain extention. Reaction (18) shows the biosynthesis of oleic acid from decanoic acid in anaerobic bacteria.[25] The steric course of the allylic rearrangement of (E)-2-decenoylthioester to (Z)-3-decenoylthioester has been investigated by specific deuterium-labelling at C-2 and C-4.[26] It is a suprafacial reaction with elimination of the *pro*-4R hydrogen and reprotonation on the *si*-face at C-2 by a proton from the solvent(19).

In most organisms oleic acid is formed by desaturation of stearic acid in the presence of oxygen, FAD, NADPH and a dehydrogenase. The mechanistic details of this fundamental reaction are not known, and are open to specula-

$$CH_3(CH_2)_7CH_2COCoA \quad + \quad CH_3COCoA \longrightarrow$$

Decanoyl CoA

$$CH_3(CH_2)_7CH_2COCH_2COCoA \xrightarrow{\text{[H]}} CH_3(CH_2)_7CH_2\overset{\overset{\displaystyle OH}{|}}{C}HCH_2COCoA$$

$$\xrightarrow{\text{- H}_2O} CH_3(CH_2)_7CH \overset{c}{=} CHCH_2COCoA \xrightarrow{\text{+ 3 Acetate}}$$

(18)

$$CH_3(CH_2)_7CH \overset{c}{=} CH(CH_2)_7COOH$$

Oleic acid

(19)

tion. We assume that the radical species 33d (section 4.3), acting on unactivated C-H bonds, abstracts a hydrogen from C^9 or C^{10} of stearic acid and
forms a carbon radical which directly reacts with FAD to *cis*-olefin and
FADH (cf. section 5.8). The 9,10-*pro*-H_R are specifically removed from stearic acid in a *syn*-manner in mammals, plants, bacteria and algae. The conversion of palmitoyl CoA to (*Z*)-11-hexadecenoyl CoA in insects follows the
same stereochemical course (20).[27]

(20)

Further desaturation of oleic acid leads to linoleic and linolenic acids.[28,29] It
is remarkable that polyunsaturated acids rarely have their double bonds in
conjugation and that they occur as the thermodynamically less stable *cis* isomer. Arachidonic acid, a precursor of the important prostaglandin hormones,
first detected in seminal plasma, is biosynthesized from oleic acid according to

Fig. 4.[30] They are reported to be widely distributed in low concentrations in insects, marine organisms and mammals. The prostaglandins possess a multitude of biological effects. They control blood pressure and renal blood flow, contractions of smooth muscle, gastric acid secretion and platelet aggregation, and have found clinical use in the regulation of pregnancy, in treatment of ulcers, heart failure, thrombosis, etc. Dietary investigations revealed that some fatty acids, e.g. linoleic acid, are essential for the normal development of mammals which are able to synthesize saturated and monounsaturated acids but not particular polyunsaturated acids, which have to be supplied in the diet. Their mechanism of action is explained, at least in part, by the fact that they are precursors of the prostaglandins.

By the action of lipoxygenase on polyunsaturated fatty acids, an array of hydroxylated fatty acids and compounds of prostaglandin type are formed. The biosynthesis of prostaglandin E_2 (section 4.3) is initiated by radical attack at C^{13} of arachidonic acid. The biosynthesis of preclavulone A, widespread in

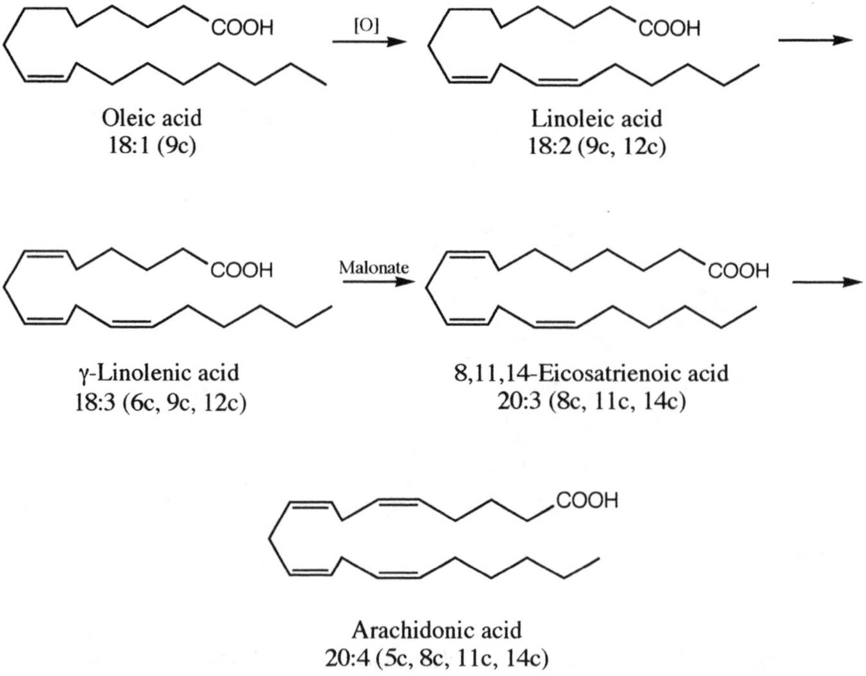
COOH
[O]
COOH
Oleic acid
18:1 (9c)
Linoleic acid
18:2 (9c, 12c)
COOH
Malonate
COOH
γ-Linolenic acid
18:3 (6c, 9c, 12c)
8,11,14-Eicosatrienoic acid
20:3 (8c, 11c, 14c)
COOH
Arachidonic acid
20:4 (5c, 8c, 11c, 14c)

Fig. 4 Biosynthesis of arachidonic acid

corals, is initiated by an attack at C^{10} with intermediate formation of the $(8R)$-hydrogen peroxide which rearranges via the allene oxide to preclavulone A.[31] This route is supported by biomimetic studies.[32] The prostanoids of this series have 8,12-*cis*-configuration whereas the PGE series has 8,12-*trans*-configuration. A shorter route, involving radical cyclization and cleavage of the C^9 hydrogen peroxide is conceivable, Fig. 5.

Lipoxygenation of linolenic acid gives a number of metabolites via hydroperoxides, Fig. 6. Abstraction of C^{11}-H leads either to the $(9S)$- or the $(13S)$-hydroperoxide.[33] Colnelenic acid is formed by a Baeyer-Villiger rearrangement of the $(9S)$-hydroperoxide and subsequent hydrolysis gives the aldehydes. The $(13S)$-hydroperoxide gives the corresponding short-chain aldehydes or cylizes to the prostanoid 12-oxyphytodienoic acid according to the mechanism shown in Fig. 5.[34]

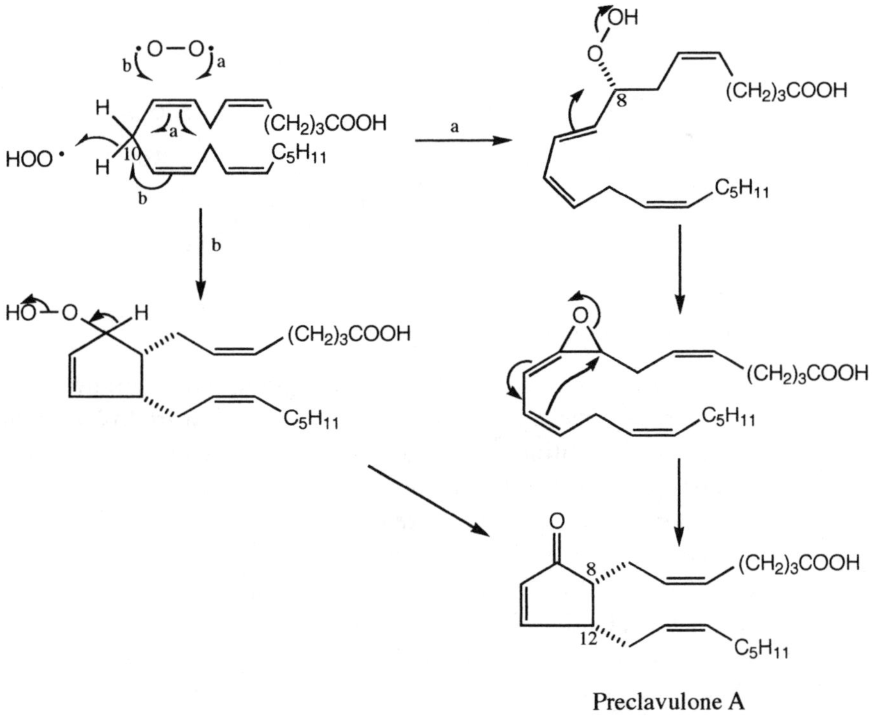

Fig. 5   Biosynthesis of preclavulone A

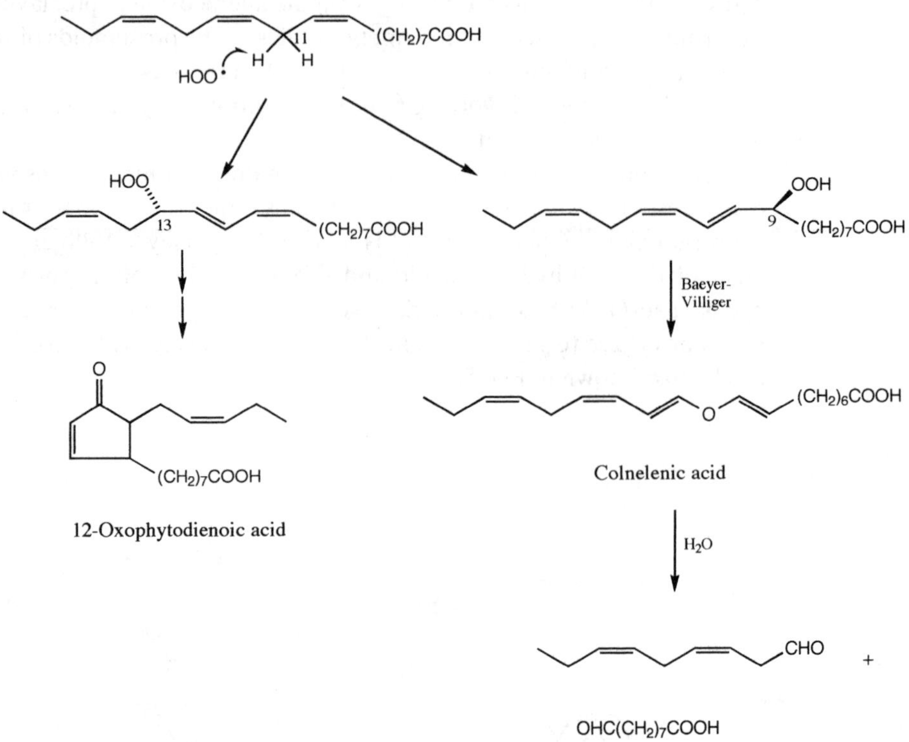

Fig. 6   Lipoxygenation of linolenic acid

5.5 Acetylenic compounds

An array of acetylenes and allenes are produced naturally, particularly in *Compositae, Umbelliferae*, and in some fungi of the group *Basidiomycetes*, albeit in small quantities.[35] They are often very unstable, sensitive to heat, light, and oxygen. Their isolation and identification have been facilitated by their characteristic UV spectra. There is convincing evidence that acetylenes are of polyketide origin arising by further desaturation of fatty acids. The structural relationship between a number of acetylenes isolated from the same plant suggests a plausible biosynthetic pathway. For example several acetylenic acids, **6–11**, have been isolated from *Santalum acuminatum*[36,37] and it appears to be a sound working hypothesis to assume that they are formed from oleic acid by sequential dehydrogenation (Table 2). The order of events is not known but as the number of known acetylenic compounds increases the biosynthetic pathway becomes clearer.

Table 2   Acetylenic acids isolated from *Santalum acuminatum*

	$CH_3(CH_2)_7CH = CH(CH_2)_7COOH$
6	$CH_3(CH_2)_5CH = CH—C \equiv C(CH_2)_7COOH$
7	$CH_3(CH_2)_3CH = CH—CH = CH—C \equiv C(CH_2)_7COOH$
8	$CH_3(CH_2)_3CH = CH—C \equiv C—C \equiv C(CH_2)_7COOH$
9	$CH_3CH_2CH = CH—CH = CH—C \equiv C—C \equiv C(CH_2)_7COOH$
10	$CH_3CH_2CH = CH—C \equiv C—C \equiv C—C \equiv C(CH_2)_7COOH$
11	$CH_2 = CH—CH = CH—C \equiv C—C \equiv C—C \equiv C(CH_2)_7COOH$

Labelling experiments with various fungi[38-40] show that the thiophene **20**, dehydromatricarianol **19**, and 10-hydroxydehydromatricariate **18** originate from oleate via linoleate **12**, crepenyate **13**, dehydrocrepenyate **14**, triynoate **15** and dehydromatricariate **17**. ω-Oxidation (**17–18**) can occur at an earlier C_{18} stage (**15–16**) but it was found that **17** is effiently incorporated into **18** which suggests that the β-oxidation (**15** to **17**) occurs before ω-oxidation as indicated in Fig. 7.

We still do not know how nature accomplishes the dehydrogenation of olefins to acetylenes and we have no suitable *in vitro* analogy to rely upon. To our dissatisfaction we are obliged to use the term desaturation for this remarkable reaction. Allylic oxidation, e.g. **15** to **16**, **17** to **18**, and **22** to **23**, often combined with allylic rearrangement, is a radical reaction. Shortening of the chain **15** to **17** proceeds via β-oxidation. The second last cleavage is coupled with a rearrangement of the original *cis*-9,10 double bond into conjugation with the carboxyl group. The formation of thiophenes and furans, which often co-occur with acetylenes, is explained by addition of hydrogen sulphide and water, respectively, to a conjugated diyne system. 1-^{14}C-labelled **17** is converted to the triol **26** exclusively labelled at C^1. Small amounts of compounds **23–25** were also detected which demonstrates that the fungus is capable of removing the terminal methyl group by oxidation and decarboxylation, reducing the ester function and *trans*-hydroxylating the double bond via an epoxide.

The calicheamicines discovered in *Microspora* spp.[41] are antibiotics with antitumour effects. They contain the highly strained medium sized enediyne ring system. The C_{15} aglycone part originates by scission from oleic acid.

Aglycone of calicheamicine, R = sugar moieties

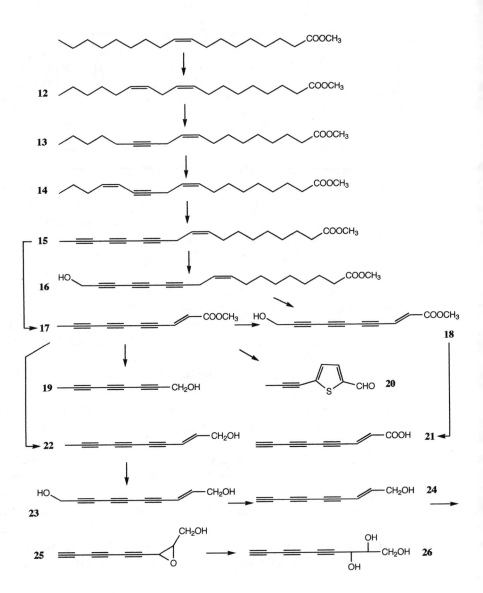

Fig. 7   Biosynthesis of acetylenic compounds

5.6 Macrolides

The macrolides are large sized (10–30-membered) lactones or lactams containing glycosidically bound sugar units. The starter unit varies widely, i.e. aromatic or short-chained aliphatic acids, and the chain is assembled from acetate, propionate and butyrate units. The enzyme system also incorporates partially assembled fragments of the metabolite which implies that the final oxidation level is attained prior to the next condensation step. Fragments of macrolide aglycones have occasionally been isolated from culture filtrates which supports the proposal that the higher polyketides are assembled in a modular manner.

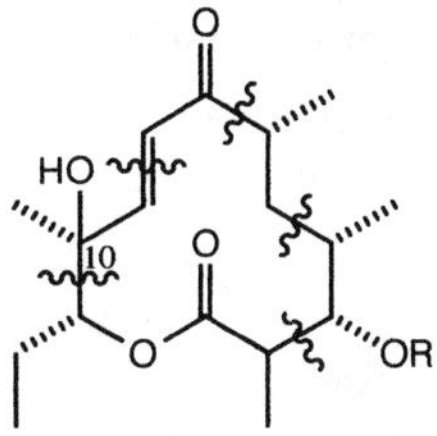

Methymycin. R = desosamine
Streptomyces venezuelae

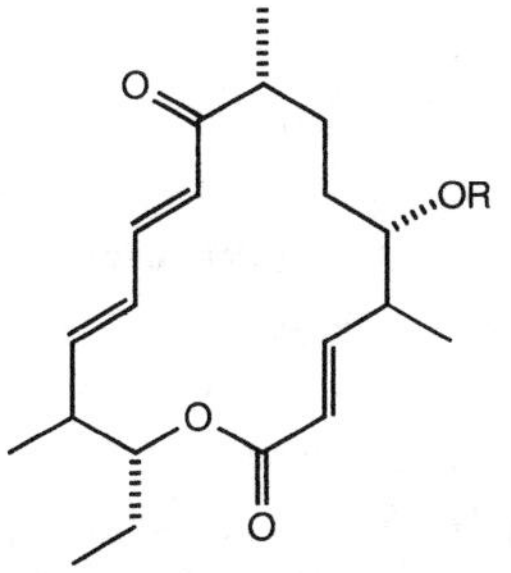

Mycinamicin. R = desosamine
Micromonospora griseorubida

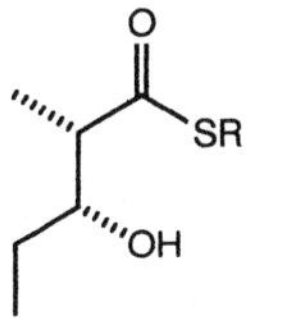

A

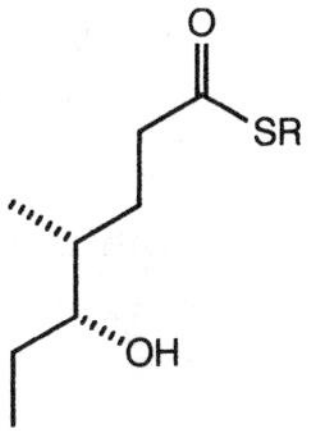

B

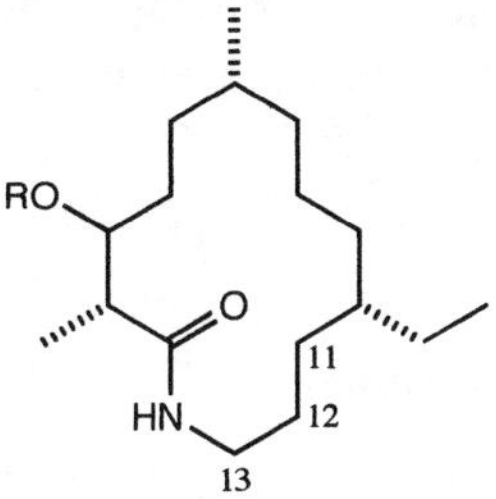

Fluvirucin A,
R = amino sugar

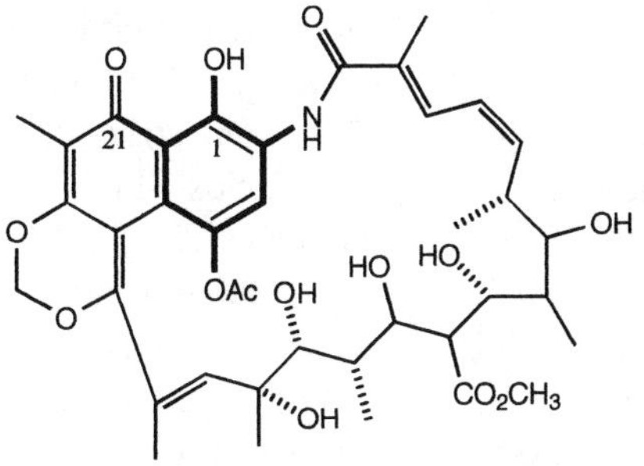

Streptovaricin C

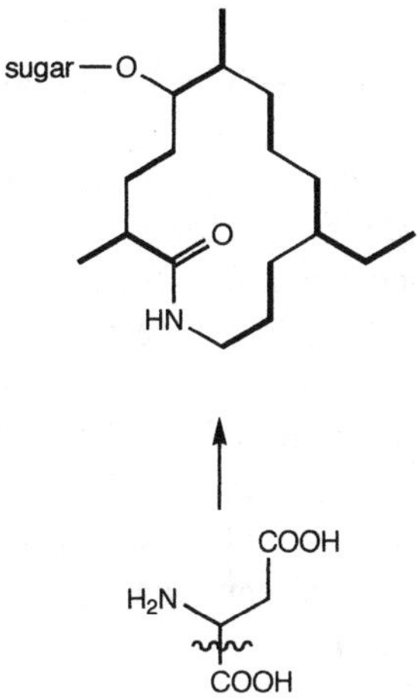

sugar

Fig. 8   Structures of macrolides

The oxygenation pattern is indicative of the original position of the carboxy groups. Visual dissection of methymycin[15] on this basis reveals the presence of one acetate and five propionates. The hydroxy group, irregularly located at C^{10}, is introduced after the assembly of the aglycone, Fig. 8. The di- and tri-ketides A and B (composed of 2 propionates, and 2 propionates + 1 acetate, respectively) are also incorporated intact and regiospecifically.

The 16-membered macrolide mycinamicin[16] is assembled from three ace-tates and five propionates. Analysis of the culture medium revealed the pre-sence of fragments clearly representing intermediates in the stepwise assem-bly of the aglycone.

Streptovaricin C is a member of the macrolactam antibiotics. The precur-sor of the starting unit is 3-amino-5-hydroxybenzoic acid which is biosynthe-sized by the shikimic acid pathway but diverted from it prior to shikimic acid because this metabolite is not incorporated. The C^1-hydroxy group derives from molecular oxygen, the C^{21} from the carboxy group of the aromatic pre-cursor and the chain is assembled from acetates and propionates.[42] Another group of 14-membered macrolactams possessing antifungal and antiviral pro-perties have been isolated from soil actinomycetes.[43] A biosynthetic investi-gation shows that fluvirucin A_1 derives from aspartic acid (C-11-13 and N), two acetates, two propionates and one butyrate.

5.7 Polyethers

Tetrahydrofuran and tetrahydropyrane rings are characteristic structural fea-tures of the polyethers. The brevetoxins are the toxic principles of the dino-flagellate *Gymnodinium breve* associated with the red tide bloom in the Gulf of Mexico. Their biosynthesis is non-classical without strict "head to tail" polyketide orientation of the acetic and propionic acids which indicates that dicarboxylic acid produced in the citric acid cycle participates.[44] Some of the marine polyethers have extraordinarily high molecular weights, i.e. maito-toxin,[45] $C_{164}H_{256}O_{68}S_2Na_2$, M.W. 3422, actually the largest monomeric organic compound known, isolated from the dinoflagellate *Gambierdiscus toxicus*. Two hydroxyl groups are esterified with sulfuric acid. It is the most toxic compound known to date with an LD_{50} in mice of 50 ng/kg and is the cause of ciguatera, a poisoning caused by ingestion of coral reef fish that have accumulated the toxin through the food chain. It affects ion transport (Na^+, K^+, Ca^{2+}) across the cell membrane. Some fragments of the polyether ring sys-tem of maitotoxin are closely related to the structure of brevetoxin and are probably assembled in the same non-classical way. Palytoxin is another high molecular weight marine polyketide, $C_{129}H_{223}N_3O_{54}$, of the same toxicity, isolated from the soft coral *Palythoa toxica*.

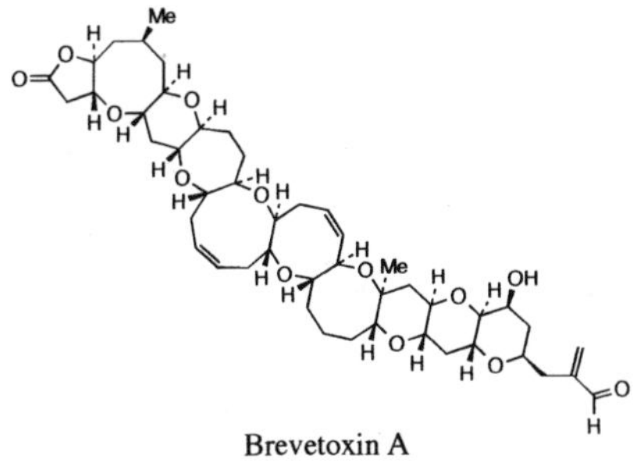

Brevetoxin A

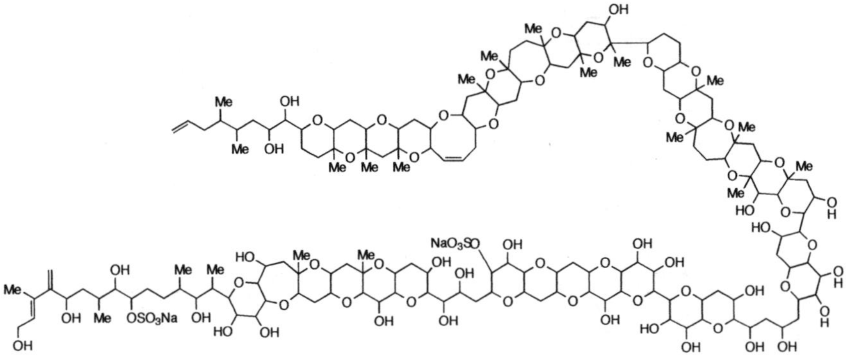

Maitotoxin

The antibiotic monensin A is assembled from five acetates, seven propionates and one butyrate. The starter unit is acetic acid but an appropriately reduced triketide is also accepted as starter. The formation of the tetrahydrofuran rings is still not well understood. The intermediacy of a triene has been suggested, which by epoxidation and subsequent cyclization could give monensin A, Fig 9. route a.[46] Experimental evidence for this sequence of reactions is still lacking, but for the findings that the ring oxygens originate from molecular oxygen. A transition metal-catalysed oxidative polycyclization of the hypothetical triene is a conceivable pathway, which avoids the intermediary epoxides. It has precedence in model reactions, route b.[47] However, it is not necessary to involve olefins or epoxides as intermediates for the formation of tetrahydrofurans. The hypothetical route c starts from the saturated hydrocarbon. This route is similar to the formation of the epoxide function in scopolamine, biosynthesis of biotin and cyclization of the penicillin skeleton.

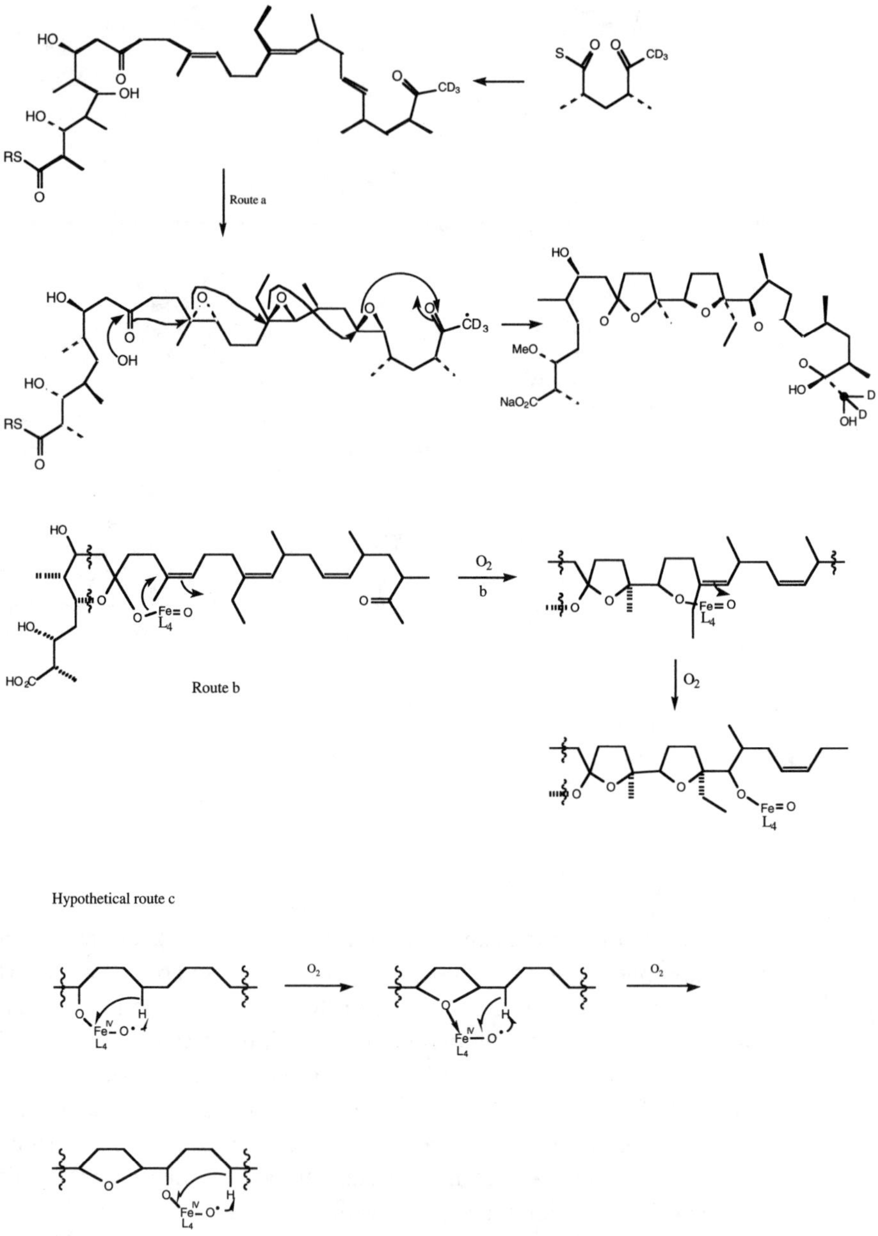

Fig. 9   Biosynthesis of monensin A

5.8 β-Oxidation

The energy stored in fatty acids is released chemically in a process called β-oxidation. At each passage through the degradative cycle one mole of acetyl CoA is generated. Palmitic acid, C_{16}, thus gives rise to eight units of acetyl CoA which eventually enter the citric acid cycle and are completely oxidized to carbon dioxide. In the first step of the sequence, the fatty acid is transported into the mitochondria and esterified with CoA (21) which activates the α-C protons, and prepares the molecule for the first FAD-mediated dehydrogenation.

$$RCOOH + ATP + CoA \rightarrow RCOCoA + AMP + PP \qquad (21)$$

The mechanism of this reaction deserves comment since it represents one of the commonest redox reactions in biosynthesis. First, it is worth noting that NADPH is reported to serve as donor in the saturation step, whereas FAD serves as oxidant in the reverse α,β-desaturation of the fatty acid. It is generally accepted that flavins act as electron transfer agents rather than hydride transmitters on the grounds that flavins can pass electrons down the respiratory chain, $FADH_2$, in contrast to NADH, is sensitive to oxygen, and finally the steps in the electron transfer mechanism are thermodynamically reasonable, e.g. for reduction of a carbonyl group (22).

(22)

It is possible to formulate the NADH reduction in terms of single electron transfers, e.g. (23), but NADH is usually regarded as a hydride shift reagent (24). In that respect the reaction has several analogies in non-enzymatic organic oxidations, e.g. the Cannizzaro (25) and the Meerwein–Ponndorf–Verley–Oppenauer reactions (26). No exchange of protons with the solution is observed in the stereospecific NADH reduction or in (25) and (26) which is taken as an argument for a hydride shift. The hydride ion is a very strong base that reacts with water with development of hydrogen. Consequently, no free hydride ion is involved in the hydride shift. It ought to be pointed out that the direct transfer of hydrogen from NADH to the oxidant by no means excludes a radical pathway because hydrogen abstraction by a radical leads to the same result. The situation is different for the FAD/FADH$_2$ couple because H–N protons equilibrate extremely rapidly with protonated solvents.

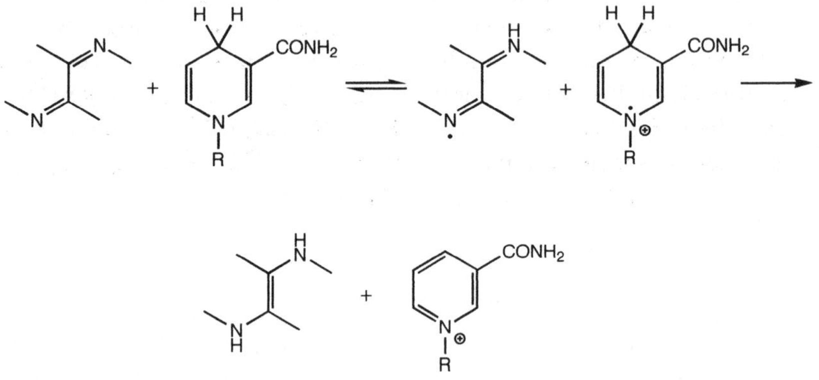

(23)

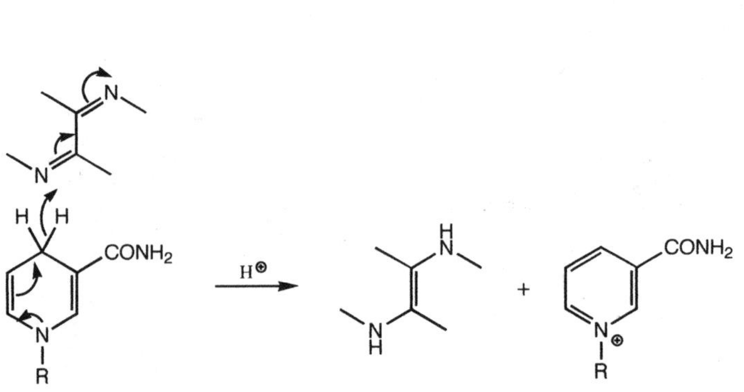

(24)

We are now ready to discuss the mechanism of the second step, the desaturation of fatty acids. A base removes the activated α-C–H of a fatty acid CoA generating the anion which is oxidized by FAD, anchored at the enzyme, to the neutral α-C-radical and flavin radical anion. H abstraction from the β-carbon completes the dehydrogenation, which is depicted in Fig. 10 as a concerted reaction. If the acid reacts in its most stable conformation and the β-C–H points axially towards FAD, the α,β-*trans* unsaturated fatty acid CoA is produced.

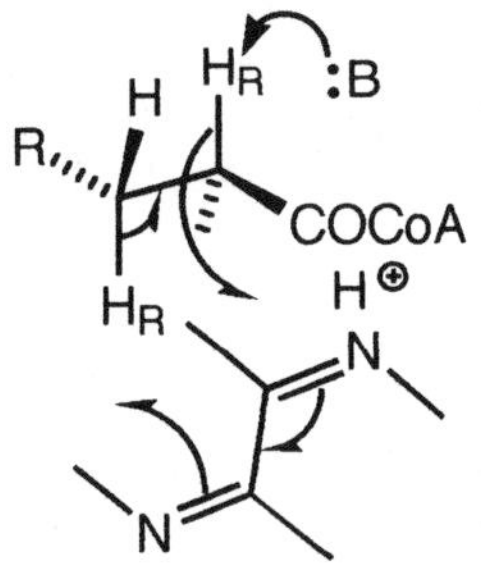

Fig. 10   Suggested mechanism for dehydrogenation of fatty acid CoA by FAD. Only the active quinonoid section of FAD is depicted

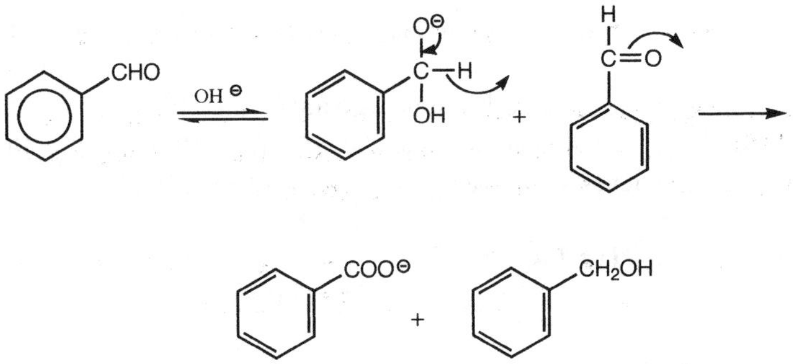

(25)

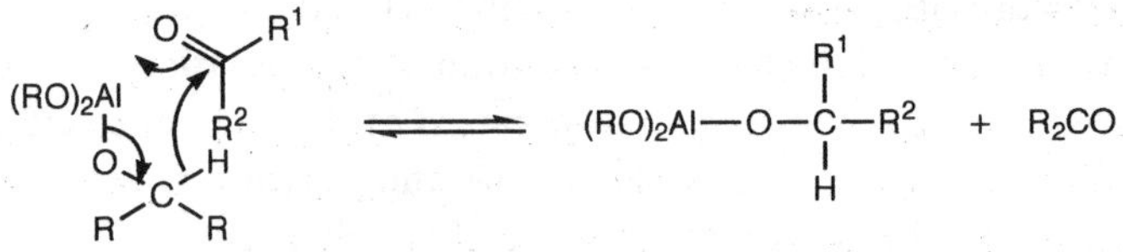

(26)

The third step consists of stereospecific hydration yielding the (3*S*)-hydroxy stereoisomer (27).

The (3*S*)-hydroxy fatty acid is oxidized by $NAD^{\oplus}$ to the 3-keto fatty acid (28). In the final step the 3-keto fatty acid undergoes thiolysis by another CoA molecule to acetyl CoA and a new fatty acid CoA, shorter by two carbon atoms (29). The overall equation for one cycle of β-oxidation starting from palmitic acid is (30).

$$\tag{27}$$

$$\tag{28}$$

$$RCOCoA \quad + \quad CH_3COCoA \tag{29}$$

$$C_{15}H_{31}COCoA + FAD + NAD^{\oplus} + CoA + H_2O \rightarrow$$
$$C_{13}H_{27}COCoA + FADH_2 + NADH + CH_3COCoA + H^{\oplus} \tag{30}$$

Part of the energy developed is thereby transformed to chemical energy in the form of ATP. If we also consider these transformations, complete degradation of palmitic acid CoA to acetyl CoA is given by (31).

$$C_{15}H_{31}COCoA + 7\ CoA + 7\ O_2 + 35\ P + 35\ ADP \rightarrow$$
$$8\ CH_3COCoA + 35\ ATP + 42\ H_2O \tag{31}$$

Odd-carbon fatty acids give propionyl CoA in the last β-oxidation step. It is eventually converted to succinyl CoA via methylmalonyl CoA in a vitamin B_{12} mediated process.

It is worth noting that the degradation occasionally can follow a non-oxidative route as is established for oxidation of *p*-coumaric acid in cell cultures of *Lithospermum erythrorhizon*. Hydration of the cinnamic acid followed by a reverse aldol reaction gives acetyl CoA and *p*-hydroxybenzaldehyde which accumulates in absence of $NAD^{\oplus}$. The aldehyde is eventually oxidized to *p*-hydroxybenzoic acid by dehydrogenase.[47a]

5.9 Cyclization of polyketides to aromatics

Reduction after each condensation step leads to fatty acids, macrolides and polyethers as described above. If further condensation occurs before reduction takes place, intermediate β-polyketoesters of various chain length are formed. These compounds are very reactive and they undergo cyclizations to form an array to aromatics. The poly-β-ketoester (polyketide) is temporarily stabilized by chelation or hydrogen bonding on the enzyme surface until the assembly is accomplished. The cyclization is then guided by the special topology of the acting enzyme. The activated methylene groups give rise to carbanions or enolates by removal of protons and the polarized carbonyl group has carbonium ion character. The tetraketide can cyclize in several ways. Path a leads to the acetophenone derivative xanthoxylin, paths b and c to the pyrone derivatives, and path d to orsellinic acid, all known from natural sources (32). Incubation of orcellinic acid synthetase with chiral [1-^{13}C-2-^{2}H] malonate combined with mass spectral analysis showed stereospecific removal of hydrogen at the enolization step.[48]

Model cyclizations of this kind have been carried out chemically in the laboratory thus mimicking the biosynthetic process.[49] Acylation of the acetoacetate dianion selectively gives the β-diketoester. By special carbonyl protection methods it is possible to repeat the procedure and synthesize poly-β-ketoesters. The unstable phenyl derivative undergoes base catalysed aldol condensation to the orsellinic acid analogue (33).

(32)

$$\text{CH}_3\text{COCH}_2\text{COOC}_2\text{H}_5 \xrightarrow{\text{2 } \overset{\ominus}{\text{NR}_2}} \overset{\ominus}{\text{CH}_2}\text{CO}\overset{\ominus}{\text{CH}}\text{COOC}_2\text{H}_5 \xrightarrow{\text{CH}_3\text{COOC}_2\text{H}_5}$$

$$\text{CH}_3\text{COCH}_2\text{COCH}_2\text{COOC}_2\text{H}_5 \xrightarrow[\text{C}_6\text{H}_5\text{COCl}]{\text{base}} \qquad\qquad (33)$$

$$\text{C}_6\text{H}_5\text{COCH}_2\text{COCH}_2\text{COCH}_2\text{COOC}_2\text{H}_5 \xrightarrow[\text{C}_2\text{H}_5\text{OH}]{\text{KOH}}$$

A great variety of metabolites are formed depending upon several factors:
1. the starter or chain initiating unit;
2. the number of acetyl CoA units involved;
3. the mode of cyclization;
4. the condensation of separately synthesized polyketides; and
5. the secondary processes, such as halogenation, alkylation, redox reactions, rearrangements, etc.

These factors determine the class of compounds biosynthesized. One distinguishes traditionally between simple benzenoids, tropolones, condensed aromatic systems, quinones, flavonoids, phenolic coupling products, and non-aromatic polyketides. A classification according to the numbers of units is also convenient; thus, one speaks of tetra-, penta-, hexaketides, etc. and the two systems overlap to a large extent.

Genetic studies show that synthetase complexes which produce aromatic polyketides largely from acetate and malonate are distinct from those which produce fatty acids, macrolides or polyethers from acetates, propionates and butyrates. The last decade has witnessed increased interest in elucidation of structure and function of polyketide synthetases.[50] Perspectives have emerged for cloning genes from one organism for synthesis of antibiotics in other related organisms, i.e. genetic manipulation in a modern sense, cf. the biosynthesis of vitamin B_{12}, section 9.5.

5.10 Confirmation of the acetate hypothesis. The use of NMR spectroscopy in biosynthetic studies

The acetate hypothesis as formulated by Birch[51] received its first confirmation by an investigation of the labelling pattern of 6-methylsalicylic acid formed by *Penicillium griseofulvum*. l-^{14}C-labelled acetic acid was added to the nutrient solution. By a special degradation technique each carbon of the molecule was separately isolated (as CO_2 etc.) and analysed for radioactivity (Fig. 11). The carbon dioxide obtained by decarboxylation and the acetic acid from Kuhn–Roth degradation were found to be active, whereas the bromopicrin was inactive. The activity of $C^{2,4}$ was calculated by difference and the relative intensity was found to be close to 1.0. In another experiment with ^{18}O-labelled acetic acid, the 6-methylsalicylic acid produced contained ^{18}O-labelled oxygens demonstrating that the original oxygens are retained throughout the biosynthesis in complete accordance with the hypothesis.[52] Some washout of ^{18}O was noted as a result of exchange with the medium.

We have seen earlier that the starter normally is acetyl CoA, but that the chain elongating units consist of malonyl CoA. This was demonstrated in an experiment with 2-^{14}C-malonate administered to *Penicillium urticae*.[53] The 6-methylsalicylic acid was isolated, degraded, and analysed. In this case activity was absent in acetic acid from the Kuhn–Roth degradation, i.e. in $C^{6,7}$, and all the activity accumulated in the bromopicrin, supporting the mechanism of polyketide formation.

One more detail in the synthesis of 6-methylsalicylic acid requires comment. The compound lacks a hydroxyl group at C–4. The reduction to alcohol occurs before cyclization since, according to experience, aromatic deoxygenation is not carried out by organisms producing the polyketides. Metabolites are frequently modified by "missing" hydroxyl functions. The exact timing of the reduction and dehydration has been the object of extensive research. 6-Methylsalicylic acid synthase is susceptible to inhibitors of *cis*-unsaturated fatty acid synthase. The inhibitor still allows NADPH reduction and synthesis of stearic acid to continue and these observations have been taken as evidence for a proposal that the reduction takes place at the 6-carbon stage rather than at the 8-carbon stage. Then follow *cis* elimination, condensation with a C_2 unit and cyclization (Fig. 11).[54]

Alternariol, produced by an enzyme extract from the mould *Alternaria tenuis*, is made up of seven acetate units and the labelling pattern conforms to theory (34).[55]

(34)

Alternariol

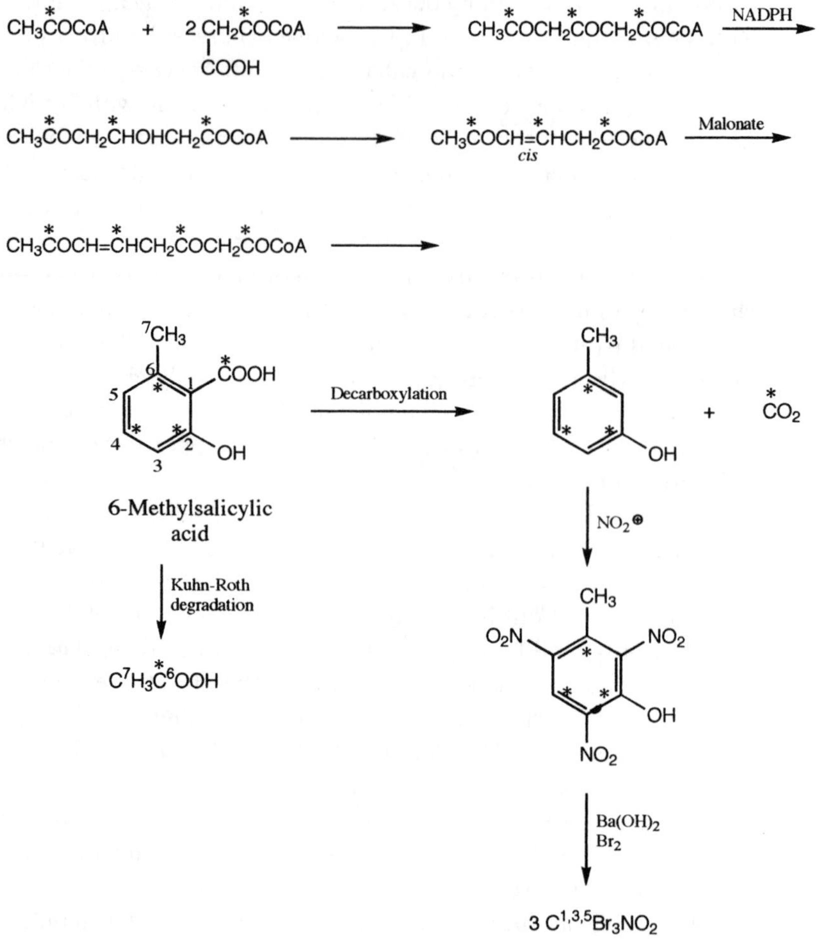
CH3COCoA
2 CH2COCoA
COOH
CH3COCH2COCH2COCoA
NADPH
CH3COCH2CHOHCH2COCoA
CH3COCH=CHCH2COCoA
cis
Malonate
CH3COCH=CHCH2COCH2COCoA
7CH3
COOH
OH
6-Methylsalicylic acid
Decarboxylation
CH3
OH
CO2
NO2⊕
Kuhn-Roth degradation
C7H3C6OOH
O2N
NO2
Ba(OH)2
Br2
3 C1,3,5Br3NO2

Fig. 11 Biosynthesis and chemical degradation of 6-methylsalicylic acid

The relative radioactivity of the starred carbon atoms is close to 1.0, the starter being somewhat higher. The reason is that the six malonate units of the chain to a larger extent are diluted by acetate (malonate) from the pool. The other positions were devoid of any activity. The uniform distribution of activity at the starred positions indicates that the whole polyketide chain is constructed in one complete sequence without leaving the protein surface. Islandicin showed the labelling pattern depicted in (35) when *Penicillium islandicum* was fed with 1-^{14}C-acetate.[56] Three secondary reactions have taken place here, decarboxylation, reduction, and hydroxylation.

In the conventional ^{14}C-labelling method it is necessary to have a degradative method allowing separation and isolation of all the carbons of the molecule. Increasing complexity of a metabolite rapidly raises serious, if not insurmountable, problems. The degradative methods, the isolation, and the purification of defined fragments become more complicated and time-consuming and small amounts of radioactive impurities seriously affect the measurements. Modern structural determinations rely heavily on spectroscopic methods and try to avoid chemical degradations as much as possible, among other things because the amounts of material available are frequently too small. As a consequence the chemistry of the compound is not sufficiently known for the application of degradation procedures.

$$(35)$$

Islandicin

Adequate NMR instrumentation for recording [2]H-, [3]H-, ^{13}C-labelled compounds became available in the sixties, which soon revolutionized the biosynthetic tracer studies.[57,58] The great advantage of NMR methods is that degradation can be eliminated. The nuclear properties of ^{3}H are ideal for NMR spectroscopy. It has a nuclear spin of 1/2, a slightly higher sensitivity than ^{1}H, gives sharp peaks and chemical shifts and coupling constants very similar to those of [1]H. The NOE effect is slightly negative and integration can therefore be used for an approximate determination of the isotopic content. The ^{3}H NMR method has lower sensitivity than conventional scintillation counting but spectra can be run at an ^{3}H enrichment level of comparably low radiation hazard.

Deuterium, with a nuclear spin of 1, has the advantage of being an inexpensive, stable isotope having low natural abundance, 0.016 per cent, allowing higher dilution of the precursor in comparison with ^{13}C-labelled compounds. It exhibits short relaxation times and no NOE effect and can therefore be accurately integrated. However, the sensitivity is low (1/100 of ^{1}H), the chemical shift scale and coupling constants are only 1/6 the value of ^{1}H and the line width is broader than that of ^{1}H. The spectral resolution is consequently low (see Fig. 13 and section 1.8).

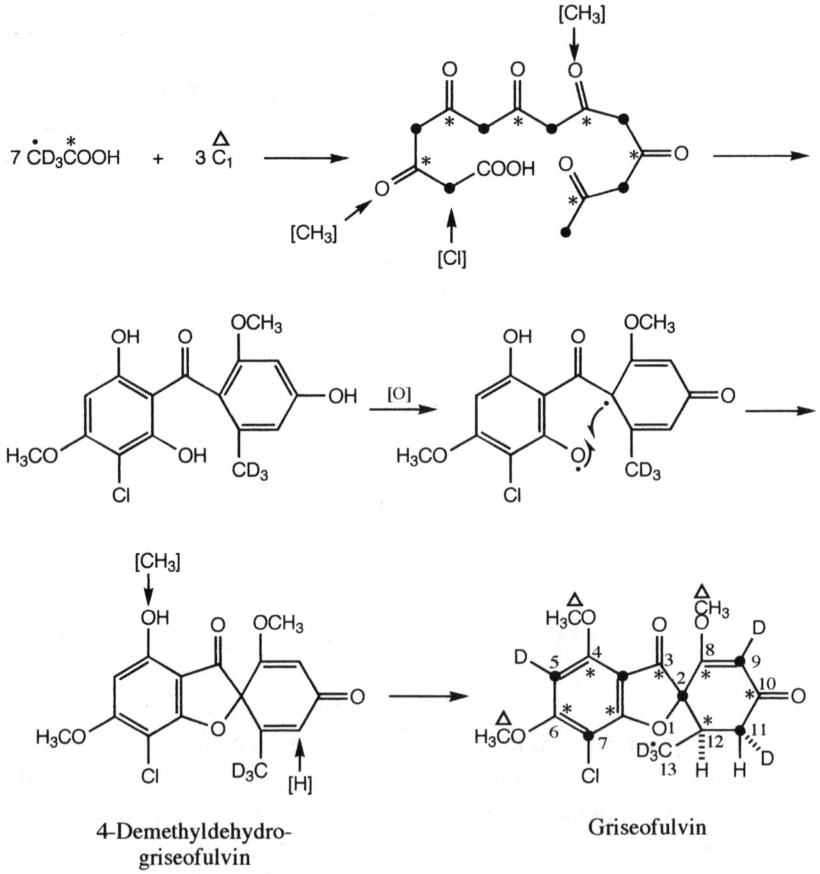

4-Demethyldehydro-
griseofulvin

Griseofulvin

Fig. 12 Biosynthesis of griseofulvin

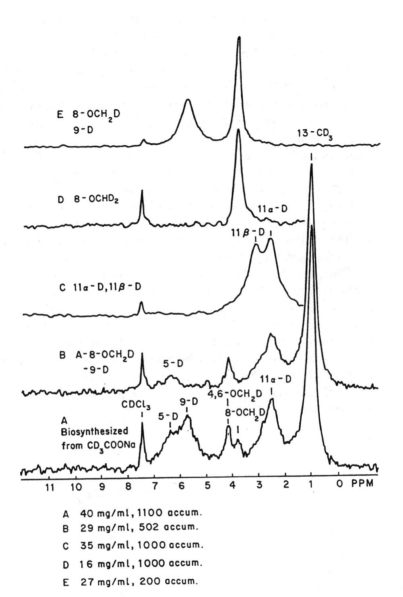

Fig. 13 A ^{2}H NMR spectrum of biosynthetically deuterated griseofulvin in CHCl$_3$ solution (4 w/v%). The lowermost signal arises from CDCl$_3$ occurring in CHCl$_3$ of natural abundance (0.02%). The assignment of ^{2}H NMR signals is straightforward to that of ^{1}H NMR since chemical-shift displacements due to isotope effects are negligible. The peak assignment is here supported with the aid of specifically deuterated griseofulvin samples, B–E. (Reproduced by permission of Pergamon Press from *Tetrahedron Lett.* **1976,** 2695)

The stable ^{13}C, natural abundance 1.1 per cent, has a nuclear spin of 1/2 and a positive NOE effect. It has a large chemical shift range (200 PPM) in comparison to ^1H (10 PPM) and gives narrow line widths. The spectra are consequently well resolved. Unfortunately, ^{13}C has a low sensitivity (1/60 of ^1H), which can in part be compensated for by proton decoupling, causing collapse of the multiplets to single sharp peaks in combination with intensity enhancement due to the positive NOE effect. Progress in the refinement of instrumental techniques and application of Fourier transformation allow direct detection of the ^{13}C nuclei in natural abundance on samples as small as 1–10 mg and an accumulation time of a few hours. By proton noise decoupling all $^{13}C-^1H$ coupling information is lost. By off-resonance technique residual couplings are observed giving information on the number of hydrogens attached to the carbon atom. Thus, a quaternary carbon appears as a narrow singlet, a methine carbon as a narrow doublet, a methylene gives a triplet and a methyl a quartet centred at the δ value of the singlet in the noise decoupled spectrum. ^{13}C NMR is used with success in tracing the fate of hydrogen in biosynthesis. In proton decoupled ^{13}C NMR spectra a carbon atom attached to one ^2H (nuclear spin 1) appears as a triplet of equal line intensity, a C^2H_2 group as a quintet (1: 2: 3: 2: 1:) and a C^2H_3 as a septet (1: 3: 6: 7: 6: 3: 1) with an upfield shift of *ca.* 0.4 PPM for each ^2H. The spectral analysis of overlapping signals can be simplified by running $^{13}C-^2H$ decoupled spectra. The weakness of the $^{13}C-^2H$ relaxation limits the use of this technique. On the other hand, reduction of the ^{13}C signal strength indicates that the carbon atom is ^2H-labelled.

(36)

Shanorellin

The biosynthesis of griseofulvin (Fig. 12) has been studied by ^{2}H NMR spectroscopy.[59] Griseofulvin produced from 2-^{2}H$_3$-acetate is expected to be deuterated at C^5, C^9, C^{11}, and C^{13}. The ^{2}H NMR spectrum (Fig. 13) shows that these positions are indeed labelled. Some washout was noted but label is also incorporated to some extent in the methoxy groups at C^4, C^6, and C^8. C^5 is exclusively labelled in the α-position demonstrating that reduction of the intermediate dehydrogriseofulvin takes place specifically in a *trans* fashion. The finding that C^{13} contained a C^2H$_3$ residue proves that this is the starter end of the C$_{14}$ chain. The ^{2}H NMR spectrum is assigned by comparison with specifically labelled derivatives. As can be seen from the spectrum, the line width is much broader than that of a corresponding ^{1}H spectrum.

Presuming that the shifts of all carbon atoms can be assigned, it is possible to identify the labelled positions by measuring the relative intensities of the decoupled ^{13}C peaks in ^{13}C enriched samples and at natural abundance. Shanorellin is a benzoquinone pigment synthesized by *Shanorella spirotricha*. By administration of 1-^{13}C-acetate, or ^{13}C-formate to the culture medium, ^{13}C enriched shanorellins are obtained (36). Table 3 shows the results of the ^{13}C NMR recordings in complete agreement with an earlier, conventional ^{14}C study.[60] Chromic acid oxidation of shanorellin labelled with 1-^{14}C-acetate gave inactive acetate (CH$_3$C^{2,6}O$_2$H), whereas samples labelled with 2-^{14}C-acetate and ^{14}CH$_3$ methionine gave acetic acid labelled in the carboxy and methyl groups, respectively. The proposal has been made that methylation and oxidation of shanorellin occur at an early stage prior to cyclization.

Table 3 ^{13}C NMR data for labelled shanorellin, δ. Positions for enhanced intensity

Position	1-^{13}C-acetate	2-^{13}C-acetate	13C-formate
C$_1$	187.7	—	—
C$_2$	—	146.8	—
C$_3$	137.8	—	—
C$_4$	—	183.3	—
C$_5$	152.0	—	—
C$_6$	—	117.4	—
CH$_2$OH	—	54.8	—
CH$_3$	—	—	12.0
CH$_3$	—	—	7.8

Loss of ^{13}C signal strenth due to deuteration was used in an investigation of the origin of skytalone produced by the fungus *Phialophora lagerbergii* (37).[61] The compound, synthesized in a culture medium containing 2-^{13}C-2-^{2}H$_3$-acetate gave enhanced signal intensities at C^2, C^4, C^5, C^7, C^{8a}, but lower intensities than expected at C^4 and C^5, indicating the presence of ^{2}H at these positions.

(37)

At C^4 a triplet was observed (0.3 PPM) upfield from the normal ^{13}C signal. In the ^2H decoupled spectrum C^5 appeared as a singlet and C^4 as a doublet establishing a $^1H^2HC^4$-labelling pattern. No deuterium was detected at C^2 and C^7, suggesting that the polyketide is folded so that $C^{4,5}$ are specifically derived from the starter acetate, whereas $C^{2,7}$ are derived from malonate which might lose ^2H more readily in the biosynthetic process. An alternative conformation of the pentaketide chain is suggested for the biosynthesis of flaviolin (section 5.11).

Chartreusin is a complex isocoumarin glycoside antibiotic produced by *Streptomyces chartreusis*. ^{13}C NMR analysis of chartreusin aglycone, biosynthesized from 1-^{13}C-, 2-^{13}C-, and $1,2$-$^{13}C_2$-acetate, revealed details of its formation which would be difficult to extract from any other method.[62] From recordings of spectra from singly labelled precursors the positions of the labelled carbons could be assigned confirming that the metabolite is acetate-derived. The carbon skeleton cannot be formed by direct cyclization of one polyketide chain. Several modifications have taken place. Condensation of two or more separate polyketides or cyclization of one decaketide, e.g. according to (38), has been suggested. The labelling pattern as determined by the intensities of the ^{13}C NMR spectrum is in agreement with the suggested folding of the decatide. The use of doubly labelled acetate in conjunction with ^{13}C–^{13}C spin–spin couplings, $^1J_{13c13c}$, also gave information as to which intact C_2 units are present. Selective ^{13}C decoupling and ^{13}C off-resonance techniques showed $^1J_{13c13c}$ couplings between $C^{3,4}$, $C^{4a,5}$, $C^{6,7}$, $C^{8,8a}$, $C^{2,1'}$, $C^{3',4'}$, $C^{5,CH3}$, and $C^{6',7'}$ superimposed on natural abundance singlets. This is in agreement with the cyclization of an undecaketide to a benzpyrene type intermediate. Decarboxylation and bond cleavages with loss of three carbon atoms, rotation around $C^{2,1'}$ and lactonization gives the aglycone, route (38b). Alternatively (38a) gives an isolated $C^{4'}$ from the decarboxylation that should not give rise to any $C^{3'4'}$ coupling, and analysis of the remaining couplings anticipated gives a pattern different from that observed. $1,2$-$^{13}C_2$-Acetate is diluted and incor-

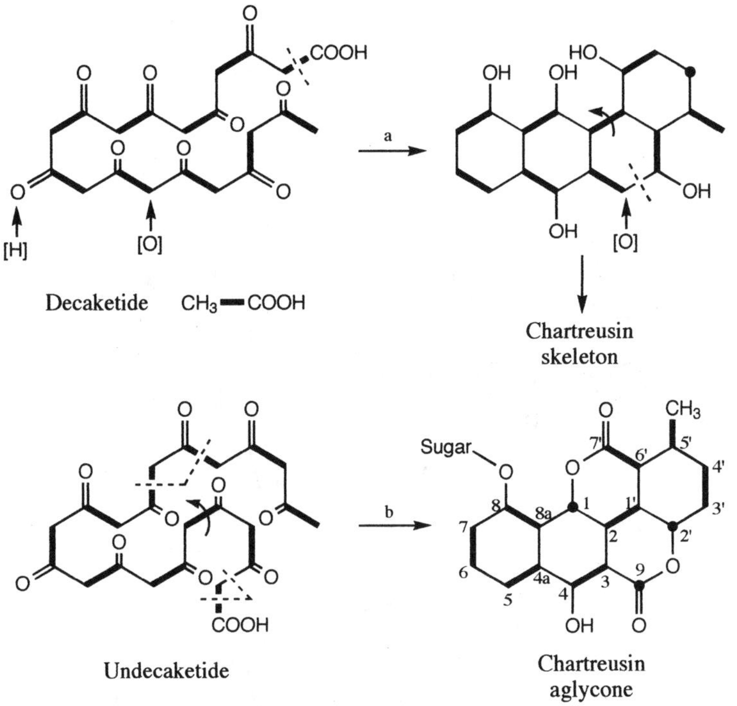

Decaketide   CH₃━COOH

Chartreusin
skeleton

(38)

Undecaketide

Chartreusin
aglycone

porated largely so that adjacent units derive from the organism's pool of un-labelled acetate. Therefore couplings are restricted to intact units and, conse-quently, the carbon atoms $C^{1,2',9}$ appear only as singlets.

The same method has been used to unveil details of the biosynthesis pro-cess leading to mollisin.[63] Condensation of two tetraketides, pathway (39a,b), has been suggested but ^{13}C NMR studies of mollisin obtained from fermen-tations of *Mollisia caesia* fed with doubly labelled ^{13}C acetate showed $^1J_{c2,12}$ 45.0 Hz, $^1J_{c13,14}$ 47.5 Hz, $J_{c6,7}$ 61.3 Hz, $J_{c3,4}$ 52.5 Hz, but no couplings at $C^{1,11}$. This eliminates pathways a and b and favours pathway c, but a single octake-tide, which is cleaved, is also conceivable, pathway d. These studies provide compelling demonstrations of how the doubly labelling technique can give in-sights into the folding processes at the enzymatic level unavailable by other methods.

Mollisin (39)

5.11 Derivation of structure

In a polyketide every second carbon is oxygenated which will give a basic pattern of *meta*-hydroxy substitution in the aromatic product. Hence, this is an indication that we are dealing with an acetate-derived metabolite. Shikimic acid derived metabolites are often characterized as having hydroxy groups in *ortho* positions. Many secondary transformations can, of course, mask the biogenesis, and a metabolite can be produced by different routes in different organisms. 2,5-Dihydroxybenzoic acid is an example of the latter case. It is produced from phenylalanine in *Primula acaulis*, but from acetate in *Penicillium griseofulvum*. As already noted (section 4.6), structure proves to be an illusory guide for biosynthesis in the naphthoquinone and anthraquinone series. Having said this, we can turn to some classical examples, where biosynthetic principles have been used successfully for structural determination. By using the oxygenation pattern and the positions of alkyl and carbonyl groups as markers, it is clear that certain structures are more likely than others. For example two structures were considered for the naphthoquinone flaviolin, **27** or **28**, from the available data. If biosynthetic considerations are taken into account, **27** is the most probable structure and this was later confirmed (Fig. 14).[64]

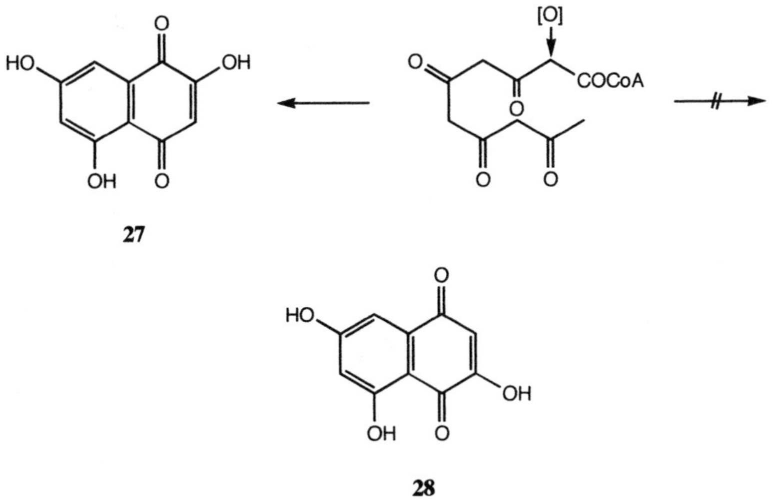

Fig. 14   Biosynthetic derivation of the most probable structure **27** for flaviolin

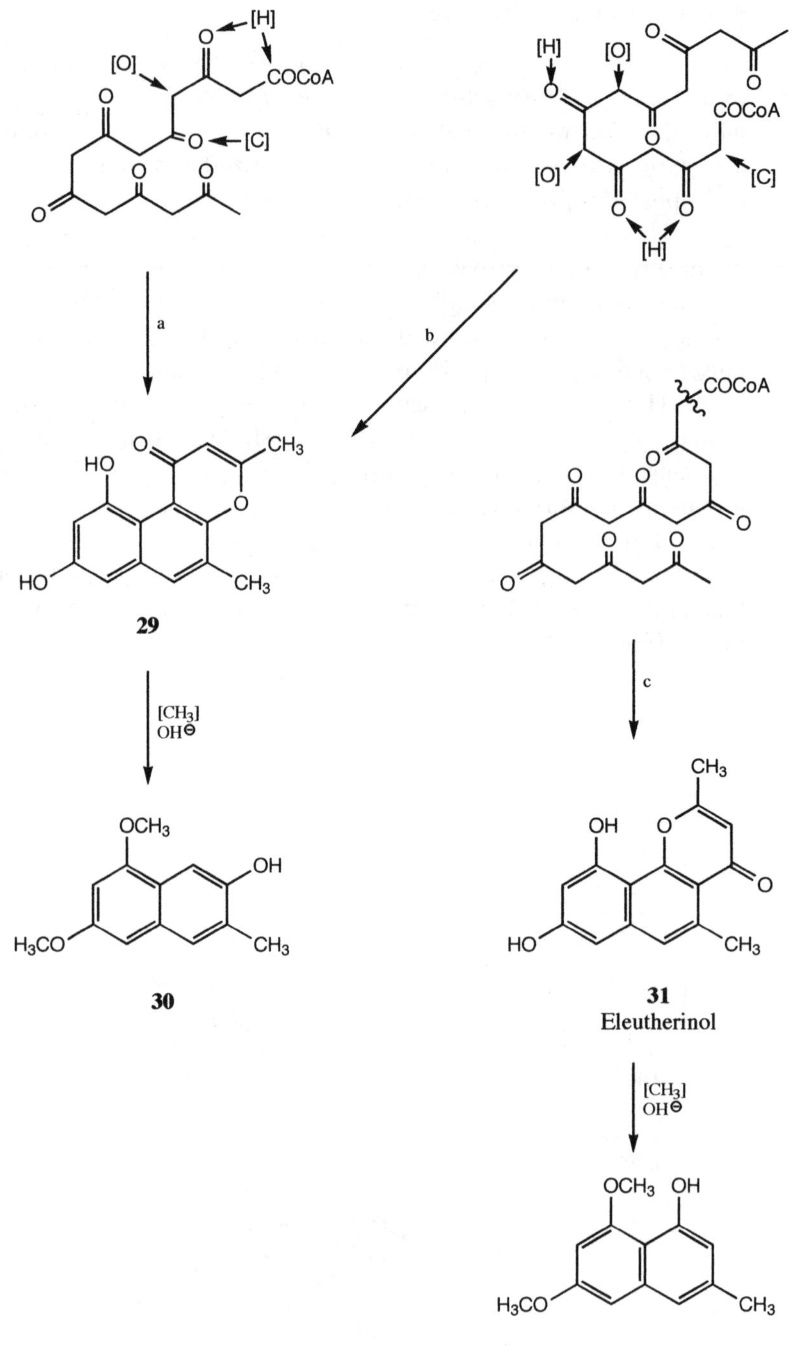

Fig. 15   Biosynthetic derivation of the structure of eleutherinol

The revision of the structure of eleutherinol illuminates the potency of the acetate rule (Fig. 15).[65] Structure **29** was first proposed for eleutherinol, based primarily on the claim that the naphthol **30** was formed on alkaline hydrolysis. It is not possible to fold a polyketide chain in harmony with structure **29** without making certain provisions. An extra carbon has to be introduced and the carboxyl group has to be reduced to methyl, route a, or the oxygenation pattern has to be drastically changed in conjunction with the introduction of one extra carbon, route b. Both possibilities are unlikely because the molecule has to undergo a series of rather unusual reactions. Suppose instead that we start with a polyketide with two more carbons, route c; decarboxylation and cyclization then give **31** which on degradation gives a different naphthol **32**. This folding directly gives the basic oxygenation pattern and correct positioning of the methyl in the naphthol. Generation of the other methyl by decarboxylation is a straightforward reaction. However, the structure of the naphthol has to be revised. Later synthetic work confirmed that **32** was indeed the correct structure for the naphthol and that **31** hence represents the correct structure for eleutherinol.

5.12 Anthraquinones, anthracyclinones and tetracyclines

Anthraquinones are the largest group of quinones. They have been used as mordant dyes, e.g. alizarin from *Rubia tinctorum*, and purgatives, e.g. emodin from *Rheum, Rumex,* or *Rhamnus* spp. They lost their importance, like so many other natural dyes, with the development of the synthetic dye industry. Anthraquinones are widely spread in lower and higher plants and occur also in the animal kingdom. They are present as glycosides in young plants.

It is difficult to systematize the biosynthesis of quinones because it reveals such a diversified picture.[66] Benzoquinones originate either from shikimic acid, polyketides, or mevalonate. Naphthoquinones may be completely synthesized from acetate, e.g. in flaviolin **27** or in the polyhydroxylated spinochromes discovered in the calcareous parts of sea urchins, or originate from mixed biosynthesis, with e.g. one ring deriving from shikimic acid and the missing carbons of the other ring coming from acetate via glutamate (section 4.6), or one ring may come from a polyketide and the other from mevalonate. The polyketide pathway leading to anthraquinones and tetracyclines is well documented. Annelation of another mevalonate to the naphthoquinone skeleton constitutes another route.

As a general rule fungal anthraquinones and plant anthraquinones with hydroxy groups are derived from polyketides, whereas plant anthraquinones devoid of hydroxy groups in one ring, e.g. alizarin, come from mixed pathways.

Emodin is constructed from one acetate and seven malonate units, but since it is questionable whether endocrocin is an intermediate, the decarboxylation may occur at an earlier stage (40). Numerous anthraquinones are pro-

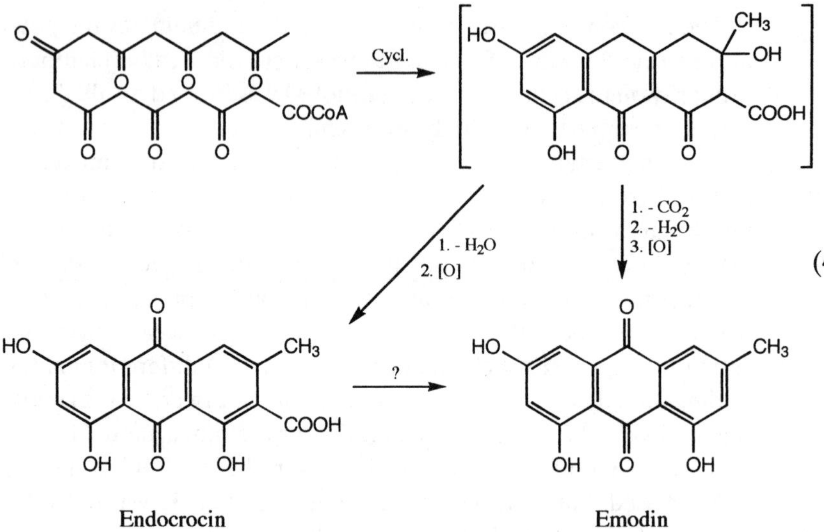

(40)

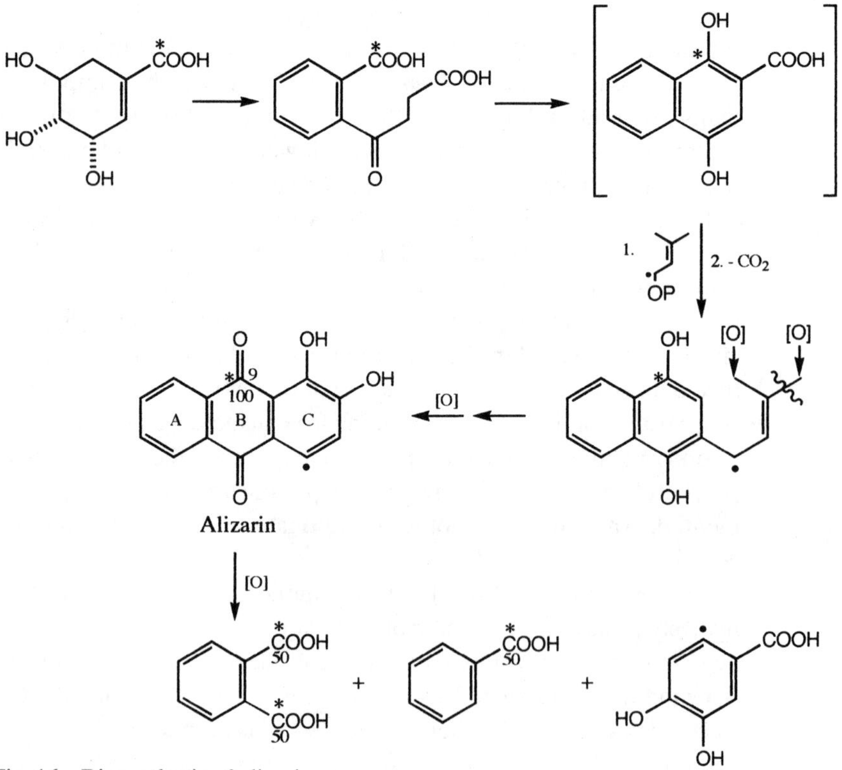

Endocrocin   Emodin

Fig. 16   Biosynthesis of alizarin

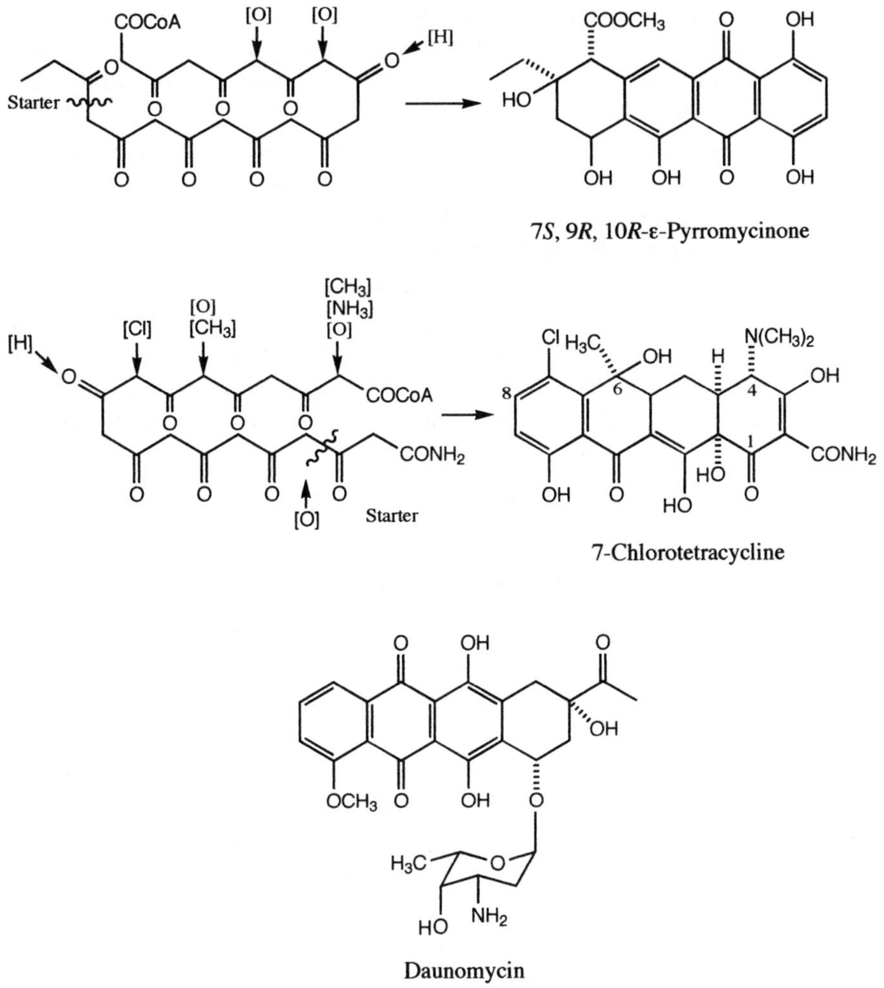

7S, 9R, 10R-ε-Pyrromycinone

7-Chlorotetracycline

Daunomycin

Fig. 17   Biosynthesis of tetracyclines

duced by the octaketide pathway conforming to the basic emodin structure. They arise via different folding, O-methylation, side chain oxidation, nuclear hydroxylation or elimination of hydroxyl groups, chlorination, dimerization via phenol oxidation, etc.

The historical dye alizarin has a more complicated biosynthesis. *Rubia tinctorum* was fed with 7-^{14}C-shikimic acid and alizarin was isolated and degraded to phthalic, veratric and benzoic acids.[67] The label was recovered only in the carboxy groups. Phthalic acid had the same specific activity as alizarin, benzoic acid 50 per cent, and veratric acid no activity, which shows that the carboxyl group of shikimic acid is located at C^9 and no symmetrical intermediate is involved during the construction of ring C. The remaining three carbon atoms of ring B come from glutamic acid. 5-^{14}C-mevalonic acid is incorporated in ring C. Dimethylallyl pyrophosphate alkylates selectively in the *meta* position relative to the labelled carbonyl group, and one of the methyls is lost as carbon dioxide (Fig. 16).

Anthracyclinones and tetracyclines belong to a group of linear tetracyclic, highly active antibiotics, some of which have found use in the treatment of cancer, e.g. daunomycin. They are produced in cultures of *Streptomyces* spp. The order of the modifications of the polyketide chain has been determined by a series of mutant studies. Amidomalonyl CoA is the starter of the nonaketide chain. Methylation at C^6, reduction at C^8, and cyclization to the fully aromatic system occur next, followed by hydroxylations at C^4, C^6 and C^{12a} and amination (Fig. 17). We have here examples of other starters, propionic acid for pyrromycinone and daunomycinone, and amidomalonic acid for tetracycline.

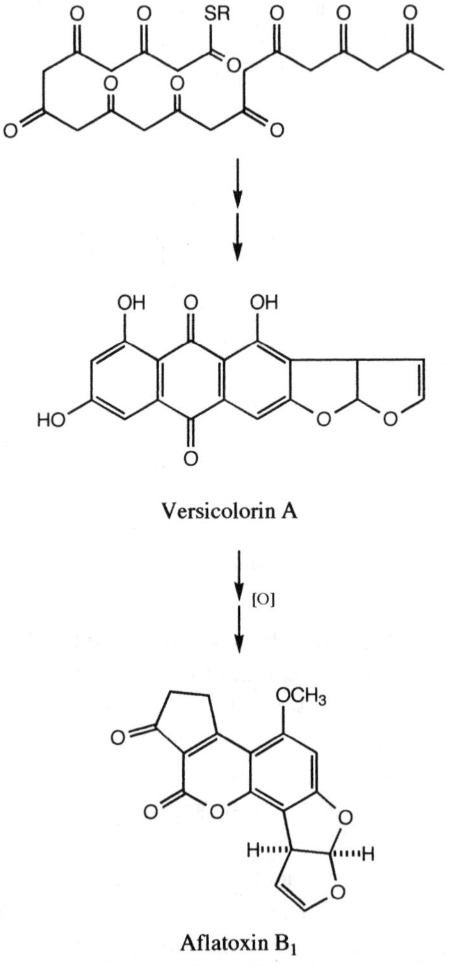

Versicolorin A

Aflatoxin B$_1$

Fig. 18   Conversion of a polyketide into aflatoxins

The carcenogenic aflatoxins (from *Aspergillus flavus*), which also act on the pulmonary and cardiovascular systems, form a group of fungal toxins characterized by two tetrahydrofuran groups. The intermediary anthraquinone nucleus is oxidatively cleaved.[68] The aflatoxins occur in mouldy food stuffs, which afflict poultry farming, Fig. 18.

5.13 Flavonoids

The flavonoids are colouring substances contributing to the beauty and splendour of flowers and fruits in nature. The flavones give yellow or orange col-

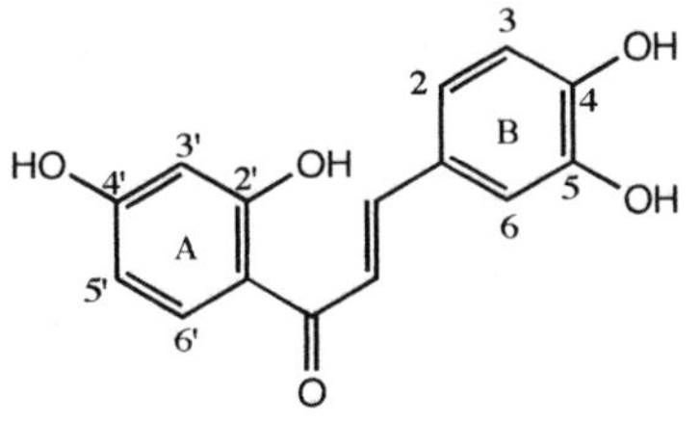

Butein
(Chalcone)

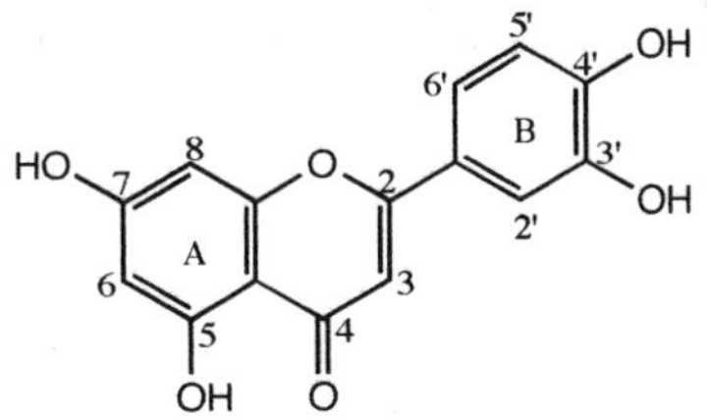

Luteolin
(Flavone)

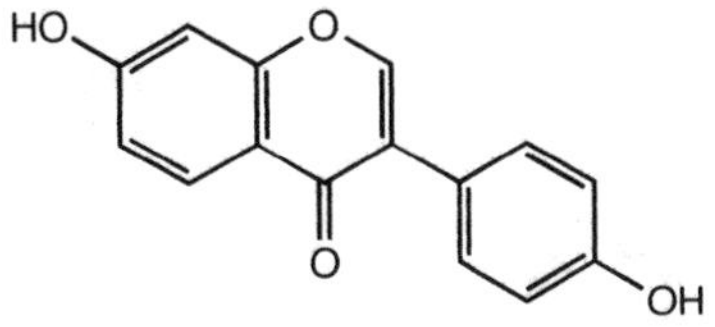

Daidzein
(Isoflavone)

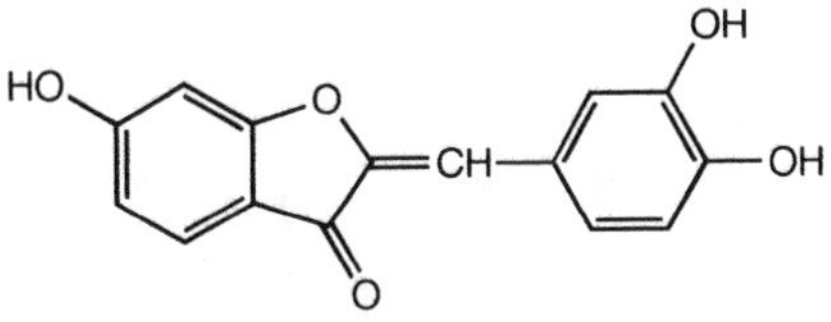

Sulphuretin
(Aurone)

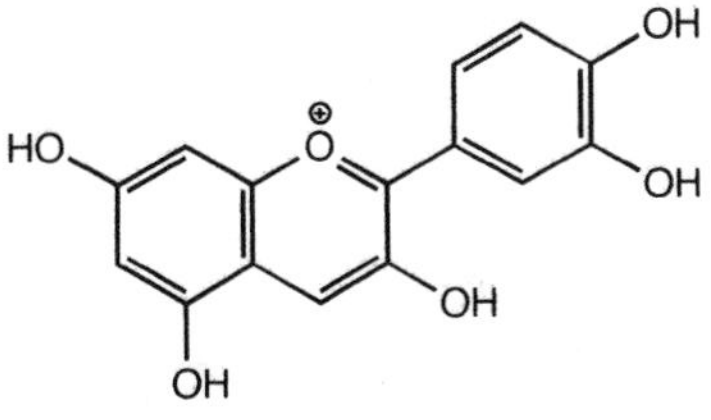

Cyanidin
(Anthocyanidin)

Fig. 19 Structures of flavonoid compounds

ours, the anthocyanins red, violet or blue colours, i.e. all the colours of the rainbow but green. The occurrence of this numerous class of oxygen hetero-cycles is restricted to higher plants and ferns. Mosses contain a few flavonoid types but they are absent in algae, fungi and bacteria. The hydroxylation and methylation patterns appear to be genetically controlled, i.e. the distribution of flavonoids is a useful auxiliary tool for classification purposes.[69] Biolog-ically the flavonoids play a major role in relation to insects pollinating or feed-ing on plants, but some flavonoids have a bitter taste, repelling certain cater-pillars from feeding on leaves, see Chapter 2.

The flavonoids are structurally characterized as having two hydroxylated aromatic rings, A and B, joined by a three carbon fragment. One hydroxyl group is often linked to a sugar. Several substructures can be distinguished; chalcones, flavones, isoflavones, aurones, and anthocyanidins (Fig. 19). With-in each group there are members at various oxidation levels.

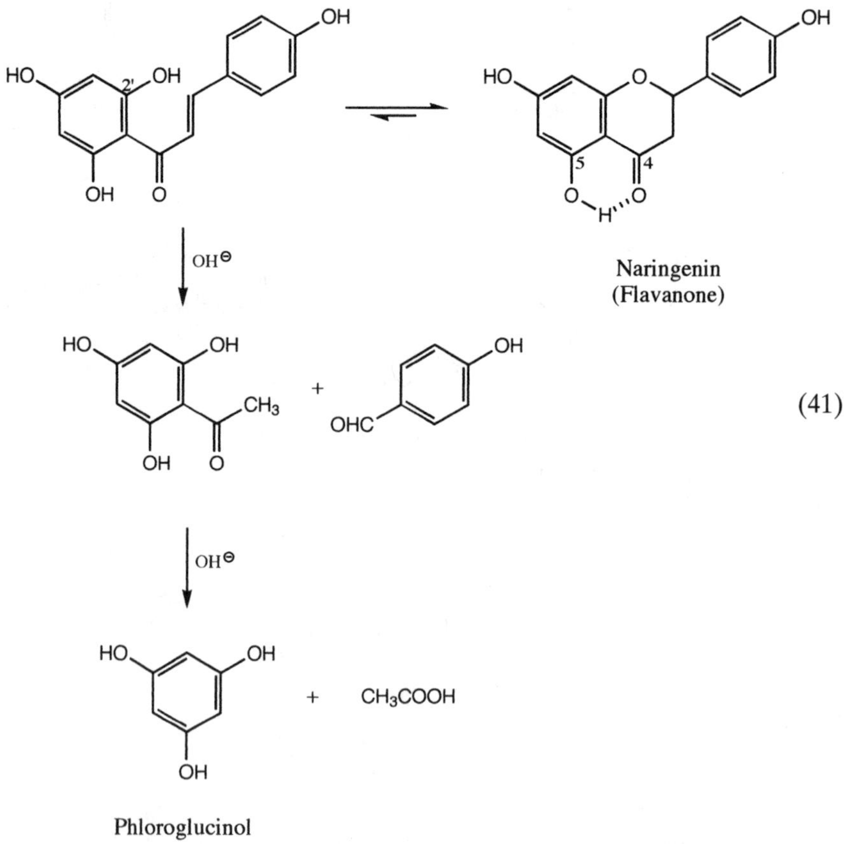

Naringenin
(Flavanone)

(41)

Phloroglucinol

2′-Hydroxy-substituted chalcones cyclize easily to flavanones, the structure of which is stabilized by hydrogen bonding at C^4O and C^5O. The structural determination was first accomplished by alkaline degradation, which gave for exemple, acetic acid, acetophlorophenone, *p*-hydroxybenzaldehyde and phloroglucinol according to a retrocondensation mechanism (41). Anthocyanidins were related to 3-hydroxyflavones by reduction of the carbonyl group followed by treatment with acid (42). The basic skeleton arises from three malonyl CoA units and a cinnamoyl CoA as starter which is incorporated intact. Specifically labelled acetic acid and phenylalanine were incorporated according to Fig. 20.[70] The cyclization of the chalcones is enzymatically controlled since the flavanones are optically active.

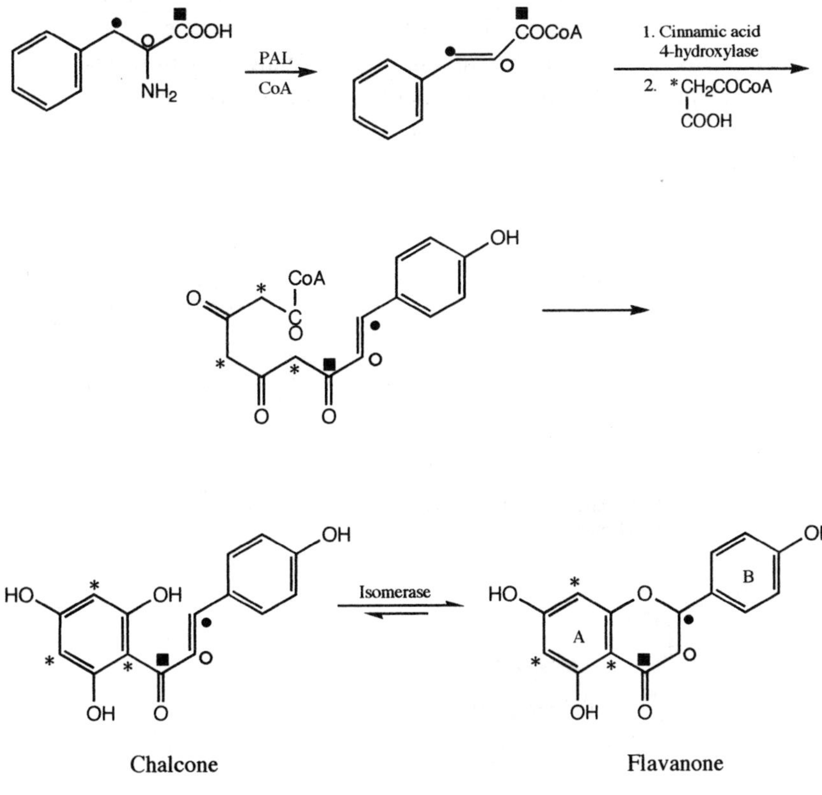

Fig. 20   Biosynthesis of the basic flavone skeleton

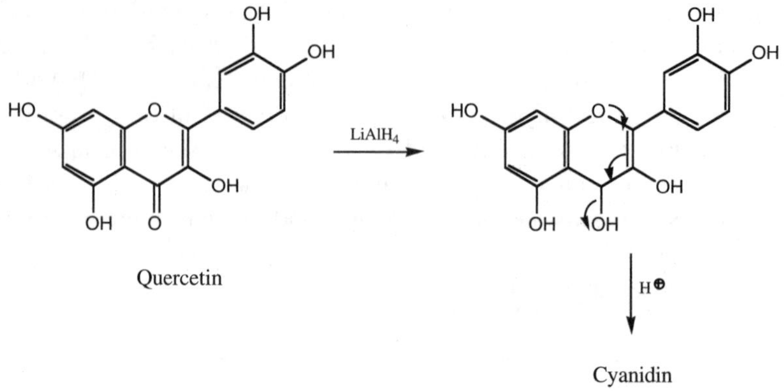

(42)

Quercetin

Cyanidin

Phenylalanine is a good precursor but cinnamic acid and p-hydroxycinnamic acid are better still; in contrast, caffeic acid is poorly incorporated implying that p-hydroxycinnamic acid is situated at a branching point of the biosynthetic pathway.[71] Several plants contain cinnamic acids which are not

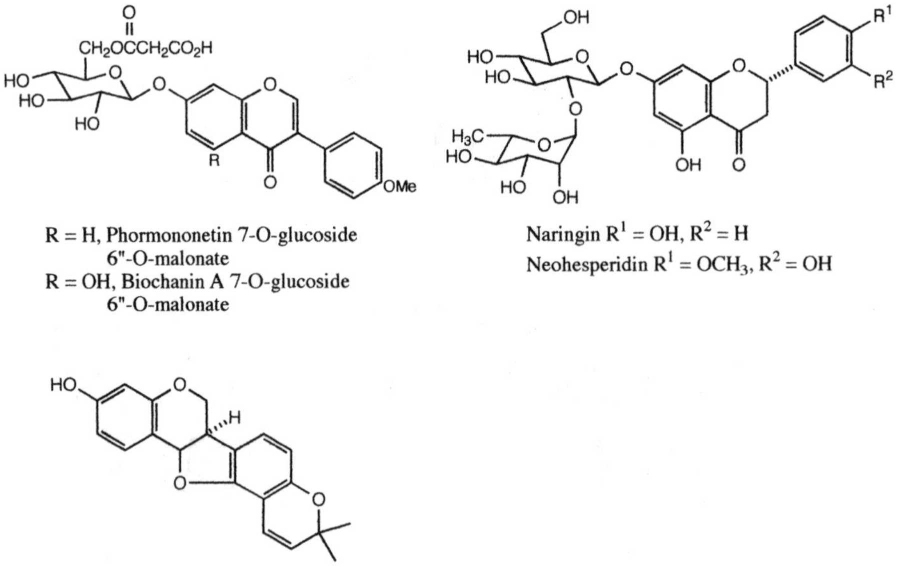

R = H, Phormononetin 7-O-glucoside
   6"-O-malonate
R = OH, Biochanin A 7-O-glucoside
   6"-O-malonate

Naringin R^1 = OH, R^2 = H
Neohesperidin R^1 = OCH$_3$, R^2 = OH

Phaceolin

Fig. 21 Secondary functionalization of flavonoids

represented in the co-occurring flavonoids. The pH of the cell seems to be critical in determining substrate specificity. In *Haplopappus gracilis* for example, *p*-hydroxycinnamic acid is utilized efficiently at pH 8.0 as precursor for eriodictyol, whereas caffeic acid is utilized for the purpose only at pH 6.5–7.0.[72] It is interesting that pH may be a factor determining the metabolic ratio in different plant cells.

O-Methylation, prenylation (Fig. 22), nuclear hydroxylation, sulphonation and glycosidation are modifications taking place at the final stage. Fig. 21 shows a few examples.

The anthocyanidins are biosynthesized from flavanones via dihydroflavonols according to (43).

The flavanones or chalcones are also precursors for the isoflavones.[70] The exact nature of this rearrangement in unknown, but a plausible mechanism is

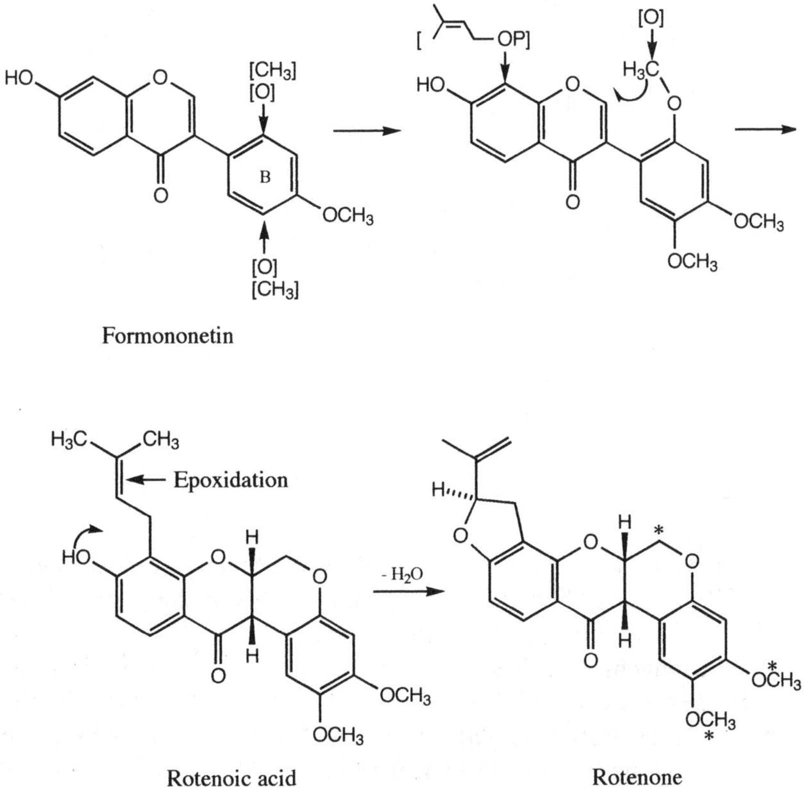

Formononetin

Rotenoic acid   Rotenone

Fig. 22 Biosynthesis of rotenone. The starred carbon atoms originate from methionine

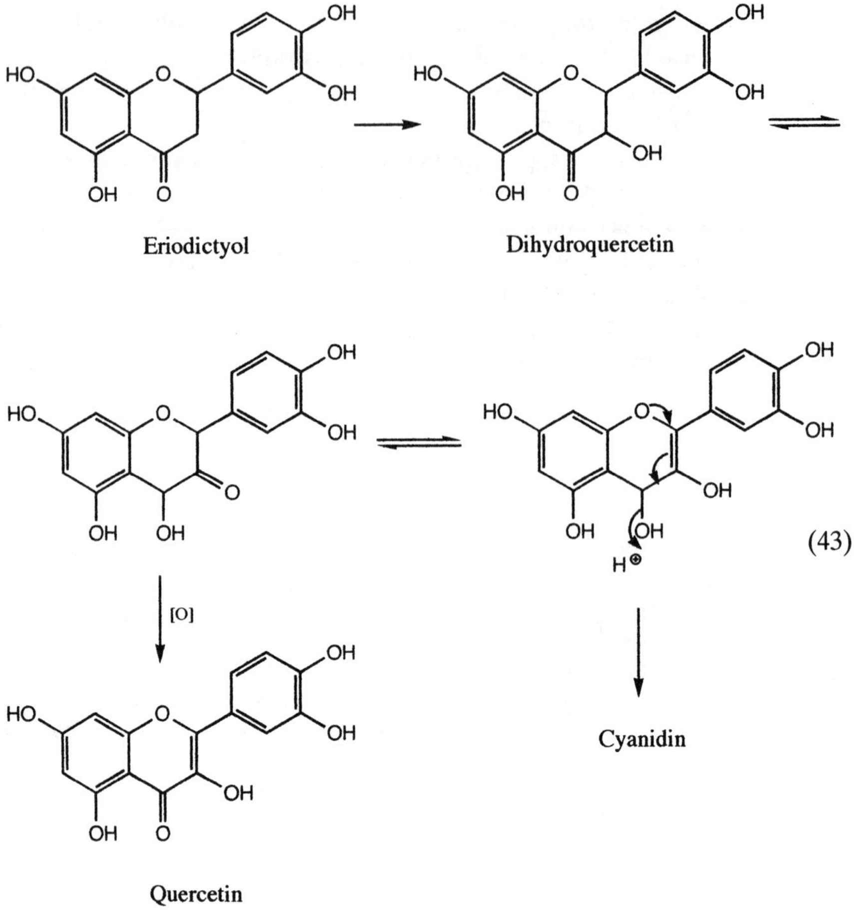

Eriodictyol

Dihydroquercetin

(43)

Cyanidin

[O]

Quercetin

oxidation of the anion to a diradical, combination to the cyclopropanoid intermediate and proton elimination (44). It is unlikely that the oxidation proceeds to a carbonium ion α to the carbonyl group followed by a carbonium ion rearrangement as has been proposed. The rotenoids, which are used as fish poisons, are structurally related to the isoflavones. Tracer experiments in seedlings of *Amorpha fruticosa* showed that rotenone is biosynthesized from formononetin by extra hydroxylation and methylation in ring B, hydrogen abstraction from the methyl (section 4.3), is followed by radical cyclization and isoprenylation to rotenoic acid. Radical Michael type addition is a favoured and common reaction. Epoxidation of the isoprenoid double bond of

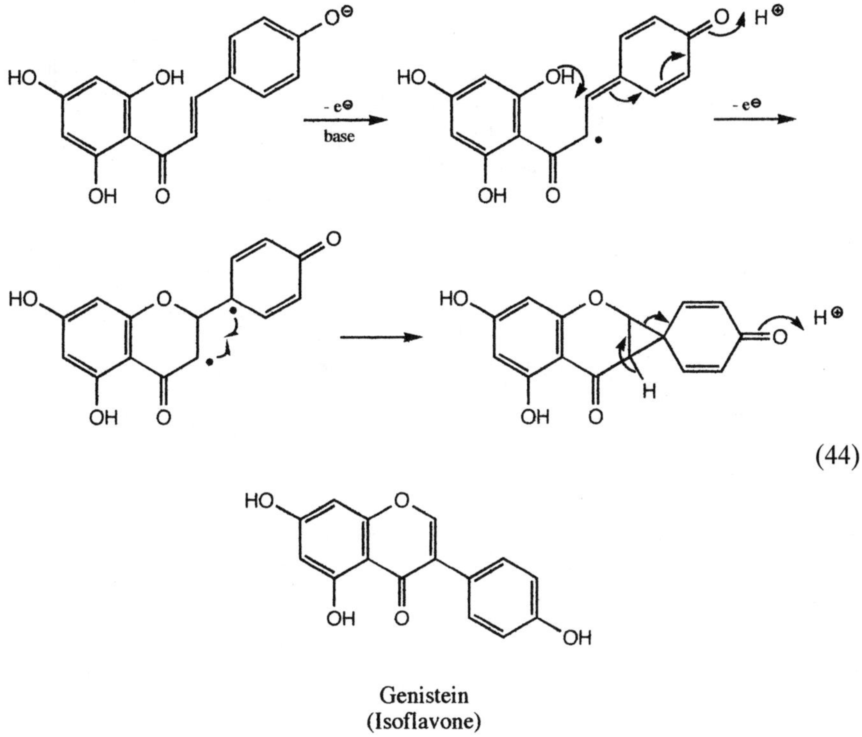

(44)

Genistein
(Isoflavone)

rotenoic acid, cyclization via opening of the epoxide, and finally dehydration of the tertiary alcohol, actually discovered as a natural constituent, give rotenone (Fig. 22).[73]

Dalrubone, isolated from *Dalea emoryi*, has an unusual oxygenation pattern in that rings A and B appear to be reversed.[74] A biomimetic synthesis (45) from co-occurring coumarin and phloroglucinol suggests that this flavonoid may arise via a different pathway.[75]

However, the reversal can be explained by a 1,3-carbonyl transposition in the chalcone in the normal biosynthesis, effected by β-oxidation, cyclization reduction and elimination of water (Fig. 23). This pathway is supported by feeding experiments in *Glycyrrhiza echinata* which produces transposed chalcones.[76]

The flavonoids are degraded by several routes, Fig. 24.[77] The first step is an elimination of the sugar moiety by a glycosidase. Ring A is usually broken down to carbon dioxide and ring B gives rise to benzoic acids.[78] The cleavage of the pyrone ring is effected by a monooxygenase at D^{4a} (route a). Oxidation at $C^{1'}$ gives rise to chromones[79] and there is indication that certain chromones are post mortem products[80] (route b).

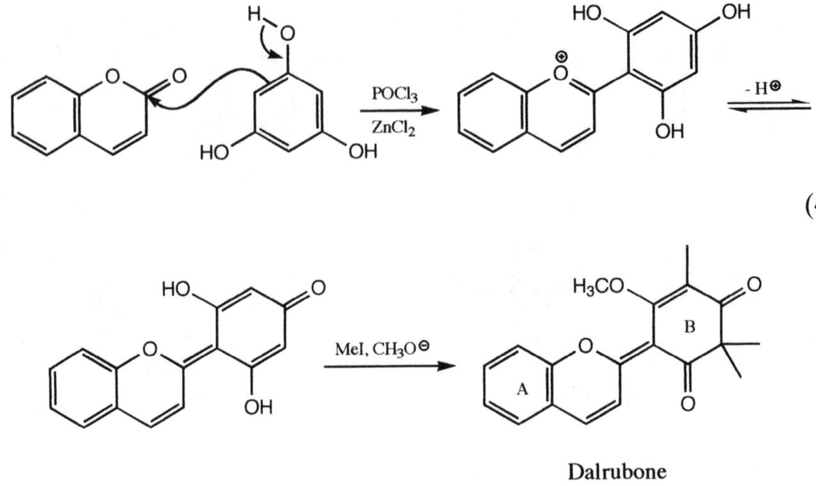

(45)

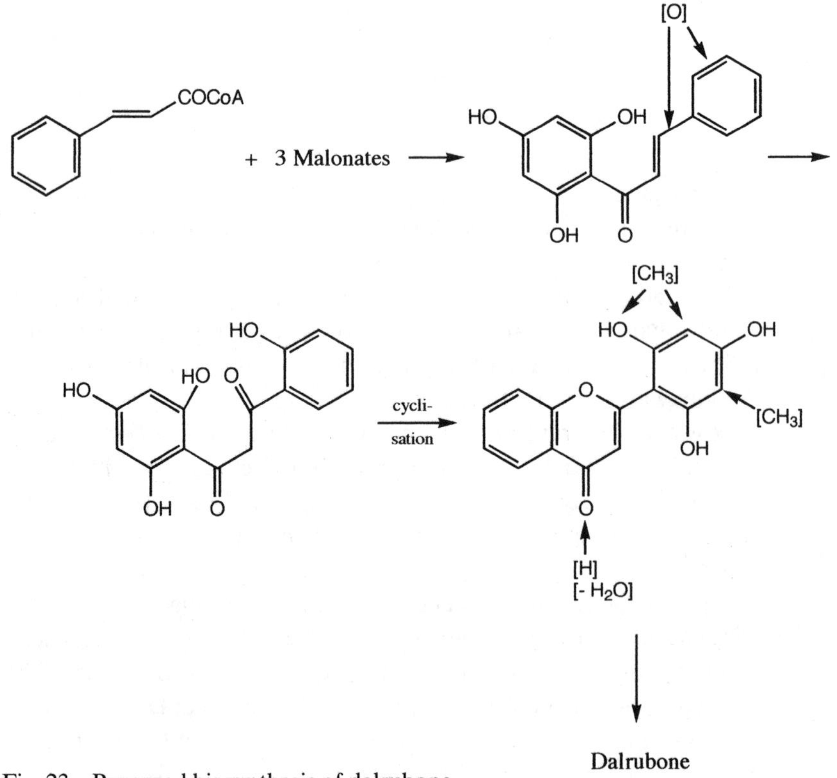

Dalrubone

Dalrubone

Fig. 23   Proposed biosynthesis of dalrubone

These findings illuminate the long standing problem concerning the true origins of a metabolite. Is it produced by a direct biosynthetic route or is it simply a secondary degradation product—an artefact? In this case it seems unnecessary to speculate on alternative routes to chromones missing a substituent at C-2.

The flavonoids are thus produced by folding the polyketide as shown in Fig. 20. A slightly different folding gives stilbenes. Feeding experiments with *Pinus resinosa* have shown that pinosylvin originates from the same C6–C3–C6 unit by cyclization according to Fig. 25.[81] Most stilbenes have lost the carboxyl group, although it remains in hydrangenol which like the antimicrobial plant constituent resveratrol derives from *p*-hydroxycinnamic acid and three malonate units.[82,83] In many plants stilbenes are formed in response to microbial attack or stress, i.e. as a phytoalexin, Chapter 2.

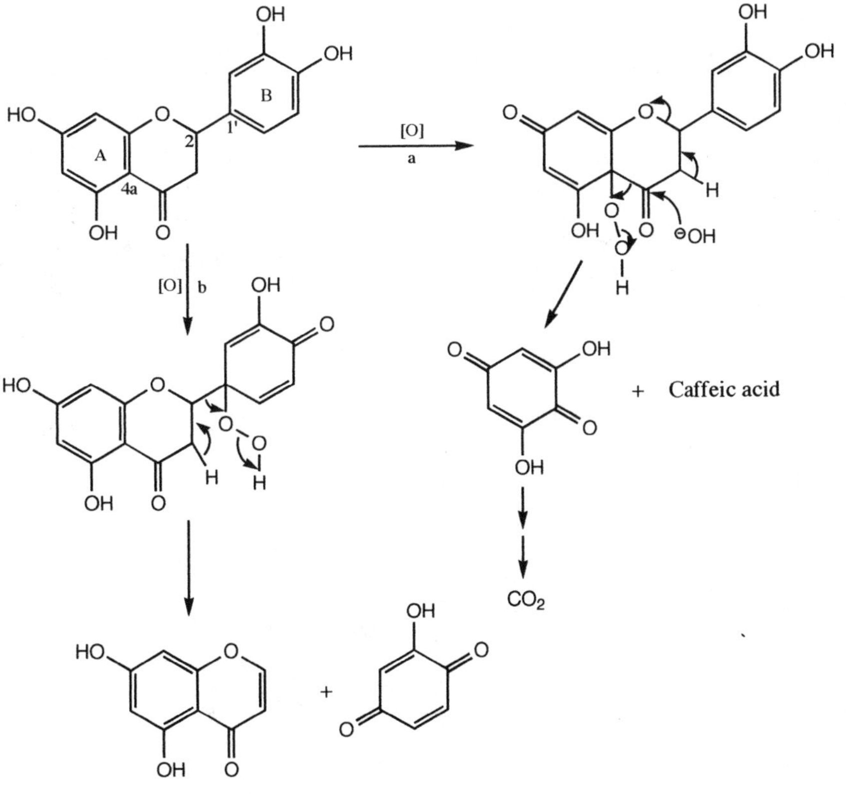

5,7-Dihydroxychromone

Fig. 24   Flavonoid metabolism

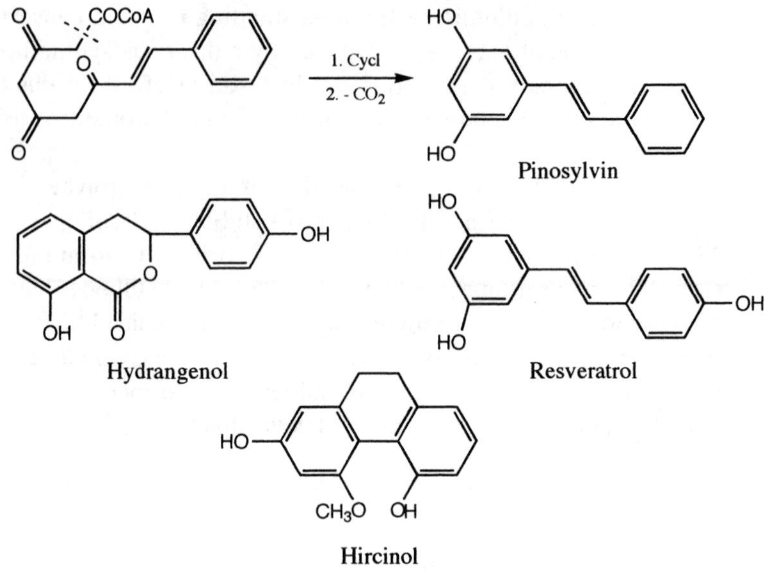
COCoA
1. Cycl
2. - CO2
Pinosylvin
Hydrangenol
Resveratrol
Hircinol

Fig. 25 Naturally occurring stilbene and phenanthrene derivatives

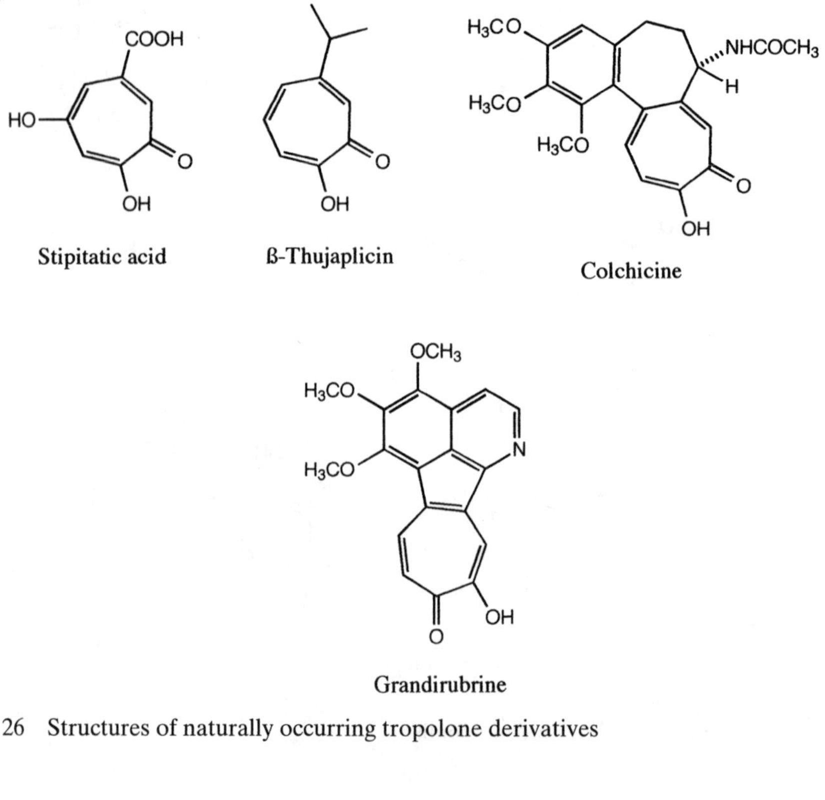
Stipitatic acid
ß-Thujaplicin
Colchicine
NHCOCH3
Grandirubrine

Fig. 26 Structures of naturally occurring tropolone derivatives

The requirements for formation of phenanthrenes are that hydroxy groups are located in the *meta*-position and that a *cis*-stilbene structure is formed or alternatively reduction of the central double bond has occurred, as in hirsinol. The two aromatic nuclei are then coupled by phenolic oxidation.

5.14 Tropolones

The non-benzenoid aromatic structure for tropolones was first put forward by Dewar in 1945 to account for the properties of stipitatic acid produced by *Penicillium stipitatum*. A number of tropolone derivatives have since been isolated from other moulds, from *Cupressaceae*, e.g. thujaplicins, from *Colchicum* spp. the antimitotic colchicine and from *Abuta grandiflora* the unusual tropoloquinoline alkaloid grandirubrine, Fig. 26.

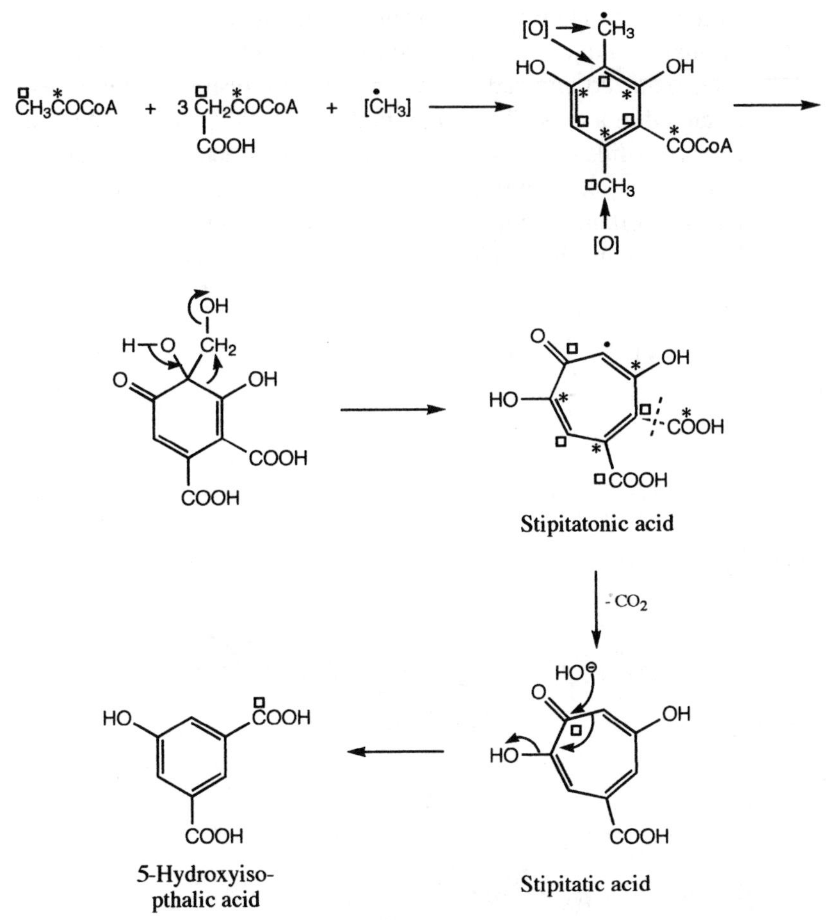

Fig. 27   Biosynthesis of tropolones

The principal biosynthetic path for stipitatic acid starts from acetate plus a one carbon unit.[84] Orsellinic acid is methylated and oxidized by a monooxygenase[85] to a cyclohexadienone. The mechanism of the ring enlargement is not fully understood but can be envisaged as a 1,2 shift facilitated by the electron donating hydroxyl group (Fig. 27). Benzilic acid rearrangement of stipitatic acid in alkaline medium gives 5-hydroxyisophthalic acid, and only the marked carbon is extruded. The mechanism of the ring enlargement in colchicine is discussed in section 8.3.

5.15 Oxidative coupling of phenols

Several groups of enzymes are capable of catalysing oxidative phenolic coupling, all of which are widespread in the plant and animal kingdoms.[86] They have iron or copper as a prosthetic group and are all able to effect one-electron transfer. Hydrogen peroxide and molecular oxygen, used as oxidants, are ultimately reduced to water, whilst the transition metal catalysts shift between their oxidized and reduced forms. Horseradish peroxidase, which has been obtained in a crystalline state, mol. wt 40 000, is non-specific in its action. It can use iodide, amines, indoles, and ascorbic acid as substrates. In spite of intense research activity and the fundamental importance of phenol oxidases, e.g. for formation of lignin, tannins, alkaloids and a wealth of microbial metabolites, the mechanism of action is not fully understood.

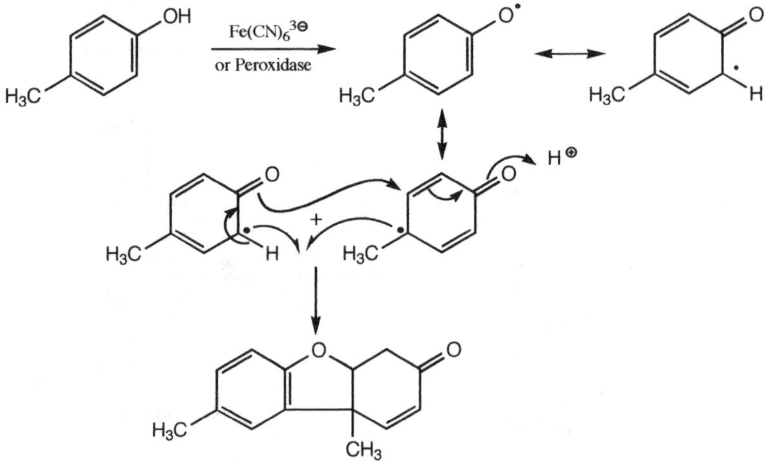

(46)

Pummerer's ketone

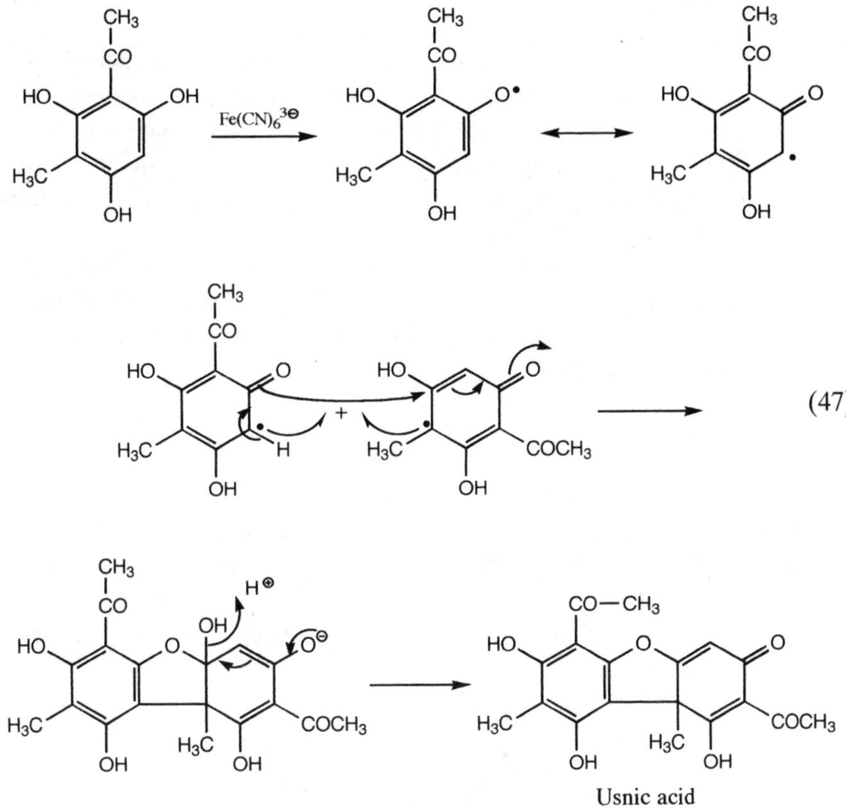

(47)

Usnic acid

As for the phenol part, the enzyme removes one electron and the phenoxy radical formed can couple in a number of ways (see section 4.7). The reaction can be simulated *in vitro* by oxidation with ferricyanide as in (46) showing the oxidation of *p*-cresol to the so-called Pummerer's ketone. When this ketone was prepared from *p*-cresol or dehydrogriseofulvin was prepared from griseophenone (Fig. 12), racemic products were obtained by the action of cell-free phenol oxidase preparations. The latter reaction is relevant to the biogenetic scheme proposed for griseofulvin. The intact organism *Penicillium griseofulvum* produces optically active griseofulvin. This suggests that the enzyme acts solely as producer of phenoxy radicals which then can couple uninfluenced by the topology of the enzyme. In the intact cell, on the other hand, the phenol oxidase must be integrated into a larger enzyme complex also catalysing the condensation and cyclization of the polyketide.[87]

It has been argued that the dimerization is not the result of a coupling of two radicals, but rather of an attack of one radical on another phenol. The new radical is then oxidized. This seems not to be the case because oxidation of a phenol in the presence of a large excess of 1,2-dimethoxybenzene does not give rise to any cross-products.

The principle of oxidative phenol coupling is convincingly demonstrated in the synthesis of usnic acid from two moles of methylphloracetophenone (47).[88] Incorporation experiments with labelled precursors later confirmed that the biosynthetic machinery followed a similar path.[89] The introduction of *C*-methyl is an early step because phloroacetophenone is not utilized as substrate in the biosynthesis.

Picrolichenic acid from the lichen *Pertusaria amara* is constructed in the same way by oxidation of two moles of 5 2,4-dihydroxy-6-pentyl-benzoic acid (48).[90] We have encountered this reaction earlier in some tannins, e.g. in ellagic acid (section 4.4).

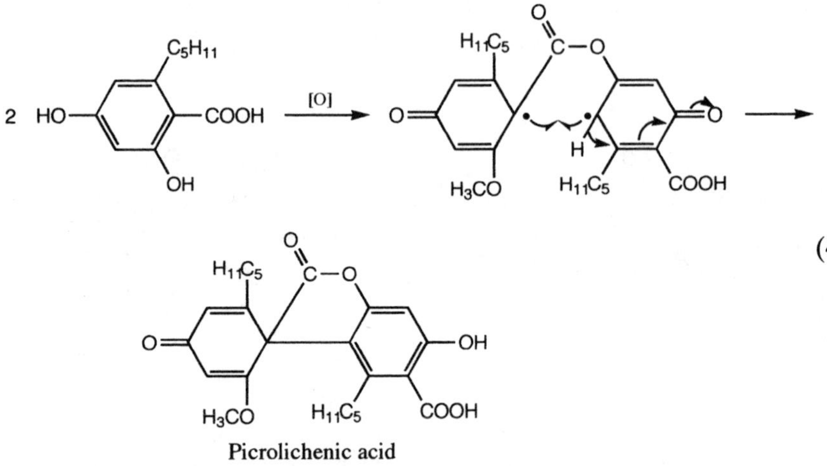

(48)

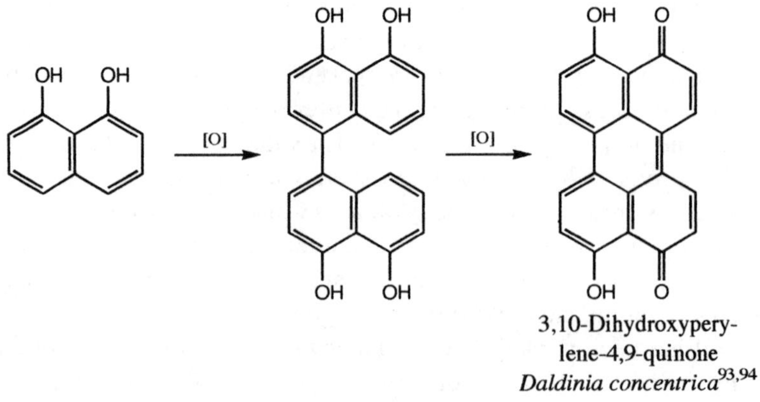

Picrolichenic acid

(49)

3,10-Dihydroxypery-
lene-4,9-quinone
Daldinia concentrica[93,94]

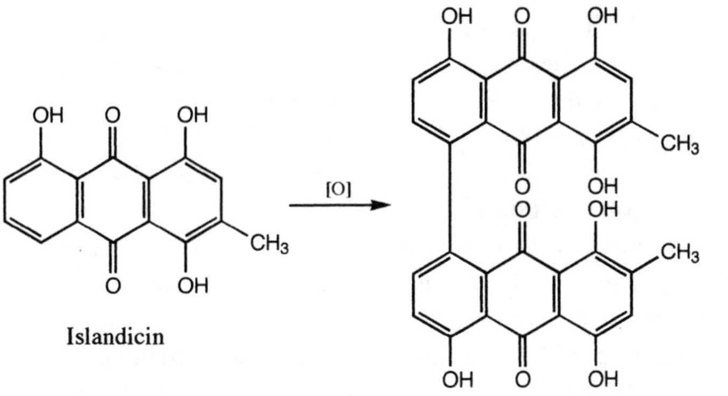

Islandicin

[O] →

Iridoskyrin
Penicillium islandicum[95,96]

(50)

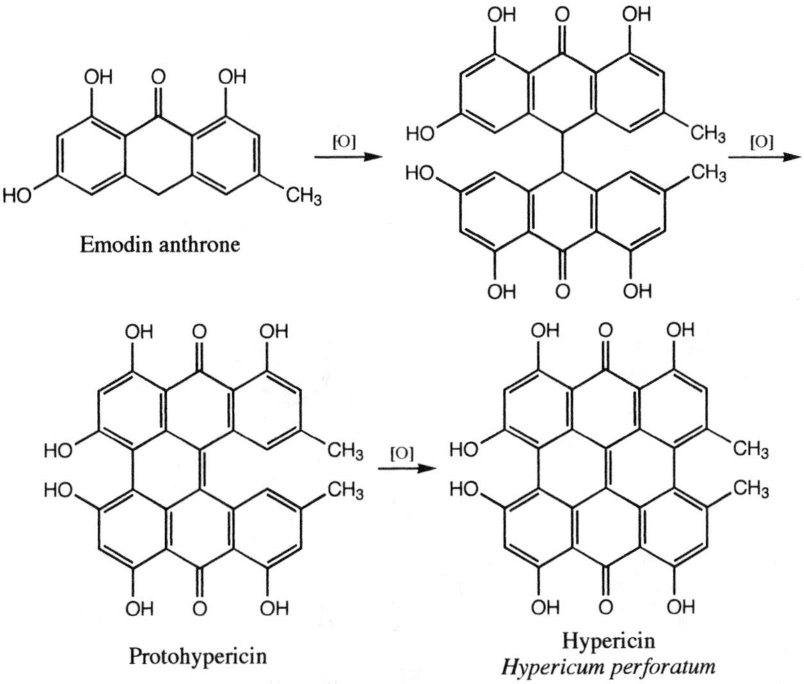

Emodin anthrone

[O] →

[O] →

(51)

Protohypericin

[O] →

Hypericin
Hypericum perforatum

A number of dimeric naphthalenes and anthracenes occur in nature which evidently arise via phenol oxidation (49–52). Insects of the Aphididae family, many of which are serious pests on cultivated crops, produce a series of pigments[91] having the dihydroxyperylenequinone nucleus. Hypericin[92] is a photodynamic pigment produced by *Hypericum perforatum*. It originates from emodin anthrone which is dimerized and further oxidized to photohypericin and finally hypericin. This process can be mimicked by passing air into an alkaline solution of emodin anthrone.

Xanthones[97] are widely distributed among several families, e.g. *Gentianaceae, Moraceae, Polygalaceae*, and are of chemotaxonomic interest. Ring A characteristically contains hydroxyl groups at C^5 and/or C^7 and originates from shikimic acid.[98] A hydroxyl group is first introduced into the *meta* position in the benzoic acid ring. Phenol oxidation and cyclization then automatically locate the OH group at C^5 or C^7 (53). It is also demonstrated that the polyhydroxybenzophenone is incorporated and co-occurrence is known in a

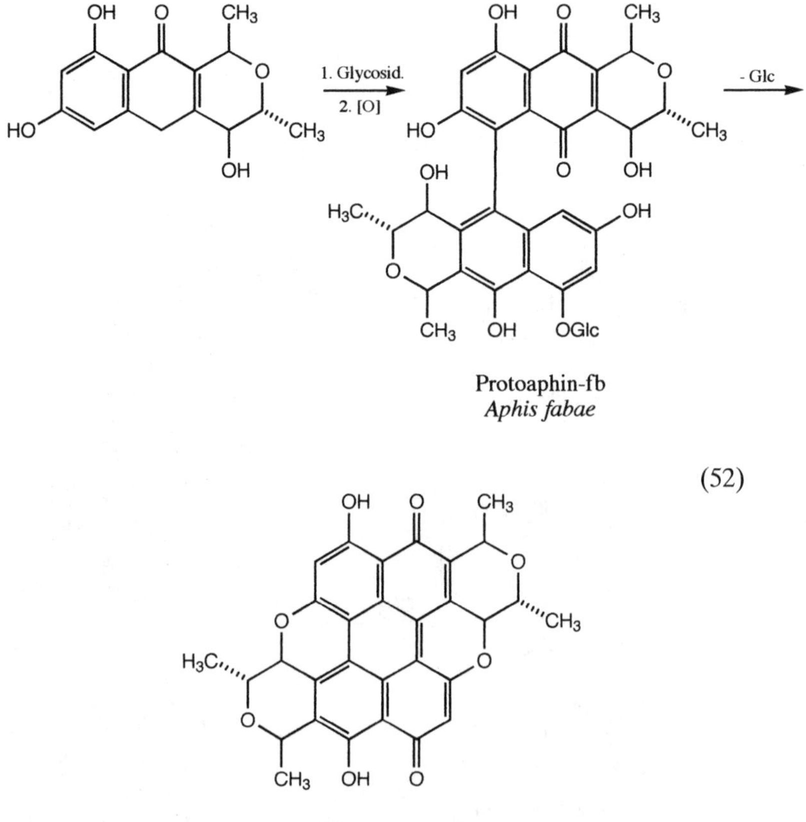

Protoaphin-fb
Aphis fabae

(52)

Erythroaphin-fb

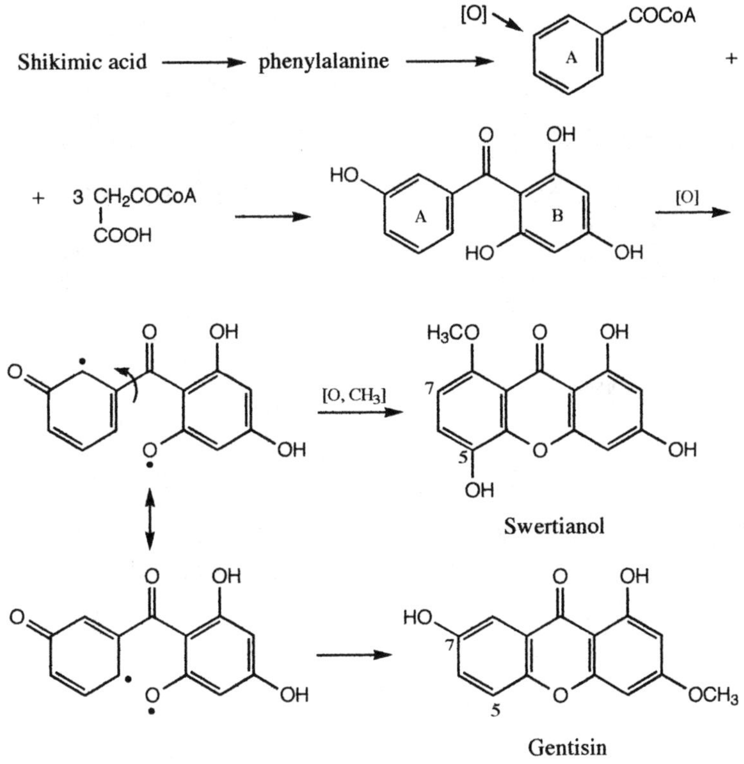

(53)

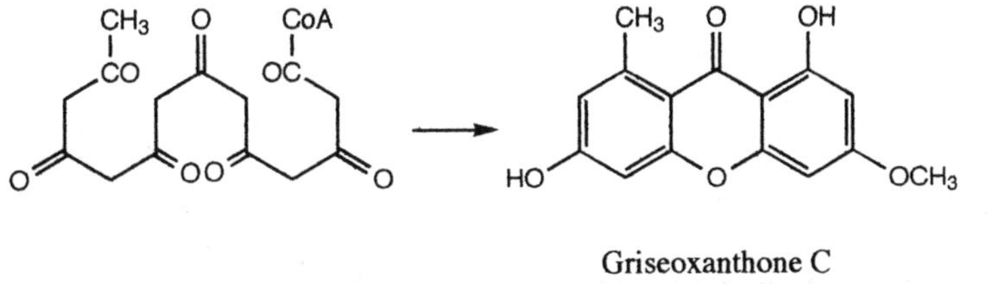

Swertianol

Gentisin

(54)

Griseoxanthone C

few cases. The *meta*-oxygenation pattern of ring B indicates its acetate origin and it has been shown by using radiolabelled acetate and ^{13}C labelled malonate that acetate solely is incorporated in ring B.[99] In some cases the whole framework is formed from one polyketide, e.g. in griseoxanthone C which is of fungal origin.[100] The oxygenation pattern is also different and no phenolic oxidation is called for in this case (54).

A third route to xanthones involves anthraquinones, which are oxidatively cleaved and subsequently cyclized. The incorporation of chrysophanol is demonstrated by using the CD$_3$ labelled compound. Oxidative cleavage, prenylation and epoxidation complete the tajixanthone biosynthesis (55).[101]

Chrysophanol   (55)

Tajixanthone

Thyroxine is an iodine-containing thyroid hormone which controls the oxygen consumption in tissues. Too high a production of thyroxine causes an increase of the metabolic rate and is recognized as Basedow's disease.

Thyroxine deficiency causes myxoedema. It is formed by electrophilic iodination of tyrosine which then couples with another iodinated tyrosine molecule with loss of a side chain, presumably in a pyridoxal phosphate mediated reaction (56).

Peroxidases are responsible for the darkening of cut fruits or vegetables, whereby catechols are rapidly oxidized forming darkly coloured polymers. Darkening of the skin by the action of sunshine is caused by formation of melanin, a polymer built up by enzymatic oxidation of tyrosine, predominantly through 3,7-linkage of the intermediate indole derivative (57).

(56)

Thyroxine

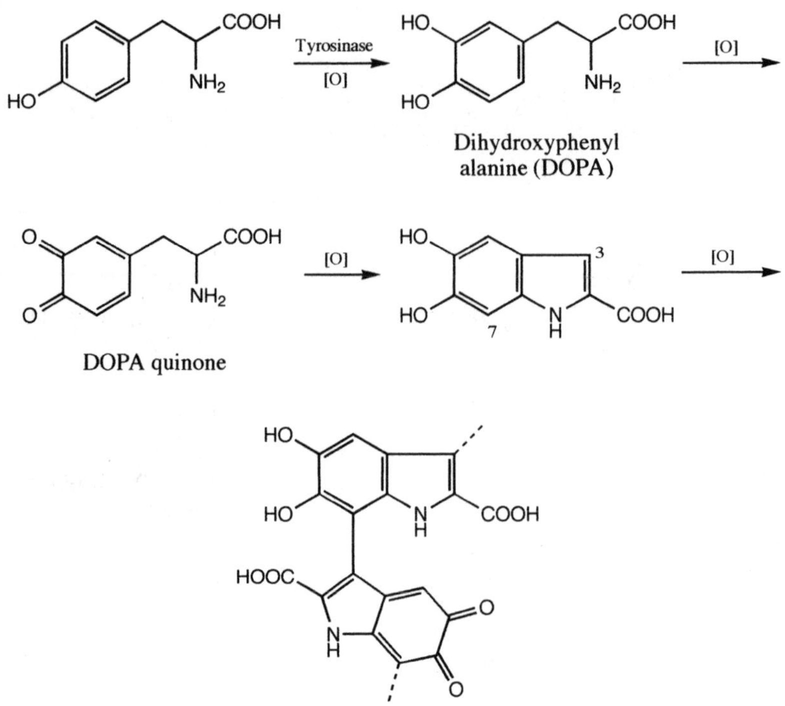

Dihydroxyphenyl
alanine (DOPA)

DOPA quinone

(57)

Melanin 3,7-polymer

5.16 Halogen compounds

The thyroid hormone thyroxine contains iodine and that represents one of the very few instances where iodine is found in terrestrial naturally occurring compounds. Terrestrial fluorine and bromine compounds are also extremely rare. The toxic principle in *Gastrolobium grandiflorum* and *Dichapetalum cymosum*, causing losses in livestock in Queensland and South Africa, was identified by Marais as fluoroacetic acid, and the toxin of the shrub *Dichapetalum toxicarium*, used as an arrow poison in Sierra Leone, was identified as ω-fluorooleic acid.[102] Fluoroacetate is metabolized in the citric acid cycle to fluorocitric acid which is an inhibitor of aconitase. Thus the organism commits suicide by carrying out lethal syntheses.

Fluorocitric acid is a naturally occurring compound found in *Acacia georginae* and also in tea, fortunately at well below toxic level. Fluorine is not incorporated in organic compounds by a fluoroperoxidase system purely for thermodynamic reasons. Nucleophilic substitution or addition of fluoride ion

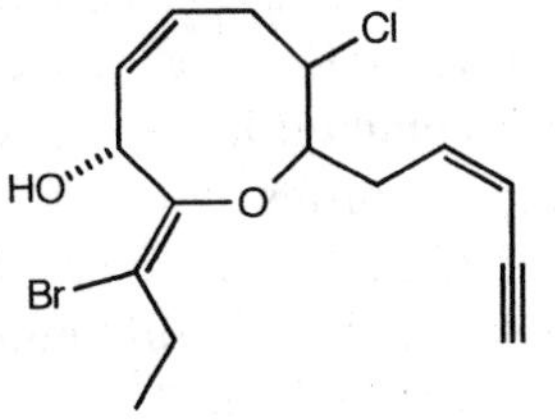

Chondriol
Laurencia yamada
(red alga)

$FCH_2(CH_2)_7CH= CH(CH_2)_7COOH$

ω-Fluorooleic acid
*Dichapetalum
toxicarium*

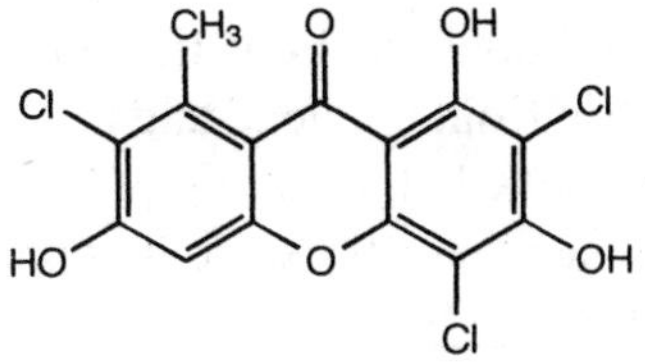

Xanthone from
Lecanora spp.

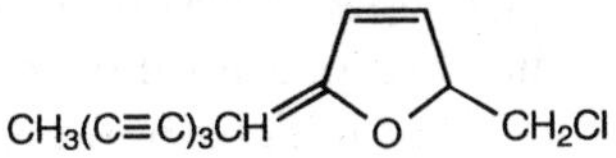

$CH_3(C≡C)_3CH$

Polyyne from
Gnaphalium sp.

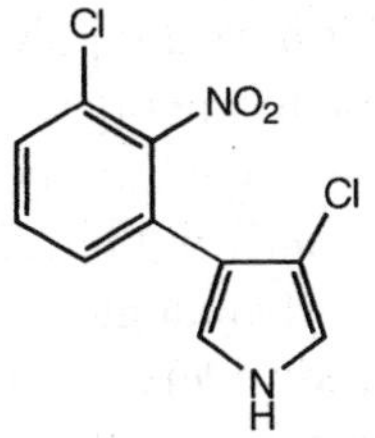

Pyrrolnitrin
Pseudomonas pyrrocinia

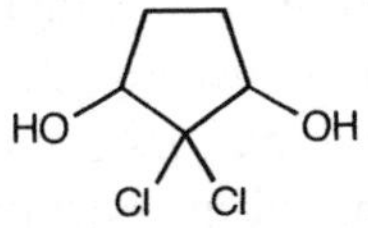

Caldariomycin
*Caldariomyces
fumago*

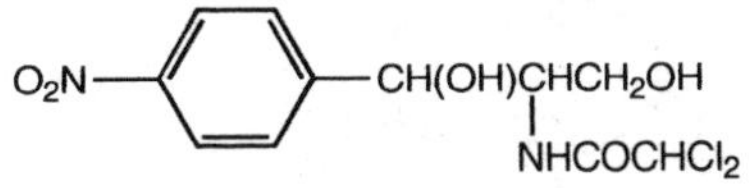

O_2N — CH(OH)CHCH$_2$OH
NHCOCHCl$_2$

Chloromycetin
Streptomyces venezuelae

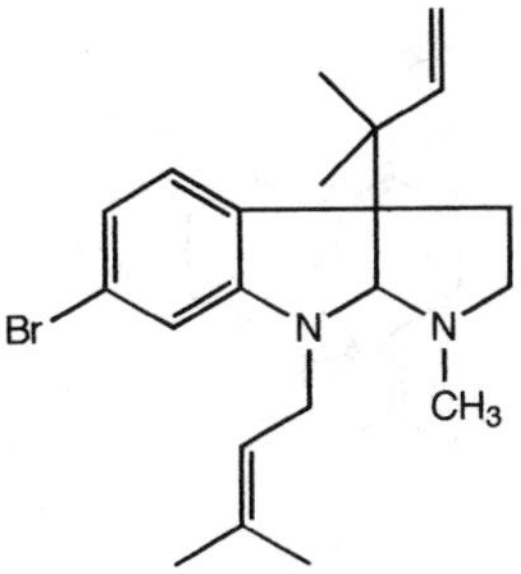

Flustramine A
Flustra foliaceae
(bryozoan)

Fig. 28   Structures of some naturally occurring halogen compounds of polyketide origin to double bonds initiated by another haloperoxidase is more likely. The ω–F-fatty acids are formed by involving fluoroacetyl CoA as starter.[102]

The chloro compounds are the most common halogenated organic compounds biosynthesized chiefly by microorganisms and various marine organisms. More than 1000 naturally occurring halogen compounds have been identified to date.[103] It has been estimated that *ca.* 5×10^6 tons of methyl chloride is produced per year in the marine environment. This is not pollution, it is production of natural products.

Both non-haem- and haem-Va and -Fe complexes catalyse the reaction of hydrogen peroxide with I⁻, Br⁻ and Cl⁻ to free halogens which, attached to the enzyme, could represent the active species.[104] The true nature of this species is a matter of controversy. In model reactions the chloroperoxidase-Cl⁻ couple forms ethylene chlorohydrin from ethylene and a 1:1 mixture of 2- and 4-chloroanisol from anisol. Enzymatic halogenation of sensitive substrates is of interest in preparative organic chemistry since it could lead to products not accessible by conventional methods.[105]

As a rule halogenation has a potentiating effect on the biological and pharmaceutical activities of the compound. The antibacterial effect of chloromycetin is 50–100 times stronger than that of the acetyl analogue. It has also been noted that many organohalogen compounds have a repulsive effect on predators or invaders, thereby offering protection to the organism producing these compounds. Fig. 28 gives further examples of naturally occurring halogen compounds. By replacing chloride with bromide in the culture medium several organisms start producing the brominated analogues, e.g. griseofulvin. In a strict sense these metabolites may not be considered natural but it is not unreasonable to argue that they do occur in nature, yet in quantities which have escaped detection.

(58)

Nidificene
Laurencia nidifica

Research on marine natural products has uncovered a great number of chlorine, bromine, and iodine containing compounds,[106–108] many of which have unusual structures, and it is quite clear that the metabolism of marine organisms has been adapted to the unique saline aqueous environment.

The genus *Laurencia* of *Rhodophyta* proved to be a rich source of structurally interesting compounds, mostly of terpenoid nature, but several C_{15} acetylenic cyclic ethers containing halogen of polyketide origin have been isolated as well. From the intestinal tract of certain gastropods, sea hares, several algal metabolites have been isolated which gave information about the dietary habits of these animals. The red alga, *Asparagopsis taxiformis*, contains an array of simple volatile polyhalogenated hydrocarbons, such as $CHBr_3$, CBr_4, $CHBrClI$, $BrCH_2COCH_2I$, $Br_2CHCH(OAc)CHI_2$, $ClCH_2COOH$, $BrCH=CBrCOCH_2Br$, $Br_2C=CH\ COOH$, etc. They were identified by GC—MS analyses and comparison with authentic samples.

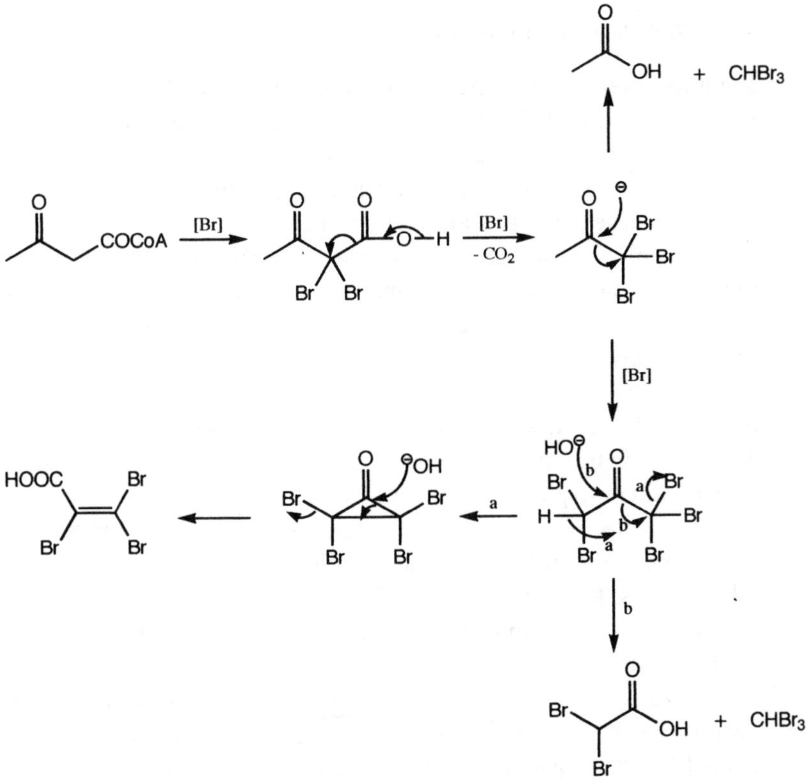

Fig. 29 Proposed biosynthesis of bromoform and halogenated acetic and acrylic acids

The relative concentration of bromine and iodine in sea water is very low in comparison to that of chlorine and contrasts sharply to the frequent occurrence especially of bromine-containing metabolites. This is explained by the capability of seaweeds to accumulate bromine and iodine and the low oxidation potential of these halogens. The halogenated acetic acids arise by halogenation of malonic acid or, together with haloforms, by the classical haloform fission of perhalogenated acetone derivatives which are formed by halogenation and decarboxylation of acetoacetic acid. The halogenated acrylic acid could be formed by a Favorsky rearrangement of halogenated acetone (Fig. 29). A route to the butenones must be still more speculative. Several chlorine-containing acetylenes of lipid origin are known in the *Compositae* family. Epoxides are present as cometabolites and the halides are formed by nucleophilic opening of the epoxide ring with HCl as in a polyyne from a *Gnaphalium* sp. (Fig. 28).[35]

Bromonium ions or their equivalents initiate cyclizations as in the biosynthesis of nidificene from farnesyl phosphate (58).

5.17 Modification of the carbon skeleton

We have already encountered the insertion of a carbon atom by methionine into the benzenoid ring to form tropolones, oxidative cleavages of the chain as in chartreusin aglycone, Wagner-Meerwein rearrangements and 1,2 shifts as in the flavone–isoflavone rearrangement or the homogentisic acid rearrangement. These processes are comparably simple to survey but rather drastic rearrangements also occur which have challenged the ingenuity of chemists. Oxidative ring cleavage of aromatics often leads to secondary products of unusual structure, the genesis of which is hard to visualize. Patulin, a metabolite of *Penicillium patulum*, was known to originate from 6-methylsalicylic acid. *m*-Cresol, *m*-hydroxybenzyl alcohol, gentisylalcohol, and gentisaldehyde were identified as cometabolites and incorporate well into patulin.[109,110] The route via toluquinol appears to be a side reaction. When $2,4,6$-$^2H_3$-*m*-cresol was added to glucose-deficient culture medium up to 57 per cent incorporation was observed.[111] The fragmentation pattern of patulin in the mass spectrometer is well understood and therefore it was possible to locate the position of the isotopes in the molecule. Fig. 30 depicts the biosynthetic pathway.

Penicillic acid produced by *Penicillium cyclopium* or *P. baarnense* derives from orsellinic acid via oxidative ring fission. 1-^{14}C-acetic acid was incorporated according to Fig. 31. If 1-^{14}C-malonic acid was administered, the methyl group stayed unlabelled. The ring fission, suggested to occur at $C^{4,5}$ was confirmed by ^{13}C NMR double labelling technique using 90 per cent enriched $1,2$-$^{13}C_2$ acetate.[112,113] Two pairs of $^1J_{13_C-13_C}$ couplings, between $C^{2,3}$ ($J = 77$ Hz, two sp^2 carbons, originally $C^{3,2}$) and $C^{5,7}$ ($J = 44$ Hz one sp^3 and one sp^2 carbon, originally $C^{6,7}$) were observed. All the ^{13}C shifts were asssigned by off-re-

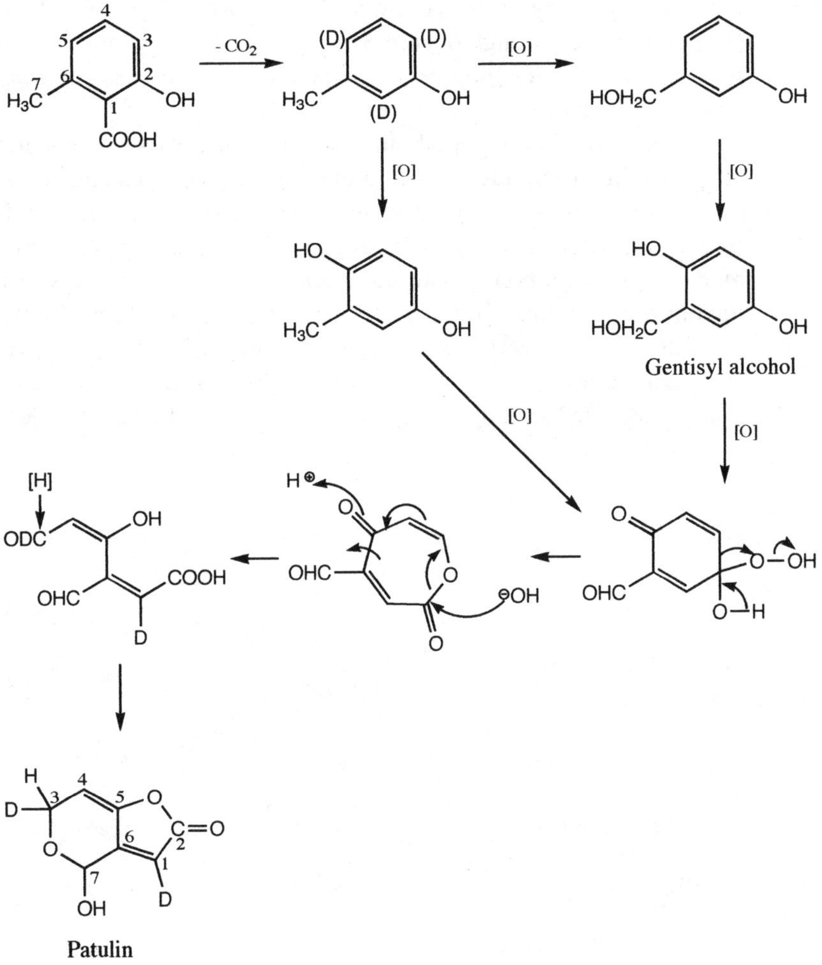

Fig. 30   Biosynthesis of patulin

sonance decoupling and by comparison with model compounds. 2,5-Di-hydroxy-3-methoxytoluene has been isolated from cultures of *Penicillium baarnense* supporting the biosynthetic scheme.[114]

The biosynthesis of penicillic acid has also been studied by ^{3}H NMR providing further details of the enzymatic processes (Fig. 32).[115] ^{3}H-Acetate labels orsellinic acid at C^{3,5,7} and penicillic acid is therefore expected to be labelled at C^{2,6,7}, which proved to be correct. C^3–^{3}H and C^5–^{3}H appear as singlets and C^7–^{3}H as a triplet, coupling geminally with two ^{1}H (orsellinic acid numbering). The very different intensities of the vinylic C^5-^3H absorptions ($C^6-^3H_2$) indicate a high degree of stereospecificity for the formation of this function. The C^7-^3H showed the highest relative intensity in accord with C^7 being the starter end of the tetraketide.

Cyclopentanoids are metabolites of varied biogenesis. The structural unit is contained in terpenoids (see Chapter 6), lipids as in prostaglandins and in ring contracted aromatics. One of several routes from aromatics can be rationalized as a benzilic acid type rearrangement of α-hydroxyketones or of intermediary *ortho*-quinones. The humulones, bitter constituents of hop, giving flavour and aroma to beer, are formed by isoprenylation of phloroglucinol. They are easily rearranged to cyclopentadiones (59). Cryptosporiopsin is derived from a tetraketide, chlorinated, oxidized, and rearranged according to (60).[116] Isotopic labelling has shown that C^6 and not C^1 is extruded.

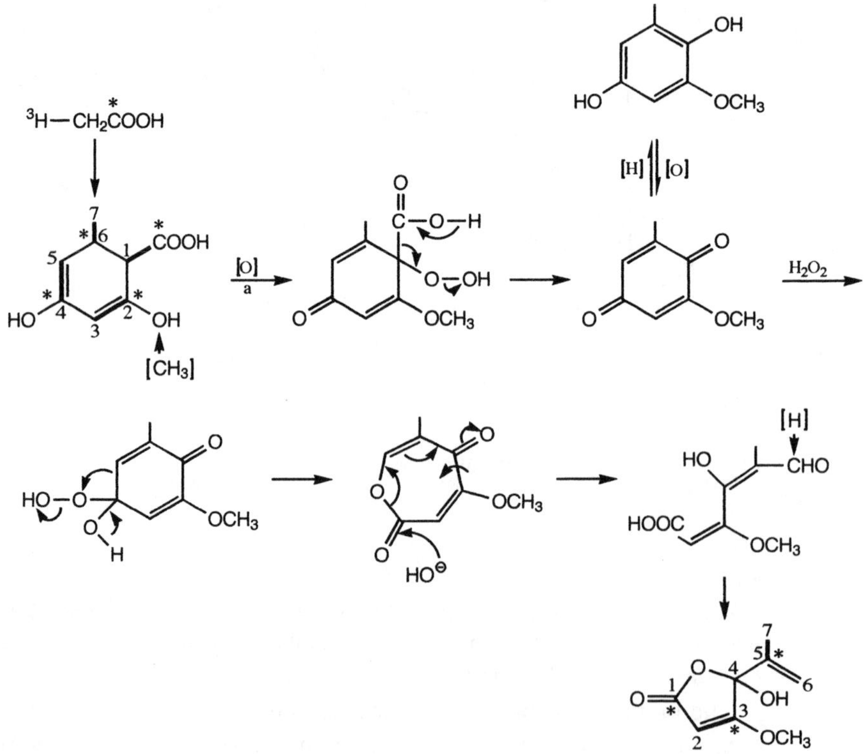

Fig. 31   Biosynthesis of penicillic acid

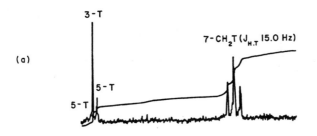

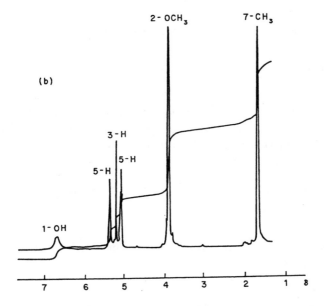

Fig. 32 (a) FT-^{3}H NMR spectrum of penicillic acid in acetone, d$_6$, 25°C, 1.12 × 10^4 pulses at 1.75 second intervals. (b) ^{1}H NMR spectrum. The numbering is that of the precursor, orsellinic acid. (Reproduced by permission of The Royal Society of Chemistry from *J. Chem. Soc. Chem. Commun.* **1974**, 220)

(59)

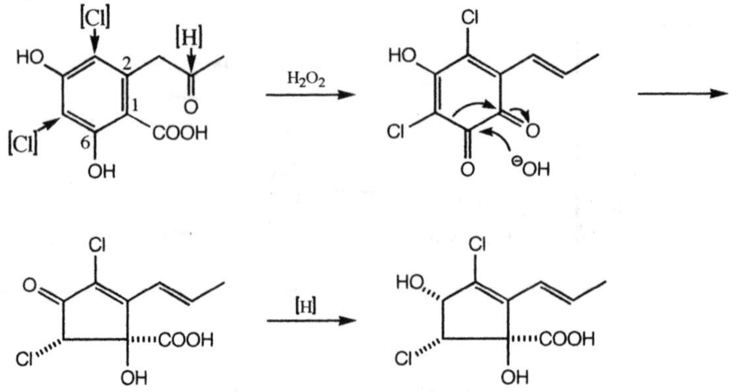

(60)

Cryptosporiopsin   Dihydrocryptosporiopsin

5.18 Problems

5.1 Suggest a biosynthetic route to resorsostatin, which represents a rare structural class isolated from *Pseudomonas* spp. Suggest also suitable experiments for proving your hypothesis. (Kato, S., Shindo, K., Kawai, H., Matsuoka, M. and Moshizuki, J. *J. Antibiotics* **46** (1993) 1024.)

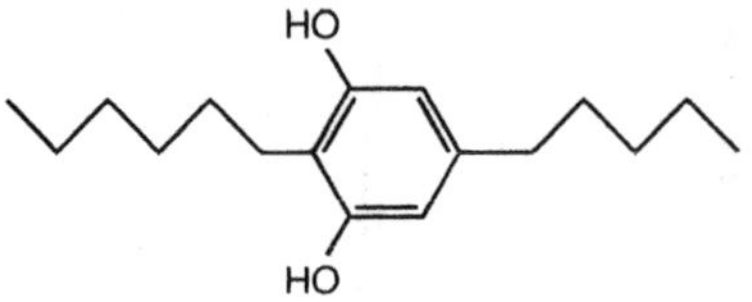

Resorsostatin

5.2 Suggest a mechanism for formation of aurones, e.g. sulphuretin, from the corresponding chalcone.

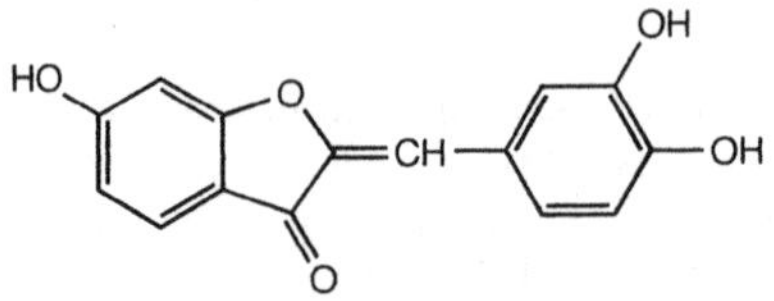

Sulphuretin

5.3 It is suggested that islandicin is assembled from eight acetate units according to equation (35). However, it can also be formed by folding the octaketide chain in the following manner:

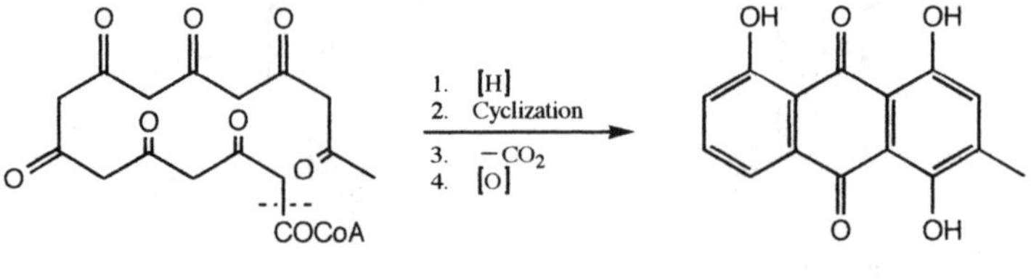

1. [H]
2. Cyclization
3. $-CO_2$
4. [O]

Islandicin

How would you settle the folding problem? (Paulick, R. C., Casey, M. L., Hillenbrand, D. F. and Whitlock, H. W. *J. Am. Chem. Soc.* **97** (1975) 5303.)

5.4 Multicolic acid, isolated from *Penicillium multicolor*, incorporated 1-^{13}C-acetate as shown below. It showed the following ^{13}C NMR data when 1,2-^{13}C-acetate was fed to the culture. It is formed from a hexaketide via ring fission

^{13}C NMR data for methyl O-methylmulticolate

Carbon	δ_C	$J_{13_C-13_C}$, Hz
1	168.3	—
2	109.7	48 (sp^2–sp^3)
3	160.6	—
4	150.2	90 (sp^2–sp^2)
5	23.4	48
6	29.7	35 (sp^3–sp^3)
7	25.5	35
8	32.2	35
9	62.2	35
10	100.9	—
11	163.8	—
CH₃OOC–	52.0	—
CH₃O–	59.5	—

of an aromatic intermediate. Discuss the biosynthetic pathway, the reaction mechanisms involved, and indicate which acetate bonds are still intact in the metabolite. (Gudgeon, J. A., Holker, J. S. E., and Simpson, T. J. *J. Chem. Soc. chem. Commun.* **1974**, 636.)

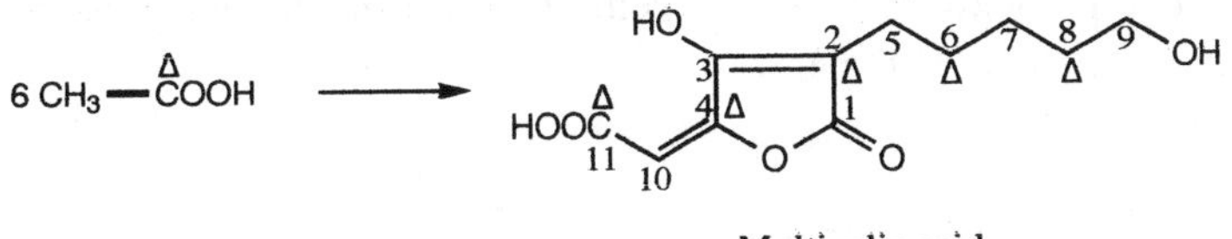

Multicolic acid

5.5 Citromycetin (1) and fulvic acid (2) are formed from seven acetate units but they cannot be formed by simple linear chain folding. No introduction of C_1 units is observed. Two pathways have been considered: (a) condensation of two separate polyketide chains; (b) oxidative ring cleavage and recondensation of a single chain heptaketide intermediate, such as fusarubin (3). Labelling experiments with 1,2-[13]C$_2$-acetate show that all seven C$_2$ units are intact in (1)–(3) and the enrichment is approximately of the same intensity at all positions. Discuss the two chain hypothesis versus the single chain hypothesis (common intermediate hypothesis). (Kurobane, I., Hutchinson, C. R. and Vining, L. C. *Tetrahedron Lett,* **1981**, 493.)

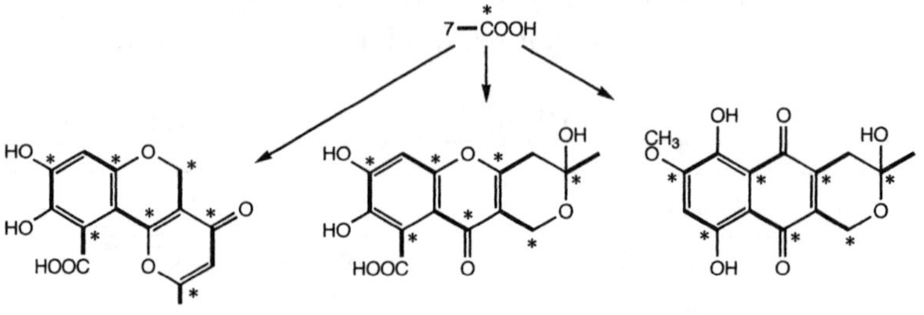

(1) (2) (3)

5.6 The antibiotic isolasalocid A has the following structure:

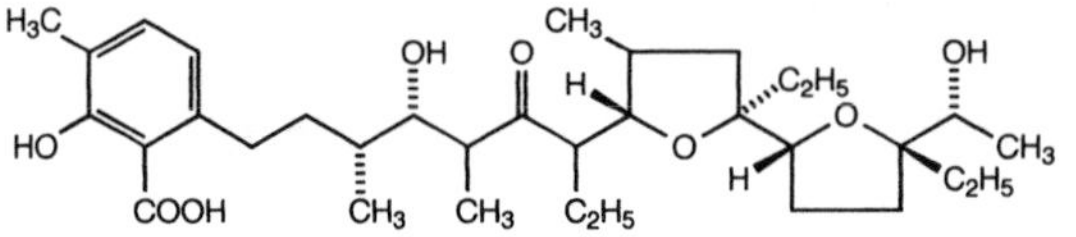

Suggest a reasonable biosynthetic pathway which accounts for the substitution, branching, and oxygenation patterns and the formation of the benzene and tetrahydrofuran rings. (Westley, J. W., Preuss, D. L. and Pitcher, R. G. *J. Chem. Soc. Chem. Commun.* **1972**, 161.)

5.7 The macrocyclic ecklonia lactone was isolated from the brown alga *Ecklonia stolonifera*. It acts as an antifeedant to deter attack from marine herbivores. Suggest a biosynthetic pathway from a suitable unsaturated fatty acid and draw the conformation of the molecule at the transition state of the cyclization step. (Kurata, K., Taniguchi, T. Shiraishi, K. and Susuki, M. *Phytochemistry 33* (1993) 155.)

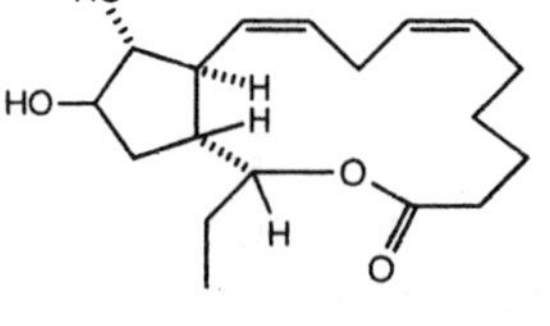

Ecklonia lactone

5.8 Suggest biosynthetic pathways for sclerin, a plant growth hormone from the phytopathogenic fungus, *Sclerotinia sclerotiorum*. Three methyls originate from the one carbon pool. Biosynthesis via one, as well as two, polyketide chains have to be considered. Suggest also suitable labelling experiments in support of your biosynthetic ideas (Barber, J., Garson, M. J. and Staunton, J. *J. Chem. Soc. Perkin I* **1981**, 2584.)

$$3 \overset{\blacktriangle}{C} + CH_3COOH \longrightarrow$$

Sclerin

Bibliography

1. For recent reviews see Simpson, T. J. *Nat. Prod. Rep.* **1** (1984) 28; **2** (1985) 321; **4** (1987) 339; **8** (1991) 578. O'Hagan, D. *Nat. Prod. Rep.* **9** (1992) 447; **10** (1993) 593; **12** (1995) 1.
2. Moss, J. and Lane, M. in *Adv. in Enzymology* **35** (1971) 321.
3. Lynen, F., Knappe, J., Lorch, E., Jutting, G. and Ringelmann, E. *Angew. Chem.* **71** (1959) 481.
4. Bonnemere, C., Hamilton, J. A., Steinrauf, L. K. and Knappe, J. *Biochemistry* **4** (1965) 240.
5. Rose, I. A., O'Connel, E. and Solomon, F. *J. Biol. Chem.* **251** (1976) 902.
6. Arnstad, K. I., Schindlbeck, G. and Lynen, F. *J. Chem. Soc. Perkin I* **1975** 52.
7. Sedgwick, B., Morris, C. and French, S. J. *J. Chem. Soc. Chem. Commun.* **1978**, 193; Schwab, J. M., Klassen, J. B. and Habib, A. *J. Chem. Soc. Chem. Commun.* **1986** 357
8. Jackowski, S. and Rock, C. O. *J. Biol. Chem.* **262** (1987) 7927; Jaworski, J. G., Post-Beittenmiller, D. and Ohlrogge, J. B. *Eur. J. Biochem.* **213** (1993) 981.
9. Schal, C., Burns, H. L., Gadot, M., Chase, J. and Blomquist, G. J. *Insect Biochem.* **21** (1991) 73.
10. Djerassi, C. and Lam, W.-K. *Acc. Chem. Res.* **24** (1991) 69.
11. Chang, S.-I. and Hammes, G. G. *Acc. Chem. Res.* **23** (1990) 363.
12. Rawlings, B. J., Reese, P. B., Rahmer, S. E. and Vederas, J. C. *J. Am. Chem. Soc.* **111** (1989) 3382; Arai, K., Rawlings, B. J., Yoshizawa, Y. and Vederas, J. C. *J. Am. Chem. Soc.* **111** (1989) 3391.
13. Eggink, G., de Waard, P. and Hujiberts, G. N. M. *FEMS, Microbiol. Rev.* **103** (1992) 159.
14. Li, Z., Martin, F. M. and Vederas, J. C. *J. Am. Chem. Soc.* **114** (1992) 1531.
15. Cane, D. E., Lambalot, P. C., Prabhakaran, P. C. and Ott, W. R. *J. Am. Chem. Soc.* **115** (1993) 522.
16. Kinoshita, K., Takenaka, S. and Hayashi, M. *J. Chem. Soc. Perkin I* **1991** 2547.
17. Staunton. J. and Sutkowski, A. C. *J. Chem. Soc. Chem. Commun.* **1991** 1110; Jacobs. A., Staunton, J. and Sutkowski, A. C. *J. Chem. Soc. Chem. Commun.* **1991** 1113.
18. Wagner, C., Eckehardt, K., Ihn, W., Schumann, G., Stengel, C., Fleck, W. F. and Tresselt, D. *J. Basic Microbiol.* **31** (1991) 223.
19. Patzelt, H. and Robinson, J. A. *J. Chem. Soc. Chem. Commun* **1993** 1258.
20. Buckner, J. S., Kolattukudy, P. E. and Rogers, L. *Arch. Biochem. Biophys.* **186** (1978) 152.

21. Cramer, U. and Spener, F. *Biochem. Biophys. Acta* **450** (1976) 261.
22. Moore, B. S., Poralla, K. and Floss, H. G. *J. Am. Chem. Soc.* **115** (1993) 5267.
23. Law, J. H. *Acc. Chem. Res.* **4** (1971) 199.
24. Buist, P. H. and MacLean, D. B. *Can. J. Chem.* **59** (1981) 828.
25. Scheuerbrandt, G. and Block, K. *J. Biol. Chem.* **235** (1962) 2064.
26. Schwab, J. M. and Klassen, J. B. *J. Am. Chem. Soc.* **106** (1984) 7217.
27. Boland, W., Frossl, C., Schottler, M. and Toth, M. *J. Chem. Soc. Chem. Commun* **1993** 1155.
28. Bloomfield, D. K. and Bloch, K. *J. Biol. Chem.* **235** (1960) 337.
29. Cherif, A., Dubacq, J. P., Mache, R., Oursel, A. and Tremalieres, A. *Phytochemistry* **14** (1975) 703.
30. Stoffel, W. *Biochem. Biophys. Res. Commun.* **6** (1961) 270.
31. Corey, E. J., D'Alarcao, M., Matsuda, S. P. T., Lansbury, Jr., P. T. and Yamada, Y. *J. Am. Chem. Soc.* **109** (1987) 289.
32. Brash, A. R. *J. Am. Chem. Soc.* **111** (1989) 1892; Corey, E. J., Ritter, K., Yus, M. and Náyera, C. *Tetrahedron Lett.* **28** (1987) 3547.
33. Crombie, L., Morgan, D. O. and Smith, E. H. *J. Chem. Soc. Perkin I* **1991** 567.
34. Crombie, L. and Morgan, D. O. *J. Chem. Soc. Perkin I* **1988** 588; **1991** 581.
35. Bohlmann, F., Burkhardt, T. and Zdero, C. *Naturally Occurring Acetylenes*, Academic Press, London, 1973.
36. Bu'Lock, J. D. and Smith, G. N. *Biochem. J.* **1962,** 35.
37. Hatt, H. H. and Szumer, A. Z. *Chem. Ind.* **1954,** 962.
38. Jones, E. R. H., Thaller, V. and Turner, J. L. *J. Chem. Soc. Perkin I* **1975** 424.
39. Jones, E. R. H., Piggin, C. M., Thaller, V. and Turner, J. L. *J. Chem. Research* **1977** (S) 68: (M) 0744.
40. Hodge, P., Jones, E. R. H. and Lowe, G. *J. Chem. Soc.* (C) **1966,** 1216.
41. Lee, M. D., Manning, J. K., Williams, D. R., Kuck, N. A., Testa, T. R. and Borders, D. B. *J. Antibiot.* **42** (1989) 1070.
42. Staley, A. L. and Rinehardt, K. L. *J. Antibiot.* **44** (1991) 218.
43. Naruse, N., Konishi, M., Oki, T., Inouye, Y. and Kakisawa, H. *J. Antibiot.* **44** (1991) 756.
44. Lee, M. S., Qin, G.-W., Nakanishi, K. and Zagorski, M. G. *J. Am. Chem. Soc.* **111** (1989) 6234; Garson, M. J. *Chem. Rev.* **93** (1993) 1699.
45. Murata, M., Naoki, N., Iwashita, T., Matsunaga, S., Sasaki, M., Yokoyama, A. and Yasumoto, T. *J. Am. Chem. Soc.* **115** (1993) 2060; Garson, M. J. *Chem. Rev.* **93** (1993) 1699; Yasumoto, T. and Murata, M. *Chem. Rev.* **93** (1993) 1897.
46. Cane, D. E., Laing. T.-C. and Hasler, H. *J. Am. Chem. Soc.* **104** (1982) 7274.
47. Townsend, C. A. and Basak, A. *Tetrahedron* **47** (1991) 2591; McDonald, F. E. and Towne, T. B. *J. Am. Chem. Soc.* **116** (1994) 7921.
47a. Yasaki, K., Heide, L. and Tabata, M. *Phytochemistry* **30** (1991) 2233.
48. Spenser, J. B. and Jordan, P. M. *J. Chem. Soc. Chem. Commun.* **1992** 646
49. Money, T. *Chem. Rev.* **70** (1970) 553.
50. Hopwood. D. A. and Sherman, D. H. *Ann. Rev. Microbiol.* **24** (1990) 37.
51. Birch, A. J. and Donovan, F. W. *Austr. J. Chem.* **6** (1953) 360; Birch, A. J. *Fortschr. Chem. Org. Naturstoffe* **14** (1957) 186, Springer, Wien.
52. Gatenbeck, S. and Mosbach, K. *Acta Chem. Scand.* **13** (1959) 1561.
53. Bu'Lock, J. D., Smalley, H. M. and Smith, G. N. *J. Biol. Chem.* **237** (1962) 1778.
54. Dimroth, P., Walter, H. and Lynen, F. *European J. Biochem.* **13** (1970) 98; Jordan, P. M. and Spenser, J. B. *J. Chem. Soc. Chem. Commun* **1990** 238.
55. Gatenbeck, S. and Hermodson, S. *Acta Chem. Scand.* **19** (1965) 65.
56. Gatenbeck, S. *Acta Chem. Scand.* **14** (1960) 296.

57. Garson, M. J. and Staunton, J. *Chem. Soc. Rev.* **8** (1979) 539.
58. Wehrli, F. W. and Nishida, T. *Prog. Chem. Org. Nat. Prod.* **36** (1979) 1.
59. Sato, Y., Oda, T. and Saito, H. *Tetrahedron Lett.* **1976,** 2695; *J. Chem. Soc. Chem. Commun* **1978,** 135.
60. Wat, C.-K., McInnes, A. G., Smith, D. G. and Vining, L. C. *Can. J. Biochem.* **50** (1972) 620.
61. Sankawa, U., Shimada, H. and Yamasaki, K. *Tetrahedron Lett.* **1978,** 3375.
62. Canham, P., Vining, L. C., McInnes, A. G., Walter, J. A. and Wright, L. J. C. *J. Chem. Soc. Chem. Commun.* **1976,** 319.
63. Seto, H., Cary, L. and Tanabe, M. *J. Chem. Soc. Chem. Commun.* **1973,** 867.
64. Birch, A. J. and Donogan, F. W. *Austr. J. Chem.* **8** (1955) 529.
65. Birch, A. J. and Donogan, F. W. *Austr. J. Chem.* **6** (1953) 373.
66. Thomson, R. H. *Naturally Occurring Quinones*, 2nd. Edn, Academic Press, London, 1971.
67. Leistner, E. and Zenk, M. H. *Tetrahedron Lett.* **1971,** 1677.
68. Townsend, C. A. and Davis, S. G. *J. Chem. Soc. Chem. Commun.* **1983** 1420.
69. Harborne, J. B. *Phytochemistry* **6** (1967) 1415, 1643.
70. Harborne, J. B. (Ed.) *The Flavonoids. Advances in Research since 1980*. Chapman and Hall, London, 1988.
71. Hrazdina, G. and Creasy, L. L. *Phytochemistry* **18** (1979) 581.
72. Saleh, N. A. M. and Fritsch, H., Kreuzaler, F. and Grisebach, H. *Phytochemistry* **17** (1978) 183.
73. Bhandari, P., Crombie, L., Kilbee, G. W., Pegg, S. J., Proudfoot, G., Rossiter, J., Sanders, M. and Whiting, D. A. *J. Chem. Soc. Perkin I* **1992** 851; Bhandari, L. C., Crombie, L., Daniels, P., Holden, I., Van Bruggen, N. and Whiting, D. A. *J. Chem. Soc. Perkin I* **1992** 839.
74. Dreyer, D. L., Munderloh, K. P. and Thiessen, W. E. *Tetrahedron* **31** (1975) 287.
75. Roitman, J. N. and Jurd, L. *Phytochemistry* **17** (1978) 161.
76. Saitoh, T., Shibata, S. and Sankawa, U. *Tetrahedron Lett.* **1975,** 4463.
77. Ellis, B. E. *Lloydia* **37** (1974) 168.
78. Hösel, W., Frey, G. and Barz, W. *Phytochemistry* **14** (1975) 417.
79. Birch, A. J. and Thompson, D. J. *Austr. J. Chem.* **25** (1972) 2731.
80. Stocker, M. and Pohl, R. *Phytochemistry* **15** (1976) 571.
81. von Rudloff, E. and Jorgensen, E. *Phytochemistry* **2** (1963) 297.
82. Pryce, R. J. *Phytochemistry* **10** (1971) 2679.
83. Rupprich, N. and Kindl, H. *Hoppe-Seyler's Z. Physiol Chem.* **359** (1978) 165.
84. Bentley, R. *J. Biol. Chem.* **238** (1963) 1889, 1895.
85. Scott, A. I. and Wiesner, K. J. *J. Chem. Soc. Chem. Commun.* **1972,** 1075.
86. *Oxidative Coupling of Phenols*. Taylor, W. I. and Battersby, A. R. (Eds.), M. Dekker, New York, 1967.
87. Rhodes, A., Somerfield, G. A. and McGonagle, M. P. *Biochem. J.* **88** (1963) 349.
88. Barton, D. H. R., DeFlorin, A. M. and Edwards, O. E. *J. Chem. Soc.* **1956,** 530.
89. Taguchi, H., Sankawa, U. and Shibata, S. *Chem. Pharm. Bull. (Japan)* **17** (1969) 2054.
90. Davidson, T. A. and Scott, A. I. *J. Chem. Soc.* **1961,** 4075.
91. Cameron, D. W. and Lord Todd in *Oxidative Coupling of Phenols*, Taylor, W. I. and Battersby, A. R. (Eds.), M. Dekker, New York, 1967, p. 203.
92. Brockmann, H. *Prog. Chem. Nat. Prods.* **14** (1957) 142.
93. Anderson, J. M. and Murray, J. *Chem. Ind.* **1956,** 376.
94. Allport, D. C. and Bu'Lock, J. D. *J. Chem. Soc.* **1958,** 4090.
95. Howard, D. H. and Raistrick, H. *Biochem. J.* **57** (1954) 212.

96. Shibata, S., Murakami, T., Kitagawa, I. and Kishi, T. *Pharm. Bull. (Japan)* **4** (1956) 111.
97. Sultanbawa, M. V. S. *Tetrahedron Reports* **84;** *Tetrahedron* **36** (1980) 1465.
98. Atkinson, J. E., Gupta, P. and Lewis, J. R. *Chem. Commun.* **1968,** 1386.
99. Floss, H. G. and Rettig, A. *Z. Naturforsch.* **19B** (1964) 1103, Bennet, G. J., Lee, H. and Das N. P. *J. Chem. Soc. Perkin I* **1990** 2671.
100. Scott, A. *Quart. Rev.* **19** (1965) 1.
101. Ahmed, S. A.; Bardshiri, E., and Simpson, T. J. *J. Chem. Soc. Chem. Commun.* **1987** 883.
102. Ward, P. F., Hall, R. J. and Peters, R. A. *Nature* **201** (1964) 611; Harper, D. B. and O'Hagan, D. *Nat. Prod. Rep.* **11** (1994) 123.
103. Naumann, K. *Chemie unserer Zeit* **27** (1993) 33; Neidleman, S. and Geigert, J. *Endevour* **11** (1987) 5.
104. Butler, A. and Walker. J. U. *Chem. Rev.* **93** (1993) 1994.
105. Laatsch, H., Budleiner, H., Pelizaeus, B. and Van Pée, K.-H. *Liebigs Ann.* **1994** 65.
106. Moore, R. E. in *Marine Natural Products*, Vol. I, Scheuer, P. J. (Ed.). Academic Press, New York, 1978, p. 44.
107. Martin, J. D. and Darias, J. in *Marine Natural Products*, Vol. I, Scheuer, P. J. (Ed.), Academic Press, New York, 1978, p. 125.
108. Faulkner, D. J. in *Environmental Chemistry*, Vol. IA, Hutzinger, O. (Ed.), Springer, Berlin, 1980, p. 229.
109. Murphy, G., Vogel, G., Krippahl, G. and Lynen, F. *European J. Biochem.* **49** (1974) 443.
110. Scott, A. I., Zamir, L., Phillips, G. T. and Yalpani, M. *Bioorg. Chem.* **2** (1973) 124.
111. Scott, A. I. and Yalpani, M. *Chem. Commun.* **1967,** 945.
112. Mosbach, K. *Acta Chem. Scand.* **14** (1960) 457.
113. Gudgeon, J. A., Holker, J. S. E. and Simpson, T. J. *J. Chem. Soc. Chem. Commun.* **1974,** 636.
114. Better, J. and Gatenbeck, S. *Acta Chem. Scand.* **B30** (1976) 368.
115. Elvidge, J. A., Jaiswal, D. K., Jones, J. R. and Thomas, R. *J. Chem. Soc. Perkin I* **1977,** 1080.
116. Holker, J. S. E. and Young, K. *J. Chem. Soc. Chem. Commun.* **1975,** 525.

Chapter 6

The mevalonic acid pathway
The terpenes

6.1 Introduction

When the number of structurally defined compounds rapidly accumulated in the late nineteenth century, Wallach noted that many compounds, especially the fragrant principles of plants—the essential oils—could be formally dissected into branched C_5 units called isopentenyl or isoprene units. These compounds, which typically possessed the molecular composition $C_{10}H_{16}$, were given the collective name terpenes, etymologically derived from the terebinth tree, *Pistacia terebinthus*, which exudes a resin. Conifers, eucalyptus trees, and citrus fruits are rich in these low molecular weight volatile terpenes. Limonene can, in principle, be synthesized from two moles of isoprene by a Diels–Alder reaction (1) but it was rapidly perceived that isoprene itself

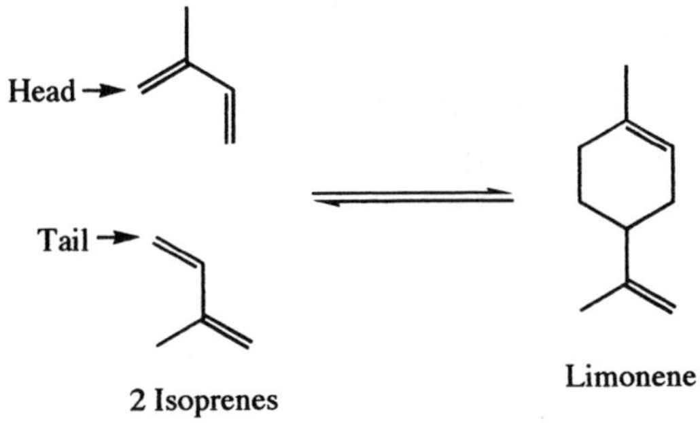

Head →

Tail →

2 Isoprenes

Limonene

(1)

could not be the functional unit used by nature. Nevertheless, the isoprene unit was a useful device for rationalizing the structures of many more complex compounds of higher molecular weight (2).

The isoprene rule[1] states that terpenes are multiples of C_5 units linked together head to tail. Several modes of cyclization are conceivable and lead to various skeletons just like the cyclization of polyketide chains. The terpenes are classified according to the number of C_5 units: monoterpenes, C_{10}; sesquiterpenes, C_{15}; diterpenes, C_{20}; sesterterpenes, C_{25}; triterpenes, C_{30}; and tetraterpenes, C_{40}. It soon appeared that neither steroids, C_{27}, nor several other related compounds obeyed the rule. Degradation had occurred, the head to tail principle was violated, and the skeleton could not be dissected into isoprene units. These changes could be rationalized on the assumption of cleavages and rearrangements of the original skeleton. In 1956 Folkers isolated the easily incorporated mevalonic acid, and subsequently Bloch, Lynen, Cornforth, Eggerer, and Popjak showed how it functioned as a building block. It was then possible to interpret the biosynthesis and secondary modifications of terpenes correctly.

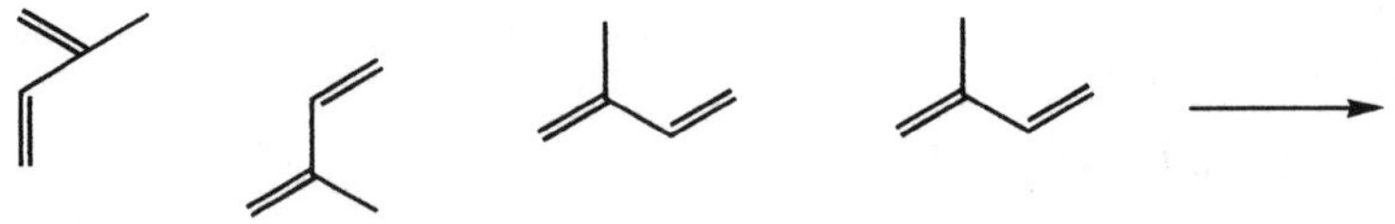

4 Isoprenes

(2)

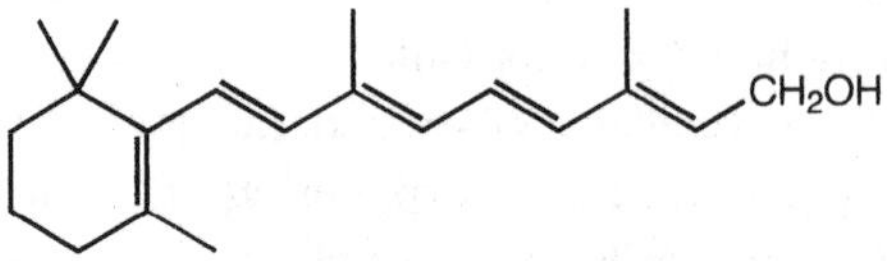

Vitamin A

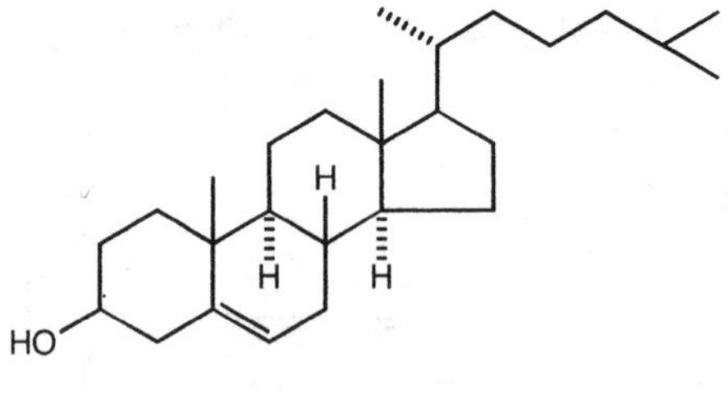

Fig. 1   Structure of cholesterol

The terpenes house a wealth of significant compounds. The perfume industry is interested in the "essential" oils, terpentine is used for painting, and most importantly, we find among the terpenes physiologically very active compounds governing the life processes, such as adrenal hormones (cortisone), sex hormones (oestrogen and testosterone), vitamins A, D, and E, etc. This has, of course, triggered intense research on all frontiers to elucidate the structures and properties of these rather complicated compounds. In the early days of terpene and steroid chemistry the chemists had to fight against difficult odds. Many asymmetric centres complicated the structures, the aliphatic nature of the compounds, with few chromophores, rendered UV spectroscopy less helpful (the only spectroscopic aid available at that time), and, finally, the ease by which rearrangements occurred in aliphatic ring systems frustrated chemists. On the other hand, hardly any group of compounds has repaid the efforts of chemists as well as the terpenes. The research has contributed immensely to the development of conformational chemistry, understanding of mechanisms, and synthesis. The need for suitable spectroscopic methods was articulated and gradually became available.

The many problems faced are reflected in the rather late date by which the final structural clarification of cholesterol (Fig. 1) was achieved by Wieland and Windaus (1932) and that was with the help of Bernal, an X-ray crystallographer who, by measuring the dimensions of the unit cell of ergosterol, indicated the most likely structure.

A few C_5 compounds occur in nature but not all of them are of isoprenoid origin (Fig. 2).

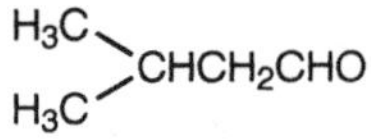

Isovaleraldehyde **1**

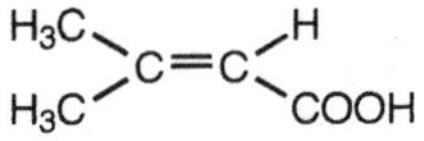

Senecioic acid **2**

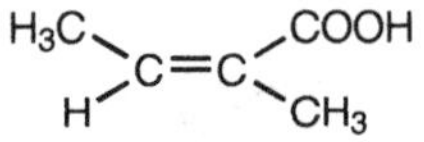

Angelic acid **3**

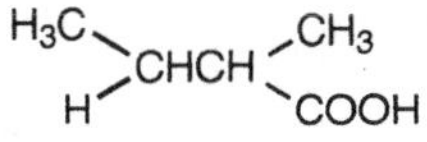

α-Methylbutyric
acid **4**

Fig. 2 Some naturally occurring C_5 compounds. **1** comes from mevalonic acid, **3** and **4** from isoleucine, and **2** can effectively be synthesized from both mevalonic acid and leucine

Single isoprene units are found in many natural compounds, so-called hemiterpenes, e.g. in furanocoumarins (section 4.5), hop constituents (section 5.17), quinones (sections 4.6 and 5.12), and in a variety of alkaloids (Chapter 8).

The formation of the very large number of terpenes from C_{10}–C_{30} precursors can be rationalized by a few basic reaction types. In the starting reaction a discrete carbonium ion is generated, either by solvolysis of an allylic pyrophosphate, opening of an epoxide or protonation or halogenation (marine organisms) of double bonds. The further reaction proceeds by electrophilic cyclization, by Wagner–Meerwein shifts (1,2 shifts), or less frequently by electrophilic substitution at the aliphatic bond (1,3 1,4 and 1,5 shifts). The reaction is terminated by proton elimination or addition of water (Fig. 3). The 1,2 shifts and 1,3 and higher shifts deserve some comment. The structure of carbonium ions[2] has been the object of considerable speculation which undoubtedly has promoted much valuable research in the field. The problem was whether the equilibrating classical carbonium ion with localized charge, or the non-classical carbonium ion, represented the structure of lowest energy. The problem has received an answer in favour of the classical structure by record-

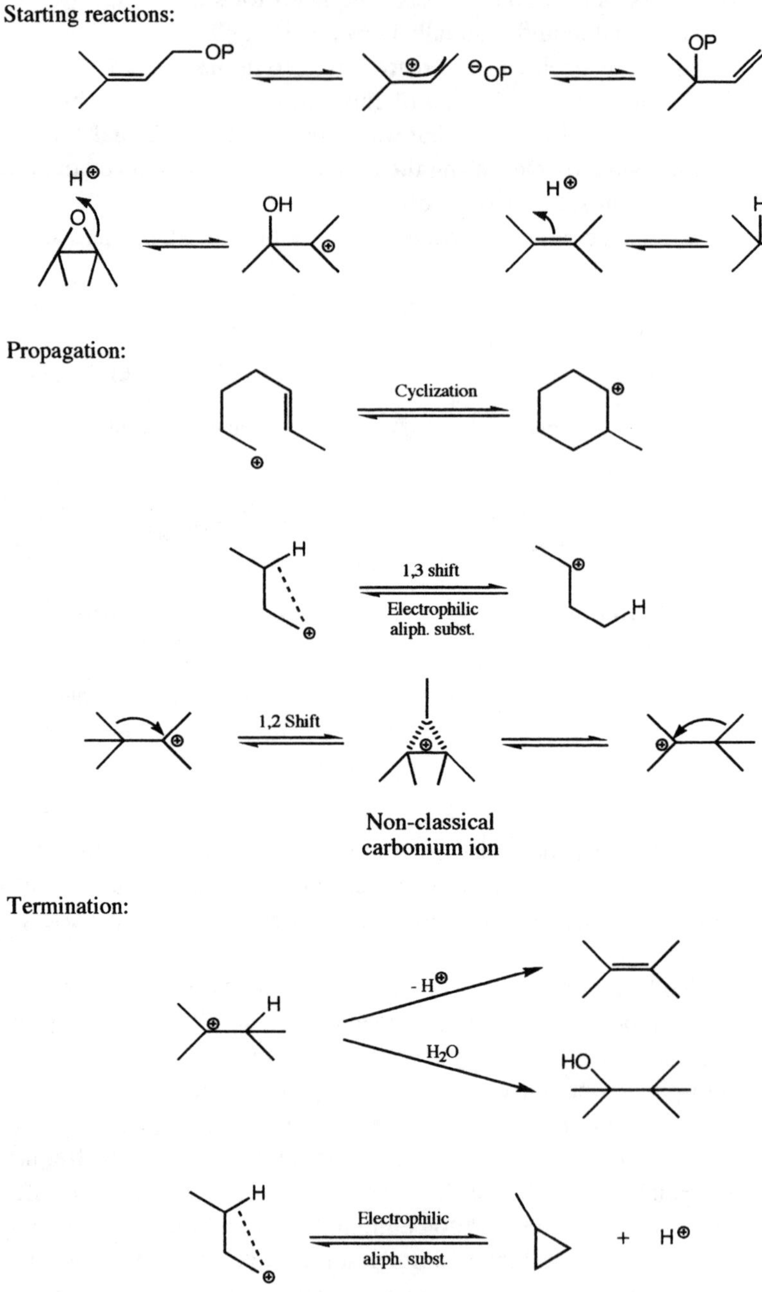

Starting reactions:

Propagation:

Termination:

Fig. 3   Reaction types in terpene chemistry

ing NMR spectra of various ions at low temperature in SbF_5–SO_2ClF solutions. It was found that the non-classical structure represents a transition state of slightly higher energy, and that the equilibrium is very rapid on the NMR time scale, $E_a \sim$ 5–10 kcal mol^{-1}. Thus, the NMR spectrum of the cyclopentyl carbonium ion shows only one peak corresponding to nine fast exchanging hydrogens.

1,3 and higher hydrogen shifts and eliminations are synonymous with electrophilic aliphatic substitution, cf. *trans*-annular reactions. The potential surface of a carbonium ion attacking an aliphatic σ-bond has a minimum in the direction perpendicular to the σ-bond. This contrasts with the radical attack which occurs in the direction of the C–H bond (Fig. 4). This implies that electrophilic aliphatic substitution in principle occurs with retention of configuration. The activation energy for the 1,3 hydrogen shift is *ca.* 10 kcal mol^{-1}.

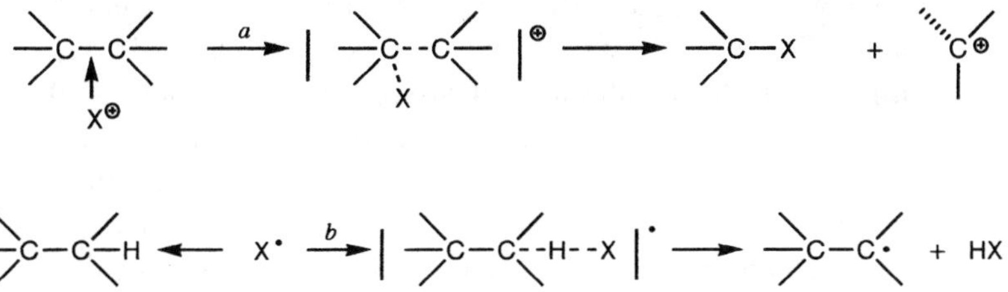

Fig. 4   (a) Preferred direction of attack in ionic reactions and (B) radical reactions

The steric structure of the product is determined by the folding of the polyprenyl chain and the principle of a synchronized antiparallel addition leading directly to the product. The few exceptions to this rule are explained by the stability of an intermediate carbonium ion or the intervention of a 'Y' group from the enzyme that is subsequently expelled by substitution and inversion of configuration (Fig. 5).

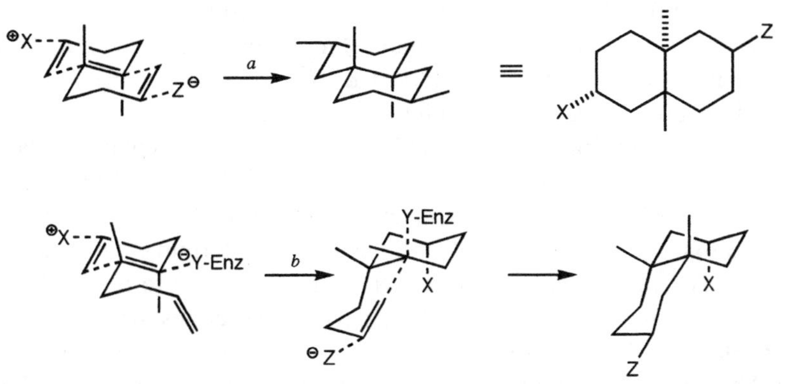

Fig. 5   (a) Antiparallel concerted cyclization. (b) Cyclization via intervention of Y group from the enzyme

6.2 Biosynthesis of mevalonic acid and the active isoprene units. The chiral methyl[3]

Early isotopic studies revealed that the carbon skeleton of terpenoids is acetate based. The hunt for active intermediates continued and a major breakthrough came with the demonstration of mevalonic acid as a general precursor which could replace acetate as an essential growth factor.[4,5] Its formation in the cell is shown in (3).[6] Acetyl CoA combines with a sulphydryl group at the active site of the enzyme and condenses with acetoacetyl CoA in a branched fashion to give 3-hydroxy-3-methylglutaryl CoA after hydrolysis. The two-step reduction with NADPH then affords mevalonic acid via mevaldic acid. Only 3R-mevalonic acid is biologically active. Phosphorylation, decarboxylation, and elimination of phosphate produces isopentenyl pyrophosphate (IPP), the active isoprene unit in the polymerization stage. Isopentenyl pyrophosphate is then reversibly isomerized to dimethylallyl pyrophosphate (DMAP), the starter unit of terpene biosynthesis. This compound, and other allylic phosphates as well, are reactive alkylating agents with phosphate being an efficient leaving group and the incipient carbonium ion being stabilized by charge delocalization. The first aldol condensation does not require malonate, unlike the straight chain condensation to fatty acids, and it is not catalysed by bicarbonate ions. If the acetyl CoA is specifically labelled at C^2, the label appears at C^2 in mevalonic acid, at C^4 in isopentenyl pyrophosphate, and the labelled methyl group appears eventually in the *trans* position[7] to the chain in dimethylallyl pyrophosphate. The two methylene protons of a CH_2XY centre are stereochemically different as a consequence of a particular orientation of the molecule in a chiral environment. Mevalonic acid contains three such so-called prochiral methylenes at C^2, C^4 and C^5. By chemical and enzymatic procedures deuterium and tritium were stereospecifically introduced in these positions. Isotopic analysis of the metabolites gave information about the steric course of the enzymatic steps. It was shown by this method that H_R at C^2 in isopentenyl pyrophosphate was lost in the isomerization,[8] and that the concomitant elimination of carbon dioxide and phosphate in mevalonic acid occurs strictly in a *trans* fashion, whereby the prochiral ^3H appears in a *cis* position to the methyl group (3,4).[9]

Two problems of considerable intricacy were formulated. Does the condensation of acetyl CoA with acetoacetyl CoA to (S)-3-hydroxy-3-methylglutaryl CoA proceed with inversion, as formulated in (3), or retention of configuration in the methyl; and second, is the terminal methylene in isopentenyl pyrophosphate protonated from the *re*-face (as formulated) or from the *si*-face? In contrast to most other enzymatic reactions, which stereospecifically lead to *cis* or *trans* isomers or to *R* or *S* enantiomers, the reactions outlined have no influence on the structure of the compound produced, they just unveil the intrinsic stereospecificity of enzymatic reactions. In other words, how

CH₃CCH₂COCoA H₂O → 3S-3-Hydroxy-3-methyl-glutaryl CoA (HMG-CoA) NADPH →

3R-Mevalonic acid ATP → − CO₂, − HOP → (3)

Isopentenyl pyrophosphate ⇌ Isomerase ⇌ ß,ß-Dimethylallyl pyrophosphate

rigidly do the reactants orient on the enzyme surface and is there time for any free rotation of the intermediate carbanion and carbonium ion? These problems cannot be solved without access to a methyl labelled with the three isotopes of hydrogen of known configuration, i.e. a chiral methyl, and an

→ ≡ (4)

Geranyl pyrophosphate IPP → all-*trans*-Farnesyl pyrophosphate analytical method for determining the absolute configuratin of such a methyl. The story of the chiral methyl is actually a close-up of enzymes in action. It is not realistic to label all methyls with ^{3}H. This will give an enormously high specific radioactivity. It appears that one can work with the usual level of tritium, *ca.* 1×10^{-6} M and merely measure the activity of the small fraction containing tritium under the supposition that all methyl groups containing tritium also contain hydrogen and deuterium. The preparation of labelled *R* and *S* acetic acids can be carried out enzymatically[10] or purely by synthetic methods[11] (Fig. 6). Phenyl acetylene was deuterated by metallation and reacton with deuterium oxide. Reduction with diimide, followed by epoxidation with *m*-chloroperbenzoic acid, and reduction with lithium borotritide gave a racemic mixture of 1-phenylethanols. Optical resolution with brucine phthalate and oxidation gave the two asymmetric acetic acids. The steric course of the chemical transformations and the absolute configuration of (+) and (−) 1-phenylethanol are well known, consequently the absolute configuration of the derived acetic acids is as depicted.

The analytical method was dependent on the action of malate synthase-fumarase, and the existence of an isotope effect discriminating between the three hydrogens.[12] Malate synthase catalyses the irreversible condensation of

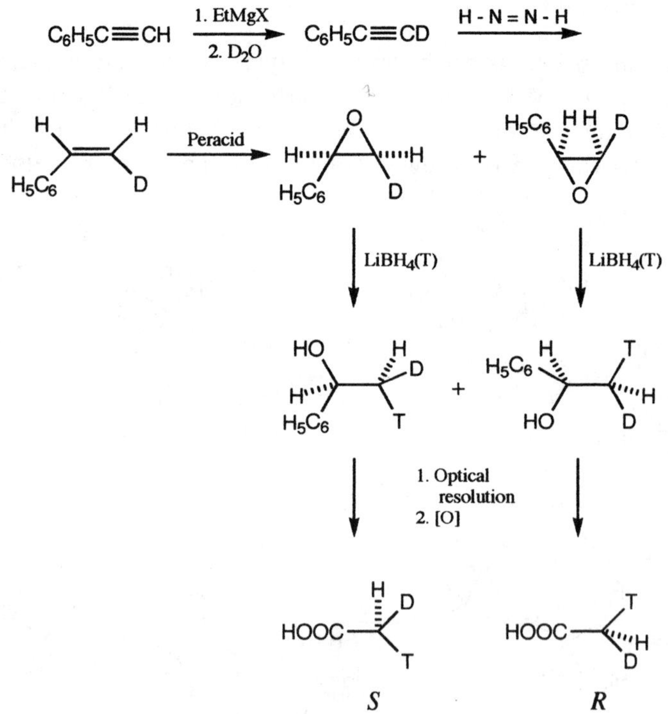

Fig. 6   Synthesis of chiral acetic acids

acetyl CoA with glyoxylate to malic acid. No fast proton exchange occurs in the methyl with the solvent in the condensation step and a normal isotope effect was observed. Fumarase eliminates water reversibly and stereospecifically from malic acid in a *trans* fashion.

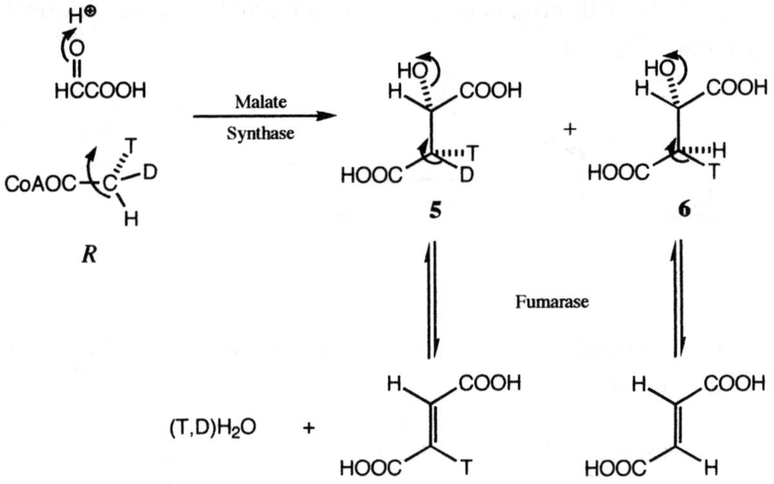

Fig. 7   Reactions of malate synthase and fumarase

Chiral acetyl CoA gives a mixture of tritium labelled malic acid **5** and **6**. If we suppose that the condensation is stereospecific, e.g. inversion of configuration as in Fig. 7, then $5/6 = k_H/k_D$. Prolonged equilibration with fumarase will exchange all D in **5** with protons from the solvent, but tritium will be retained. On the other hand, **6** will lose all of its tritium. Thus measurement of the activity of malic acid produced from chiral acetic CoA and the activity of malic acid resulting from a sample incubated with fumarase will give us the relative amounts of **5** and **6** and consequently, k_H/k_D. Malate from R acetyl CoA was found to retain 69 per cent of its activity. When the S acetate was used 31 per cent of the activity was retained, e.g. $k_H/k_D = 2.2$. This is a normal isotope effect and the malate condensation proceeds thus with inversion. Without regard to the isotope effect, this series of enzymatic transformations constitutes a method to determine the chirality of a given acetic acid. When dimethylallyl pyrophosphate was degraded and the acetic acid formed was put through the malate synthase-fumarase procedure, it turned out that the steric course followed (3), i.e. protonation occurs from the *re* side of IPP.[13] Isoprenylation, on the other hand, proceeds with the opposite configurational outcome (4). The incoming isoprenyl group and the released C^2-H_R are here located on the same side and the *E*-double bond is generated in this process. It has been suggested that loss of H_S typically should give rise to *Z*-double bonds as is the case for the synthesis of the *cis*-polyisoprene chain in rubber. However, there is no obligatory correlation between the geometry of the double bond formed and the prochiral proton loss.[14]

By using chiral acetyl CoA it was also demonstrated that its condensation with acetoacetyl CoA to 3-hydroxy-3-methylglutaryl CoA proceeded stereo-specifically with inversion of configuration and involved a normal hydrogen isotope effect (5).[15]

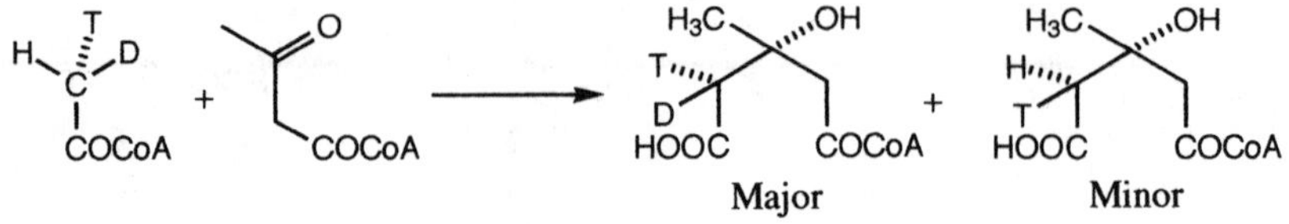

(5)

$$\text{Major} \qquad \text{Minor}$$

Two mechanisms have been advanced to account for the stereochemistry of the prenylation reaction:

1. consecutive *trans*-1,2-addition and *trans*-1,2-elimination involving formation of an intermediate σ-bond with the enzyme (6);[9]
2. ionization of the allylic pyrophosphate synchronized with alkylation and proton release: stereoelectronic and topological factors control the steric course of the reaction (7).[16]

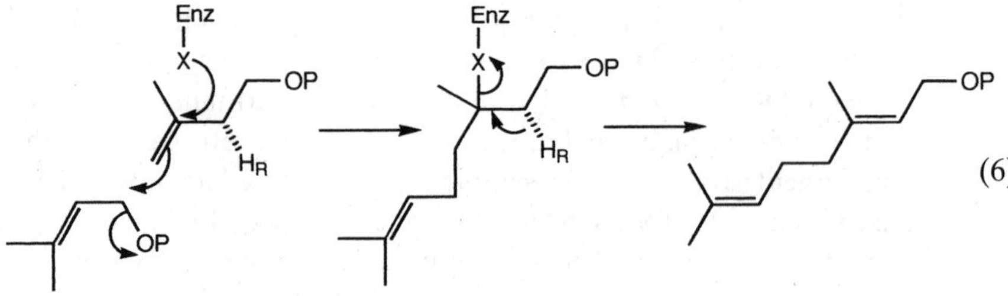

(6)

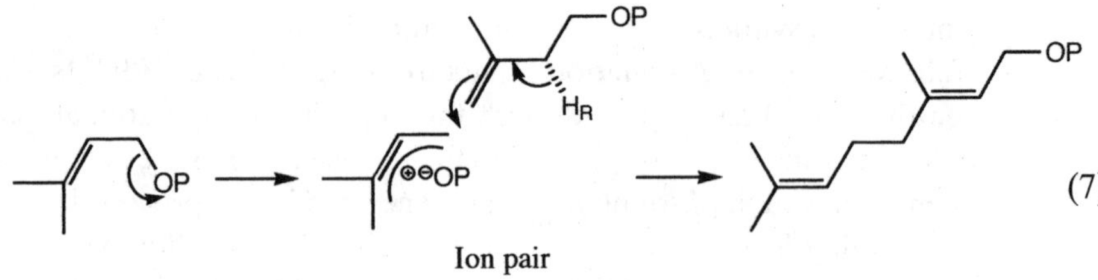

Ion pair

(7)

(8)

A decision favouring mechanism 2 was reached on the basis of incubation experiments with 2-fluoroisopentylpyrophosphate (8). It was argued that a σ-bonded R 2-F-isoprenylpyrophosphate will be irreversibly attached to the enzyme since it lacks a proton in an antiparallel position. The activity of the enzyme was indeed reduced, but restored again after dialysis. This suggests a competitive absorption at the active site but no participation of an X-group of the enzyme. 2-Fluorofarnesylpyrophosphate was identified as a metabolite from incubation of geranyl pyrophosphate and (R,S)-2-fluoroisopentenyl pyrophosphate. In another series of experiments it was found that the relative rate for geranylation and 2-fluorogeranylation of IPP is nearly identical to the relative rate for solvolysis of the geranyl and 2-fluorogeranyl methanesulphonates ($ca.10^3$). The electron withdrawing effect of fluorine retards both ionization and prenylation but it has practically no effect on the rate of S_N2 displacements at C^1. This implies that the C-OP bond breaking is far advanced before alkylation of the double bond occurs. The steric integrity at C^1 of the allyl carbonium ion is maintained by a rotational barrier of 28 kcal.[17] The experiments also show that the substrate specificity of prenyl transferase is not very stringent.

6.3 Monoterpenes

Geranyl pyrophosphate is the parent compound for the monoterpenes. Labelling experiments in higher plants have shown that mevalonic acid is incorporated preferentially in the part derived from isopentenyl pyrophosphate. The reason for the imbalance is not clear, but it is suggested that leucine participates in the biosynthesis of monoterpenoids by being primarily converted to dimethylallyl pyrophosphate through an alternative route[18] or that there is a pool of DMAP which primarily is metabolized. In contrast to the situation in higher plants the fungus *Ceratocystis monoliformis* produces monoterpenes with symmetrical incorporation of labelled mevalonic acid.[19]

Before cyclization can occur, *trans*-geranyl pyrophosphate must isomerize to the *cis*-form, either to the tertiary R- or S-allylic linalyl pyrophosphate or to neryl pyrophosphate, Fig. 8.[20,21] This is accomplished by ionization of the pyrophosphate to a tight ion pair with full integrity of the stereo structure at C^1.

The rotation barrier about the $C^{2,3}$ bond becomes smaller the tighter the pyrophosphate ion is associated with the tertiary center in the ion pair. The subsequent cyclization occurs with net retention of configuration at C^1. Neryl and linalyl pyrophosphates are exellent precursors but they seem not to be obli-

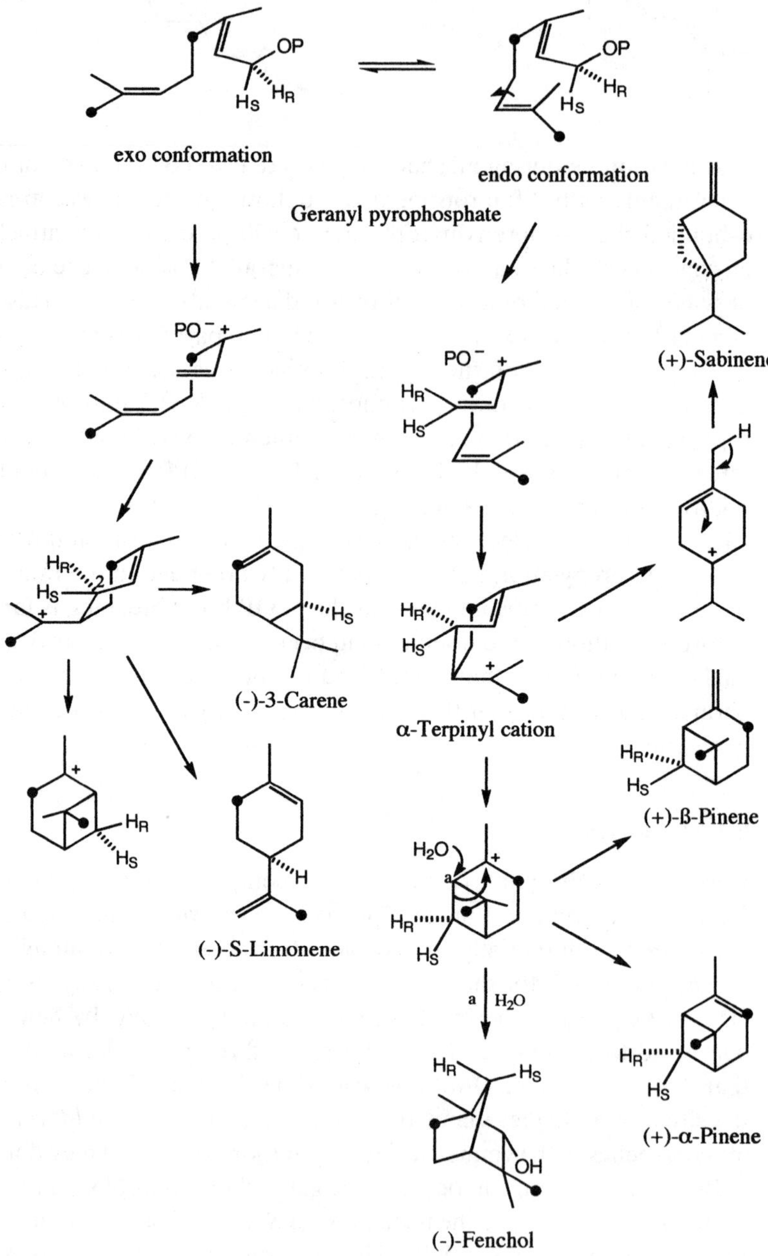

Fig. 8   Biosynthesis of monoterpenes

gatory intermediates on the pathway to monoterpenoids. It has also been demonstrated on several occasions that geranyl, neryl and linalyl pyrophosphate
(and the corresponding *trans-trans, cis-trans*-farnesyl and nerolidyl pyrophosphates in the sesquiterpene series) are cyclized to the same products but no
interconversion is observed. This finding can be taken as evidence for formation of a common asymmetric ion pair which cyclizes faster than it rearranges.

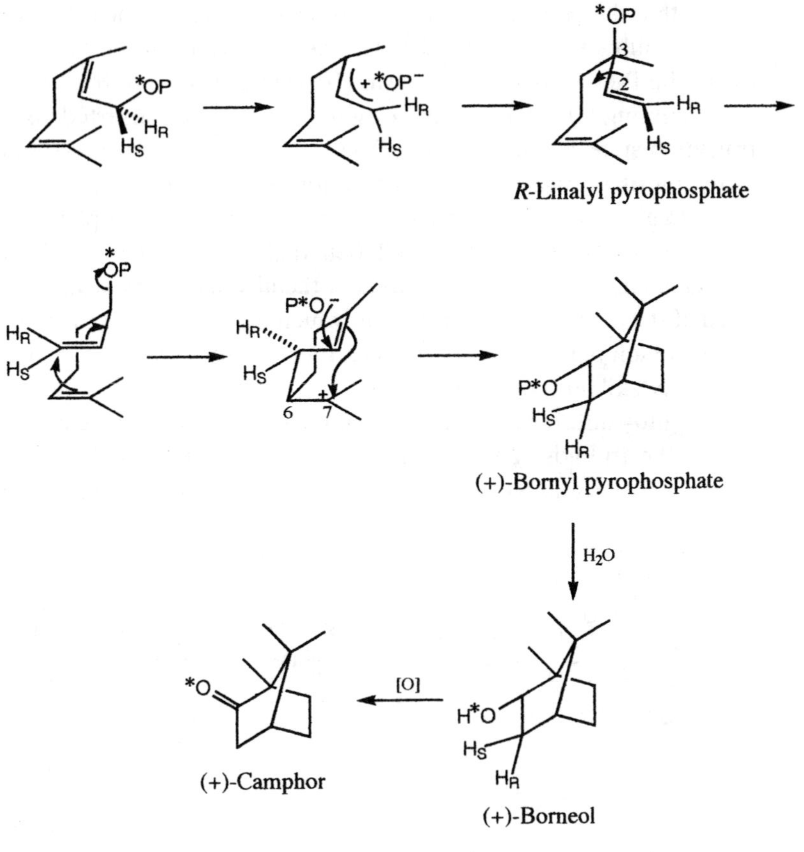

R-Linalyl pyrophosphate

(9)

(+)-Bornyl pyrophosphate

H_2O

(+)-Camphor ⟵ [O] (+)-Borneol

The initial endo-conformation of geranyl pyrophosphate has the advantage
that only minor conformational changes are required to accomplish subsequent cyclizations. This is manifested in the integrity of the prochiral methyl
groups at C^7 of geranyl pyrophosphate. The *E*-methyl is specifically located
anti to the olefin function in the α- and β-pinenes, Fig. 8.[22]

(-)-3-Carene can be formed from the exo-conformation under the supposition that the *syn* C^2-H_R undergoes electrophilic aliphatic substitution with the
terpinyl cation. Accoding to labelling experiments the double bond is shifted
to the 3-position.[22a]

Limonene cyclase from *Citrus sinensis* shows high regioselectivity for the proton elimination from the *cis*-terminal methyl group in the biosynthesis of limonene.

The allylic rearrangement (9) exhibits some interesting features. It is established to proceed with *syn*-stereochemistry. C^1-^{18}O labelled geranyl pyrophosphate gave an enrichment of the intermediate C^3-^{18}O labelled linalyl pyrophosphate and bornyl pyrophosphate practically identical to that of the precursor implying a 1,3-sigmatropic rearrangement mechanism[23] (10 c). This means that the phosphate ion remains tightly paired to the substrate throughout the multistep reaction with no space and time allowed for free rotation about the P-O-P bond. Mechanism (10 a) requires 1/3 of the original activity and tumbling (10 b) 1/6 of the original activity. A concerted six-membered transition state requires loss of activity (10 d).[20] The terminal step in terpene biosynthesis is either proton elimination to give the olefin or reaction with water to give the alcohol. In the borneol case the pyrophosphate is first formed which on hydrolysis gives the labelled alcohol by fission of the P-O bond. C^1-^{18}O labelling of farnesol shows that the allylic tertiary alcohol carries onethird of the activity of the precursor, which indicates that a carbonium/pyrophosphate ion pair is formed in which there is sufficient time for free rotation (10 a) but still no time for tumbling.

A slightly different folding of geraniol gives rise to the cyclopentene skeleton of the iridoids. They are widely distributed in the plant kingdom and often glucosidically bound. They have also been isolated from the secretion of

(10)

several insects where they play a role in chemical defence and communication (Chapter 2). The name originates from ants of the genus *Iridomyrmex* from which they first were isolated. 8-Hydroxygeraniol and 8-hydroxynerol are found to be on the pathway. Oxidation of $C^{1,8}$ to the dial precedes the cyclization leading to the iridoid skeleton, Fig. 9.[24,25] At one stage of the subsequent oxidation carbon atoms 8 and 9 of geraniol become equivalent. Loganin is a precursor for the monoterpene and indole alkaloides. Oxidative fission of the $C^{7,8}$ bond gives secologanin.

The biosynthesis of some irregular monoterpenes[26], e.g. artemisia, yomogi, and santolina alcohols and chrysanthemic acid, a component of the efficient pyrethrum insecticides, is started by condensation of two molecules of dime-

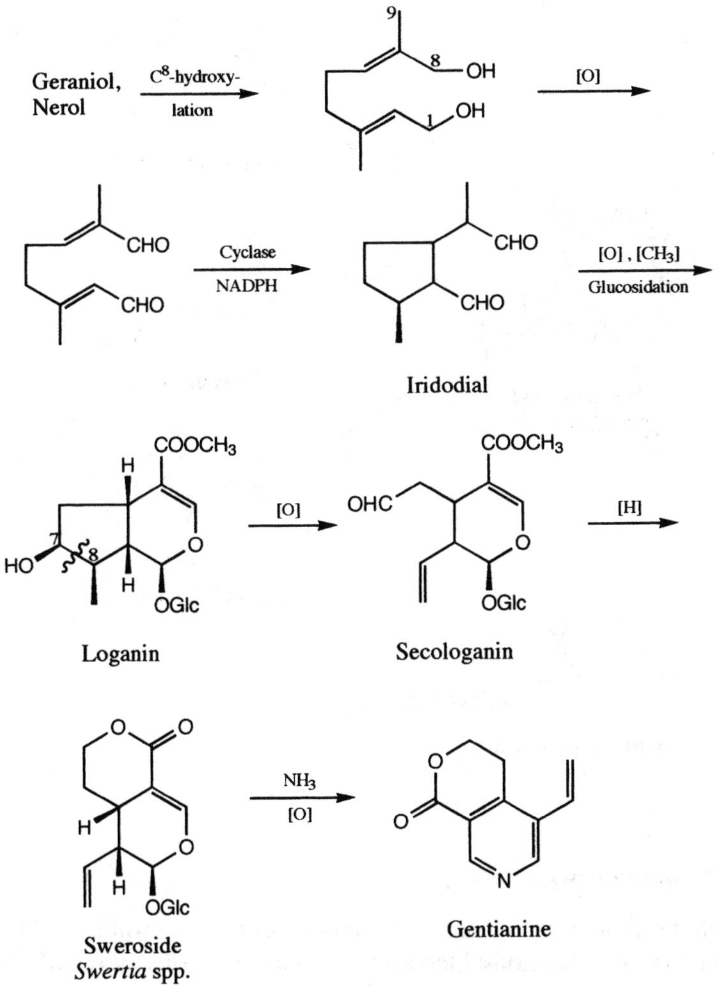

Fig. 9   Biosynthesis of iridoids

thylallyl pyrophosphate. Nucleophilic aliphatic substitution gives chrysan-themyl pyrophosphate and opening of the ring gives the various alcohols in a reaction initiated by solvolysis of the pyrophosphate (11). Geraniol and nerol are not obligatory precursors for these monoterpenes because extensive scrambling of label was observed when these alcohols were incorporated into artemisia ketone by *Artemisia annua*.[27]

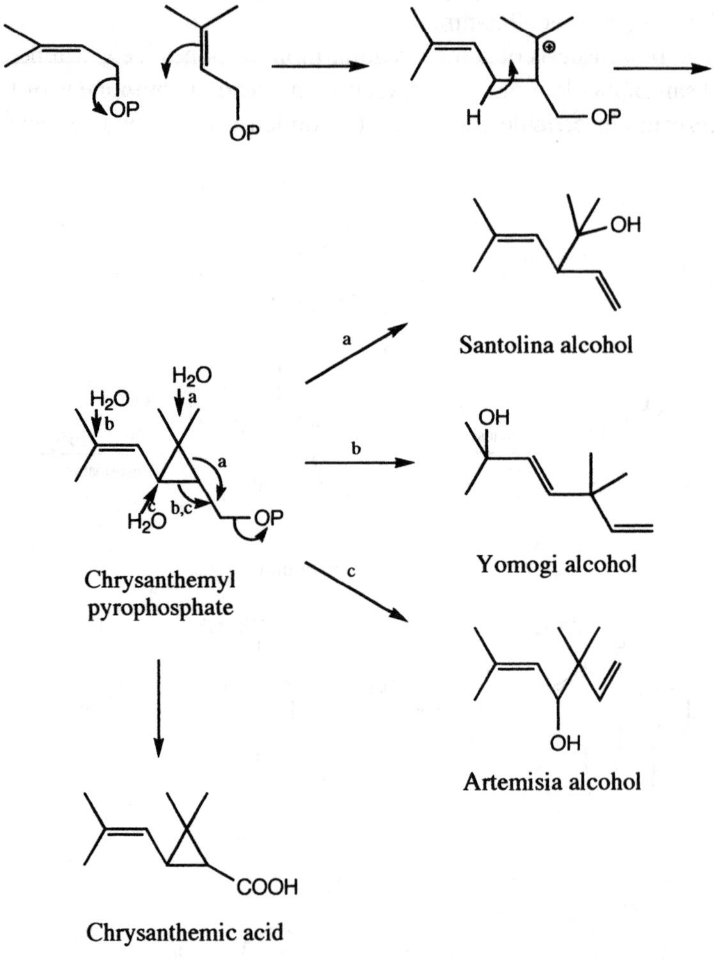

Santolina alcohol

(11)

Yomogi alcohol

Chrysanthemyl
pyrophosphate

Artemisia alcohol

Chrysanthemic acid

6.4 Sesquiterpenes

When the chain length increases, the number of conceivable cyclizations and secondary modifications increases dramatically. This is manifested in the large number of sesquiterpenes isolated from all parts of the plant and animal kingdoms. The structural variation is impressive, especially within the sesquiterpene and diterpene series. None the less the skeletons can be derived by suitable folding of the chain and by applying the basic reactions outlined in Fig. 3. The success of this semi-theoretical approach and the success of many syntheses inspired by biosynthetic principles lend strength to our conceptions of enzymes at work.

A few acyclic sesquiterpenes are known (Fig. 10). Dehydration of farnesol gives farnesenes. Furanofarnesenes are detected in such different phyla as marine sponges, ants, and sweet potato infected by *Ceratocystis fimbriata*. It is suggested that the ethyl groups in the juvenile hormone come from propionic acid by the synthesis of the methyl homologue of mevalonic acid.[28] The juvenile hormone prevents the metamorphosis of larvae and hampers the development of insects into adults. Compounds of this type have interest as insecticides.

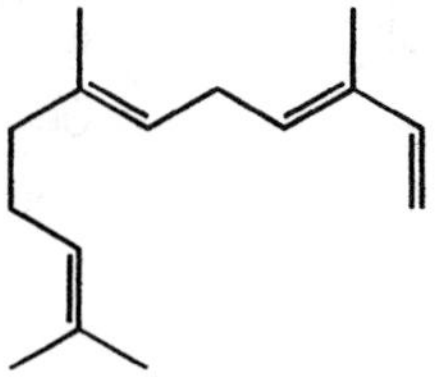

ß-Farnesene
Populus balsamifera

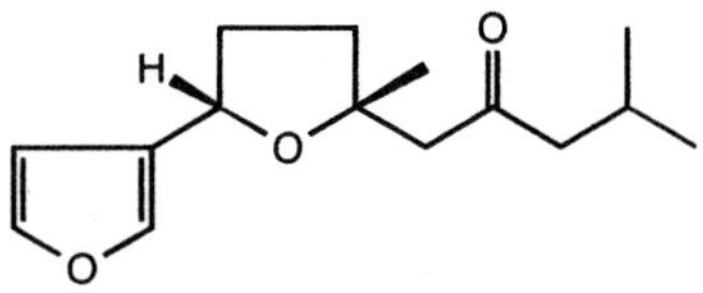

Ipomoeamarone
Ipomoea batatas

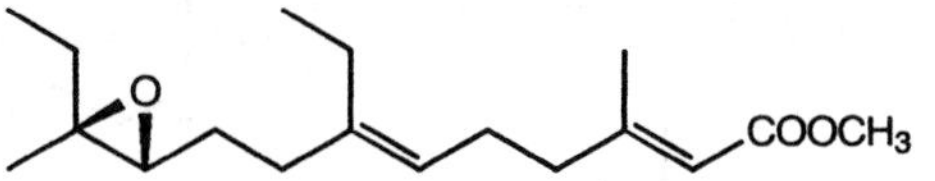

Juvenile hormone
Platysamia cecropia

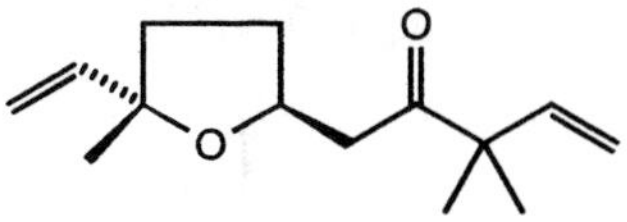

Artemone
Artemisia pallens

Fig. 10   Linear sesquiterpenes

When *cis-trans*-farnesyl pyrophosphate reacts analogously to neryl pyrophosphate, we obtain initially the corresponding menthane, bornane, pinane and carane skeletons (Fig. 11). Sirenin has a unique function as sperm attractant in *Allomyces*, a marine mould. The folding of the chain was demonstrated for the menthane derivatives, γ-bisabolene and paniculide B by feeding tissue cultures of *Andrographis paniculata* with 1,2-^{13}C-acetic acid.[29] The positions of the labels were located by measuring the ^{13}C NMR shifts, intensities, and $^1J_{13c13c}$, and were found to be in agreement with the biosynthetic scheme, Figs 11, 12. The biosynthesis of ovalicin proceeds via β-bergamotene.[30] [13]C_2 labelled acetic acid is incorporated in agreement with predictions. *cis-trans*-Farnesol, but not *cis-cis*-farnesol is incorporated.

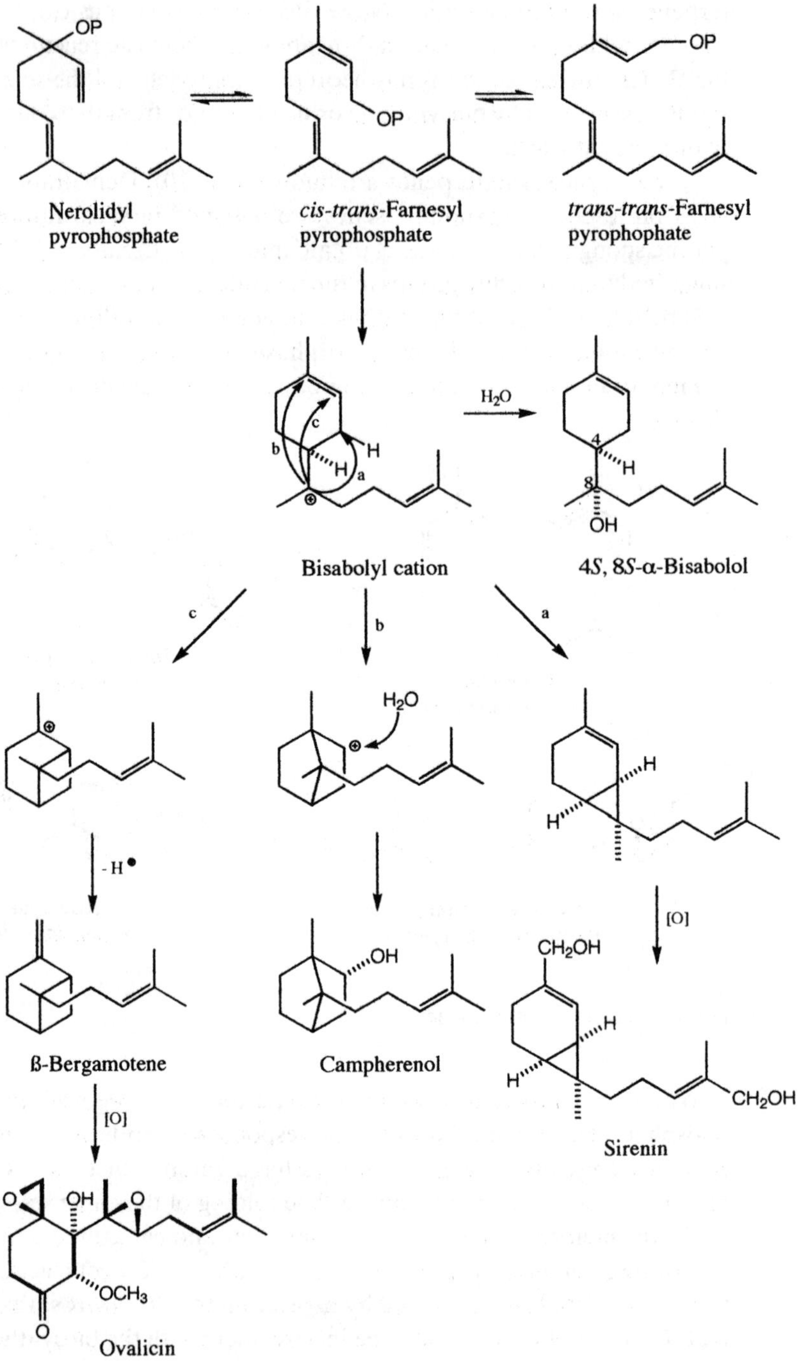

Fig. 11   Generation of menthane, bornane, pinane, and carane skeletons from *cis–trans*-farnesyl pyrophosphate

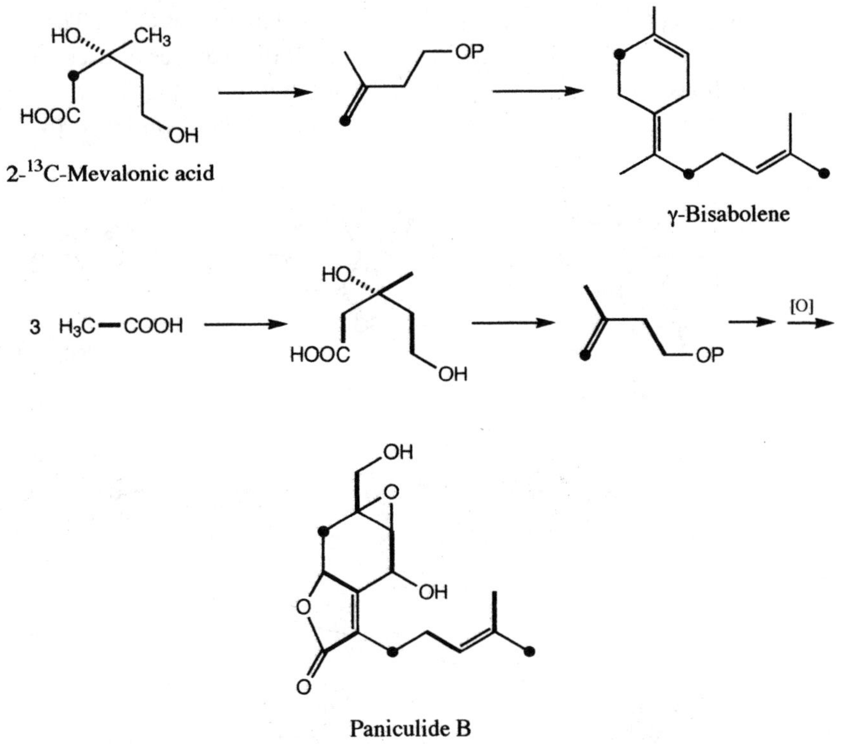

2-^{13}C-Mevalonic acid

γ-Bisabolene

3 H$_3$C—COOH

Paniculide B

Fig. 12   Biosynthesis of γ-bisabolene and paniculide B

If the C^{10} double bond of the bisabolyl cation is involved in a second cyliza-ton, we obtain the skeletons of the cuparane-widdrane families, Fig. 13. *cis*-Alkylation of the central double bond with C^1 and C^{11} and 1,4-hydrogen shift from C^6 to C^{10} create a new carbonium ion at C^6. Wagner-Meerwein shifts of the methyl groups *trans* to the leaving C^6–H give the trichothecane skeleton, route b, whereas a 1,2-C^7, C^6 shift leads to ring enlargement and the chami-grane spiroskeleton, route a. A second ring enlargement, C^5–C^7, followed by a homoallylcyclopropylcarbinyl rearrangement, gives thujopsene and finally widdrol.

It was shown by deuteration of C^1 (starred carbon atom) that this carbon atom had switched position with C^2 in widdrol. Widdrol can, as a matter of fact, be dissected into isoprene units, linked head to tail, but as it appears from Fig. 13, these fragments do not any longer represent the original build-ing blocks.

The biosynthesis of trichothecane mycotoxins and the related fungal qui-none helicobasidin have been studied in detail. The folding of farnesyl pyro-phosphate was established by incorporation of 2-^{13}C-mevalonic acid in trichothecolone.[31] *cis–trans*-Farnesol will be labelled at $C^{4,8,12}$. The label was observed by ^{13}C NMR spectroscopy at the dotted centres of trichothecolone and it showed that the C^{12} methyl had migrated to C^7, Fig. 13. The 1,4-hydrogen shift was established by incorporation of 6-^3H-farnesyl pyrophosphate.

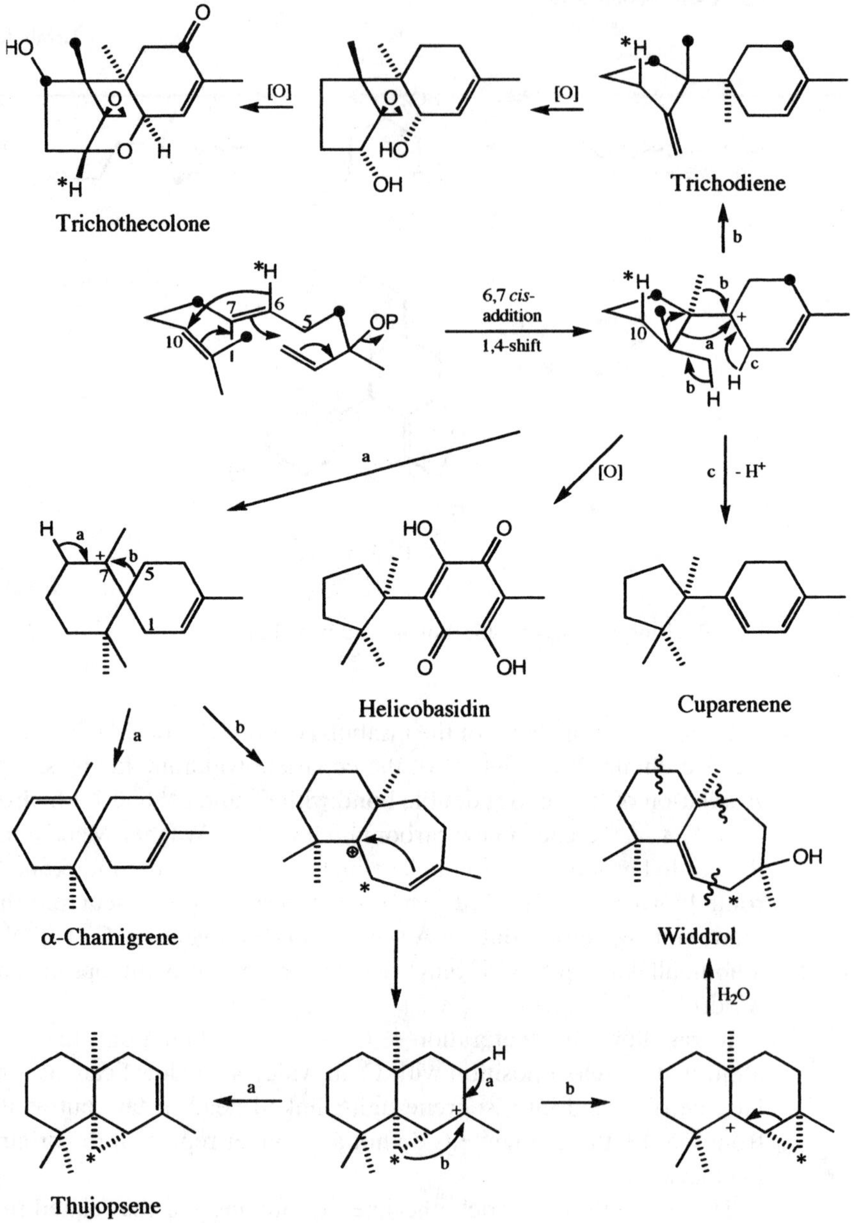

Fig. 13   Derivation of some sesquiterpene skeletons

Tritium was recovered with no loss at C^{10} which showed that the subsequent hydroxylation at C^{10} occurs with retention of configuration.[32] It is found that trichodiene, but not γ-bisabolene, is a precursor for trichothecolone.[33]

The biosynthesis of acoranes and cedranes[34] involves an initial formation of the bisabolyl cation followed by 1,2-hydrogen shift and cyclization C^6 to C^{10}. Deprotonation gives α-acoradiene, whereas a second cyclization, C^{11} to C^2, leads to the tricyclic α-cedrene, a constituent of *Juniperus virginiana* (Fig. 14).

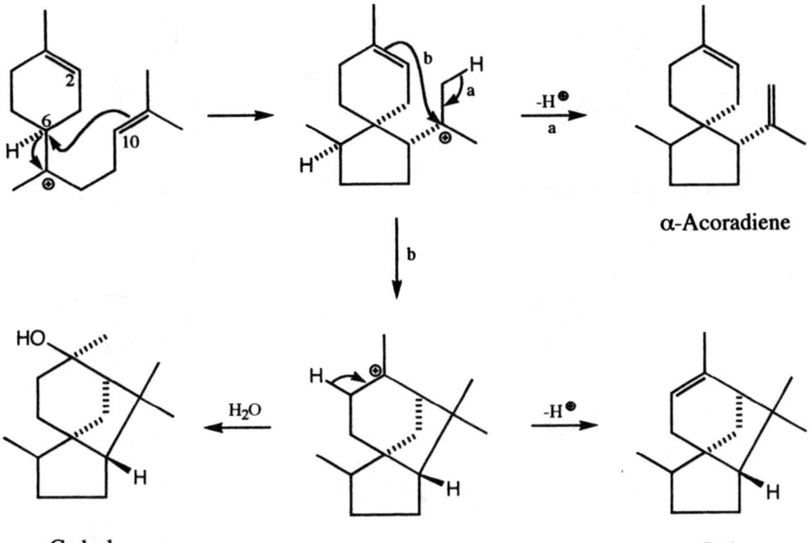

Fig. 14   Biosynthesis of the acorane and cedrane skeletons

Both *cis–trans*- and *trans–trans*-farnesyl pyrophosphate may undergo direct ring closure with the terminal double bond (Fig. 15). The *trans*-2,3 double bond isomerizes at some instant to a *cis* configuration, probably before cyclization in conjunction with the solvolysis of the pyrophosphate as discussed above. Various decalins and hydroazulenes are formed via the ten-membered germacrane ring system. The formation of α-cadinol involves an attack of C^1 at C^{10} followed by a 1,3 hydrogen shift from C^1 to regenerate the reactive centre at C^1, and a second ring closure C^1 to C^6. The direct 1,3 shift is proved by C^1–H labelling. A double 1,2 shift is eliminated by retention of the C^{10}–H.[35,36] Both *cis* and *trans* fusion of the rings can occur depending upon the folding of the chain. The spermotoxic and antitumour agent gossipol contained in the seed of the cotton plant *Gossypium arboreum* belongs to the cadinane type of sesquiterpenes. The decaline ring system is aromatized, hydroxylated, dimerized via phenol oxidation and finally methylated.[37] Fig. 15 shows the labelling pattern from incorporated $^{13}C_2$-acetate. *trans–trans*-Farnesol gives an

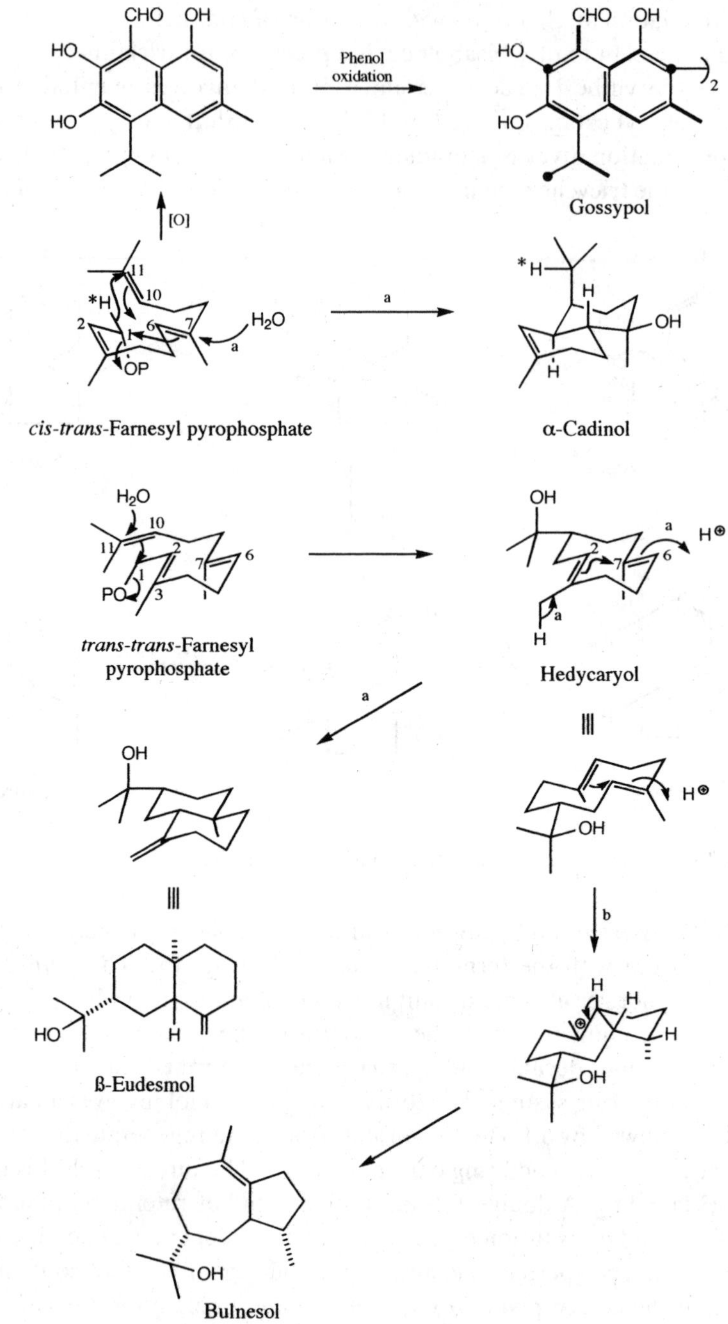

Fig. 15   Biosynthesis of decalin and hydroazulene derivatives

isomeric cation by cyclization of C^1 to C^{10}. It is neutralized by water to hedycaryol. Further protonation at C^6 and cyclization of C^2 to C^7 in the Markovnikov sense gives β-eudesmol, route a. Anti-Markovnikov protonation at C^7 and cyclization of C^2 to C^6 lead to the hydroazulene derivative, bulnesol, route b.

Cyclization of *trans–trans*-farnesyl pyrophosphate C^1 to C^{11} leads to the humulyl cation which gives humulene by proton elimination (Fig. 16, route a) or undergoes a further cyclization to caryophyllene, route b. Cyclization of *cis–trans*-farnesyl pyrophosphate C^1 to C^{11} followed by a 1,3 hydrogen shift from C^1 to C^{10} and cyclization of C^1 to C^6 gives himachalol on neutralization with water (Fig. 16).

The preceding cyclizations are initiated by an enzyme-mediated solvolysis of the phosphate group. A category of sesquiterpenes is formed in an entirely different manner not involving the phosphate group (Fig. 17). The cycliza-

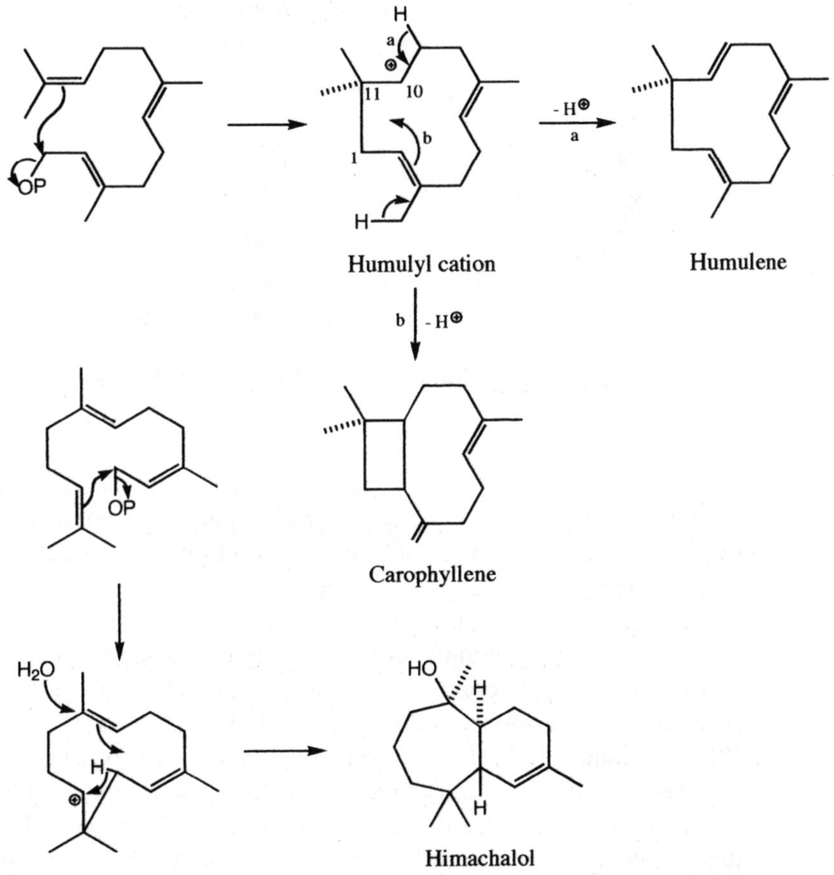

Fig. 16 Biosynthesis of humulanes, caryophyllanes, and himachalanes

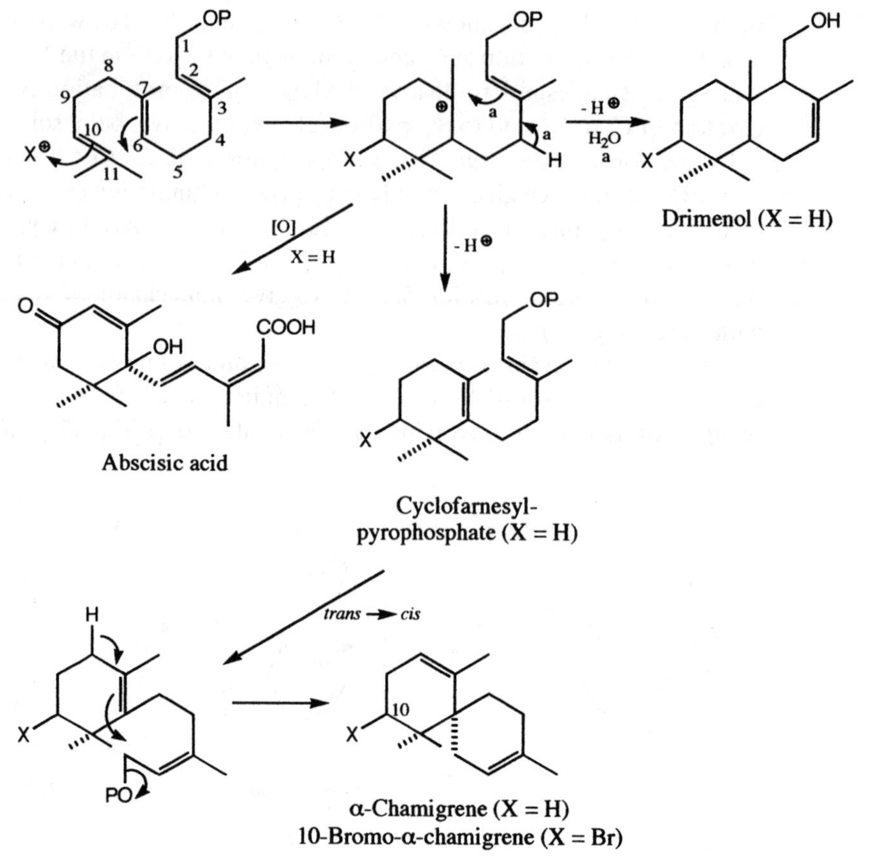

Fig. 17   Sesquiterpene cyclizatons initiated by electrophilic attack at C^{10}

tion of C^{11} to C^6 is initiated by an attack of an electrophile ($X^\oplus = H^\oplus$, halogen or epoxide) at C^{10}. A cyclofarnesyl pyrophosphate is formed in the first step by proton elimination. A second concerted cyclization leads either to a chamigrene or to a decalin derivative, drimenol.

There is some ambiguity concerning the biosynthesis of the chamigrene skeleton. It can be generated either by an initial solvolysis of the phosphate, followed by a double cyclization and a cuparene–chamigrene rearrangement (Fig. 13), or by a double cyclization initiated by an electrophile at C^{10} (Fig. 17). Brominated and chlorinated chamigrene derivatives occur frequently in the seaweeds of the genus *Laurencia*, together with a number of halogenated cuparenes, bisabolenes, and monocyclofarnesenes.[38] It is clear that at least in this particular environment the cyclization is initiated by the peroxide–bromide couple at C^{10}. There are also indications that chamigrenes rearrange to cuparenes in *Laurencia* spp.

Abscisic acid, Fig. 17, is a plant hormone controlling the shedding of leaves.[39] It appears to be formed by oxidative cleavage of the carotinoid violaxanthin in higher plants. In fungal systems the compound derives from farnesyl pyrophosphate by cyclization and oxidation.[40]

6.5 Diterpenes

Geranylgeranyl pyrophosphate is the precursor of the diterpenes. Phytol, 6,7,10,11,14,15-hexahydrogeranylgeraniol, forming the lipophilic side chain of chlorophyll, is the most prominent member of the linear diterpenes. A most unusual functional group, the isonitrile group, is produced by some marine sponges and some fungi and its origin is still a mystery. Incorporation of labelled cyanide into marine tertiary isonitrile compounds has been observed.[41] In this case it is not possible to introduce the nitrogen atom by transamination. In some fungal metabolites the nitrogen atom and the carbon framework arise from an amino acid. Tyrosine is the precursor for xanthocillin X isolated from *Penicillum notatum* but the origin of the carbon unit is still unknown.[42] Geranyllinalyl isonitrile, formamide and isothiocyanate are synthesized by *Halicondria* spp.[43] The formamide and thiocyanate are formed from the isonitrile unit but the reverse reactions do not occur. The isonitrile seems to protect the sponge from predators. The enol acetate trifarin has been isolated from the green alga *Caulerpa trifaria* (Fig. 18).[44]

Xanthocillin

The biosynthesis of the majority of diterpenes is initiated by electrophilic attack at the terminal double bond. This triggers a series of cyclizations leading to mono- (rare), di-, tri-, and tetracyclic derivatives.[45,46] The most important member of the monocyclic diterpenes is vitamin A or retinol, essential in the process of vision. It is manufctured by oxidative fission of β-carotene in the intestine and stored in the liver. Since man cannot synthesize vitamin A, it has to be supplied in his food. Vegetables, such as carrots, spinach, and lettuce, are rich in carotenes (section 6.10).

The cyclization normally progresses directly to the bicyclic decalin system which is subsequently discharged by addition of water or proton elimination,

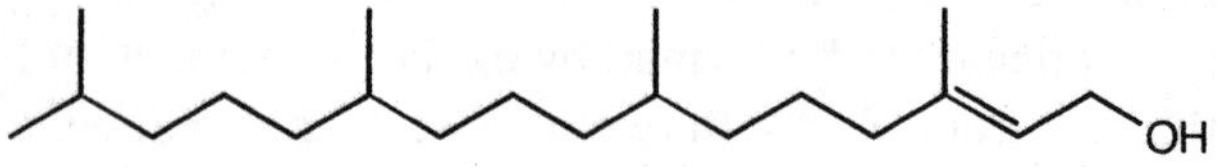

Phytol

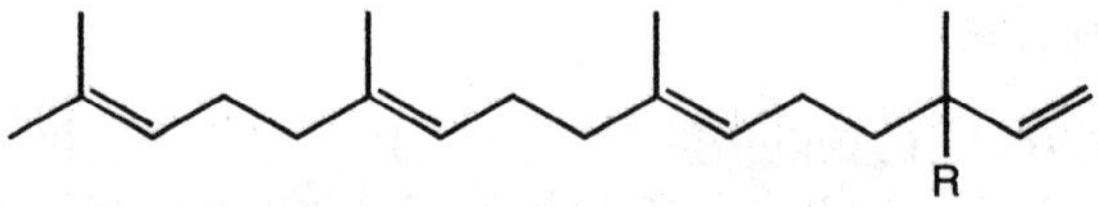

Geranyllinalyl isonitrile   R = N=C
                  formamide   R = NHCHO
                  isothiocyanate   R = NCS

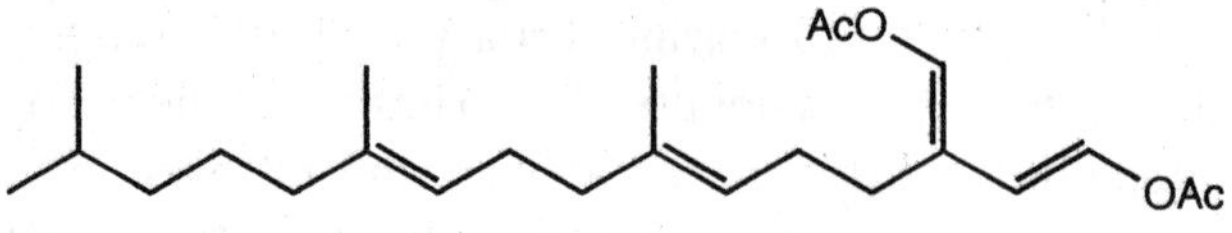

Trifarin

Fig. 18   Linear diterpenes

often via a series of Wagner–Meerwein shifts. The configuration of $C^{5,8,9,10}$ (steroid numbering) is determined by the folding of the *trans–trans*-geranyl-geraniol chain. A chair–chair conformation gives the labdane skeleton (Fig. 19). Hydration of the C^8-carbonium ion and allylic rearrangement in the side-chain give sclareol. Deprotonation to form the C^8 exocyclic double bond produces labdadienol pyrophosphate which is an important intermediate for bio-synthesis of tri- and tetracyclic diterpenes. A five-step concerted Wagner–Meerwein shift with inversion at each centre, route a, leads to the skeleton of hardwickiic acid. Modifications in the side chain are less stereospecific. It is characteristic for diterpenes that often both enantiomeric forms are pro-duced, occasionally in the same species. The cyclizations have been imitated with great success *in vitro* and consequently the mode of cyclization is not en-tirely determined by the topology of the enzyme but is stereoelectronically controlled.[47,48]

Solvolysis of labdadienyl pyrophosphate gives, via the 8-pimarenyl cation, pimarane and abietane derivatives, widely distributed in *Coniferae*. Forma-tion of rosenonolactone is rationalized by a double Wagner–Meerwein shift, $C^9 \rightarrow C^8$ and $C^{10}CH_3 \rightarrow C^9$, route a, giving a carbonium, ion at C^{10} which fin-ally is neutralized either by water or by direct lactonization. The scheme was verified by incorporation of $(4R)$-4-^{3}H, 2-^{3}H$_2$ and 5-^{3}H$_2$ mevalonic acid in rose-nonolactone by the fungus *Trichothecium roseum*[49]. 4-^{3}H-Mevalonate will be *trans-trans*-geranylgeranyl pyrophosphate

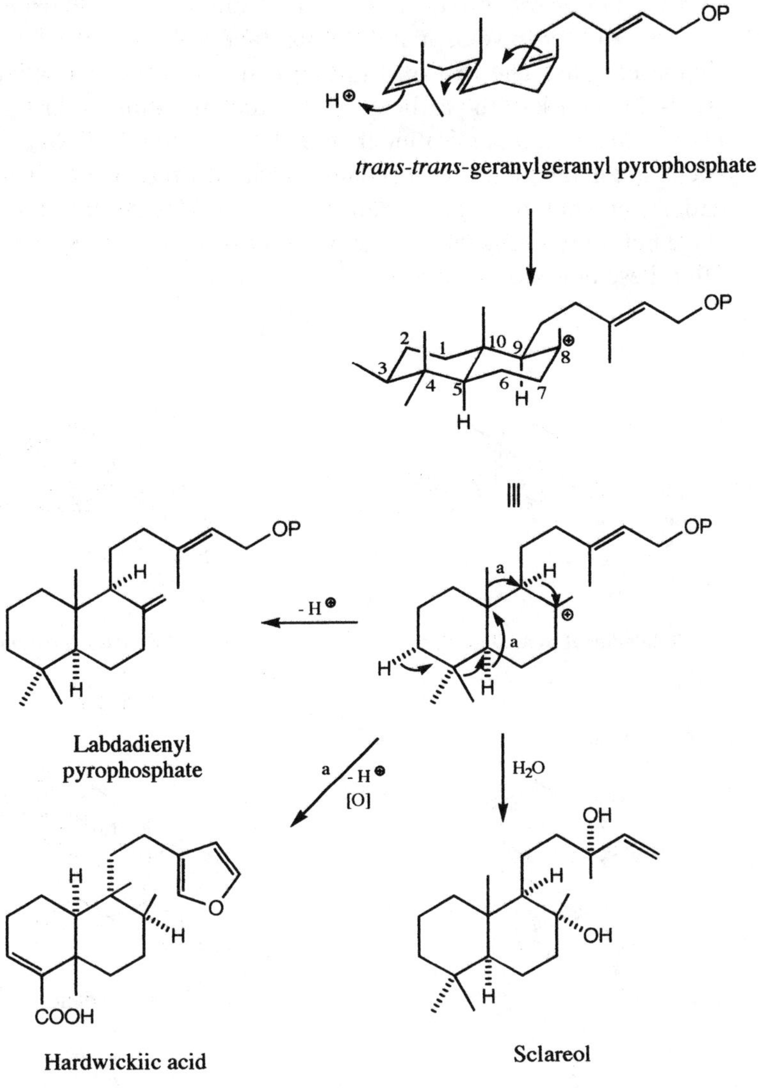

Fig. 19 Cyclization of all-*trans*-geranylgeranyl pyrophosphate in a chair–chair conformation to bicyclic diterpenes

Labdadienyl
pyrophosphate

Hardwickiic acid

Sclareol

incorporated at $C^{5,9}$ of the pimarane skeleton and ^3H was recovered at $C^{5,8}$ in rosenonolactone which confirms the 1,2 shift C^9–$H \rightarrow C^8$ and excludes a $C^{5,10}$-ene as intermediate. C^1 and C^6 will be ^3H-labelled by 2-$^3H_2$- and 5-$^3H_2$-mevalonic acid, respectively. Both hydrogens were recovered in rosenonolactone and exclude $C^{1,10}$- or $C^{5,6}$-enes on the pathway thus supporting either hydroxylation or *cis*-lactonization of the C^{10} centred carbonium ion (Fig. 20). Concerted Wagner–Meerwein shifts proceed as a rule in a *trans* fashion, i.e.

by inversion. In order to keep to the rule, water should add from the rear side or one could conceive of an interfering stabilization of the C^{10} carbonium ion by the enzyme. The hydroxyl group or the enzyme is finally displaced by front–side attack of the carboxyl group. Incorporation experiments with *Trichothecium roseum*, indicating that oxidation of the C^4-β-CH_3 group is a late event, are in favour of this mechanism. On the other hand, the direct lactonization, implying that the oxidation of the β-CH_3 group occurs at an early stage before rearrangement, is supported by an *in vitro* synthesis, mimicked after these biogenetic principles.[50]

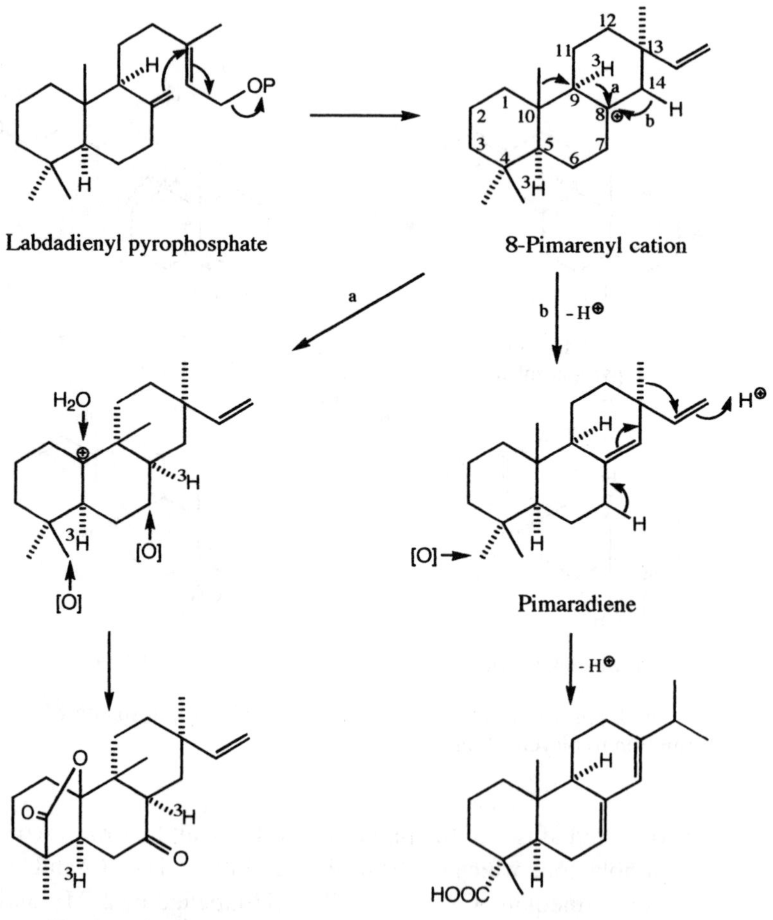
OP
Labdadienyl pyrophosphate
8-Pimarenyl cation
a
b
- H⊕
H₂O
[O]
Pimaradiene
H⊕
Rosenonolactone
Abietic acid
HOOC

Fig. 20 Biosynthesis of tricyclic diterpenes

Oxidation of ring C of the abietanes leads to a number of aromatic diterpenes. C^{14}–O opening of an assumed epoxide at $C^{13,14}$ accounts for the C^{13} to C^{14} migration of the isoprenyl group in totarol (Fig. 21).

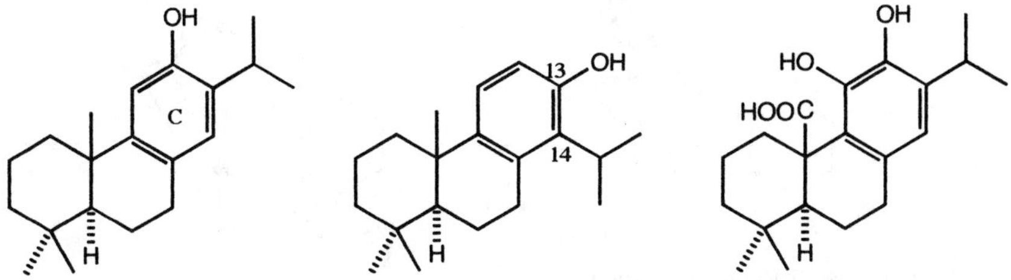

Ferruginol
Podocarpus ferruginea

Totarol
Podocarpus ferruginea

Carnosic acid
Rosmarinus officinalis

Fig. 21 Aromatic diterpenes

The tetracyclic gibberellins are of importance as plant growth hormones. They control cell elongation and were first isolated from the fungus *Gibberella fujikuroi*, a parasite of rice causing reduced straw stiffness as a result of formation of excessively long, limp straw cells. This feared infection causes great losses of rice crops. It was later found that gibberellins are produced universally by plants in small quantities as natural plant growth hormones. The chemistry and biosynthesis of gibberellins have been studied extensively.[51,52] The immediate precursor is *ent*-kaurene derived from *ent*-labdadienyl pyrophosphate = copalyl pyrophosphate (Fig. 22). C^{19} of *ent*-kaurene is oxidized to a carboxylic group followed by 7-α-hydroxylation, 6-β-H-abstraction and ring contraction, possibly via a diol-one rearrangement, to gibberellin A_{12} aldehyde, the first detectable derivative with a gibbane skeleton. The aldehyde is then oxidized to the dicarboxylic acid.[53–56] The order of the following oxidative events is dependent on the metabolizing organism. There is consensus that hydroxylation occurs first at either C^3 or C^{13} followed by oxidative elimination of C^{10}-CH_3 as carbon dioxide with concomitant formation of the lactone bridge. There is no loss of oxygen from ^{18}O-labelled C^{19}-carboxylate and no loss of hydrogen atoms from adjacent centres. These findings suggest a peracid as a plausible intermediate (12). Dehydrogenation of $C^{1,2}$ is a late event.

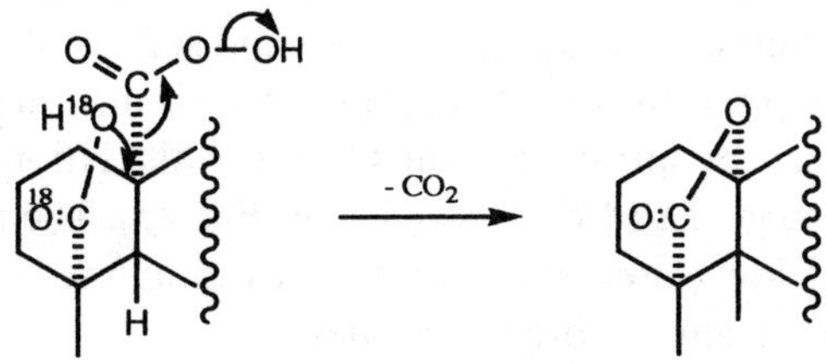

(12)

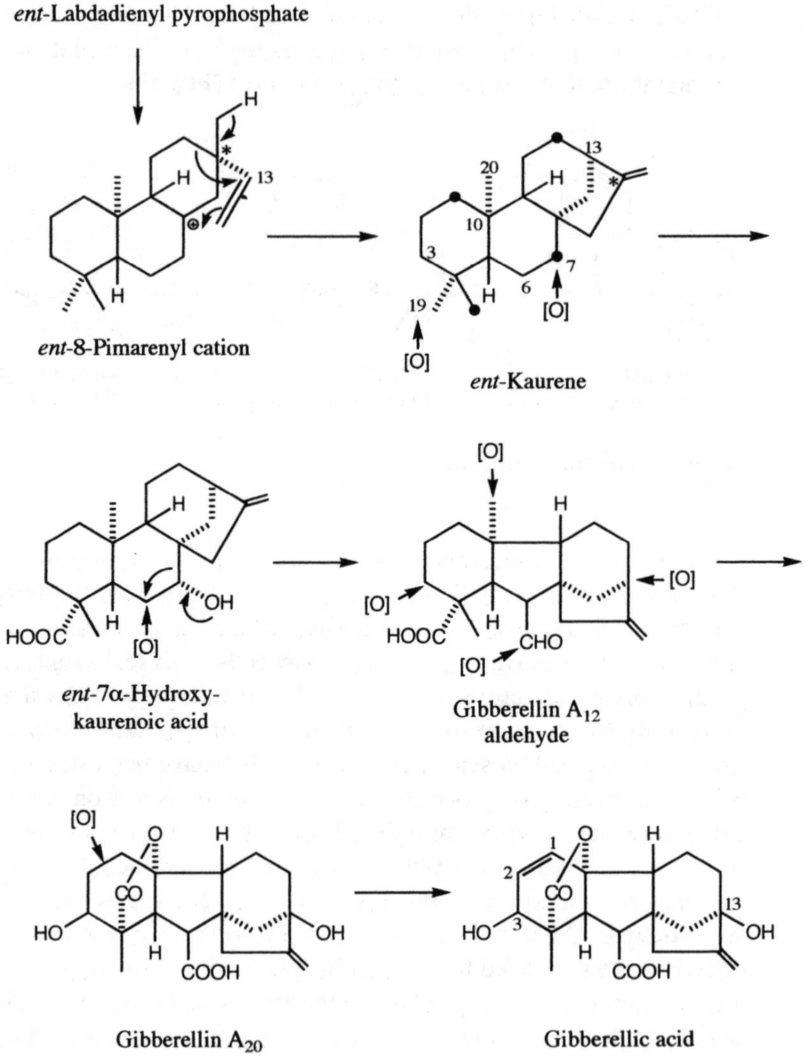
ent-Labdadienyl pyrophosphate
ent-8-Pimarenyl cation
ent-Kaurene
[O]
ent-7α-Hydroxy-kaurenoic acid
Gibberellin A12 aldehyde
Gibberellin A20
Gibberellic acid
HOOC
CHO
COOH
HO
OH

Fig. 22 Biosynthesis of gibberellic acid

Macrocyclic diterpenes,[57] cembranes, have been isolated from *Pinus* spp., from marine coelenterates such as sea fans and soft corals, and also from the termite *Nasutitermes exitiosus* where they act as scent-trail pheromones. They are biosynthesized by intramolecular C^1 to C^{14} alkylation of geranylgeranyl pyrophosphate to give a 14-membered ring (Fig. 23). The research on constituents of the shallow water coelenterates was stimulated by the tumour inhibiting effects of certain cembrane lactones.

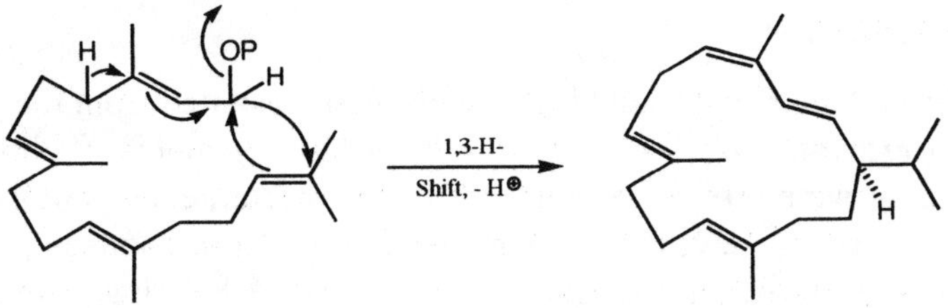

Cembrene
Pinus sibirica

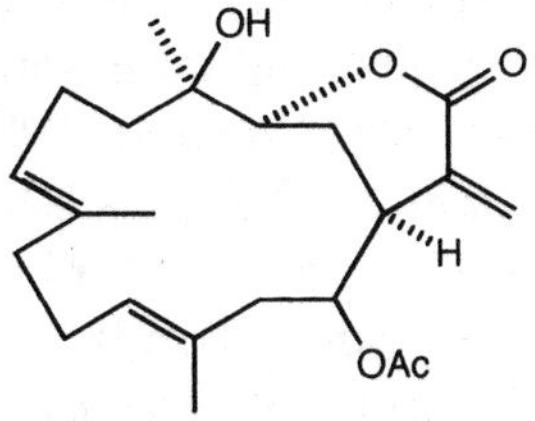

Crassin acetate
Pseudoplexaura porosa

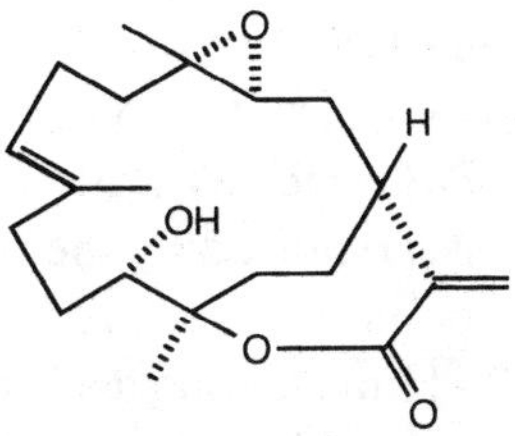

Sinulariolide
Sinularia flexibilis

Fig. 23   Macrocyclic diterpenes

The biosynthesis of taxoids is closely related to that of the cembranes. A slightly different folding of geranylgeranyl pyrophosphate gives rise to the tricyclic taxane skeleton (13) and extensive secondary oxygenation involving molecular oxygen leads to the tumour inhibiting taxol, a metabolite of the yew tree, *Taxus baccata*. We are still awaiting experimental support for the proposed biosynthetic pathway.[57a]

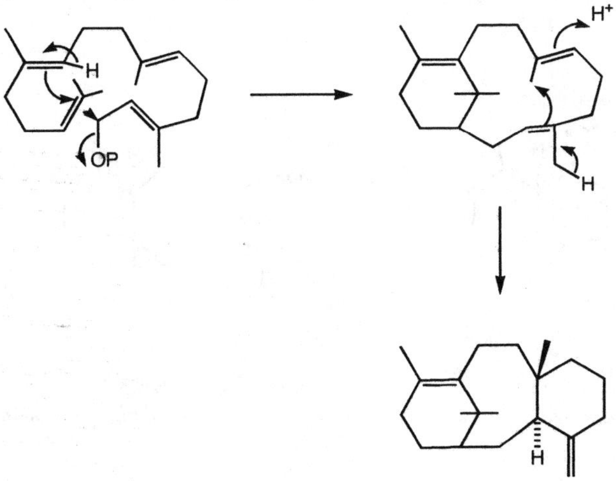

(13)

Taxene
skeleton

6.6 Sesterterpenes[58]

The parent linear compound is geranylfarnesol, isolated from the wax of the insect *Ceroplastes albolineatus*.[59] Until recently members of this group of compounds were rare. Several linear furanosesterterpenes as well as tetracyclic sesterterpenes have now been isolated from marine sponges.[60] A group of tricyclic sesterterpenes, the ophiobolanes, with a 5–8–5 ring-system was isolated from phytopathogenic fungi (Fig. 24). To account for the biosynthesis of ophiobolin A,[61] all-*trans*-geranylfarnesyl pyrophosphate is rearranged to the 2,3-*cis* isomer and folded according to Fig. 25. The cyclization is initiated by alkylation of C^{11} with C^1 phosphate and C^{10} reacts concertedly with C^{14} thus creating a carbonium ion at C^{15}. One of the C^8 hydrogens is transferred in a 1,5 shift to C^{15}. The 5–8–5 ring structure is finally accomplished by $C^{2,6}$ cyclization and *trans* attack by water at C^3. This scheme, which initially leads to ophiobolin F, is supported by the finding that C^3OH derives from water but C^{14}–O in ophiobolin A is derived from atmospheric oxygen.[62] By administration of (R)-2-^3H mevalonic acid to cultures of *Cochliobolus miyabeanus* and degradation of ophiobolin A, it was demonstrated that the label was retained at C^{15} thus proving that the 8-α-H (= (R)-2-^3H of mevalonic acid) is shifted stereospecifically. Apparently the 8-α-H lies closer to the carbonium ion centre at C^{15} than the 8-β hydrogen.[63]

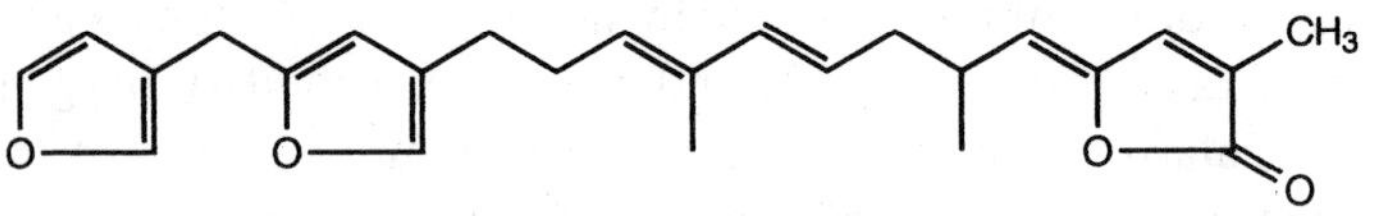

Furanosesterterpene from *Thorecta marginalis*

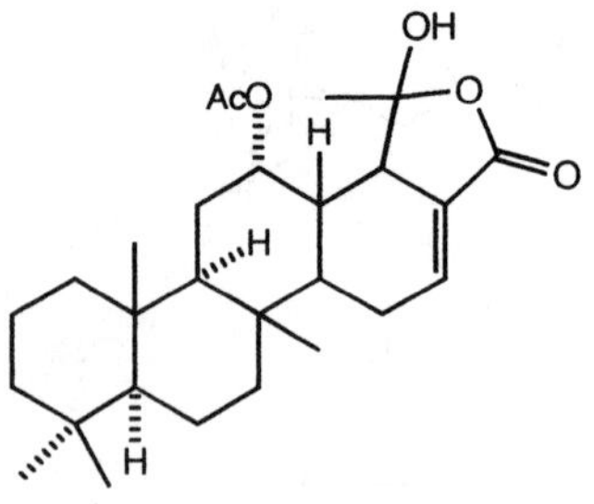

Scalarin
Cacospongia scalaris

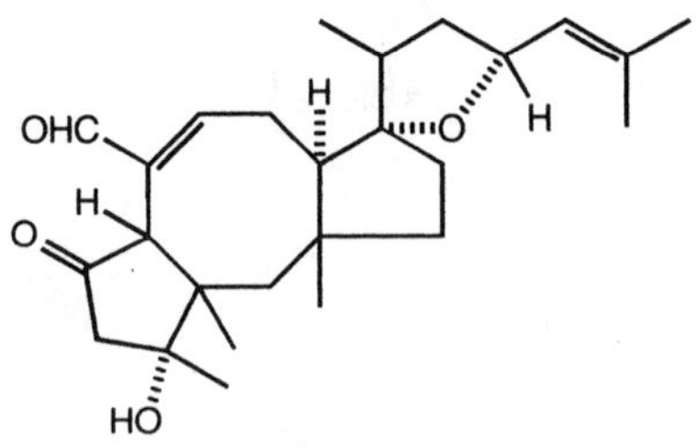

Ophiobolin A
Cochliobolus miyabeanus

Fig. 24   Structure of some sesterterpenes

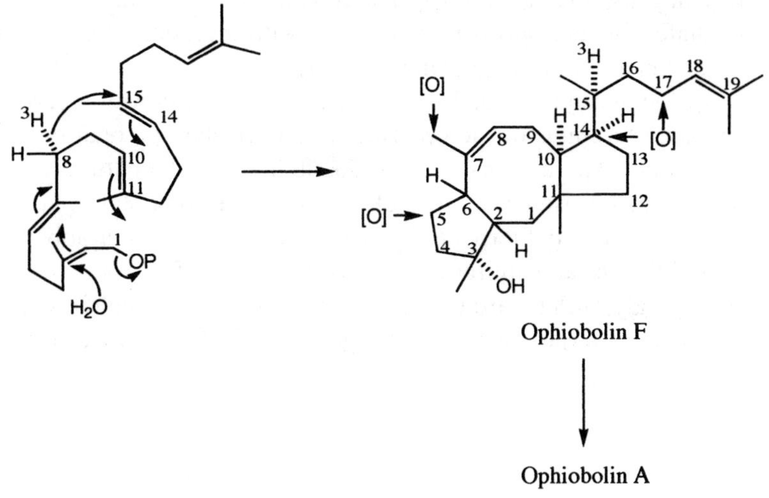

Fig. 25   Folding of geranylfarnesyl pyrophosphate for the biosynthesis of the ophio-bolane skeleton

6.7 Squalene. Triterpenes[64]

Parallel studies on the different aspects of terpene and steroid chemistry gradually focused the interest around a rare C_{30} hydrocarbon, squalene, as a conceivable progenitor of the higher terpenoids. Squalene was first isolated from shark liver, *Squalus* spp., but was later found to be ubiquitously distributed. By folding this compound in certain modes one can construct the basic triterpenoid skeleton with the angular methyls and side chain in correct positions. Experiments soon revealed that squalene was indeed efficiently incorporated into cholesterol.

Squalene consists of two all-*trans* farnesyl groups joined tail to tail. The mechanism of this puzzling coupling remained unsolved for a long period. It was first noted that *one* proton of the central unit was derived from NADPH by using an isotopic labelling technique. The solution to the problem came eventually with the isolation of an intermediate, presqualene pyrophosphate, having a cyclopropane structure.[65,66] The biosynthesis of presqualene pyrophosphate initially follows the same sequence as the synthesis of chrysanthemyl pyrophosphate (section 6.3). The 2,3 double bond of one farnesyl pyrophosphate molecule is alkylated by another farnesyl pyrophosphate with inversion of configuration, and its C^1H_S hydrogen is stereospecifically eliminated thus giving rise to the cyclopropane moiety of presqualene. In the absence of NADPH this compound accumulates. It is formed more rapidly than squalene in yeast microsomes, which is consistent with its function as a true inter-mediate. The cyclopropane rearranges with inversion of configuration at C^4 to the linear all-*trans*-squalene (Fig. 26).

Squalene can be folded in a number of ways both with regard to ring size, start and terminus of the cyclization and prechair–preboat conformation of the chain.[67] There exists also the possibility that the naturally occurring all-*trans* form may isomerize at an olefinic centre at some stage of the cyclization. In the sesqui- and diterpene series the enzymes have admirably demonstrated their diligence and aptitude for the construction of a variety of skeletons, but surprisingly, with regard to the long multifunctional squalene chain, there are few skeletal variations in the triterpene series. With few exceptions the A, B,

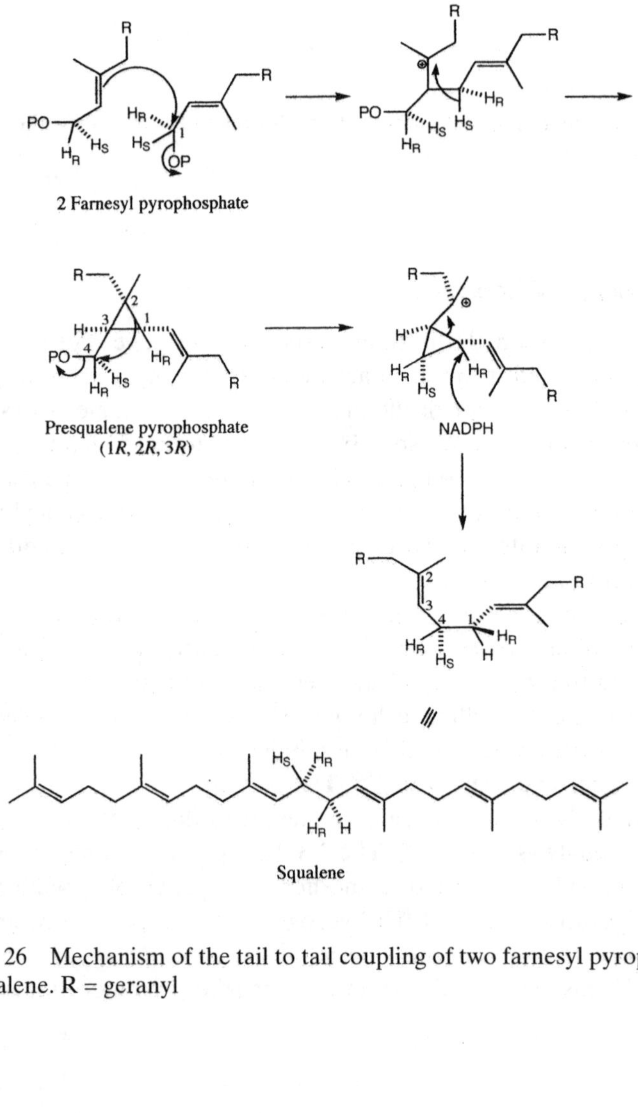

2 Farnesyl pyrophosphate

Presqualene pyrophosphate
(1R, 2R, 3R)

NADPH

Squalene

Fig. 26   Mechanism of the tail to tail coupling of two farnesyl pyrophoshates to squalene. R = geranyl

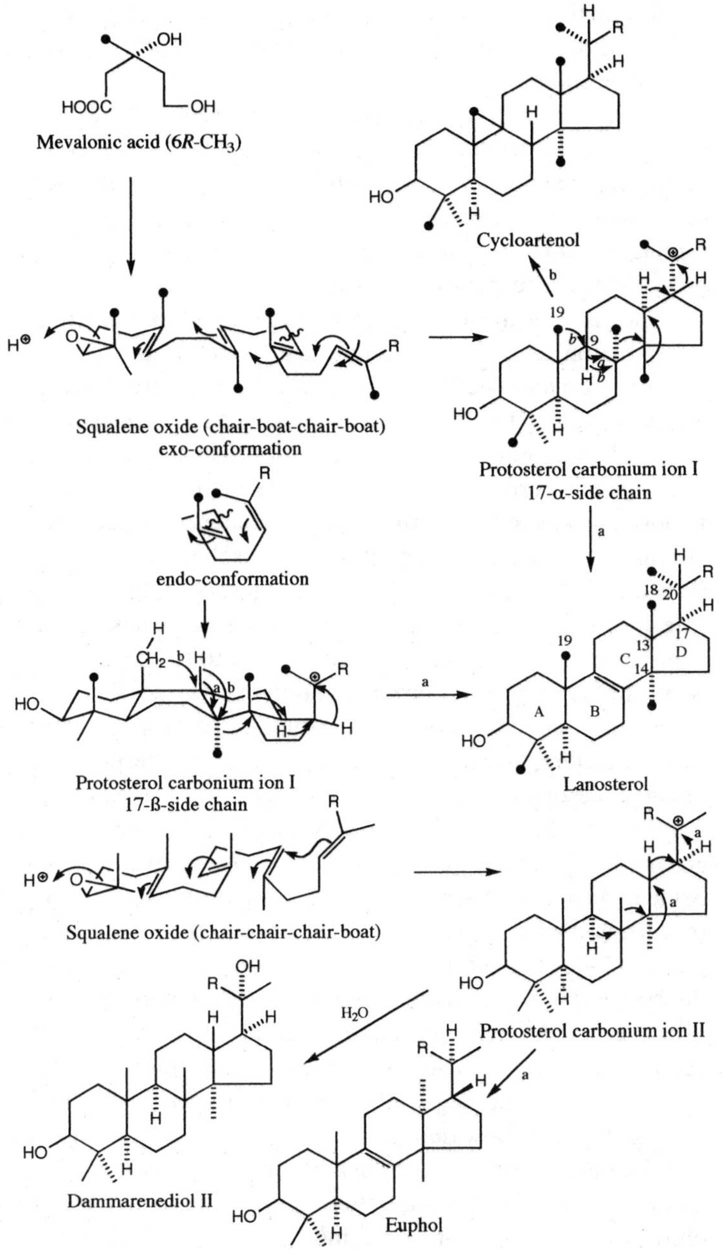

Fig. 27 Biosynthesis of tetracyclic triterpenes. Incorporation of mevalonic acid, labelled in the methyl group, into lanosterol

and C rings are six-membered rings and the cyclization is always initiated at the terminal double bond. However, there are numerous secondary modifications such as oxidations, dehydrogenations and Wagner–Meerwein shifts within the framework.

A salient feature of nearly all triterpenes is the equatorial hydroxy group at C^3. The process leading to this functionality was first formulated as an attack at the terminal double bond of $OH^{\oplus}$ or some less clearly defined biochemical equivalent, which triggers the multiple cyclization. This species is highly improbable in a biological environment but nevertheless useful for rationalization of many triterpenoid structures.

A mechanistically more satisfactory process that leads to the same products is the proton catalysed opening of an epoxide. Epoxide formation is a common reaction mediated by an epoxidase that requires NADPH, FAD and molecular oxygen. Whether cytochromes partake is questionable since carbon monoxide does not affect the epoxidation. The reduction of oxygen to activated hydrogen peroxide can very well be mediated by $FADH_2$ (section 4.3). Lanosterol was indeed formed when squalene-2,3-oxide was incubated with enzyme preparations,[68,69] and it was later isolated from a number of plants. Its formation required molecular oxygen and NADPH.[70] Consequently, it is a true intermediate on the path to the cyclic triterpenes. Squalene is epoxidized at either one of the terminal double bonds to the 3S-isomer indicating that it is assembled on another enzyme and then released.[71]

One can recognize two main groups of triterpenes, the tetracyclic triterpenes including the sterols, and the pentacyclic triterpenes. The conformation of squalene oxide on the enzyme surface and the extent of backbone rearrangement which follows upon cyclization, give rise to a number of subgroups.[46,72] The widely distributed lanosterol derives from (3S)-squalene-2,3-oxide in a chair–boat–chair–boat conformation via the hypothetical protosterol carbonium ion I and a four-step Wagner–Meerwein 1,2 shift with elimination of C^9-H (Fig. 27, route a). Taking the required skeletal movements for the cyclization into consideration, it seems unlikely that the whole process is fully concerted.[73] The enzymatic cyclization is more adequately viewed as involving formation of a series of discrete carbonium ions, the fate of which is governed by the topology of the enzyme. It is worth noting that chemical model experiments show that the opening of the epoxide ring receives considerable anchimeric assistance from adjacent double bonds. The six-membered A and B rings are formed via a thermodynamically favoured Markovnikov addition whereas the C ring is formed via an anti-Markovnikov cycloaddition. The formation of the five-membered E ring in the sterol series follows the Markovnikov rule. The stereoelectronically favoured *trans*-addition to the double bond is a rule in the biosynthesis of triterpenes, in contrast to the biosynthesis of many sesqui- and diterpenes. Hence, there exists a fine balance between stereoelectronic and enzymatic effects. Non-enzymatic acid catalysed cyclization of epoxisqualene gives typically the 6,6,5-fused tricyclic products as expected from thermodynamic considerations.

A large rotation, *ca.* 120°, of the $C^{17,20}$ bond is required for the formation of lanosterol from (3*S*)-squalene-2,3-oxide in a chair-boat-chair-boat-exo conformation via the protosterol carbonium ion I in order to account for the configuration of C^{20} and the presumed *trans*- anti-*trans* mode of the Wagner–Meerwein shifts.[74] The same result can in fact be achieved also from the endo conformation via a least motion pathway and a *syn*-$C^{13,17}$ H-shift, Fig. 27.[75]

It was possible to excise the C^{18} methyl moiety of lanosterol as acetic acid and determine the absolute configuration of the methyl by the malate synthase–fumarase method.[76] *R*-acetic acid was isolated from lanosterol, biosynthesized from mevalonic acid with a chiral methyl (6*R*–CH₃, Fig. 27) which shows that the $C^{14,13}$ methyl shift proceeds stereospecifically with retention of configuration. Experiments with various labelled squalenes confirm in full these 1,2 shifts.[77] One further shift of $C^9H_\beta \rightarrow C^8H_\beta$ and C^9 alkylation by C^{18} leads to cycloartenol containing a cyclopropane ring (route b).

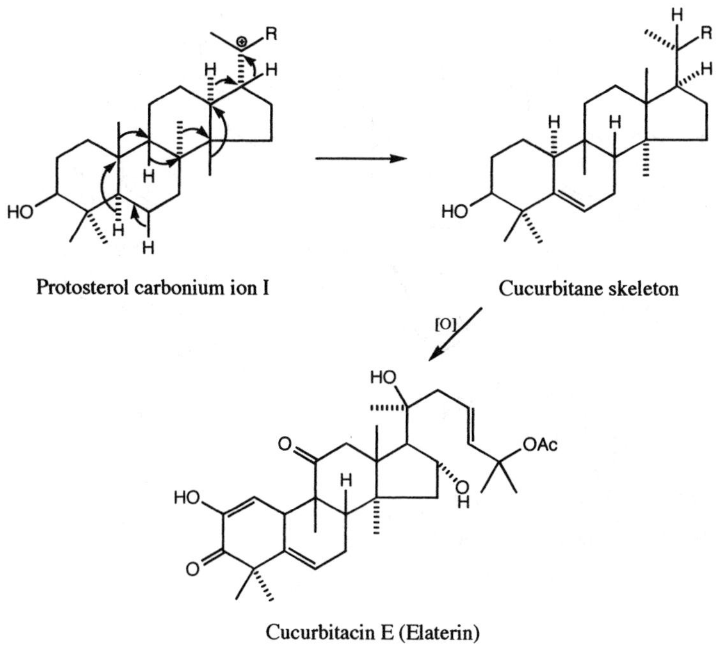

Protosterol carbonium ion I   Cucurbitane skeleton

Cucurbitacin E (Elaterin)

Fig. 28   Variations on the cucurbitane skeleton

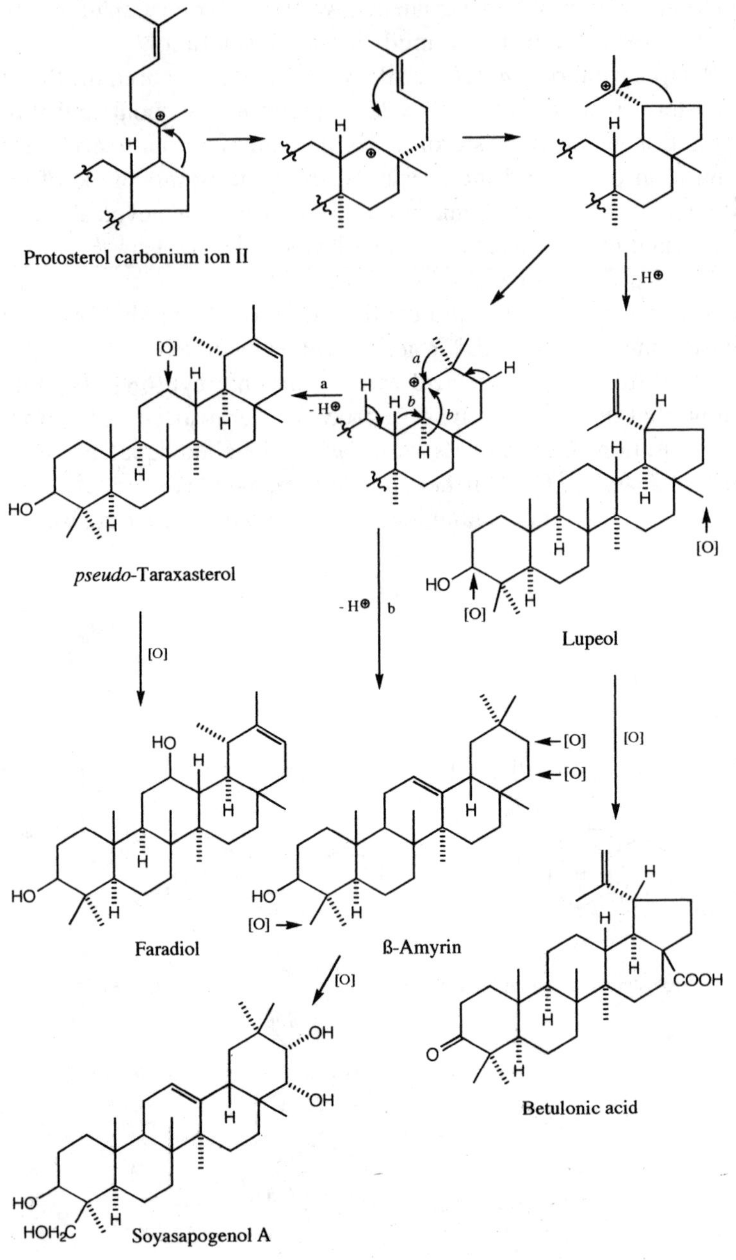

Fig. 29 Biosynthesis of pentacyclic triterpenes with no backbone rearrangements

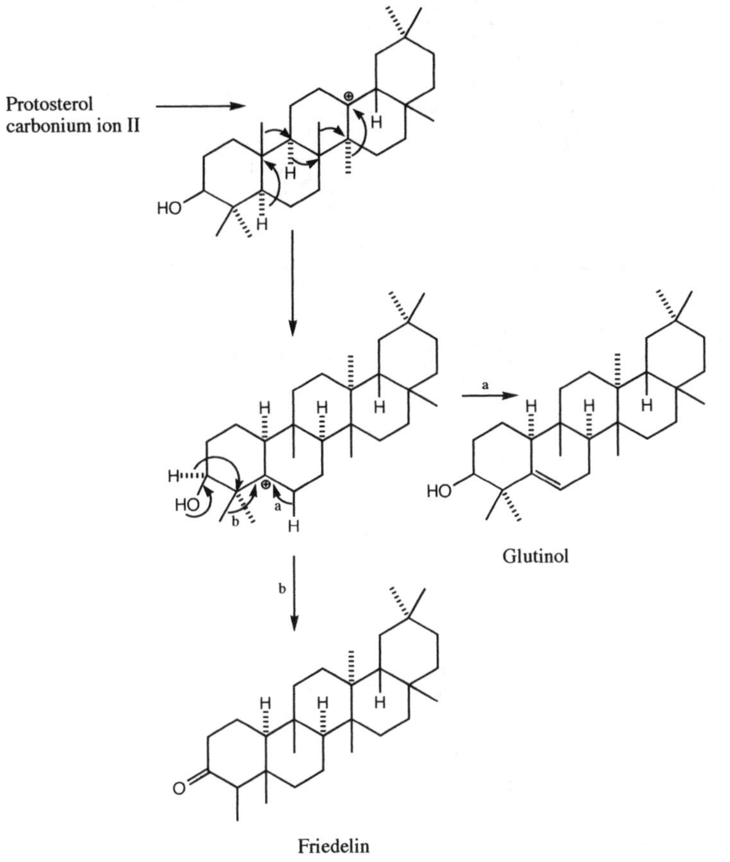

Fig. 30   Biosynthesis of pentacyclic triterpenes with extended backbone rearrangement

A still more deep-seated rearrangement of protosterol carbonium ion I leads to the cucurbitanes,[78] the toxic principles of cucurbitaceous plants, a highly oxygenated group of tetracyclic triterpenes used in medieval medicine and still attracting interest because of their toxicity (Fig. 28).

Most pentacyclic triterpenes derive from protosterol carbonium ion II by expansion of the D ring and cyclization with the side chain R. If we keep the rearrangements within the D and E rings, we arrive at lupeol, β-amyrin and pseudotaraxasterol and their oxygenated derivatives (Fig. 29). A large group of pentacyclic triterpenes are formed by backbone rearrangements, i.e. multi-step Wagner–Meerwein 1,2 shifts leading to friedelanes, glutinanes, and taraxeranes etc. (Fig. 30). These arrangements are not solely the result of

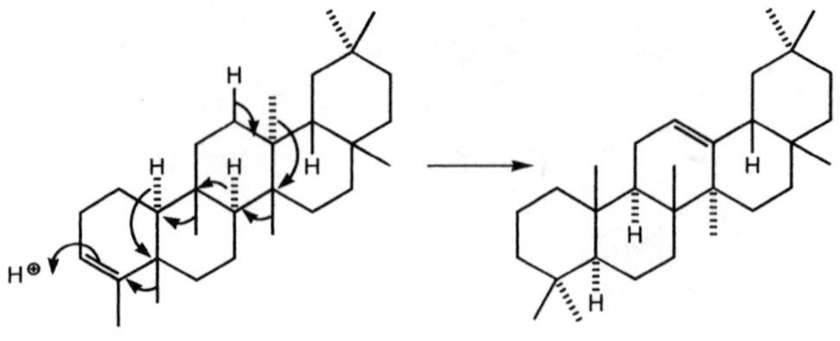

Δ^3-Friedelene Δ^{12}-Oleanene

Fig. 31   Acid induced retrobiosynthetic backbone rearrangement

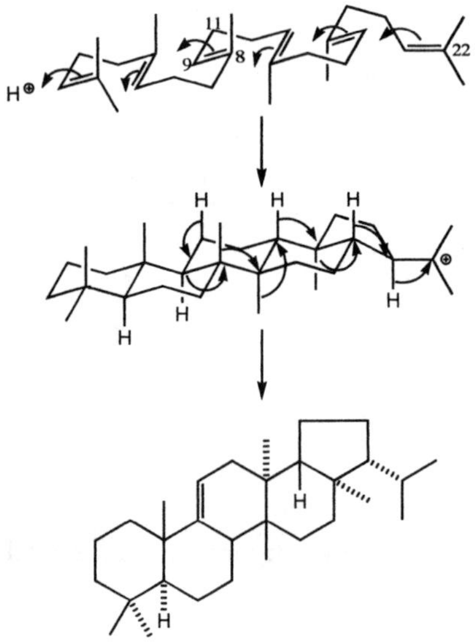

Fern-9-ene

Fig. 32   Biosynthesis of fern-9-ene

specific enzyme action. In these highly condensed aliphatic ring systems, 1,3-diaxial steric interactions, conformational factors and stereoelectronic effects contribute to the driving force of the rearrangements. A great many biosynthetic rearrangements can be imitated in thermodynamically controlled reactions by acid treatment of the substrate.[49] Chemically induced retrobiogenetic rearrangements occur when the metabolite possesses high free energy. This is demonstrated by the concerted retrobackbone rearrangement of Δ3-friede-lene to the more stable Δ12-oleanene[79] (Fig. 31). Cyclizations initiated by epoxide cleavage are rather uncommon in lower terpenoids, whereas this is the rule for the triterpenoids. A small number of triterpenes are derived by an initial protonation, e.g. the filicenes and the fernenes from ferns. The penta-cyclic system is generated from squalene in an all-chair conformation by a se-ries of concerted cyclizations generating a carbonium ion at C^{22} followed by a seven-step concerted Wagner–Meerwein shift that is terminated by elimina-tion of the C^{11} proton (Fig. 32). In accordance with this non-oxidative route to triterpenes is the finding that squalene but not squalene-2,3-oxide is incorpor-ated in the fern-9-ene.[80]

6.8 Secondary modifications of triterpenes

Wagner–Meerwein shifts, introduction of additional hydroxyl and olefinic groups, oxidation of alcohol functions to carbonyl groups and side chain alkylation by *S*-adenosyl methionine are common modifications. The cucur-bitanes (Fig. 28) are characterized by extensive oxidation, but with retention of the basic triterpenoid skeleton. In other cases extensive degradation has taken place as for the quassinoids,[81] the bitter principles of *Simaroubaceae*, for the limonoids, bitter principles of citrus species, belonging to the *Rutaceae* family, and the meliacins of *Meliaceae*. The compounds are typical for these botanical families. The quassinoid skeleton is formed by oxidative cleavage of the side chain and opening of ring D of a tetracyclic triterpene, e.g. apotiru-callol. Feeding experiments with 2-^{14}C, (4*R*)-^{3}H- and 5-^{14}C-mevalonic acids in the seeds of *Simarouba glauca* gave labelled glaucarubinone (Fig. 33). The β-4-methyl is oxidatively decarboxylated via formation of an intermediary 3-β-keto ester as shown by loss of 3-^{3}H and the formation of inactive acetic acid from $C^{4,10,13}$ methyls on Kuhn–Roth oxidation of glaucarubinone derived from (4*R*)-^{3}H- and 2-^{14}C mevalonic acids.[82] Hydrolysis of active glaucarubinone gave inactive 2-hydroxy-2-methylbutyric acid which is derived from isoleuci-ne.[83] Selective degradation confirmed the labelling pattern expected from the tetracyclic triterpene precursor. The finding that 9-^{3}H is retained excludes a precursor with a C^8 double bond.

In the limonoids the side chain is transformed to a furane and ring A is fur-ther oxidized to a lactone. The constituents of *Meliaceae* are closely related to the limonoids but have an intact ring A (Fig. 33).

The fundamental secondary modificaton that leads to the steroids is selec-tive C^4 and C^{14} demethylation. This reaction is separately discussed in the next section.

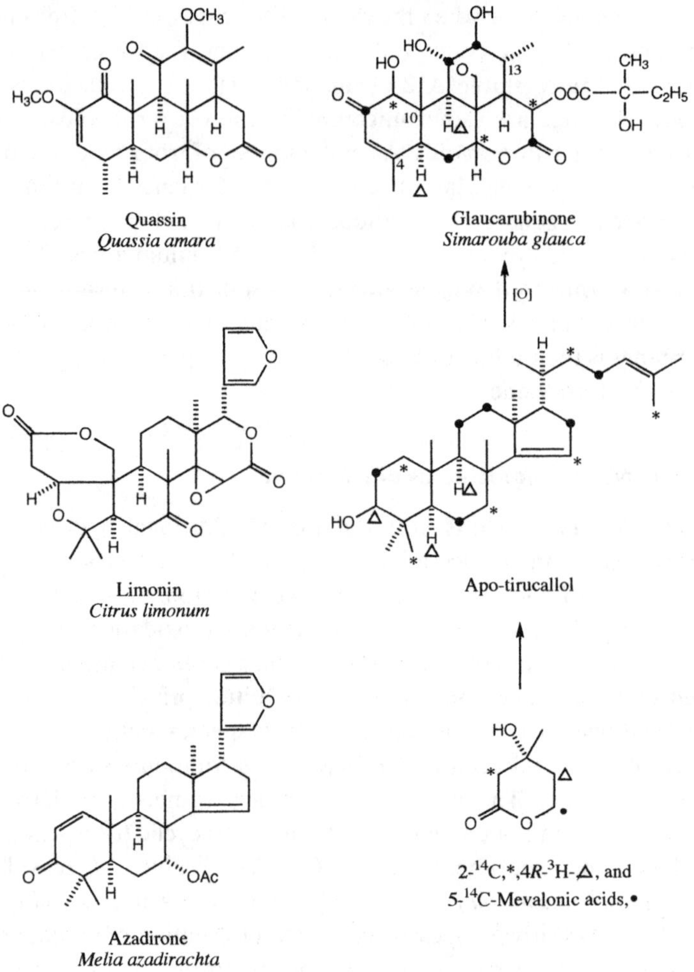

Quassin
Quassia amara

Glaucarubinone
Simarouba glauca

[O]

Limonin
Citrus limonum

Apo-tirucallol

2-^{14}C,*,4R-^{3}H-△, and
5-^{14}C-Mevalonic acids,•

Azadirone
Melia azadirachta

Fig. 33   Partially degraded triterpenes of *Simaroubaceae, Rutaceae* and *Meliaceae*

6.9 Steroids

The steroids comprise a large number of ubiquitous compounds which are divided into subgroups, chiefly according to side chain functionality: sterols, sapogenins, cardiac aglycones, bile acids, adrenal steroids, and sex hormones, all of vital importance for life (Fig. 34). They have a tetracyclic ring system derived from lanosterol, by convention drawn with the axial 18- and 19-methyls on the near face, called the β-side. Substituents on the opposite side are α. The configuration of chiral centres in the side chain are described by the *R,S* system.

The most common sterol of animal origin is cholesterol[84] (section 6.1, Fig. 1). Gall stones are mainly made up of cholesterol. The precursor of cholesterol and animal sterols in general, is lanosterol (Fig. 27) which undergoes $C^{24,25}$ reduction, $C^{4,14}$ demethylation and transposition of the C^8 double bond to C^5.

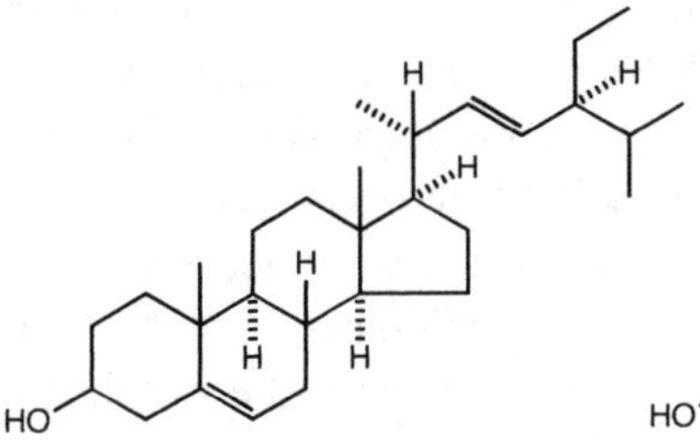

Stigmasterol (Phytosterol)

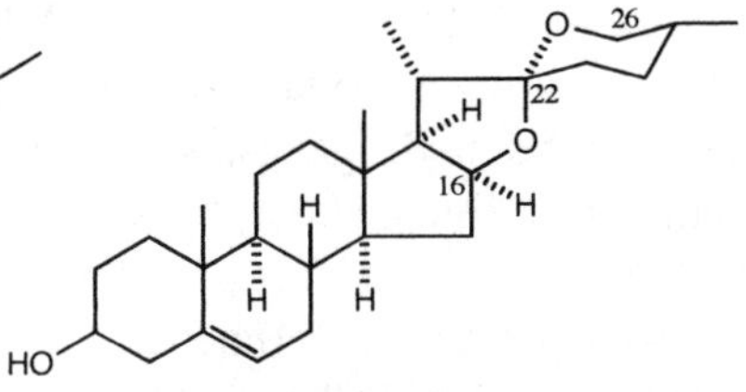

Diosgenin
(Sapogenin)

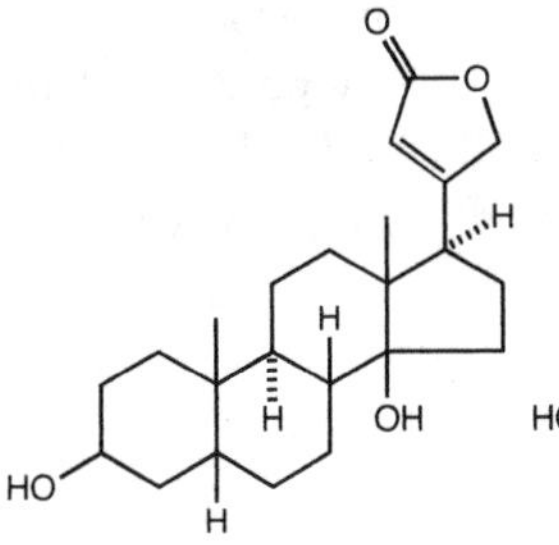

Digitoxigenin
(Cardiac aglycone)

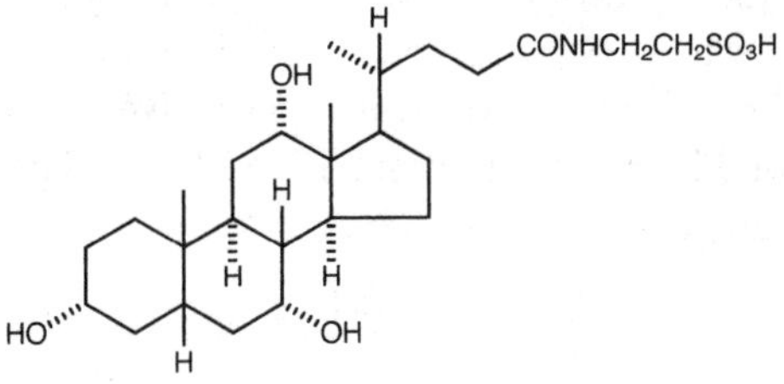

Tauro-cholic acid (Bile acid)

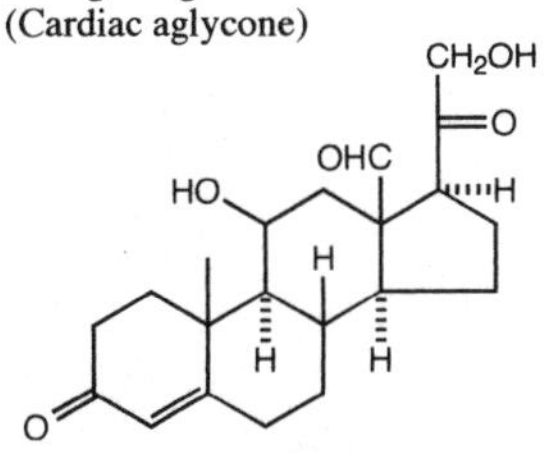

Aldosterone (Adrenal steroid)

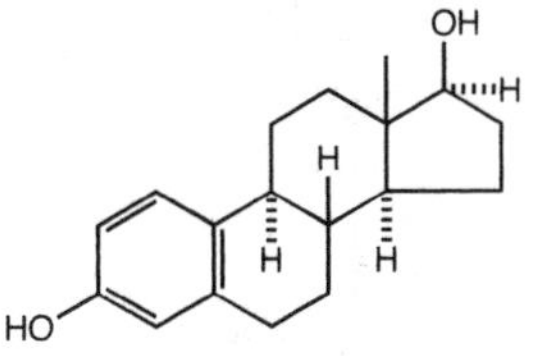

ß-Oestradiol
(Female sex hormone)

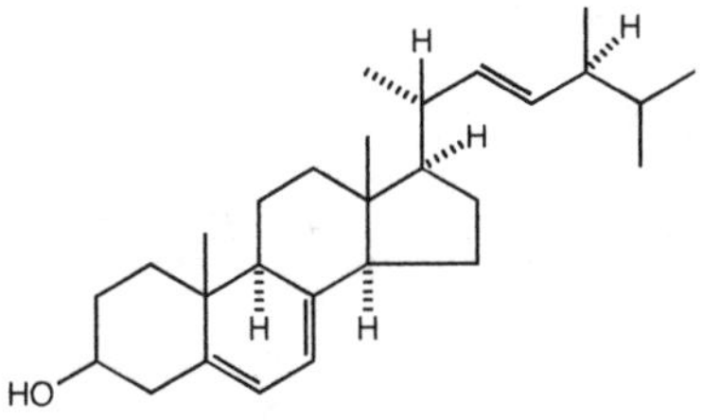

Ergosterol
(Precursor of vitamin D)

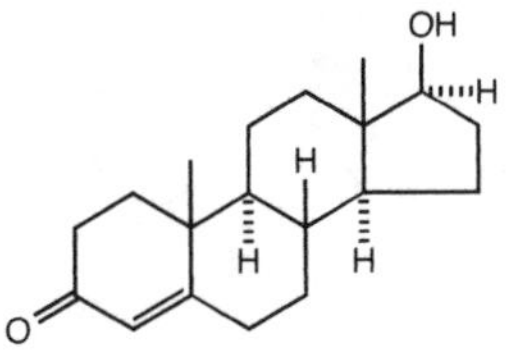

Testosterone
(Male sex hormone)

Fig. 34   Representative steroids

The stage at which the C^{24} double bond reduction occurs, may vary with the organism. The C^{24}–H of lanosterol is derived from 4-pro-*R* H of mevalonic acid and can be labelled with $(4R)$-4-^{3}H-mevalonic acid by incubation of a rat liver enzyme preparation (14). The reduction requires NADPH and its steric course was determined by isolation of the labelled cholesterol, the side chain of which was cleaved by a bovine adrenal enzyme preparation to form 3-^{3}H-4-methyl-pentanoic acid.[85] A Barbier–Wieland degradation followed by Baeyer–Villiger oxidation gives 1-^{3}H-2-methylpropanol, which was subjected to a yeast dehydrogenase known to selectively remove the pro-*R* H. The oxidation to isobutyraldehyde proceeds without loss of activity. This proves that C^{24}–^{3}H of cholesterol is pro-*R*, i.e. the reduction proceeds from the back side = *si*-side at C^{24} (14). The next problem concerns the prochirality at C-25. Cholesterol biosynthesized from 2-^{14}C-mevalonic acid was enzymatically oxidized by a mouse liver preparation to 26-OH-cholesterol shown by X-ray crystallography to be 25*S*.[86] The hydroxymethyl group was furthermore shown to be radioactive, i.e. it is originally *trans* positioned in lanosterol. These facts demonstrate conclusively that the reduction of the C^{24} double bond involves an overall *cis* addition. The C^{24} hydrogen comes from the medium and C^{25}–H from NADPH.

(14)

The next step involves the oxidative demethylation of the C^{14}-α-methyl as formic acid, taking place before C^4 is demethylated. This reaction is mediated by an NADH dependent cytochrome system as shown by accumulation of lanosterol, $C^{24,25}$ dihydrolanosterol and 4,4-dimethyl sterols in the presence of carbon monoxide during cholesterol biosynthesis.[87] Abstraction of a hydrogen from the C^{14} methyl by $PorFe^{IV}O^•$ (33 d, section 4.3) initiates the oxidation to a C^{14} formyl group which is eliminated by the mechanism suggested in (15).[88] This last oxidative step produces formic acid with incorporation of molecular oxygen. The oxidation proceeds in three steps, each using one molecule of oxygen and one NADPH. It was observed that the C^{15}-α-hydrogen is lost in a *cis* elimination during cholesterol formation and the $C^{8,14}$ diene is an intermediate on the pathway. Contrary to the C^{14}-methyl the C^4-methyls are lost as carbon dioxide[89], starting with a three-stage oxidation of the α-methyl involving oxygen and NAD(P)H. The decarboxylation is mediated by oxidation of 3-OH to a β-keto acid. The oxidative enzyme system has rigid steric requirements.[90]

$$(15)$$

Thus, the 4-α-ethyl and 4-β-ethyl homologues, 4-β-hydroxymethyl isomers and the 4-β-monomethyl derivatives are inactive. This actually requires a rearrangement of 4-β-methyl to 4-α-methyl, probably via the 3-oxo derivative, prior to demethylation. The stepwise removal of the C^4-methyl groups is shown in (16).[91]

The transposition of double bonds to C^5 involves a C^{14} double bond reduction of the C^{14}-demethylation product (15) to the C^8-ene prior to C^4-demethylation[92]. C^8-ene → C^7-ene → $C^{5,7}$-diene → C^5-ene transpositions complete the biosynthesis of cholesterol which is the precursor for most other steroids.

Sapogenins are aglycones of a class of steroid glycosides, the saponins,[93] often containing a spiroketal side chain (Fig. 34). They acquired their name from their property for forming soapy emulsions in water, a property they share with the cardiac glycosides and the bile acids. They cause haemolysis by destroying the membranes of the erythrocytes. These compounds contain a hydrophobic and a hydrophilic part which make them surface active. Diosgenin (Fig. 34), a sapogenin from yams, a *Dioscorea* sp., is used as a valuable starting material for steroid hormone synthesis. It has the same skeletal configuration as cholesterol from which it is biosynthesized by oxidation of C^{22} to a carbonyl function and introduction of hydroxyl groups at $C^{16,26}$, which then ketalize the carbonyl group. The bile acids emulsify lipids thereby promoting the absorption through the intestinal wall.

(16)

In the cardiac glycosides, e.g. digitoxigenin (Fig. 34), the side chain has been converted to an α,β-unsaturated γ-lactone. It appears that the side chain of cholesterol rather unexpectedly is cleaved first at C^{20} in a cytochrome P450 catalysed reaction (O_2, NADPH) to give 4-methylpentanal and pregnenolone which subsequently is oxidized to progesterone, and then condensed with an acetate unit (17).[94] Both of these pregnane derivatives are incorporated into digitoxigenin by *Digitalis lanata*. An alternative route to the butenolide ring is followed in *D.purpurea*.[95] It involves cleavage of the side chain at C^{23} to give norbile acids followed by oxidation and lactonization. Rings A,B and C,D are *cis* fused in digitoxigenin and C^3–OH is coupled to various sugars in the glycosides. Pregnenolone and progesterone are precursors for the sex hormones and are produced in the ovaries.

The cardiac glycosides have a powerful and specific action on the heart muscle and are most valuable agents in the treatment of heart ailments. These glycosides are also the active principle in African arrow poisons from *Strophanthus* spp.

In the bile acids[96] additional hydroxyl groups are introduced in the nucleus and the side chain is shortened to a 24-oic acid. The order of events is not completely settled but nuclear hydroxylation occurs prior to side chain cleav-

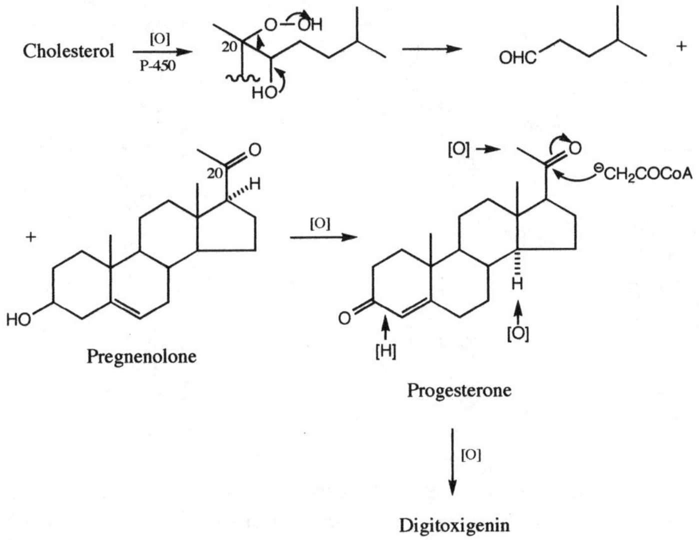
Cholesterol
[O]
P-450
20
O–OH
HO
OHC
+
O
20
H
[O]
[O]
CH2COCoA
HO
Pregnenolone
O
H
[H]
[O]
Progesterone
[O]
Digitoxigenin

(17)

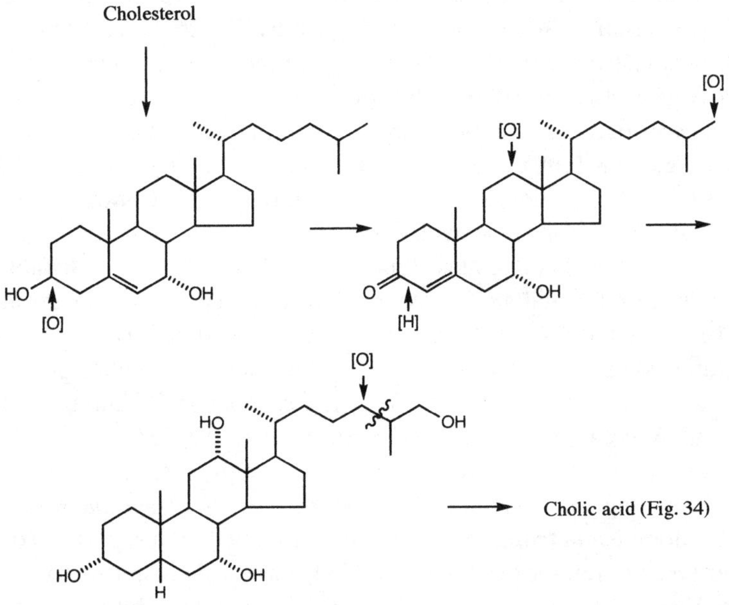
Cholesterol
[O]
[O]
HO
OH
[O]
O
OH
[H]
[O]
HO
OH
Cholic acid (Fig. 34)
HO
H
OH

(18)

age (18). The bile acids, conjugated with taurine or glycine, occur in the bile. Oxidation of the C_5 side chain of the bile acids to a 2-pyrone ring leads to bufadienolides which are the toxic principles in the skin glands of certain toads.

(19)

The adrenal hormones, e.g. aldosterone (Fig. 34), are formed from cholesterol by side chain cleavage at C^{20} followed by C^{17} and C^{21} hydroxylations. These hormones are secreted by two small glands adjacent to the kidneys. They regulate the metabolism of sugars and proteins, the salt balance in the animal organism, and are used in medicine for the treatment of inflammations and rheumatoid arthritis. Further cleavage of the side chain leads to the sex hormones, the C_{19} steroids (Fig. 34), controlling the development of male and female characteristics, lactation and the menstruation cycle, etc. The microbiological cleavage of the $C^{17,20}$ bond in progesterone to testosterone proceeds via hydroperoxidation of C^{20} (route 19a), rather than by a Baeyer–Villiger rearrangement (19b) because C^{17}–H is lost in the pocess.[97] Oestradiol is produced in the ovaries and testosterone in the testes. Contraceptives are modified sterols suppressing ovulation.

Cycloartenol (Fig. 27) is considered to represent the first stable tetracyclic product from squalene-2,3-oxide cyclization in plants on the grounds that it is efficiently incorporated into phytosterols and that it is widely distributed in plants compared with lanosterol.[98,99] The sterols of higher plants, algae, and fungi are characterized as having extra carbons at C^{24} and often a C^{22} double bond. Several marine sterols[100] with methyls at C^{22}, C^{23}, C^{24}, C^{26} and C^{27} are known as well as some containing a cyclopropyl side chain.

Labelling experiments with *Saccharomyces cerevisiae* using deuterated methionine and tritiated mevalonic acid show that in ergosterol only two deuteriums are retained in the C^{24} methyl, and C^{24}–H is shifted to C–25. The biosynthesis thus proceeds by way of a C^{24} methylene intermediate and this is also the case for the methyl sterol produced by *Hordeum vulgare* (20).[101] The ethyl sterol contained four deuteriums indicating an ethylidene intermediate. Various mechanisms are operating in the C^{24} alkylation since labelling experiments show that methyl and ethyl sterols of *Chlorella* contain three and five deuteriums consistent with loss of C^{24}–H or C^{25}–H, respectively.

Ergosterol, first isolated from ergot, but more readily available from yeast, is transformed by light to precalciferol which is rearranged to calciferol, vitamin D_2 (21).[102] Vitamin D deficiency causes rickets, a weakening of the bone structure that can be prevented by intake of fish liver oil, a rather distasteful experience for children in the old days. Since irradiation by light is of importance for the production of calciferol in the skin, rickets is more common in areas where the winter is long.

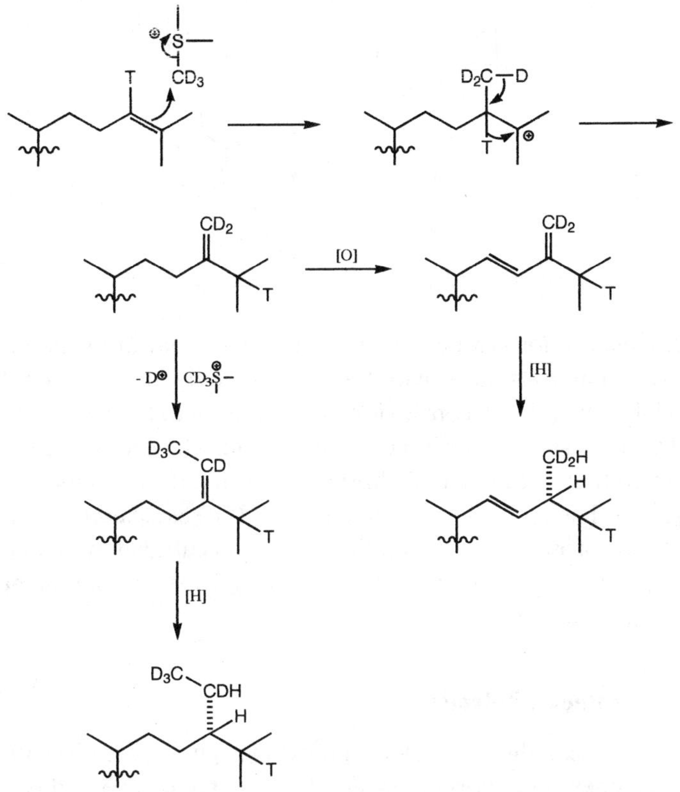

(20)

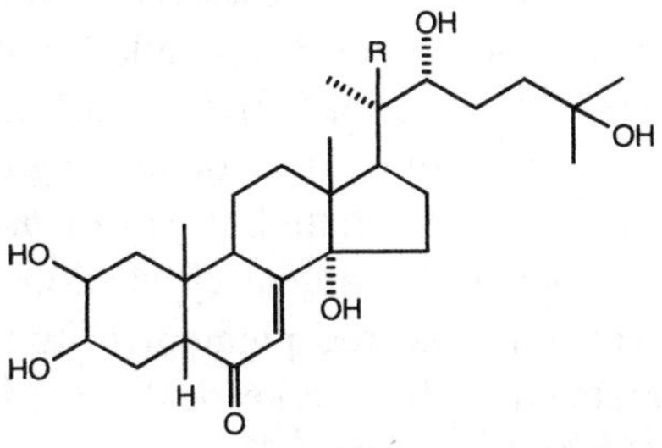

Fig. 35 Structure of insect moulting hormones. R = H, ecdysone; R = OH, ecdysterone

Ergosterol

hv

Precalciferol

Calciferol

(21)

Insects do not synthesize steroids *de novo* but are capable of processing suitable materials taken with the diet to provide for hormonal functions. Of special interest is the conversion of cholesterol to the moulting hormones, the ecdysones, essential for insect development. The ecdysones have an opposite effect to that of the juvenile hormones in that they stimulate metamorphosis (Fig. 35). The name derives from ecdysis (Greek, shedding), the entomological term for moulting. Several ecdysones occur also in higher plants, where they presumably play an ecological role in affecting the metabolism of phytophagous insects.

6.10 Carotenes. Polymers

Tail-to-tail coupling of geranylgeranyl pyrophosphate gives the C_{40} terpenes or carotenes, an important group of yellow-red conjugated polyene pigments with widespread occurrence in nature.[103,104] They are the chief pigments of egg yolk, yellow corn, carrots, tomatoes, pansy flowers, yellow autumn leaves, algae, etc. The biosynthesis (Fig. 36) parallels that of squalene. A cyclopropanoid prephytoene[105,106] is formed first which rearranges, either assisted by NADPH to lycopersene (route a) or, depending upon which of the prochiral hydrogens, H_A or H_B, is eliminated, to *cis*- or *trans*-phytoene, route b. Further dehydrogenations and terminal cyclizations lead to a variety of carotenes, such as lycopene, the red pigment in tomatoes, or to β-carotene, a widespread pigment which by symmetrical oxidative fission at $C^{15,15'}$ produces two molecules of vitamin A (retinol) or retinal (22), R = CH_2OH and CHO respectively. They play an important role in vision. The aldehyde forms a light sensitive Schiff's base with opsin, a protein present in the retina. Violaxanthin

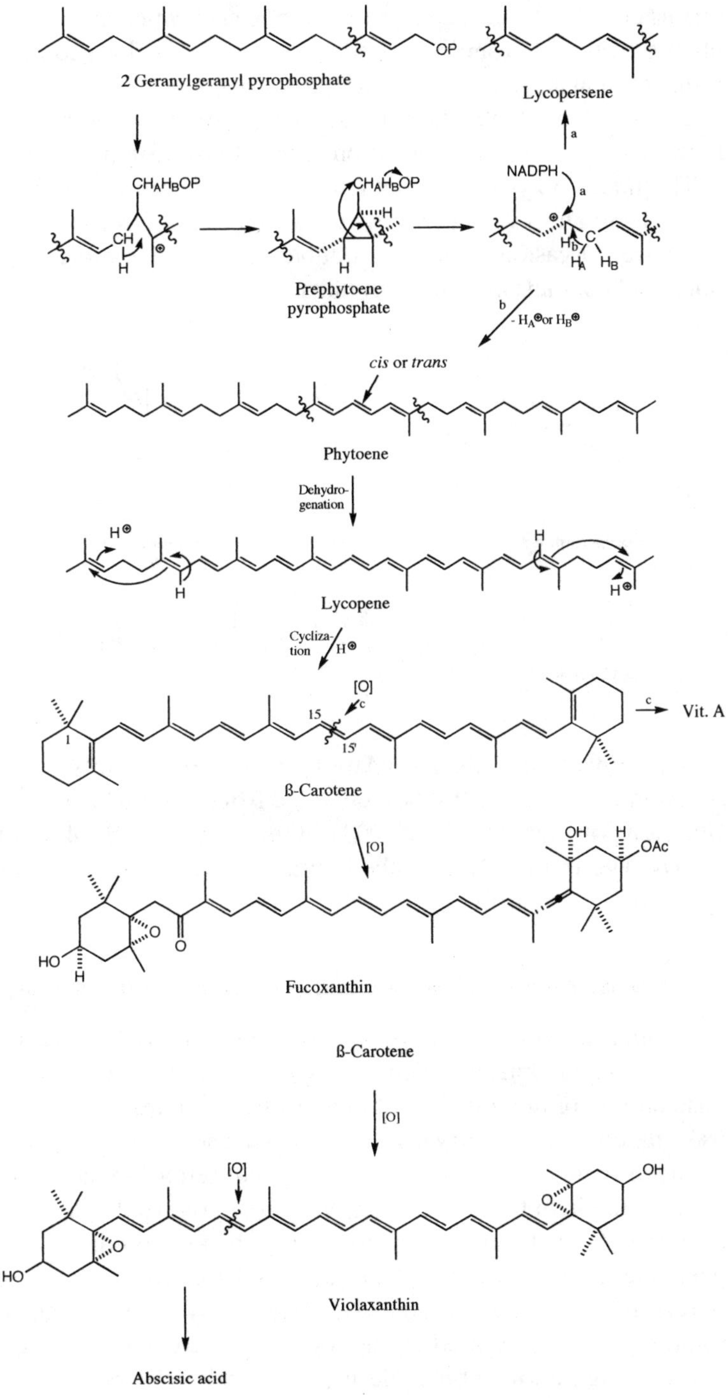

Fig. 36   General biosynthetic pathway to carotenes

is considered to be a precursor for abscisic acid which has a role in regulating plant growth. Fucoxantin, a common carotene confined to the marine environment in algae, has the unusual allene function.

The carotenoids function as supplementary light receptors in photosynthesis, transmitting their excitation energy to chlorophyll.

The linear polyprenols containing *ca.* ten or more isoprene units bound tail to head and with varying *cis-trans* isomeric ratios are widely distributed in nature. Occasionally several double bonds in the molecule are reduced. Solanesol contains nine all-*trans* units.

Polyprenol, n = 7-20   Solanesol   (22)

β-Carotene $\xrightarrow{[O]}$ 2

In continuation of the prenylation process some plants produce polymers as a white fluid, latex. Rubber, obtained from the rubber tree, *Hevea brasiliensis*, is a polymer with about 2000 isoprene units. Nearly all the double bonds are *cis*. The *trans* polymer called gutta-percha is obtained from *Palaquium* spp. and is more horny.

6.11 Optical rotatory dispersion and circular dichroism. The octant rule

Chiroptical methods[107,108] have been applied for more than a century for characterization of optically active compounds. It was already recognized by Pasteur, one of the founders of modern stereochemistry, that optical activity was related to asymmetry of the molecule. The rotatory power measured by the polarimeter has its origin in the different refractive indices of the optically active medium for left and right circularly polarized light. It increases with decreasing wavelength of the incident light in a predictable way and this phenomenon is defined as optical rotatory dispersion, ORD. In the early days of research it was tried more or less successfully to relate sign and molecular rotatory power, measured at one wavelength, commonly the sodium D line, to the configuration of the molecule, i.e. chirality. However, it turned out not to be advisable to rely on the sign for assigning the structure. This point is illustrated by the data obtained from the *o*-, *m*-, and *p*-iodophenyl ethers of

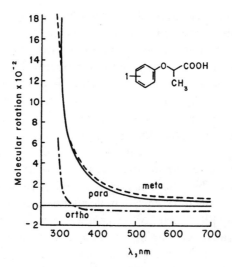

Fig. 37   ORD curves of *o, m-*, and *p*-iodophenyl ethers of lactic acid. (From *Optical Rotatory Dispersion* by C. Djerrassi. Copyright © 1960, McGraw-Hill. Used with the permission of McGraw-Hill Book Company)

lactic acid. The *m*- and *p*-derivatives show positive, the *o*-derivative negative rotation at the D line, from which one could be tempted to conclude that the *o*-derivative may possess the opposite configuration at C^2. The positive curvature of ORD, however, demonstrates that all three compounds have the same configuration (Fig. 37). A plain ORD curve can thus occasionally cross the *x*-axis. ORD curves are useful for detection of optical activity in compounds showing very small rotation values at the D line. As a consequence

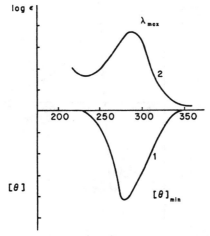

Fig. 38   CD curve, 1, showing a negative Cotton effect, and the UV absorption curve, 2

less substance is required for determination of activity. Instrumental improvements, stimulated by the demand from organic chemists for simple spectroscopic methods for determination of absolute configuration and conformation of optically active natural products, have considerably facilitated rapid automatic recordings of rotatory power as a function of wavelength. A novel valuable tool became available to the chemist. It turned out that chiroptical techniques were of immense importance for structural elucidations of terpenoids containing many chiral centres.

If the compound has a chromophore which is asymmetrically perturbated by a chiral centre, left and right circularly polarized light is absorbed to different extents. This phenomenon is called circular dichroism, CD. The CD curve graphs the difference in molecular extinction coefficients, $\Delta\epsilon = \epsilon_L - \epsilon_R$, or ellipticity $[\theta] = 3300 \cdot \Delta\epsilon$, against wavelength. This difference can be positive or negative, depending upon the chirality at the active centre. Its maximum, $\Delta\epsilon_{max}$, coincides with the maximum, λ_{max}, of the absorption curve (Fig. 38). These conditions are visible as a drastic anomaly in the ORD curve, the Cotton effect. A single positive Cotton effect gives rise to a sharp maximum (peak) which suddenly drops to a minimum (trough) (Fig. 39). If the peak in the ORD curve occurs at shorter wavelength than the trough, the curve is defined as a negative Cotton effect curve. These anomalous curves provide more information than the plain curves and are important for structural determinations. A molecule containing several chromophores gives rise to a complex spectrum with multiple Cotton effects.

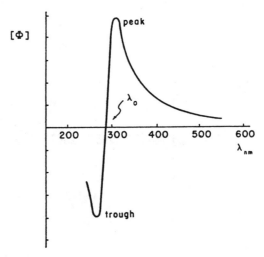

Fig. 39   ORD curve showing positive Cotton effect. λ_0 corresponds closely to λ_{max} of the UV absorption curve

The carbonyl function was found to be most suitable for rotatory dispersion studies. It is a common function in natural products and hydroxyl groups can easily be oxidized to carbonyl groups. It absorbs in a readily accessible spectral region, it has a low extinction coefficient, which does not interfere with ORD measurements, and most importantly it is electronically perturbated by proximate chiral centres. The terpenes offered numerous compounds of known structure with the carbonyl function located in a variety of molecular environments, such as different ring sizes and proximities to chiral centres of different configuration and conformation. In condensed cyclic structures the number of conformers is considerably reduced and they are controlled in a predictable way. Correlation between sign and magnitude of the Cotton effect and the absolute configuration of chiral centres and their disposition relative to the carbonyl group were noted, and eventually these systematic studies culminated in the formulation of the semi-empirical *octant rule*. Cyclohexanone in its chair form, the most stable form, is oriented in a three-dimensional coordinate system so that the yz plane is bisecting the ring through the carbonyl group, the xz plane passes through

the $O=C\overset{\displaystyle -C}{\underset{\displaystyle \diagdown C}{}}$ atoms and the xy plane bisects the C=O bond (Fig. 40). The

octant rule states that substituents lying in the different octants contibute to the Cotton effect with the sign shown. Substituents located in the planes and located distant from the chromophore make a negligible contribution. Only rarely do substituents fall into the front octants.

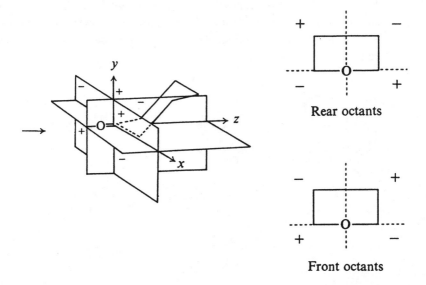

Fig. 40   The octant rule. A projection of cyclohexanone in a three-dimensional coordinate system and simplified front views along the CO bond (along the arrow)

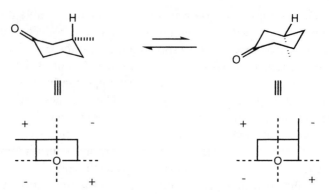

Fig. 41 The preferred conformation of *R*(+)-3-methylcyclohexanone has an equatorial methyl. The Cotton effect is positive

Two examples will clearly demonstrate the usefulness of the octant rule. (+)-3-Methylcylohexanone has been shown to have *R* configuration and a positive Cotton effect. Consequently the methyl groups must be located in a positive octant and that is only possible if the molecule acquires a conformation with an equatorial methyl group (Fig. 41). It is known from infrared spectroscopic studies that the energy difference between axial and equatorial methylcyclohexane is 1.7 kcal, i.e. *ca.* 95 per cent of the compound has the equatorial conformation. Provided with this information it is possible, on the other hand, to determine the absolute cofiguration of (+)-3-methylcyclohexanone from the ORD curve.

Reduction of podocarpic acid gave two ketones exhibiting positive and negative Cotton effects. Application of the octant rule shows that the compound exhibiting positive Cotton effect is *trans* fused, whereas the other must be *cis* fused (Fig. 42).

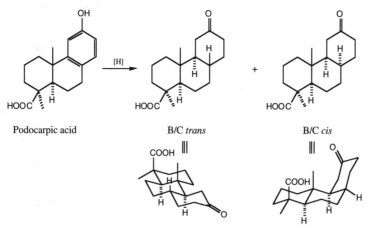

Fig. 42 Reduction of podocarpic acid. The projections of *trans* and *cis* products fall in positive and negaive octants, respectively

6.12 Problems

6.1 Several biosynthetic schemes have been proposed for artemisia ketone from IPP and DMAP (Banthorpe, D.V., Charlwood, B. V. and Francis, M. J. O. *Chem. Revs.* **72** (1972)115). A research group wishes to test the correctness of the following scheme:

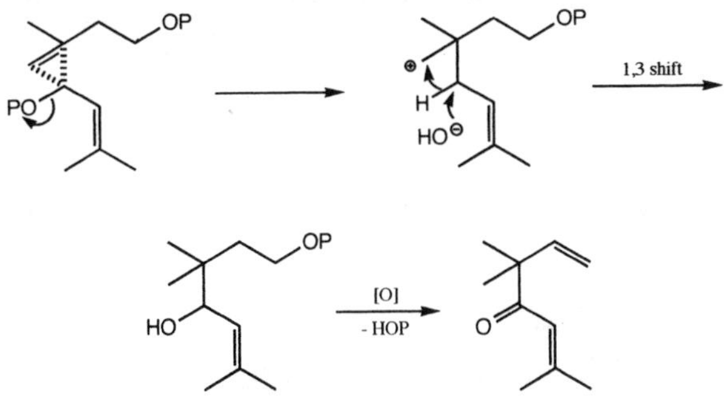

Artemisia ketone

How would you attack the problem of verifying the proposed pathway?

6.2 The biosynthesis of a large number of sesquiterpenes is initiated by cyclization of *trans–trans*-farnesyl pyrophosphate with the terminal double bond. Suggest a biosynthetic pathway to eremophilone (Coates, R. M. *Prog. Chem. Org. Nat. Prod.* **33** (1976) 74).

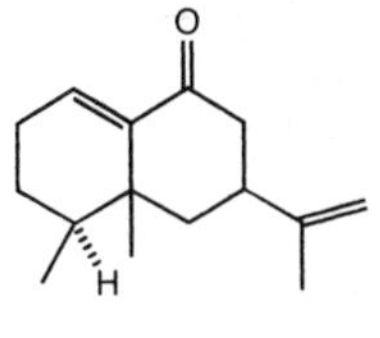

Eremophilone

6.3 The coral *Pacifigorgia adamsii* produces an ichthyotoxic sesquiterpene with the structure shown below. Suggest a plausible biosynthetic pathway to the compound (Izac. R. R., Poet. S. E. and Fenical, W. *Tetrahedron Lett.* **1982** 3743).

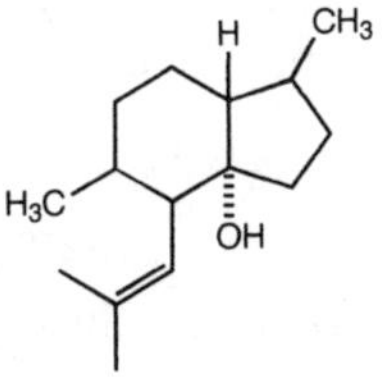

6.4 One can arrive at the hirsutane skeleton from farnesyl pyrophosphate via the humulyl cation, see Fig. 16, along three possible cyclization routes. The correct pathway was determined by fermentation of *Coriolus consors* in the

presence of 1,2-^{13}C-acetate which produced 5-dihydrocoriolin. The $^1J_{13c13c}$ of the ^{13}C NMR spectrum of labelled 5-dihydrocoriolin indicated the presence of six intact acetate units in the ring and four intact units in the side chain. Discuss the biosynthetic pathway for the metabolite. The ^{13}C NMR data are summarized in the table (Tanabe, M., Suzuki, K. T. and Jankowski, W. C. *Tetrahedron Lett*, **1974**, 2271).

Table 1 ^{13}C NMR data for dihydrocoriolin C (as triacetate)

Position, δ, multiplicity	$^1J^{13}C^{13}C$
C^1 (80.0,d)–C^2 (51.0,d)	42
C^3 (47.1,s)–C^{12} (13.1,q)	38
C^4 (64.5,s)–C^{13} (44.9,t)	30
C^5 (71.7,d)	
C^6 (60.9,d)–C^7 (73.5,s)	26
C^8 (72.7,d)	
C^9 (41.3,d)–C^{10} (37.6,t)	33
C^{11} (43.9,s)–C^{15} (21.4,q)	34
C^{14} (26.5)	
C$^{1'}$ (170.0,s)–C$^{2'}$ (72.4,d)	65
C$^{3'}$ (31.3,t)–C$^{4'}$ (25.2,t)	34
C$^{5'}$ (28.8,t)–C$^{6'}$ (31.5,t)	35
C$^{7'}$ (22.5,t)–C$^{8'}$ (14.0,q)	34
CH$_3$CO (20.5, 20.5, 20.9, 170.0, 170.5, 170.8)	

The multiplicity (s = singlet, d = doublet, t = triplet, q = quartet) refers to off-resonance residual couplings

6.5 The coelenterates (Octocorallia), forming a prominent part of the biomass in the tropical reefs, produce a number of terpenoids of unusual structures. Xenicin, I, was isolated from the Australian soft coral *Xenia elongata* (Alcyonaria), and from the Okinawan soft coral, *Alcyonium* sp., alcyonolide II was isolated. Suggest plausible biosynthetic pathways for the metabolites. Lead: the highly oxygenated skeletal frameworks could be formed by cleavage of a caryophyllene type precursor (Vanderah, D. J., Steudler, P. A., Ciereszko, L. S., Schmitz, F. J., Ekstrand, J. D. and van der Helm, D. *J. Am. Chem. Soc.* **99** (1977) 5780; Kobayashi, M., Yasuzawa, T., Kobayashi, Y., Kyogoku, Y. and Kitagawa, I. *Tetrahedron Lett.* **1981**, 4445).

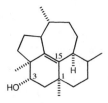

I   II

6.6 The termite *Nasutitermes rippertii* produces a gluey defence excretion from which 3α-hydroxy-15-rippertene was isolated. Suggest a biosynthetic scheme for this tetracyclic terpenoid (Prestwich, G. D., Spanton, S. G., Lauber, J. W. and Vrkoč, J. *J. Am. Chem. Soc.* **102** (1980) 6825).

3α-Hydroxy-15-rippertene

Bibliography

1. Ruzicka, L. *Experientia* **9** (1953) 357.
2. McManus, S. P. and Pittman, Jr., C. U. in *Organic Reactive Intermediates*, McManus, S. P. (Ed.), Academic Press, New York, 1973, p. 193.
3. Floss, H. G. and Lee, S. *Acc. Chem. Res.* **28** (1993) 116.
4. Tavormina, P. A., Gibbs, M. H. and Huff, J. W. *J. Am. Chem. Soc.* **78** (1956) 4498.
5. Wolf, D. E., Hoffman, C. H. Aldrich, P. E., Skeggs, H. R., Wright, L. D. and Folkers, K. *J. Am. Chem. Soc.* **78** (1956) 4499.
6. Lynen, F. and Henning, U. *Angew. Chem.* **72** (1960) 820.
7. Stone, K. J., Roeske, W. R., Clayton, R. B. and van Tamelen, E. E. *Chem. Commun.* **1969**, 530.

8. Cornforth, J. W., Cornforth, R. H., Donninger, C. and Popjak, G. *Proc. Roy. Soc. (B)* **163** (1965) 492.
9. Cornforth, J. W., Cornforth, R. H., Popjak, G. and Yengoyan, L. *J. Biol. Chem.* **241** (1966) 3970.
10. Lüthy, J., Rétey, J. and Arigoni, D. *Nature* **221** (1969) 1213.
11. Cornforth, J. W., Redmond, J. W., Eggerer, H., Buckel, W. and Gutschow, C. *Nature* **221** (1969) 1212.
12. Cornforth, J. W. *Chem. Brit.* **6** (1970) 431.
13. Clifford, K. H., Cornforth, J. W., Mallaby, R. and Phillips, G. T. *Chem. Commun.* **1971**, 1599.
14. Banthorpe, D. W., Bunton, C. A., Cory, O. and Francis, M. J. O. *Phytochemistry* **24** (1985) 251; Suga, T., Hirata, T., Aoki, T. and Kataoka, T. *J. Am. Chem. Soc.* **108** (1986) 2366.
15. Cornforth, J. W., Phillips, G. T., Messner, B. and Eggerer, H. *Eur. J. Biochem.* **42,** (1974) 591.
16. Poulter, C. D. and Rilling, H. C. *Acct. Chem. Res.* **11** (1978) 307.
17. Allinger, N. L. and Seifert, J. H. *J. Am. Chem. Soc.* **97** (1975) 752.
18. Banthorpe, D. V., Charlwood, B. V. and Francis, M. J. O. *Chem. Revs.* **72** (1972) 115; Tange, K., Okita, H., Nakao, Y., Hirata, T. and Suga, T. *Chem. Letters* **1981,** 777.
19. Lanza, E. and Palmer, J. K. *Phytochemistry* **16** (1977) 1555.
20. Cane, D. E. *Tetrahedron* **36** (1980) 1109; *Acc. Chem. Res.* **18** (1985) 220.
21. Croteau, R. *Chem. Rev.* **87** (1987) 929.
22. Banthorpe, D. V., Ekundayo, O. and Njar, V. C. O. *Phytochemistry* **23** (1984) 291.
22a. Banthorpe, D. V. and Ekundayo, O. *Phytochemistry* **15** (1976) 109.
23. Croteau, R., Shaskus, J., Renström, B., Felton, N. M., Cane, D. E., Saito, A. and Chang, C. *Biochemistry* **24** (1985) 7077.
24. Uesato, S., Ikeda, H., Fujita, T. Inouye, H. and Zenk, M. H. *Tetrahedron Lett.*, **28** (1989) 97.
25. Jensen, S. R., Kirk, O. and Nielsen, B. J. *Phytochemistry* **28** (1989) 97.
26. Poulter, C. D. *Acc. Chem. Res.* **23** (1990) 70.
27. Banthorpe, D. V., Charlwood, B. V., Greaves, G. M. and Voller, C. M. *Phytochemistry* **16** (1977) 1387.
28. Peter, M. G. and Dahm, K. H. *Helv. Chim. Acta* **58** (1975) 1037.
29. Overton, K. H. and Picken, D. J. *J. Chem. Soc. Chem. Commun.* **1976,** 105.
30. Cane, D. E. and McIlwain, D. B. *Tetrahedron Lett.* **28** (1987) 6545.
31. Hanson, J. R., Marten, T. and Siverns, M. *J. Chem. Soc. Perkin I* **1974,** 1033.
32. Achilladelis, B. A., Adams, P. M. and Hanson, J. R. *J. Chem. Soc. Perkin I* **1972,** 1425.
33. Machida, Y. and Nozoe, S., *Tetrahedron* **28** (1972) 5113.
34. Marshall, J. A., Brady, S. F. and Andersen, N. H. in *Prog. Chem. Org. Nat. Prod.* **31** (1974) 283.
35. Corbella, A., Gariboldi, P. and Jommi, G. *J. Chem. Soc. Chem. Commun.* **1972,** 600.
36. Arigoni, D. *Pure Appl. Chem.* **41** (1975) 219.
37. Stipanovic, R. D., Stoessl, A., Stothers, J. B., Altman, D. W., Bell, A. and Heinstein, P. *J. Chem. Soc. Chem. Commun.* **1986** 100; Akhila, A. and Rani, K. *Phytochemistry* **33** (1993) 335.
38. Martin, J. D. and Darias, J. in *Marine Natural Products*, Scheuer, P. J. (Ed.) Vol. I, p. 125. Academic Press, New York, 1978.
39. Hirai, N., in *Chemistry of Plant Hormones*. Takehashi, N. (Ed.) CRC Press, Boca

Raton, Florida, 1986, p. 201; Creelman, R. A. *Physiol. Plantarum* **75** (1989) 131.
40. Okamoto, M., Hirai, N. and Koshimizu, K. *Phytochemistry* **27** (1988) 2099, 3465.
41. Bornemann, V., Patterson, G. M. L. and Moore, R. E. *J. Am. Chem. Soc.* **110** (1988) 2339.
42. Edenborough, M. S. and Herbert, R. B. *Nat. Prod. Rep.* **5** (1988) 229; Aschenbach, H. and Griesebach. H. *Z. Naturforsch.* **20 B** (1965) 137.
43. Burreson, B. J., Christophersen, C. and Scheuer, P. J. *Tetrahedron* **31** (1975) 2015.
44. Blackman, A. and Wells, R. J. *Tetrahedron Lett.* **1978,** 3063.
45. Hanson, J. R. in *Prog. Chem. Org. Nat. Prod.* **29** (1971) 395.
46. Coates, R. M. in *Prog. Chem. Org. Nat. Prod.* **33** (1976) 73.
47. Johnson, W. S. *Acc. Chem. Res* **1** (1968) 1.
48. Goldsmith, D. in *Prog. Chem. Org. Nat. Prod.* **29** (1971) 363.
49. Achilladelis, B. and Hanson, J. R. *J. Chem. Soc.* (C) **1969,** 2010.
50. Hancock, W. S., Mander, L. N. and Massy-Westropp. R. A. *J. Org. Chem.* **38** (1973) 4090.
51. Mander, L. N. *Chem. Rev.* **92** (1992) 573.
52. MacMillan, J. and Pryce, R. J. in *Phytochemistry*. Vol. III p. 283, Miller, L. P. (Ed.), Van Nostrand Reinhold, New York, 1973.
53. Graebe, J. E., Bowen, D. H. and MacMillan, J. *Planta* **102** (1972) 261.
54. Graebe, J. E., Hedden, P. and MacMillan, J. *J. Chem. Soc. Chem. Commun.* **1975,** 161.
55. Evans, R. and Hanson, J. R. *J. Chem. Soc. Perkin I* **1975,** 633.
56. Bearder, J. R., MacMillan, J. and Phinney, B. O. *J. Chem. Soc. Perkin I*, **1975,** 721.
57. Weinheimer, A. J., Chang, C. W. J. and Matson, J. A. *Prog. Chem. Org. Nat. Prod.* **36** (1979) 285.
57a. Kingstone, D. G. I., Molinero. A. A. and Rimoldi, J. M. *Prog. Chem. Org. Nat. Prod.* **61** (1993) 1.
58. Cordell, G. A. *Phytochemistry* **13** (1974) 2343.
59. Rios, T. and Perez, C. S. *Chem. Commun.* **1969,** 214.
60. Minale, L. in *Marine Natural Products*, Scheuer, P. J. (Ed.), Vol. I, p. 174. Academic Press, New York, 1978.
61. Nozoe, S., Morisaki, M., Tsuda, K., Iitaki, Y., Takahashi, N., Tamura, S., Ishibashi, K. and Shirasaka, M. *J. Am. Chem. Soc.* **87** (1965) 4968.
62. Nozoe, S., Morisaki, M., Tsuda, K. and Okuda, S. *Tetrahedron Lett.* **1967,** 3365.
63. Canonica, L., Fiecchi, A., Galli Kienle, M., Ranzi, B. M. and Scala, A. *Tetrahedron Lett.* **1967,** 4657.
64. Harrison, D. M. *Nat. Prod. Rep.* **5** (1988) 387; **7** (1990) 459.
65. Epstein, W. W. and Rilling, H. C. *J. Biol. Chem.* **245** (1970) 4597.
66. Popjak, G., Edmond, J. and Wong, S.-M. *J. Am. Chem. Soc.* **95** (1973) 2713.
67. Abe, I., Rohmer, M. and Prestwich, G. D. *Chem. Rev.* **93** (1993) 2189.
68. Corey, E. J., Russey, W. E. and de Montellano, P. P. O. *J. Am. Chem. Soc.* **88** (1966) 4750.
69. van Tamelen, E. E., Willett, J. D., Clayton, R. B. and Lord, K. E. *J. Am. Chem. Soc.* **88** (1966) 4752.
70. Yamamoto, S. and Bloch, K. *J. Biol. Chem.* **245** (1970) 1670.
71. Ebersole, R. C., Godtfredsen, W. O., Vangedal, S. and Caspi, E. *J. Am. Chem. Soc.* **95** (1973) 8133.
72. Mulheirn, L. J. and Ramm, P. J. *Chem. Soc. Revs.* **1** (1972) 259.
73. van Tamelen, E. E. and James, D. R. *J. Am. Chem. Soc.* **99** (1977) 950.
74. Cornforth, J. W. *Angew. Chem. Int. Ed.* **7** (1968) 903.
75. Corey, E. J. and Virgil, S. C. *J. Am. Chem. Soc.* **113** (1991) 4025.

76. Clifford, K. H. and Phillips, G. T. *European J. Biochem* **61** (1976) 271.
77. Popjak, G., Edmond, J., Anet, F. A. L. and Easton, Jr. N. R. *J. Am. Chem. Soc.* **99** (1977) 931.
78. Lavie, D. and Glotter, F. *Prog. Chem. Org. Nat. Prod.* **29** (1971) 307.
79. Courney, J. L., Gasciogne, R. M. and Szumer, A. Z. *J. Chem. Soc.* **1958,** 881.
80. Barton, D. H. R., Mellows, G. and Widdowson, D. A. *J. Chem. Soc.* (*C*) **1971,** 110.
81. Polonsky, J. *Prog. Chem. Org. Nat. Prod.* **30** (1973) 101.
82. Moron, J., Merrien, M.-A., and Polonsky, J. *Phytochemistry* **1971,** 585.
83. Moron, J. and Polonsky, J. *European J. Biochem.* **3** (1968) 488.
84. Sabine, J. R. *Cholesterol*, Dekker, New York, 1977.
85. Greig, J. B., Varma, K. R. and Caspi, E. *J. Am. Chem. Soc.* **93** (1971) 760.
86. Duchamp, D. J., Chidester, C. G., Wickramasinghe, J. A. F., Caspi, E. and Yagen, B. *J. Am. Chem. Soc.* **93** (1971) 6283.
87. Gibbons, G. F. and Mitropoulos, K. A. *Biochem. J.* **132** (1973) 439.
88. Akhtar, M., Alexander, K., Boar, R. B., McGhie, J. F. and Barton, D. H. R. *Biochem. J.* **169** (1978) 449; Tuck, S. F., Robinson, C. H. and Silverton, J. V. *J. Org. Chem.* **56** (1991) 1260.
89. Gautschi, F. and Bloch, K. *J. Biol. Chem.* **233** (1958) 1343.
90. Nelson, J. A., Kahn, S., Spencer, T. A., Sharpless, K. B. and Clayton, R. B. *Bioorg. Chem.* **4** (1975) 363.
91. Rahimtula, A. D. and Gaylor, J. L. *J. Biol. Chem.* **247** (1972) 9.
92. Kawata, S., Trzaskos, J. M. and Gaylor, J. L. *J. Biol. Chem.* **260** (1985) 6609.
93. Hostettmann, K. and Marston, A. *Chemistry and Pharmacology of Natural Products: Saponins.* Cambridge University Press, New York 1995.
94. Hume, R., Kelly, R. W., Taylor, P. L. and Boyd, G. S. *Eur. J. Biochem.* **140** (1984) 583.
95. Maier, M. S., Seldes, A. M. and Gros, E. G. *Phytochemistry* **25** (1986) 1327.
96. *The Bile Acids* Vol. 2, Nair, P. P. and Kritchevsky, D. (Eds.), Plenum Press, New York, 1973.
97. Milewich, L. and Axelrod, L. R. *Arch. Biochem. Biophys.* **153** (1972) 188.
98. Goodwin, T. W. in *Rec. Adv. Phytochemistry*, Vol. 6, p. 97, Runeckles, V. C. and Mabry, T. J. (Eds.), Academic Press, New York, 1973.
99. Goad, L. J. and Goodwin, T. W. in *Progress in Phytochemistry*, Vol. 3, p. 113, Reinhold, L. and Liwschitz, Y. (Eds.), J. Wiley, London, 1972.
100. Goad, L. J. in *Marine Natural Products*, Vol. II, p. 75. Scheuer, P. J. (Ed.), Academic Press, New York, 1978.
101. Goad, L. J., Lenton, J. R., Knapp, F. F. and Goodwin, T. W. *Lipids* **9** (1974) 582.
102. Jones, H. and Rasmusson, G. H. *Prog. Chem. Org. Nat. Prod.* **39** (1980) 63.
103. Liaan-Jensen, S. *Prog. Chem. Org. Nat. Prod.* **39** (1980) 123.
104. Britton, G. *Plant Biol.* **2** (1986) 125.
105. Altman, L. J., Ash, L., Kowerski, R. C., Epstein, W. W., Larsen, B. R., Rilling, H. C., Muscio, F. and Gregonis, D. E. *J. Am. Chem. Soc.* **94** (1972) 3257.
106. Qureshi, A. A., Barnes, F. J. and Porter, J. W. *J. Biol. Chem.* **247** (1972) 6730.
107. Crabbé, P. *Optical Rotatory Dispersion and Circular Dichroism in Organic Chemistry*, Holden-Day, San Francisco, 1965.
108. Scopes, P. M., *Prog. Chem. Org. Nat. Prod.* **32** (1975) 167.

Chapter 7

Amino acids, peptides and proteins

7.1 Introduction

The title compounds are typical first order metabolites indispensable for life at all levels. They are the starting materials for an array of secondary products such as simple amines, alkaloids, and aromatic *N*-heterocycles, as well as phenylpropanoids, i.e. the C_6C_3 compounds, where the amino groups is lost (Chapter 4). The amino acid, leucine, is also a precursor of isopentenyl pyrophosphate. Ever since Hofmeister and Fischer proposed at the turn of the century that amino acids are the structural units of proteins, biochemists have extensively devoted their attention and skill to this domain of chemistry. This research was highlighted by such events as the crystallization of the first enzyme, urease, and the proof that it was a protein by Sumner in 1926, the automatic amino acid analyser introduced by Stein and Moore which made rapid protein analysis possible, the high resolution X-ray analysis of enzymes by Kendrew and Perutz in the late 1950s and the unravelling of the mechanisms of action of genes by Jacob, Monod, Nirenberg, Khorana, Ochoa, and others in the 1960s. The biosynthesis, degradation and reactions of amino acids and proteins are well covered in textbooks of biochemistry. Nevertheless, it is appropriate to include a few common transformations here, simply because in a predictable way they illustrate principles of organic reaction mechanism. The elucidation of the biogenetic pathways of amino acids was a difficult task, but of utmost importance for the development of biochemistry, physiology and medicine. It so happens that the classification of amino acids as essential and non-essential has its counterpart in the different complexities of formation of the two groups. The biosynthesis of the essential amino acids is more complex than the biosynthesis of non-essential amino acids, which in principle are available in a few steps by enzymatic reductive amination of α-keto acids originating from the citric acid cycle or from glycolysis.

7.2 Amino acids. Classification, structure and properties

The amino acids, derived from proteins, are all L-α-amino acids, the configuration of which is defined by the Fischer projection in Fig. 1. L-(—)-Serine is related to L-(—)-glyceraldehyde by reactions of known stereochemistry.

Fig. 1 Fischer projection of L-α-amino acids and L-(—)-glyceraldehyde. L-(—)-Serine = S-serine, R = CH_2OH

Table 1 Structure of the commonest amino acids derived from proteins. Name of amine derived by decarboxylation

Amino acid	Structure	Abbreviated symbol	Amine
Alanine N,1	$CH_3CHCOOH$ $\mid$ NH_2	Ala	Ethylamine
Arginine E,3	$H_2N^{\oplus} = CNH(CH_2)_3CHCOOH$ $\mid \qquad\qquad\quad \mid$ $NH_2 \qquad\qquad NH_2$	Arg	Agmaline (4-Guanidobutyl-amine)
Aspartic acid N,4	$^{\ominus}OOCCH_2CHCOOH$ $\mid$ NH_2	Asp	β-Alanine
Asparagine N,2	$NH_2COCH_2CHCOOH$ $\mid$ NH_2	Asn	β-Alanyl amide
Cysteine N,2	$HSCH_2CHCOOH$ $\mid$ NH_2	Cys	2-Mercaptoethyl amine
Glutamic acid N,4	$^{\ominus}OOCCH_2CH_2CHCOOH$ $\mid$ NH_2	Glu	γ-Aminobutyric acid (GABA)
Glutamine N,2	$H_2NCOCH_2CH_2CHCOOH$ $\mid$ NH_2	Gln	γ-Aminobutyr-amide
Glycine N,2	CH_2COOH $\mid$ NH_2	Gly	Methylamine

Table 1   (*continued*)

Amino acid	Structure	Abbreviated symbol	Amine
Histidine E,3	![structure] —CH$_2$CHCOOH with NH$_2$	His	Histamine
Isoleucine E,1	C$_2$H$_5$CHCHCOOH with H$_3$C NH$_2$	Ile	2-Methylbutyl-amine
Leucine E,1	(CH$_3$)$_2$CHCH$_2$CHCOOH with NH$_2$	Leu	3-Methylbutyl-amine
Lysine E,3	H$_2$N(CH$_2$)$_4$CHCOOH with NH$_2$	Lys	Cadaverine, 1,5-Diamino-pentane
Methionine E,1	CH$_3$S(CH$_2$)$_2$CHCOOH with NH$_2$	Met	3-Methylmercapto-propylamine
Phenylalanine E,1	![phenyl]—CH$_2$CHCOOH with NH$_2$	Phe	Phenylethylamine
Proline N,1	![pyrrolidine] COOH	Pro	Pyrrolidine
Serine N,2	HOCH$_2$CHCOOH with NH$_2$	Ser	Ethanolamine
Threonine E,1	CH$_3$CHCHCOOH with HO NH$_2$	Thr	2-Hydroxypropyl-amine
Tryptophan E,1	![indole] CH$_2$CHCOOH with NH$_2$	Try	Tryptamine
Tyrosine N,2	HO—![phenyl]—CH$_2$CHCOOH with NH$_2$	Tyr	Tyramine
Valine E,1	(CH$_3$)$_2$CHCHCOOH with NH$_2$	Val	*i*-Butylamine

Requirement by man: E, essential; N, non-essential. Polarity of the chain; 1, non-polar; 2, neutral polar; 3, positively charged; 4, negatively charged

The first amino acid isolated was the only symmetrical and hence optically inactive amino acid, glycine, obtained by Braconnot in 1820 from a hydrolysate of gelatine. Besides the 20 common acids from proteins (Table 1) there are about 300 non-protein amino acids found in nature. Some of them have D-configuration. D-Alanine was found free in insects[1] and also in some plants, e.g. *Pisum sativum*.[2] It occurs as the dipeptide, D-alanyl-D-alanine, in the leaves of *Nicotiana tabacum*.[3] D-Glutamic acid is a constituent of the glycopeptides in the cell wall of many bacteria.[4] A few are β-, γ- or δ-amino acids. β-Alanine, which arises by decarboxylation of aspartic acid, is a building block of coenzyme A (section 1.6). γ-Aminobutyric acid, which is formed by decarboxylation of glutamic acid, acts as a relay compound for transmission of nerve impulses.[5] Ornithine, α,γ-diaminovaleric acid, was first isolated by Jaffe in 1877 in excretion products from birds fed with benzoic acid. Birds and reptiles excrete benzoic acid as the dibenzoate of ornithine, ornithuric acid, in contrast to man, who excretes benzoic acid as benzoyl glycine, hippuric acid. If 3 g of benzoic acid is given orally to a man, hippuric acid can readily be extracted in substantial amounts from his urine after 12 h. Conjugation of toxic substances with glycine, glutamine, or cysteine, thereby rendering them more water soluble, is a frequently encountered detoxification mechanism. Sulphatization of hydroxy compounds and acetylation are other detoxification procedures used by the body. Ornithine, a carrier in the urea cycle, is the precursor of the pyrrolidine and tropane alkaloids. It decarboxylates to evil smelling

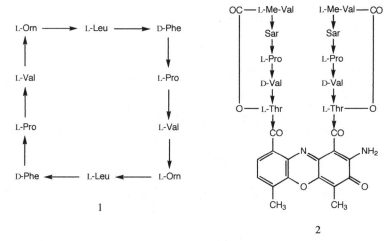

Fig. 2   1, Structure of the antiobiotic gramicidin S, a cyclic decapeptide containing non-protein D-phenylalanine and L-ornithine. The arrows indicate the N→C direction of the peptide bond. 2, Structure of actinomycin D from *Streptomyces* spp., a lactone peptide acting as a DNA inhibitor. It contains the non-protein D-valine and *N*-methylated L-valine and *N*-methylated glycine = Sarcosine = Sar. The carbonyl group of L-Me-Val is lactonized with the OH group of L-Thr. The two cyclic peptides are bound to the phenoxazine ring via amide bonds

1,4-diaminobutane, putrescine. Citrulline, or carboxamido-ornithine, homo-cysteine, homoserine, and β-cyanoalanine, are other non-protein amino acids serving as precursors and intermediates in metabolism.

Bacteria produce several peptide antibiotics,[6,7] e.g. the actinomycins from *Streptomyces* spp. and gramicidin S from *Bacillus brevis*, the latter containing the non-protein amino acids D-phenylalanine and L-ornithine (Fig 2.). Gramicidin S acts—like the crown ethers—by its property to form fat-soluble salts.

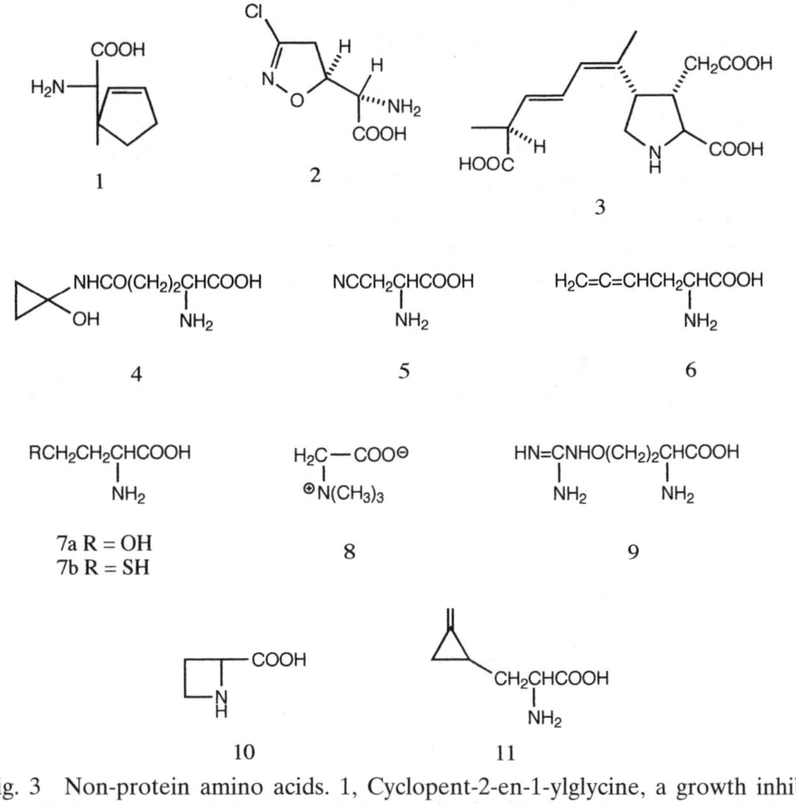

Fig. 3 Non-protein amino acids. 1, Cyclopent-2-en-1-ylglycine, a growth inhibitor from *Hydnocarpus anthelminthica*; 2, 3-chloroisoxazolin-5-ylglycine, a metabolite with antitumour activity from *Streptomyces sviceus*; 3, domoic acid, an anthelminthic from the red alga *Chondria armata*; 4, coprine, a metabolite from the mushroom *Coprinus atramentarius* with antabuse activity; 5, β-cyanoalanine, a constituent of *Lathyrus* and *Vicea* spp.; 6, allenic antibiotic amino acid from *Amanita solitaria*; 7a, homoserine and 7b, homocysteine are intermediates in amino acid metabolism; 8, betaine widespread zwitterionic amino acid; 9, canavanine, constituent of the *Leguminosae* family, phyto-toxic amino acid containing the unique guanidoxy function; 10, azetidine-2-carboxylic acid, proline antagonist, almost ubiquitously distributed; 11, hypoglycine A, blood sugar depressing amino acid from *Blighia sapida*.

The metabolic pathways to these non-protein amino acids are not known with certainty yet, but in several cases it is rather obvious that they are con-structed in a straightforward manner from mainstream metabolic units discussed in Chapters 3, 4, and 5. The biosynthesis of cyclopent-2-en-1-ylgly-cine,[8] 1 (Fig. 3) is not known but is related to the biosynthesis of chaulmoogric acid, which is present in the same plant. The isoxazoline,[9] 2, acivicin, origin-ates from ornithine via *N*-hydroxyornithine. The ring closure involves loss of C^3-H_s, and C^2-H of ornithine is exchanged in the process.[9] Domoic acid,[10] a shellfish toxin, 3, can be dissected into glutamic acid and a monoterpene. It is probably synthesized by *N*-alkylation of glutamic acid with geranyl phos-phate followed by ring closure. The exact timing of the formation of the car-boxyl group in the terpenoid side chain is difficult to assess. Domoic acid oc-curs in several *cis–trans*-isomeric forms. Coprine,[11] 4, is an adduct of cyclo-propanone, the origin of which is 1-aminocyclopropane carboxylic acid and glutamic acid. It is shown by radioactive labelling experiments that the bio-synthesis of cyanoalanine, 5, takes place from L-cysteine by elimination of hy-drogen sulphide and Michael addition of cyanide to the intermediate α-ami-noacrylic acid.[12] β-Cyanoalanine is hydrolysed to asparagine,[13] decarboxyla-ted to β-aminopropionitrile and reduced to 2, 4-diaminobutyric acid (Fig. 4).

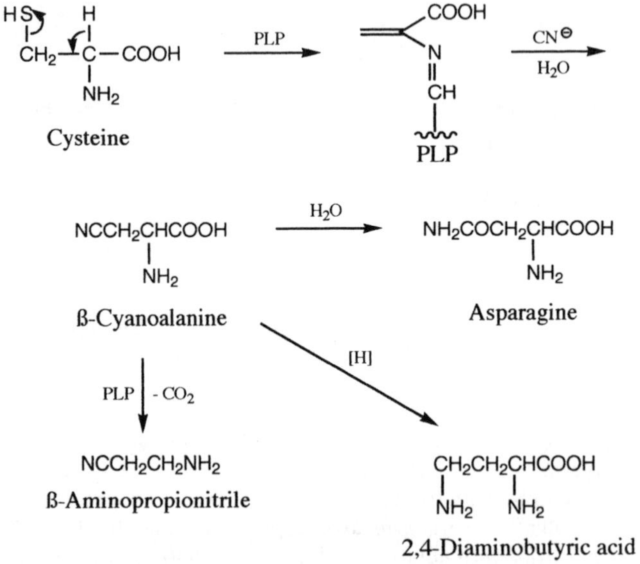

Fig. 4   Biosynthesis and metabolism of β-cyanoalanine. PLP, pyridoxal phosphate

Several unsaturated amino acids of the type 6 (Fig. 3) are known, acetylenic as well as olefinic. The amino group of amino acids has been found to be al-kylated to various degrees, methylation being especially common. Glycine gives via sarcosine and dimethylglycine the fully methylated and widely dis-tributed betaine 8 (Fig. 3), which gave its name to this class of dipolar com-pounds; the betaines. Another biosynthetic route to betaine starts with serine, which is phosphorylated, decarboxylated, methylated, and finally oxidized (Fig. 5). It was isolated by Huseman in 1863 from *Lycium barbarum* and by Scheibler from the sugar beet, *Beta vulgaris*, in 1869, containing up to 5 per cent betaine in dried leaves. The corresponding arsenobetaine has been isolated from lobster.[13a] Arsenic is metabolized especially by marine organisms.

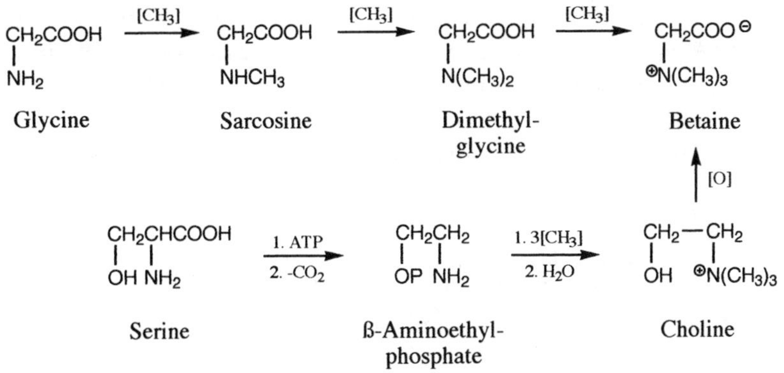

CH₂COOH — [CH₃] → CH₂COOH — [CH₃] → CH₂COOH — [CH₃] → CH₂COO⁻

$$CH_2COOH \xrightarrow{[CH_3]} CH_2COOH \xrightarrow{[CH_3]} CH_2COOH \xrightarrow{[CH_3]} CH_2COO^{\ominus}$$

Glycine — Sarcosine — Dimethylglycine — Betaine

$$\underset{NH_2}{|}CH_2COOH \xrightarrow{[CH_3]} \underset{NHCH_3}{|}CH_2COOH \xrightarrow{[CH_3]} \underset{N(CH_3)_2}{|}CH_2COOH \xrightarrow{[CH_3]} \underset{\overset{\oplus}{N}(CH_3)_3}{|}CH_2COO^{\ominus}$$

$$\underset{OH\ NH_2}{| \quad |}CH_2CHCOOH \xrightarrow[2.\ -CO_2]{1.\ ATP} \underset{OP\ NH_2}{| \quad |}CH_2CH_2 \xrightarrow[2.\ H_2O]{1.\ 3[CH_3]} \underset{OH \quad \overset{\oplus}{N}(CH_3)_3}{| \quad \quad |}CH_2-CH_2$$

Serine — β-Aminoethylphosphate — Choline

Fig. 5   Biosynthesis of betaine

The functions of most of these secondary amino acids are unknown but circumstantial evidence indicates that their presence is not meaningless. Several non-protein amino acids are highly toxic or have inhibitory effects on insects, microorganisms, and plants. Canavanine, 9 (Fig. 3), inhibits growth of yeast and seedling growth of *Lathyrus* spp., whereas it has no effect on seedlings of *Vicea bengalensis*, which actually synthesizes canavanine. It is biosynthesized from homoserine via *O*-aminohomoserine which is guanidated by carbamylphosphate and ammonia (*cf.* formation of citrulline from ornithine, section 7.4). Azetidine-2-carboxylic acid, 10, shows also phytotoxic effects, inhibiting seedling growth in species which do not produce the amino acid. It is biosynthesized from methionine via 2,4-aminobutyric acid and 4-amino-2-oxobutyric acid. Homoarginine and pipecolic acid occurring in *Acacia* species inhibit feeding in the locust, *Locusta migratoria*, but have no effect on the nymphs of *Anacridium melanorhodon* feeding on leaves of *Acacia* spp. It is therefore clear that non-protein amino acids exert an influence of ecological significance.[14] It is not uncommon to find that certain free non-protein amino acids account for 1–10 per cent of the dry weight of the tissue. Seeds of *Dioclea megacarpa* contain up to 13 per cent canavanine, and leaves of *Convallaria majalis* contain 3 per cent azetidine-2-carboxylic acid. An accumulation of this order suggests that they have a storage role. Several amino acids occurring in fodder plants are toxic and could cause death. The unusual cyclopropanoid amino acid hypoglycine A, 11, occurs in the fruits of the akee tree, *Blighia sa-*

pida. It causes hypoglycaemia and occasionally death among people in the West Indies. Mimosine (see Fig. 11), a metabolite of the tropical legume *Leucaena glauca*, causes loss of hair and is of great concern to sheep breeders in Australia. Certain plants when growing on selenium rich soil can metabolize selenium and substitute selenium for sulphur. The SeCH$_3$ analogue of cysteine, produced by *Astragalus bisulcatus* in the Western United States, the so-called locoweeds, causes madness among grazing livestock.

Amino acids are classified in various ways. Since the polarity of the R group (Table 1) controls the stability of various conformations of the free acid or still more important controls the conformation and coiling of the peptide chain with consequences for the topology of the enzyme surface, it is meaningful to base a classification on the polarity of the R group:

1. non-polar,
2. neutral polar,
3. positively charged and
4. negatively charged amino acids (see Table 1).

Another classification is based on the nutritional requirements of amino acids. *E. coli* can utilize ammonia alone for the biosynthesis of all of its amino acids and derivatives, whereas most vertebrates including man require some amino acids in their diet. Of the twenty amino acids required for the synthesis of proteins man can manufacture ten, the other ten are called the essential amino acids.

$$\underset{\overset{|}{\oplus}NH_3}{RCHCOOH} \xrightarrow{-H^\oplus} \underset{\overset{|}{N}H_2}{RCHCOOH} \rightleftharpoons \underset{\overset{|}{\oplus}NH_3}{RCHCOO^\ominus} \xrightarrow{OH^\ominus} \underset{\overset{|}{N}H_2}{RCHCOO^\ominus} \qquad (1)$$

Amino acids are bifunctional compounds having an acid carboxyl group and a basic amino group. Consequently they are better represented by their dipolar ionic or zwitterionic structure (1). They act either as acids or bases and as such are called amphoteric compounds. Because of their zwitterionic nature they are high-melting solids, soluble in water and insoluble in non-polar solvents. Having an asymmetric centre α to the carboxyl group they racemize in strong acidic or alkaline media. Actually even at room temperature and under neutral conditions an extremely slow racemization takes place that has been utilized for dating fossils.[15] Different amino acids have different half-lives, and so it has been possible to cover times from *ca*. 100 to 100 000 years. Investigations of proteins from the human lens and tooth enamel show that racemization is related to age and provided that the notorious old Caucasians have not lost their teeth, it should be possible to check their claim of longevity.[16] The conclusions rest on the assumption that conversion of some trapped proteins is low in the organism. It has been proposed that ageing is coupled with racemization of chiral centres in proteins thus making them malfunction. The absolute configuration of amino acids not possessing interfering chromophores can be deduced from their CD curves at the 210 nm n–π^* transition. The L-stereoisomers give a positive Cotton effect,[17] but care must be exercised because conformational factors affect the sign. A sextor rule based on the octant rule is proposed on the basic assumption that the N-C$^\alpha$-COO atoms are coplanar in solution.[18]

Ninhydrin

Resonance stabil-
ized ion

(2)

Ninhydrin

bluish-red

Amino acids are detected by a sensitive reaction with ninhydrin which gives a bluish-red colour (2)[18a]. This reaction is closely related to the Strecker degradation of amino acids and the pyridoxal mediated enzymatic decarboxylation of amino acids. The reaction has been used in forensic medicine for the detection of fingerprints on paper. The minute amount of sweat exuded from the pores of a fingertip contains enough amino acids to give a coloured reproduction of the fingerprints by spraying the paper with ninhydrin and heating it to 100°C.

Few groups of organic compounds have been the object of so much qualitative and quantitative analytical research as the amino acids and peptides. Different kinds of chromatography based on either the principles of partition or absorption have been worked out for complete separation of complicated mixtures, e.g. a hydrolysed protein. These include one- or two-dimensional paper chromatography, silica gel thin-layer chromatography (TLC), electrophoresis on different carriers, ion-exchange chromatography and column chromatography for larger quantities. The countercurrent technique developed by Craig is based solely on the principle of partition of a solute between immiscible liquids and chromatographic separations on carriers are to varying extents based on the same principle. On the surface of, for example, cellulose fibres of filter paper, a film of solvent and hydrated polysaccharides is formed, different from the bulk of solvent, and the amino acids are partitioned between these two media. Absorption phenomena have a substantial influence on the partition which can be varied widely by change in solvent composition. The spots or bands are usually developed by ninhydrin. By making amino acids more volatile, by methylation or silylation, it is possible to apply gas chromatographic procedures combined with highly sensitive mass spectrometric recordings. The analytical procedures have continual been refined and are now automated and carried out with considerable speed and precision.[19]

7.3 Reactions of amino acids promoted by pyridoxal phosphate

Glutamic acid and glutamine occupy a central position in group transfer metabolism for amino acids. Glutamine is the carrier of ammonia in an unreactive form, and it is released from the amido function by enzymatic hydrolysis. Glutamic acid serves as donor of the amino group in transamination. The α-amino group of glutamic acid is introduced by reductive amination of α-keto glutaric acid via the imine, a reversible reaction catalysed by glutamate dehydrogenase which uses NADH or NADPH as reductanct (3). It was shown by tracer experiments that the prochiral H_s of NADH was transferred stereospecifically.

$$NH_3 \;+\; HOOCCH_2CH_2COCOOH \;\underset{}{\overset{-H_2O}{\rightleftharpoons}}\; \underset{NH}{HOOCCH_2CH_2\overset{\|}{C}COOH} \;\underset{NAD^\oplus}{\overset{NADH}{\rightleftharpoons}}$$

$$\underset{NH_2}{HOOCCH_2CH_2\overset{|}{C}HCOOH} \tag{3}$$

L-Glutamic acid

This represents one of the major pathways to amino acids directly from ammonia and α-keto acids. The other is transamination with pyridoxal phosphate, PLP, as a coenzyme.[23] PLP is anchored as a Schiff base at the active site

$$\tag{4}$$

PLP-Enz-aldimine
(Schiff base)

by a lysine residue as shown by sodium borohydride reduction of the enzyme followed by hydrolysis (4). Glutamic acid displaces the lysine reside forming a new aldimine in which H$^\alpha$ is doubly activated and can be removed by a strategically located basic function on the enzyme (Fig. 6). The aldimine rearranges to the ketimine which on hydrolysis gives α-ketoglutarate and pyridoxamine phosphate. This constitutes the first half of the transamination, the mechanism of which originally was proposed by Snell and Braunshtein. The H$^\oplus$ transfer takes places intramolecularly to some extent. The second half is a reversal of the first part with another keto acid generating a new amino acid

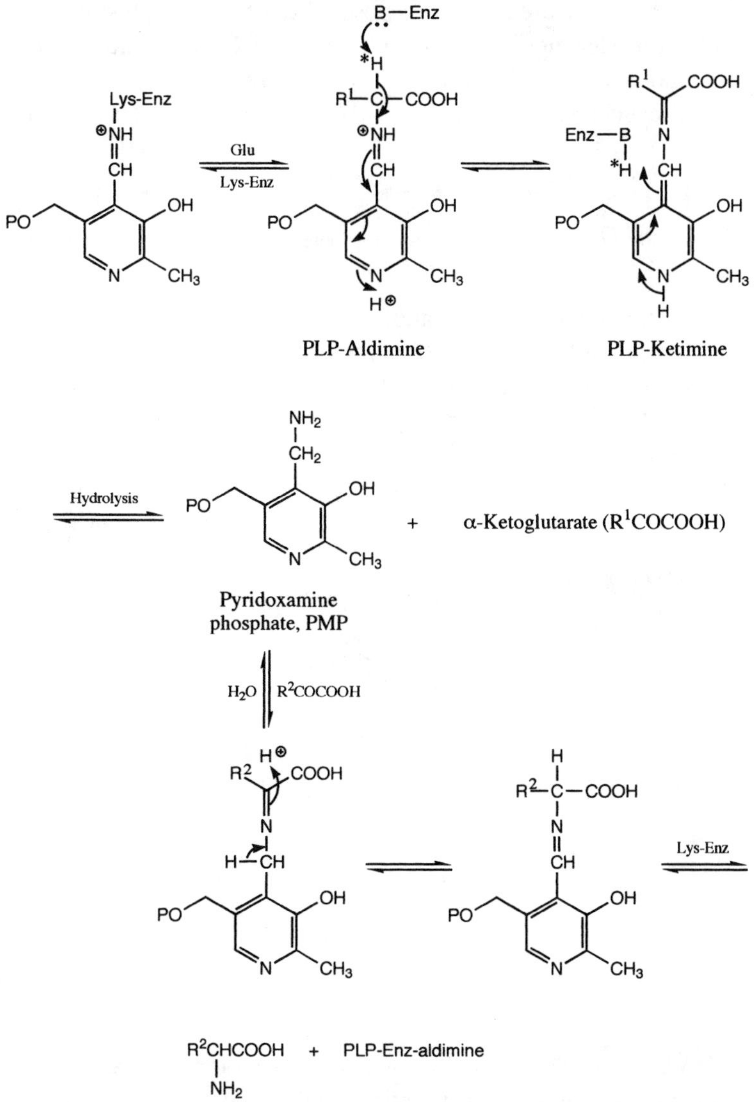

Fig. 6   Mechanism of transamination with glutamic acid as donor of the amino group

and pyridoxal phosphate etc. The two steps can be summarized as in (5). The mechanism of action of PLP dependent racemization is readily envisioned according to the scheme above.

$$\text{Amino acid}^1 \;+\; \text{Keto acid}^2 \;\rightleftharpoons\; \text{Keto acid}^1 \;+\; \text{Amino acid}^2 \qquad (5)$$

The purpose of the coenzyme is thus to activate the C^α–H bond to such a degree that the proton can be removed by a weak base. The anion formed is stabilized as a result of efficient delocalization. These two effects are of course interrelated. But the possibility of forming a stabilized anion affects also the cleavage of the C^α–COOH and C^β–C^α bonds. Decarboxylation of amino acids is a common and important reaction for the formation of amines. A direct cleavage of the amino acids according to (6) is unlikely since they are

very stable and α-amino anions are unfavourably high in energy. Decarboxylation of the PLP aldimine, on the other hand, leads to the charge delocalized anion (7). It is rewarding for theoreticians to see that enzymes acknowledge predictions made by simple principles of orbital overlap.

Whether H^α abstraction or decarboxylation will take place is decided by the position of the basic functions at the active site of the enzyme and how the substrate is oriented on the enzyme surface. Decarboxylation is not preceded by any H^α exchange. The product from the decarboxylation of tyrosine contains one deuterium when the decarboxylation is run in D_2O and it was found that it proceeds with retention of configuration.[24]

Generation of the stabilized C^α-anion of PLP-glycerine by cleavage of a C^β–C^α bond occurs in the retro aldol cleavage of serine and threonine cata-

$$(8)$$

lysed by serine hydroxymethylase (8). Formaldehyde is not set free but scavenged by tetrahydrofolate for later usage. The presence of tetrahydrofolate is found to increase the rate of cleavage as a result of fomaldehyde being removed from the equilibrium. Experiments with chirally labelled glycine[25] show that the enzyme exclusively abstracts the H_S and that substitution takes place with retention of configuration at C^2 of glycine, e.g. reversal of (8). The mechanism is analogous to that for decarboxylation of amino acids. No C^α–H exchange occurs with the solvent implying that dehydroalanine is not on the pathway and tetrahydrofolate does not directly participate in the displacement as formulated in (9). Consistent with the transient formation of formaldehyde are results with serine, chirally labelled with ^3H at C^3. Partial loss of chirality was observed in the CHT fragment of N^5,N^{10}-methylene tetrahydrofolate in accordance with the intermediary formation of formaldehyde,

$$(9)$$

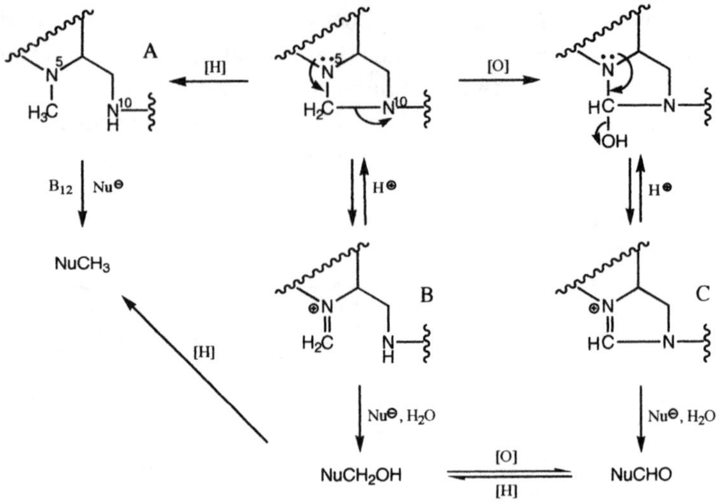

Fig. 7   Structures of active C_1 transferring tetrahydrofolate at different oxidation levels

free long enough to rotate.[26] The cleavage of threonine is independent of tetrahydrofolate.[27] N^5,N^{10}-methylene tetrahydrofolate is one of the most important donors of the one carbon fragment; the other is S-adenosylmethionine which actually receives its methyl from N^5, N^{10}-methylene tetrahydrofolate. The C_1 unit can be used directly at the formaldehyde oxidation level or be oxidized or reduced to active formic acid or methyl, respectively. The active species in formylation is the amidinium ion C and in hydroxymethylation the immonium ion B (Fig. 7). It is more difficult to reconcile the methylating ability of the N^5–CH_3 derivative A with the known sluggishness of trialkylamines as alkylating agents in regular organic reactions. Hence, it seems more plausible that B is the active species in combination with a reductant. It ought to be mentioned that the methylation is mediated by vitamin B_{12} and it is suggested that Co–CH_3 bonding is involved.[28] The reaction pattern of organocobalt compounds is not well understood and does not follow the traditional mechanistic principles organic chemists are used to handling. An example of an unusual skeletal rearrangement is the methylmalonyl-CoA– succinyl-CoA rearrangement which is mediated by vitamin B_{12} (10). There are indications that the rearrangement is radical in nature[29] but the mechanism of $-CH_2-$*H homolysis and *H transfer is not clear. The action of vitamin B_{12} will be discussed in greater detail in section 9.4.

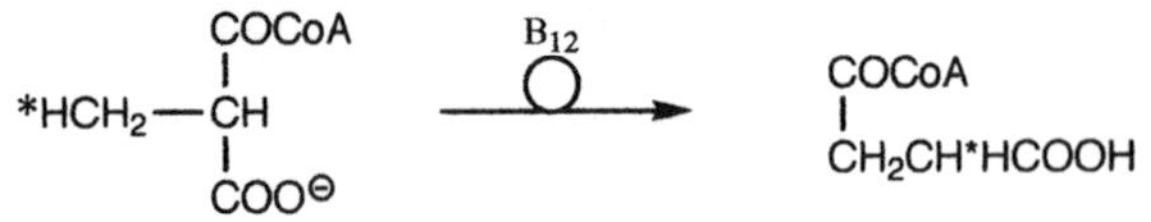

$$\begin{array}{c} \text{COCoA} \\ | \\ \text{*HCH}_2\text{—CH} \\ | \\ \text{COO}^{\ominus} \end{array} \quad \xrightarrow{\quad B_{12} \quad} \quad \begin{array}{c} \text{COCoA} \\ | \\ \text{CH}_2\text{CH*HCOOH} \end{array} \qquad (10)$$

The first part of the mechanism in (11) represents an α,β-elimination which is observed for serine and other amino acids with good leaving groups at C^3, e.g. 3-chloroalanine and cysteine. The intermediate α-iminoacrylate gives either pyruvic acid on hydrolysis (11a), or if it finds a good nucleophile in the vicinity of the active site of the enzyme, it undergoes β-substitution (11b). We have already encountered such cases in the biosynthesis of tryptophan (section 4.2) and cyanoalanine $Nu^\ominus = CN^\ominus$

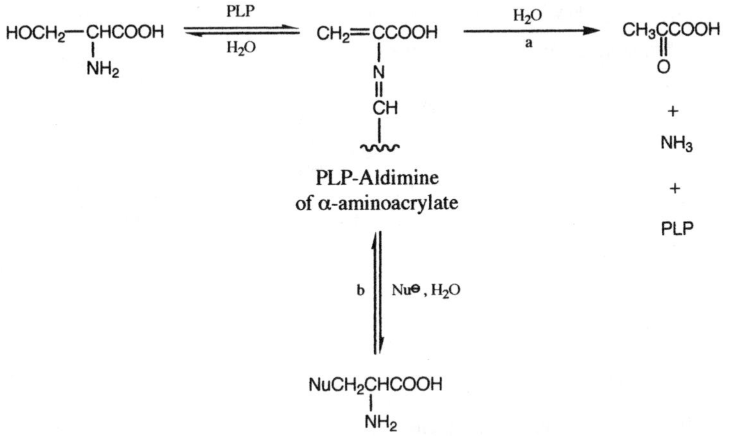

PLP also activates reactions at γ-positions as exemplified by its role as cofactor for cleavage of cystathionine to cysteine and α-ketobutyric acid, the last step in the biosynthesis of cysteine (Fig. 8). In support of the depicted β,γ-elimination mechanism which presupposes β-ionization are experiments carried out in D_2O showing incorporation of one $3S$-^2H in α-ketobutyrate.[30]

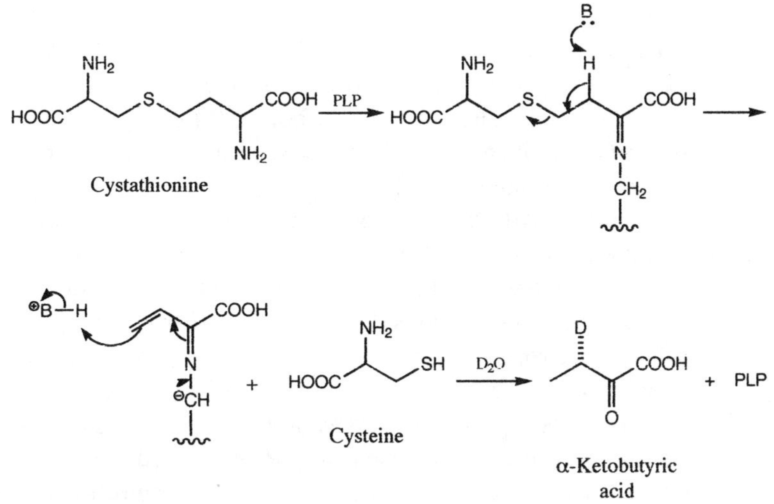

Fig. 8   Cleavage of cystathionine

7.4 The guanidino function

The guanidino function appears in several compounds from widely different sources among which marine metabolites are well represented. Octopine from *Octopus* spp. is, for structural and mechanistic reasons, assumed to be formed from arginine and pyruvic acid[31] (Fig. 9). It is known that pyruvic acid acts in the role of cofactor in transaminations. Reduction of the Schiff base with NADPH completes the biosynthesis. Octopine was also detected in tissue cultures of *Nicotiana tabacum* and the amount was increased by induction with *Agrobacterium tumefaciens.*[32]

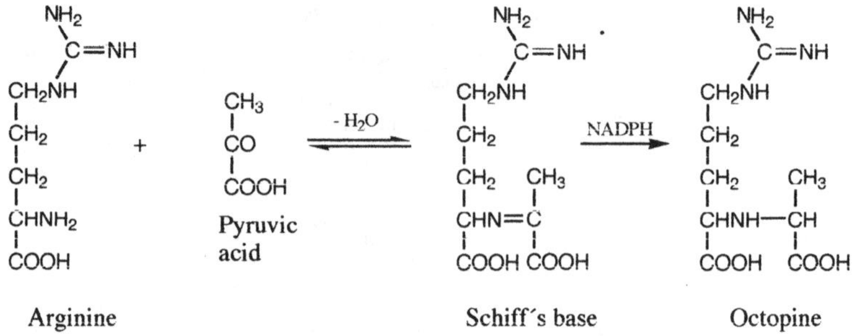

Fig. 9   Biosynthesis of octopine

The formation of the guanidino function is mechanistically interesting as an illustration of the action of ATP and utilization of its energy. This is discussed in conjunction with a presentation of the urea cycle uncovered by Krebs and Henseleit in 1932 (Fig. 10). The urea cycle solves for the organism the problem of how to dispose of its excess of ammonia which is excreted in the urine as urea.

The first step is carbamoylation of ornithine with carbamoyl phosphate (12). This reactive species is formed from ammonia, bicarbonate and ATP[33] (13). The bond energy of the P–O bonds of the triphosphate chain is utilized for formation of an unstable activated carbonate, an acyl phosphate, which is able to carboxylate ammonia with displacement of phosphate. Support for the intermediate hydroxyacyl phosphate was given by trapping experiments with diazomethane which indicated the presence of a trimethyl ester[34] (14).

Ammonia is released from glutamine by the action of glutaminase. Inactivation of glutaminase stops the formation of carbamoyl phosphate but the capacity of carbamoyl-P synthetase can be regenerated by addition of free ammonia which shows that the glutaminase activity is located on another subunit of the enzyme. One mole of carbamoyl phosphate thus requires two moles of ATP. A γ- or β-cleavage releases more energy (ΔG *ca.* 7 kcal) than α-cleavage (ΔG *ca.* 3 kcal) which means that mono- and diphosphorylation occur most frequently. The acyl phosphate has two reactive sites. If the nucleophile,

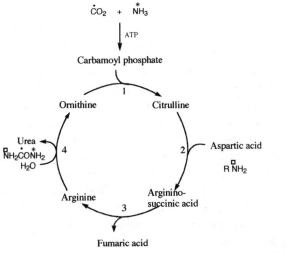

(12)

Fig. 10   The urea cycle

as in this case, attacks the polarized carbon, acylation occurs. The nucleophile could also attack phosphorus and the cleavage on the other side of the oxygen leads to phosphorylation. These mixed anhydrides are both good acylating and phosphorylating agents. The mode of action depends on how the enzyme orients the reactants.

The second step in the urea cycle is formation of argininosuccinic acid from citrulline and aspartic acid. This reaction requires another mole of ATP for activation of the carbamoyl group of citrulline. In the third step the synthesis of a guanidine group is completed by elimination of fumaric acid. The last step is a hydrolysis to urea and ornithine by the action of arginase, an enzyme requiring $Mn^{2\oplus}$ as cofactor.

$$(13)$$

Acyl phosphate   Carbamoyl phosphate

$$(14)$$

7.5 Secondary products from serine and cysteine

The biosynthetic versatility of serine was pointed out earlier in connection with the reactions of pyridoxal phosphate. Serine is located at a branching point in the secondary metabolism of amino acids. Michael addition of various N-, S- and C-nucleophiles to the derived PLP aldimine leads to β-substituted alanines[35] (Fig. 11). The phenylalanine derivative is especially interesting because from structural evidence it may be concluded that it is derived from shikimic acid as are other phenylalanines. Tracer experiments indicate that the aromatic nucleus is derived from orsellinic acid and the side chain from serine,[36] cf. the biosynthesis of thyroxine (section 5.15).

It it actually O-acetylserine which is the key intermediate in the synthesis of cysteine and other β-substituted alanine derivatives. Acetylation makes the hydroxyl group easier to eliminate. The corresponding O-phosphoserine does

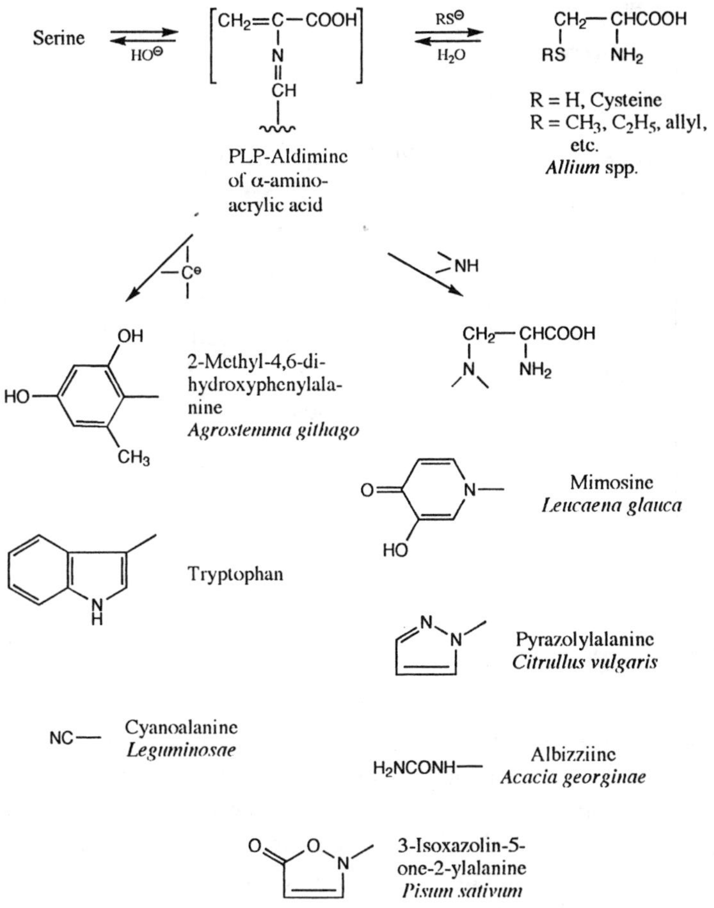

Serine $\underset{HO^\ominus}{\rightleftharpoons}$ $\begin{bmatrix} CH_2=C-COOH \\ | \\ N \\ \| \\ CH \\ | \\ \sim\sim\sim \end{bmatrix}$ $\underset{H_2O}{\overset{RS^\ominus}{\rightleftharpoons}}$ $\begin{array}{c} CH_2-CHCOOH \\ | \qquad | \\ RS \qquad NH_2 \end{array}$

PLP-Aldimine
of α-amino-
acrylic acid

R = H, Cysteine
R = CH₃, C₂H₅, allyl,
etc.
Allium spp.

$-\overset{|}{\underset{|}{C^\ominus}}$

$\geq NH$

**2-Methyl-4,6-di-
hydroxyphenylala-
nine**
Agrostemma githago

$\begin{array}{c} CH_2-CHCOOH \\ | \qquad | \\ N \qquad NH_2 \end{array}$

Mimosine
Leucaena glauca

Tryptophan

Pyrazolylalanine
Citrullus vulgaris

NC— Cyanoalanine
Leguminosae

Albizziine
H₂NCONH— *Acacia georginae*

3-Isoxazolin-5-
one-2-ylalanine
Pisum sativum

Fig. 11 Secondary derivatives of serine

not seem to have any metabolic function. Like other thiols, cysteine is sensitive to oxidation. It dimerizes easily to the disulphide cystine, a reaction of great consequences for the folding of peptide chains. The thiol can also be oxidized to sulphinic and sulphonic acids[37] (Fig. 12). Decarboxylation of the sulphonic acid gives taurine, which is widespread in nature. Oxidation of alkylated cysteines gives sulphoxides, e.g. alliin, common in the *Liliaceae, Cruciferae* and *Mimosaceae* families. *S-trans*-propenyl-L-cysteine functions as the progenitor of the characteristic lachrymatory principle in onions.

Cysteine has also been invoked as precursor for the coenzyme biotin but since it turned out that the biosynthesis proceeds via dethiobiotin,[38] it is more likely that alanine reacts with pimeloyl CoA in a pyridoxal promoted reaction. Labelling experiments demonstrate that adjacent protons at $C^{2,3,5}$ are not engaged in the sulphuration process, which at C^4 proceeds with retention of

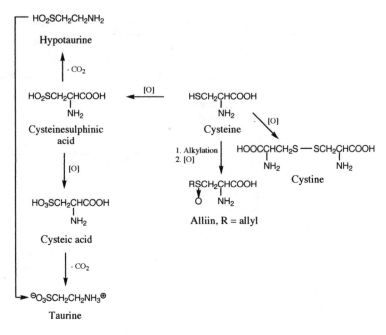

Fig. 12   Oxidations of cysteine

configuration.[39]. Evidently sulphur is not introduced by a Michael addition. A possible mechanism for the S-insertion is shown in Fig. 13. The hydrogen atom is abstracted by a non-haem, S-ligated $L_4(RS)Fe^{IV}O^{\cdot}$ radical, cf. section 4.3, and the S-ligand (conceivably cysteine) migrates subsequently to the C-centred radical in an inner sphere reaction. The cysteinyl moiety is eventually cleaved to give the thiol and the process is repeated at C^4.

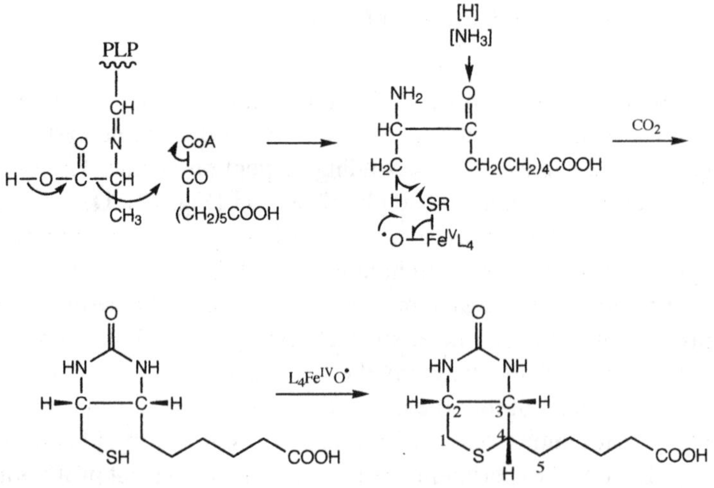

Fig. 13   Biosynthesis of biotin

7.6 Secondary products from valine, leucine and isoleucine

The biosynthesis of these three essential amino acids is similar. It embraces some general mechanistic principles and is therefore discussed here.

We have first the thiamine catalysed decarboxylation of pyruvic acid leading to acetaldehyde which condenses with an α-keto acid (Fig. 14). Pinacolone type rearrangement and NADPH reduction lead to the β-branched α-keto acid which either is directly aminated to valine and isoleucine or un-

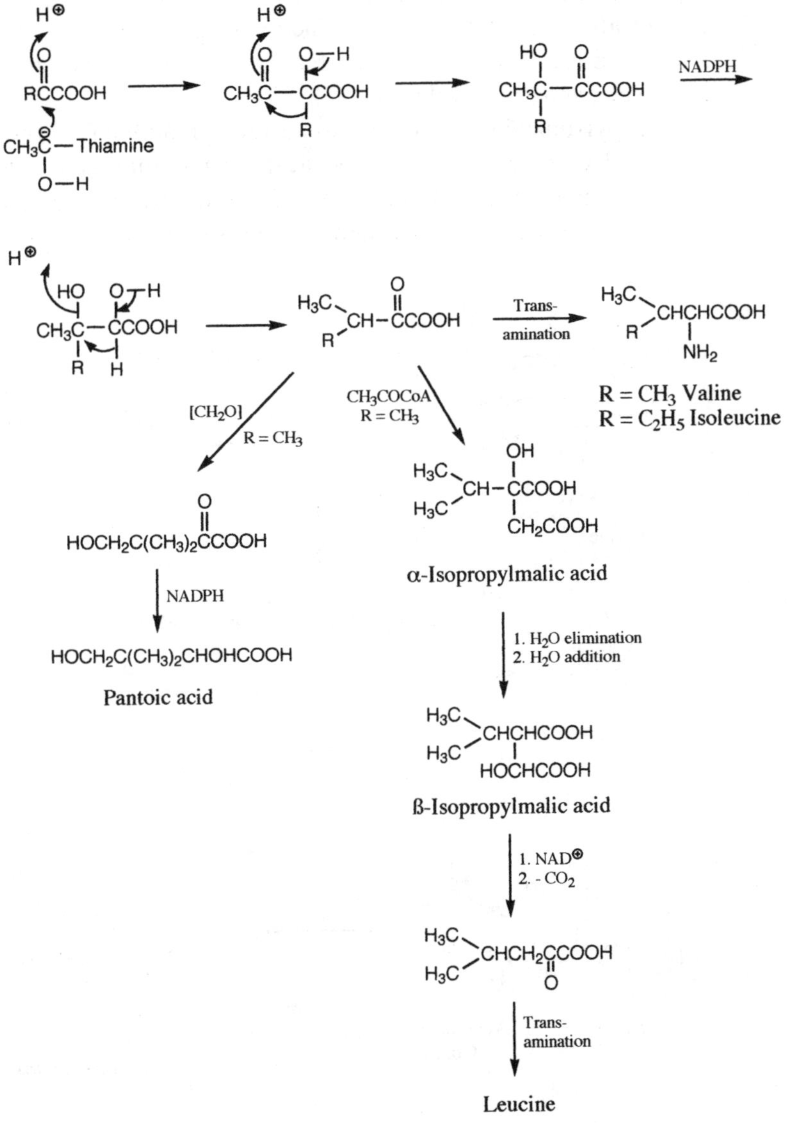

Fig. 14   Biosynthesis of valine, isoleucine, leucine and pantoic acid

dergoes chain lengthening to leucine (R = CH$_3$). This sequence, which has several analogies, starts with a condensation with acetyl CoA to form an α-substituted malic acid, cf. the biosynthesis of mevalonic acid. Elimination and readdition of water gives β-substituted malic acid, cf. the citric acid cycle. Oxidation of the hydroxy group and decarboxylation of the β-keto acid formed eventually leads to the homologous α-keto acid and, after transamination with glutamic acid, to leucine.

Pantoic acid, one of the building blocks in coenzyme A, is synthesized from the intermediary α-ketoisovaleric acid by condensation with N^5,N^{10}-methylenetetrafolic acid followed by reduction (Fig. 14). Erythroskyrin and tenuazonic acid are formed by mixed biogenesis from valine and isoleucine, respectively, and polyketide (Fig. 15).

The pyrrolizidine alkaloids are esters of aminoalcohols, necines, and branched carboxylic acids, necic acids, e.g. echimidinic acid and senecic acid, originally thought to be derived from acetate. Tracer experiments have shown, however, that they originate from simple amino acids. Echimidinic[40]

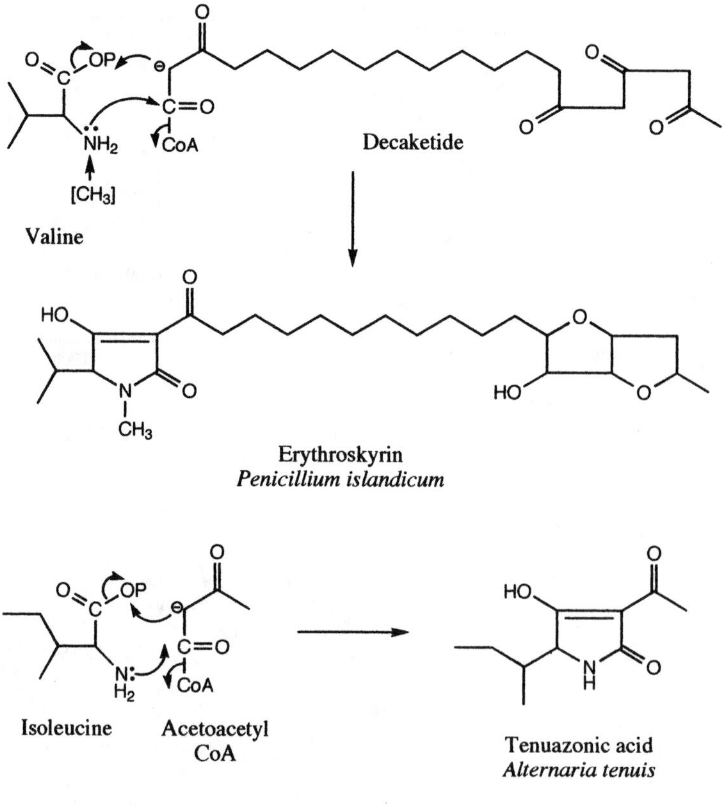

Decaketide

Valine

Erythroskyrin
Penicillium islandicum

Isoleucine   Acetoacetyl
CoA

Tenuazonic acid
Alternaria tenuis

Fig. 15   Secondary products from valine and isoleucine

acid originates from valine, senecioic acid[41] from leucine or mevalonic acid and senecic acid[42] by decarboxylation and dimerization of two isoleucine units (Fig. 16).

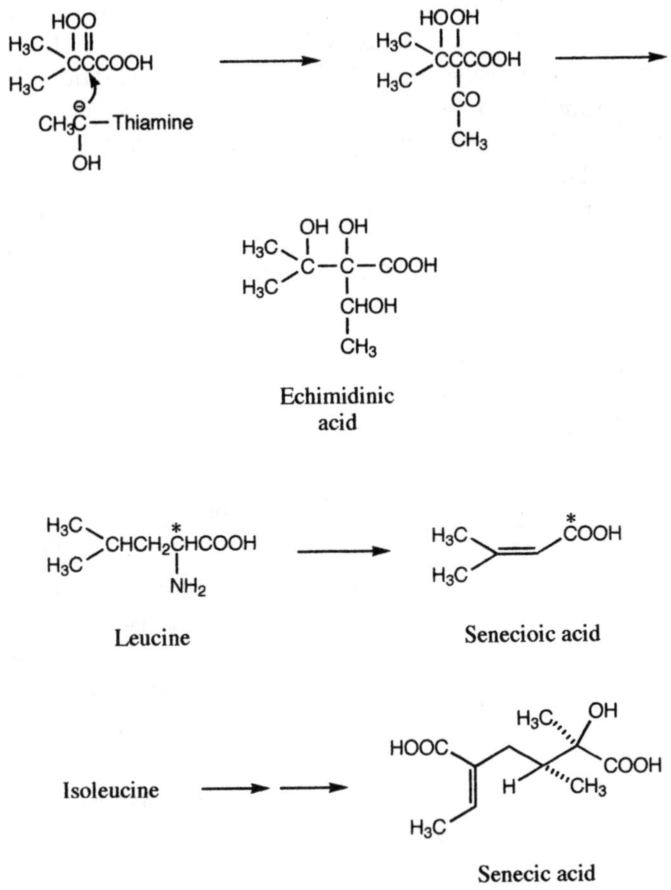

Echimidinic
acid

Leucine   Senecioic acid

Isoleucine   Senecic acid

Fig. 16   Pathway for echimidinic, senecioic and senecic acids

7.7 Cyanogenic glycosides and glucosinolates

The cyanogenic glycosides are the principal precursors of hydrocyanic acid in plants. Strangely enough some plants are able to metabolize the highly toxic cyanide ions known to effectively block the action of ferroporphyrins by stable ligand formation. In higher plants cyanide reacts with cysteine and serine (section 7.5), forming cyanoalanine which can be further hydrolysed to asparagine. This reaction is believed to serve as a detoxification mechanism for cyanide ions. It is well established that the cyanogenic glycosides are derived from amino acids. Experiments with labelled precursors show that the carboxyl group is lost as carbon dioxide, the cyano group is formed from C^2 and the amino group and C^3 becomes hydroxylated (15). There is evidence for the production of *N*-hydroxyvaline, isobutyraldoxime and isobutyronitrile from valine in flax seedlings as well as for conversion of these intermediates to glycosides that suggests a Beckmann type dehydration and α-hydroxylation to cyanohydrins as the most likely pathway.[43] Molecular oxygen is introduced by P450-dependent oxygenases into aldoximes and cyanohydrins,[44] and a β-glucosyltransferase has been partially purified from *Sorghum* seedlings exhibiting activity for cyanohydrins.[45] The occurrence of cyanohydrins is not limited to the plant kingdom. The millipede *Harpaphe haydeniana* uses hydrogen cyanide, stored as mandelonitrile, as a defensive weapon.[46]

$$R_1R_2\overset{\triangle}{C}H\overset{O}{C}HCOOH \underset{\substack{\text{NADPH, -CO}_2}}{\overset{2\,O_2,\,Fe^{II}}{\longrightarrow}} R_1R_2CHCH=NOH \overset{-H_2O}{\longrightarrow}$$
$$\underset{*NH_2}{}$$
$$\text{Oxime}$$

$$\underset{\text{Nitrile}}{R_1R_2CHCN} \underset{\text{NADPH}}{\overset{O_2,\,Fe^{II}}{\longrightarrow}} \underset{\text{Cyanohydrin}}{R_1R_2C\overset{OH}{\underset{CN}{\Big\langle}}} \underset{\text{glucose}}{\overset{UDP}{\longrightarrow}} R_1R_2\overset{\triangle}{C}\overset{OGlc}{\underset{CN}{\Big\langle}}_* \qquad (15)$$

$R_1=R^2=CH_3$ (from valine)

Linamarin

Linum usitatissimum

$R_1=H$; $R_2=p$-OH-phenyl

(from tyrosine)

Dhurrin

Sorghum vulgare

$$\underset{\text{Glucosinolate}}{R-C\overset{SGlc}{\underset{NOSO_3^-}{\Big\langle}}} \underset{\text{Enz}}{\overset{H_2O}{\longrightarrow}} \underset{\text{Isothiocyanate}}{R-NCS} \qquad (16)$$

Degradation products of glucosinolates give a characteristic smell to various plants of the *Cruciferae, Capparidaceae, Euphorbiaceae, Phytolaccaceae, Resedaceae* and *Tropaeolaceae* families. Destruction of the compartmental organization of the cell by rubbing the tissue brings glycolytic enzymes in contact with glucosinolates. Isothiocyanates or mustard oils (16) are formed by a Lossen type rearrangement. A common pathway to the aldoxime has been

$$\text{(17)}$$

Glucosino-
late

Thiohydrox-
imic acid

proposed for cyanogenic glycosides and glucosinolates. Characteristic features of the biosynthesis of glucosinolates are preservation of the labelling pattern from the parent amino acid, formation of aldoximes as intermediates, glucosidation of the thiol function and the observation that amines are poor precursors. The formation of the aldoxime can mechanistically be formulated as an oxidation of the *N*-hydroxyamino acid to the nitroso compound with concomitant decarboxylation. The oxidative sulphuration of the aldoxime is speculative (R^l conceivably cysteinyl) but the reaction has analogies in organic synthesis (17).

7.8 Peptides, β-lactam antibiotics and proteins

Peptides and proteins are polycondensation products of amino acids. The carbonyl group of one amino acid is joined to the amino function of another amino acid thus forming an amide bond or a peptide bond. The term peptide refers usually to polymers of molecular weight lower than *ca.* 5000. The molecular weight of proteins ranges from *ca.* 5000 to several millions but there is no clear distinction between the two groups. A peptide can be a partially hydrolysed protein, but also a compound containing characteristic non-protein amino acids. These are frequently found in animals, plants and fungi. A dipeptide is derived from two amino acids, a tripeptide from three amino acids, etc. Cyclization of dipeptides leads to the widespread 2,5-dioxopiperazines and related compounds,[47] frequently synthesized by microorganisms. *Aspergillus echinulatus* produces echinulin,[48,49] a dipeptide derived from L-tryptophan, L-alanine, and three isoprene units (Fig. 17) as shown by efficient incorporation of labelled amino acids and mevalonic acid. Cyclo-L-Ala-L-Try is also incorporated indicating isoprenylation as the last step. The isoprenyl group at C^2 has an unusual orientation. Model studies support the idea that an *N*-alkylated intermediate rearranges in a Claisen reaction to the isolated product.[50] A direct S_N2' alkylation or a rearrangement from C_3 is less likely. Several other mould metabolites are, in fact, known to contain the pro-

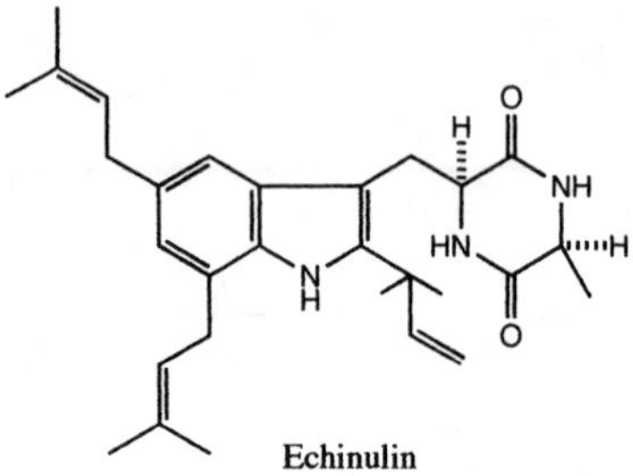

Echinulin
Aspergillus echinulatus

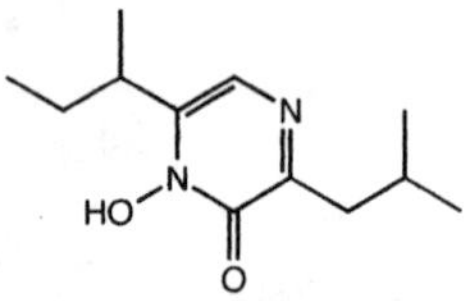

Aspergillic acid
Aspergillus flavus

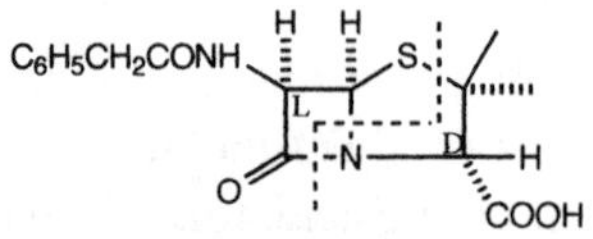

Benzylpenicillin
Penicillin G
Penicillium chrysogenum

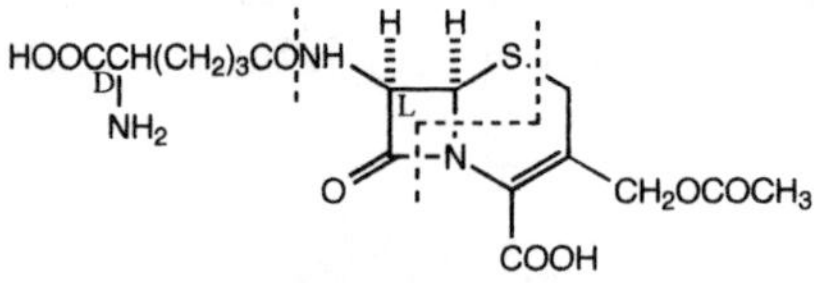

Cephalosporin C
Cephalosporium acremonium

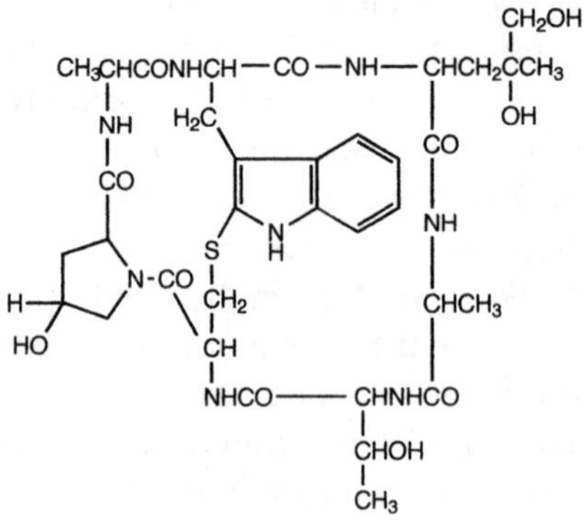

Phalloidin
Amanita phalloides

Fig. 17 Structure of naturally occuring peptides

posed *N*-isoprenyl grouping. Gliotoxin and aranotin, which were discussed in section 4.3, also belong to the dioxopiperazine group. Other *Aspergillus* spp. produce the dehydrogenated aromatic pyrazine nucleus, e.g. aspergillic acid[51] which is derived from leucine and isoleucine.

The important antibiotics penicillin and cephalosporin[52,53] produced by *Penicillium* and *Cephalosporium* spp. are cyclic tripeptides. It is established that their mode of action is related to an inhibition of the biosynthesis of the bacterial cell wall. They were the first antibiotics produced by microorganisms used in medicine. The biosynthesis is discussed in some detail in order to illustrate the mechanistic problems encountered. The penicillins can visually be dissected into L-cysteine, D-valine, and a substituted acetic acid, and the cephalosporins into L-cysteine, valine, and D-α-aminoadipic acid. The configurational identity of valine in cephalosporins is lost by double bond formation.

The L-amino acids were found to be more efficiently incorporated than the D-amino acids. This is noteworthy because it shows that inversion takes place at some late stage in the synthesis. D-Valine actually caused inhibition[54] of the production of penicillin, indicating competitive absorption, but poor conversion of this isomer, at the active site. Tracer experiments demonstrated phenylacetic acid as the immediate precursor of the benzylpenicillin and this side chain can be varied by adding other appropriate acids to the culture. α-Aminoadipic acid derives from α-ketoglutaric acid by chain elongation

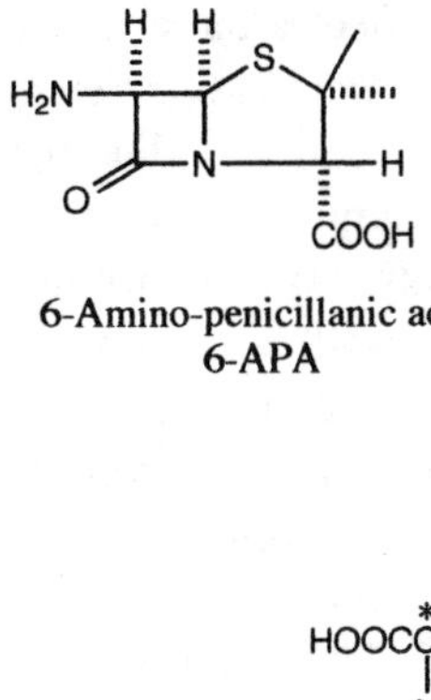

6-Amino-penicillanic acid,
6-APA

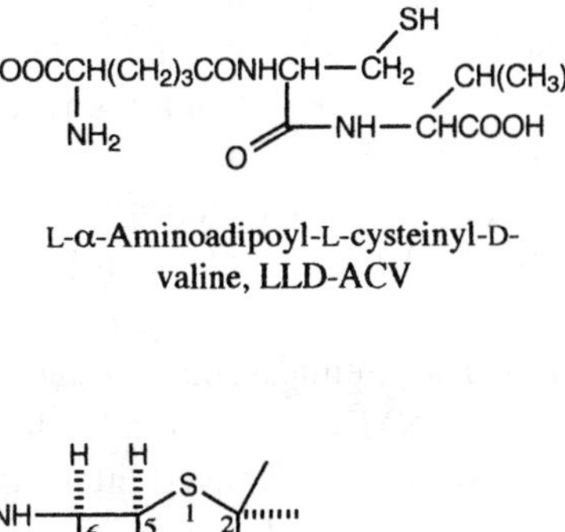

L-α-Aminoadipoyl-L-cysteinyl-D-
valine, LLD-ACV

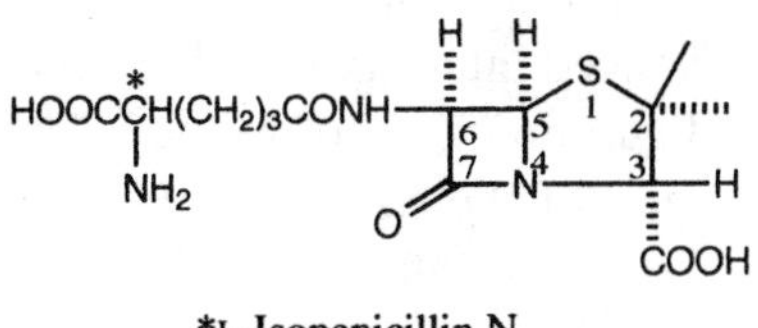

*L-Isopenicillin N
*D-Penicillin N

Fig. 18   Intermediates in the biosynthesis of β-lactam antibiotics

$$\text{Isopenicillin N} \xrightarrow[\text{Acyltransferase}]{C_6H_5CH_2COOH} \text{Penicillin G} \tag{18}$$

(section 7.6) and transamination. By multiple labelling it was shown that the amino acids are incorporated intact into both penicillins and cephalosporins. The building blocks are thus readily established but the mechanism of cyclization proved to be an intricate problem. Few intermediates of relevance to biosynthesis have been identified. The tripeptide L-α-aminoadipoyl-L-cysteinyl-D-valine, LLD-ACV, L-isopenicillin N, D-penicillin N and 6-aminopenicillanic acid, 6-APA, have been isolated, Fig. 18. No dipeptides have been observed as intermediates. Enzymatic experiments point to isopenicillin N as the most likely precursor for penicillins with non-polar side chains (18). 6-APA appears as a transient in this reaction and it accumulates in the fermentation broth of *P. chrysogenum* in the absence of other side chain precursors. Its appeerence indicates the stage at which exchange of acyl groups occurs in the side chain. The tripetide, LLD-ACV, is the most interesting intermediate and only this stereoisomer is converted into β-lactam antibiotics by *C. acremonium*. The valine moiety is thus inverted at an early stage of the biosynthesis, whereas the α-aminoadipoyl moiety becomes inverted first after the ring closures. The inversion of valine in the tripeptide is accompanied by exchange of its C^2 proton with protons from the solvent.[55] Involvement of 2,3- and 3,4-dehydrovaline in the cyclization process could be eliminated by isotopic labelling experiments, which excludes thiazolidine ring closure via intramolecular addition of the thiol to an olefinic bond. The methyl protons and the C^2-H of the D-valine moiety are all retained in the penicillin molecule. Proof that LLD-ACV is a true intermediate on the road to penicillins was provided by observation of ^{13}C and ^{1}H NMR spectra directly in the NMR probe of the tripeptide incubated with a cell-free extract of *C. acremonium*. The characteristic peaks originating from the D-valine moiety diminish with time and corresponding signals characteristic of isopenicillin N increase.[56] The stereochemical course of thiazolidine ring formation was determined by processing chiral (*2R, 3R*)-4-^{13}C-valine. This amino acid was converted to phenoxymethylpenicillin by *P. chrysogenum* and to cephalosporin C by *C. acremonium*. The chemical shifts of the germinal methyl groups were known from earlier ^{13}C NMR studies and the labelling pattern from incorporated labelled valine showed enchanced intensity of only one signal consistent with complete retention of configuration at C^3 of valine.[57] This result was verified by studying the NOE effects of the *cis*-located H^5 and C^2-α-CH$_3$ protons. Cephalosporin C became stereospecifically labelled in the SCH$_2$ group, Fig. 19.

The formation of the β-lactam ring has been investigated extensively. We know from experiments with 2-^{3}H-cysteine and (*3R*)-^{3}H-cysteine that the labels are retained in benzylpenicillin. These results imply that 2,3-dehydrocysteine does not appear on the pathway and that cyclization occurs with retention of configuration at C^3. A Michael type cyclization is therefore excluded.

Examination of tripeptides, in which the terminal valine moiety is replaced by other D-amino acids, as substrate for penicillin synthetase showed that a

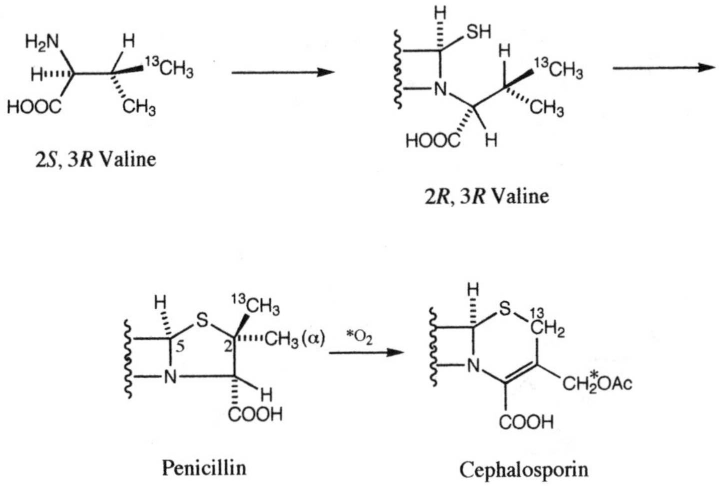

2*S*, 3*R* Valine

2*R*, 3*R* Valine

Penicillin   Cephalosporin

Fig. 19   Stereochemical course of valine incorporation

radical reaction mechanism best accounted for the formation of the bicyclic ring system,[52,53] and that it was related to the oxidative conversion of dethiobiotin to biotin,[58] Fig. 13. The general principles developed for biological hydroxylation can be applied, section 4.3. Reaction (19) shows the formation of the β-lactam. The cysteinyl H_S-hydrogen is selectively abstracted by the S-ligated iron-oxo species $L_4(RS)Fe^{IV}$-O·. Single electron transfer and cyclization complete the reaction. The Fe^{III} ion is reoxidized by H_2O_2, derived from molecular oxygen and α-ketoglutarate (or NADPH), and in the next step the ligated S-atom is oxidatively inserted into the valine-C^3-H bond with retention of configuration in an inner sphere reaction. The ring enlargement to the cephalosporins involves loss of one hydrogen from the β-methyl, a radical S-1,2-shift, dehydrogenation and eventually hydroxylation of the methyl group (19, 20). The oxidation state of the iron atom is not known with certainty. It could be a ligated Fe^V ion as shown here or a Fe^{IV} ion as previously suggested. The question about which ring is formed first is solved in favour of the β-lactam ring by measuring relative reaction velocities of LLD-ACV, LLD-A(3,3-²H)CV and LLD-AC(3-²H)V incubated with isopenicillin synthethase under both competitive and non-competitive conditions. It was shown that cleavage of the C^3-H bond of cysteine was the first irreversible step.[59] A biomimetic conversion of the tripeptide into a *cis*-β-lactam has been reported.[60]

One of the toxic principles of the mushroom *Amanita phalloides* was shown to be an unusual bicyclic heptapeptide, phalloidin, derived from cysteine, *allo*hydroxyproline, threonine, oxindolylalanine, γ,δ-dihydroxyleucine and two alanines, all with L-configuration[61] (Fig. 17). The toxic principles

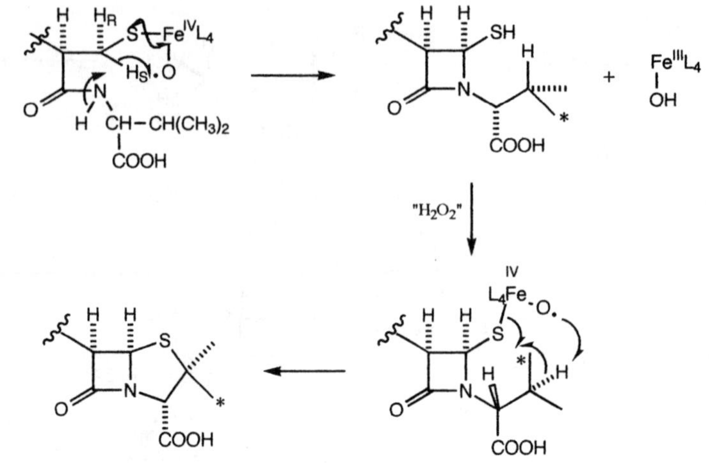

(19)

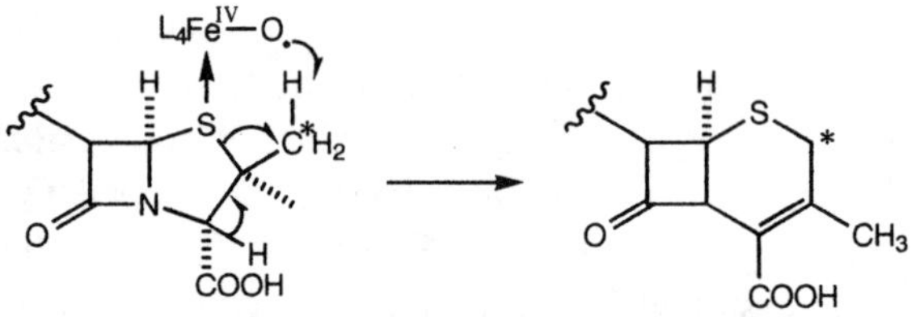

(20)

in the venoms from snakes and bees consist of single polypeptide chains containing cysteine cross-linkages. The neurotoxin of the spectacled Indian cobra, *Naja naja*, contains 71 residues, the sequence of which has been established,[62] see also section 7.2, Fig. 2.

The peptide bond of protein is synthesized by genetically controlled reactions on ribosomes, the details of which are best presented in textbooks of biochemistry.

(21)

$$RCHCOOH \ + \ ATP \ \rightleftharpoons \ RCHCOAMP \ + \ P_2$$
$$\hspace{0.4cm} | \hspace{5.3cm} |$$
$$\hspace{0.3cm} NH_2 \hspace{4.9cm} NH_2$$

Small peptides and non-protein peptides are produced enzymatically by activation of the amino acid with ATP according to (21) and by formation of a thioester (22) with the enzyme. Next to this site another amino acid, which fits the topology of the enzyme, enters, forms a thioester and reacts with the first amino acid (23). Subsequent reactions lead to polypeptides.

$$\text{Aminoacyl-AMP} \quad + \quad \text{Enz-SH} \quad \rightleftharpoons \quad \underset{\underset{NH_2}{|}}{\text{Enz}-S-\overset{\overset{O}{\|}}{C}-CHR} \tag{22}$$

$$\tag{23}$$

$$\begin{array}{cc}
\underset{|}{\overset{NH_2}{|}} & \\
\underset{|}{CHR^1} & H_2NCHR^2 \\
O=\underset{|}{C} & \underset{|}{C}=O \\
S & S \\
\hline
\end{array} \qquad \begin{array}{cc}
\underset{|}{\overset{NH_2}{|}} & \\
R^1CHCONHCHR^2 & \\
& \underset{|}{C}=O \\
HS & S \\
\hline
\end{array}$$

$$\text{Enz} \qquad\qquad \text{Enz}$$

7.9 Problems

7.1 Aspartic acid undergoes β-decarboxylation with formation of alanine. The reaction is catalysed by a PLP-dependent enzyme. It was found that the α-H is exchanged in the process. Occasionally pyruvic acid is formed instead of alanine. Suggest a mechanism explaining the conversions.

7.2 It has been suggested that glucosinolates are formed from the corresponding amino acids by oxidation of the amino group to a nitro group as the first step. Rearrangement of the nitro group assisted by ATP, cysteine, and uridine diphosphoglucose (UDPG) as reagents could then lead to the natural product. How would you account for the rearrangements? (Ettlinger, M. and Kjaer, A., in *Recent Advs. in Phytochemistry*, Mabry, T. J., Alston, R. E., and Runeckles, V. C. (Eds.), Appleton-Century-Crofts, New York 1, 1968, p. 58.)

7.3 The concentration of γ-aminobutyric acid, GABA, a neurotransmitter compound, is regulated in the body partly by its rate of formation via decarboxylation of glutamic acid, and partly by its rate of removal in a transamination as succinic semialdehyde mediated by PLP. It is shown that the structurally related gabaculine, irreversibly inactivates the transaminase by competitive formation of a stable aromatic derivative with PLP. Suggest a structure for this derivative and formulate a mechanism. Gabaculine can be regarded as a kind of suicidal substrate.

Gabaculine

7.4 1-Aminocyclopropanecarboxylic acid I derives from *S*-adenosyl methionine and is shown to be the precursor for ethylene, a hormone affecting the ripening of fruits. Suggest mechanisms for the formation of I and its fragmentation into ethylene which occurs under oxidizing conditions. Design experiments which support these transformations.

$$H_2N \quad\quad COOH$$

I

7.5 A hypothetical hexapeptide gave Gly_2, Asn, Ala, Leu, and Tyr on complete hydrolysis. N-Terminal analysis of the peptide with 2,4-dinitrofluorobenzene followed by hydrolysis gave 2,4-dinitrophenylalanate and treatment with carboxypeptidase released the first glycine. Partial hydrolysis gave, among others, Ala-Gly and Tyr-Leu. Give structures consistent with the data.

7.6 Fusaric acid is formed by mixed biogenesis from an amino acid and a polyketide. Suggest a biosynthesis route.

Fusaric acid

HOOC   N

7.7 Holomycin is an antibiotic isolated from *Streptomyces griseus* (Ettlinger, L., Gäumann, E., Hütter, R., Keller-Schierlein, W., Kradolfer, F., Neipp, L., Prelog, V. and Zähner, H. *Helv. Chim. Acta* **42** (1959) 563). Suggest precursors for this unusual structure. How would you design experiments to prove your working hypothesis?

NHCOCH$_3$

7.8 *Streptomyces refuineus* produces the antibiotic anthramycin (Hurley, L. H., Zmijewski, M., and Chang, C-J. *J. Am. Chem. Soc.* **97** (1975) 4372). ^{14}C-labelled methionine was incorporated at the starred positions and 1-^{14}C-labelled tyrosine at the dotted position. Dopa was metabolized with the same efficiency as tyrosine. It was shown that the aromatic ring of anthramycin originated from tryptophan. When 3,5-^{3}H$_2$ tyrosine was fed to the culture, one ^{3}H was retained in the side chain. Suggest a metabolic pathway.

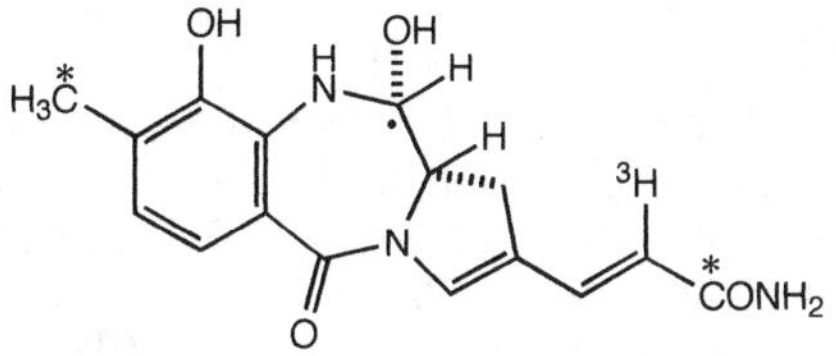

7.9 2-Methyl propionic acid is incorporated intact into asparagusic acid by *Asparagus officinalis* plants when administered by the cotton-wick method. 3,4-³H-Isobutyrate showed nearly complete retention of tritium whereas administration of 2-³H-isobutyrate resulted in nearly complete loss of label. Methacrylate was also effectively incorporated. Suggest a biosynthetic pathway for asparagusic acid and compare the result with the pathway established for lipoic acid. (Parry, R. J. *J. Am. Chem. Soc.* **99** (1977) 6464; Parry, J. R., Mitsuzawa, A. E. and Riccardione, M. *J. Am. Chem. Soc.* **104** (1982) 1442.)

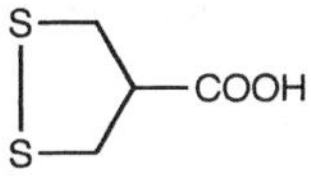

Asparagusic acid

7.10 Pyrroloquinoline quinone, PQQ, is a cofactor used by a copper containing amine oxidase. It derives from one aromatic and one aliphatic amino acid which are incorporated intact. Identify the building units and suggest a biosynthetic pathway. (van Kleef, M. A. G. and Duine, J. *FEBS Lett.* **237** (1988) 91.) Formulate a plausible mechanism for the oxidation of a primary amine to the aldehyde involving PQQ and with the Cu^I-H_2O_2 couple as oxidant.

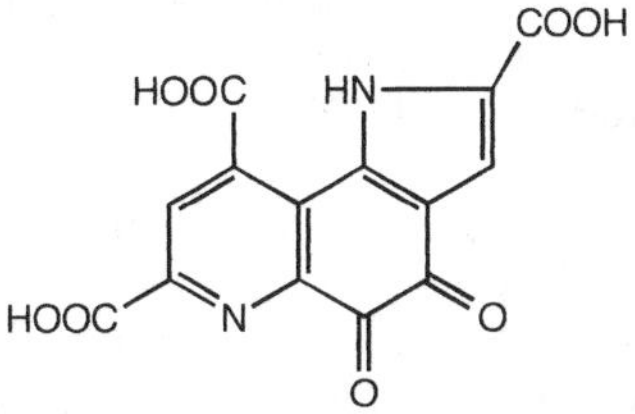

PQQ

7.11 Glycine and a suitably substituted naturally occurring phenylalanine derivate are precursors for the amino acid moiety of the antibiotic obafluorin. Give a rationale for the formation of the C_4 side chain of the amino acid. (2-²H)-Phenylalanine showed loss of label when it was incorporated into obafluorin. (Herbert, R. B. and Knaggs, A. R. *Tetrahedron Lett.* **1990** 7515.)

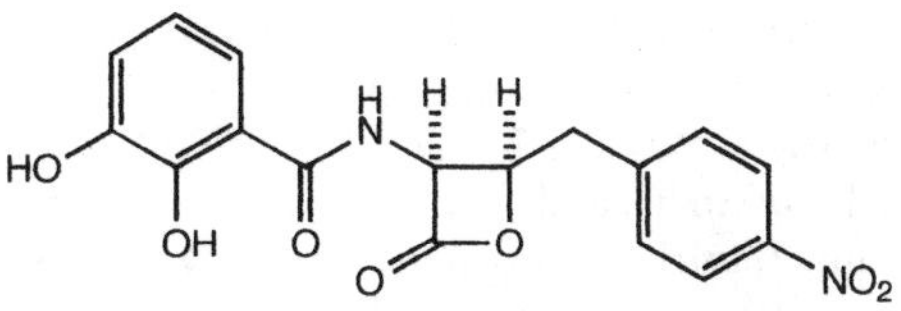

Obafluorin

Bibliography

1. Auclair, J. L. and Patton, R. L. *Rev. Can. Biol.* **9** (1950) 3.
2. Ogawa, T., Fukuda, M. and Sasaoka, K. *Biochim. Biophys. Acta* **297** (1973) 60.
3. Noma, M., Noguchi, M. and Tamaki, F. *Agric. Biol. Chem. Japan* **37** (1973) 2439.
4. Marshall, R. D. *Ann. Rev. Biochem.* **41** (1972) 673.
5. Fonnum, F. (Ed) *Amino acids as Chemical Transmitters*, Plenum, New York, 1978.
6. Walter, R. and Meienhofer, J. (Eds.) *Peptides. Chemistry, Structure and Biology*, Ann Arbor Science Publishers, Ann Arbor, 1975.
7. Ovchinnikov, Yu. A. and Ivanov, V. T. in *Heterodetic Peptides, Int. Review of Science, Organic Chemistry, Series Two* **6** (1976) 183.
8. Cramer, U. and Spener, F. *European J. Biochem.* **74** (1977) 495.
9. Gould, S. J. and Ju, S. *J. Am. Chem. Soc.* **114** (1992) 10166.
10. Wright, J. L. C., Falk, M., McInnes, A. G. and Walter, J. A. *Can. J. Chem.* **68** (1990) 22.
11. Lindberg, P., Bergman, R. and Wickberg, B. *J. Chem. Soc. Chem. Commun.* **1975,** 946.
12. Castric, P. A. and Conn, E. E. *J. Bacteriol.* **108** (1971) 132.
13. Castric, P. A., Farnden, K. J. F. and Conn, E. E. *Arch. Biochem. Biophys.* **152** (1972) 62.
13a. Edmonds, J. S., Francesconi, K. A. and Stick, R. V. *Nat. Prod. Rep.* **11** (1993), 311.
14. Bell, E. A. *Endeavour* **4** (1980) 102.
15. Bada, J. L. *Earth Planet Sci. Letters* **15** (1972) 273.
16. Helfman, P. M. and Bada, J. L. *Proc. Natl. Acad. Sci. USA* **72** (1975) 2891; Barret, G. C. in *Amino acids, Peptides and Proteins*, Specialist Periodical Reports, Chemical Soc., London **8** (1976) 21.
17. Fowden, L., Scopes, P. M. and Thomas, R. N. *J. Chem. Soc.* (C) **1971,** 833.
18. Jorgensen, E. G. *Tetrahedron Lett.* **1971,** 863.
18a. Joullié, M. M. and Thompson, T. R. *Tetrahedron Rep.* **1991,** 8791.
19. Spackman, D. H., Stein, W. H., and Moore, S. *Anal. Chem.* **30** (1958) 1190.
20. Miller, S. L., Urey, H. C. and Oro, J. *J. Mol. Evol.* **9** (1976) 59.
21. Shevlin, P. B., McPherson, D. W. and Melius, P. *J. Am. Chem. Soc.* **103** (1981) 7007.
22. Wolman, Y., Haverland, W. J. and Miller, S. L. *Proc. Natl. Acad. Sci. USA* **69** (1972) 809.
23. Walsh, C. *Enzymatic Reaction Mechanisms.* Freeman and Co., San Francisco, 1979, and refs. therein. Gives a thorough mechanistic account of PLP mediated reactions.
24. Belleau. B. and Burba, J. *J. Am. Chem. Soc.* **82** (1960) 5751, 5752; Stevenson, D. E., Akhtar, M. and Gani, D. *Tetrahedron Lett.* **1986,** 5661.
25. Jordan, P. and Akhtar, M. *Biochem. J.* **116** (1970) 277.
26. Tatum, C. M., Benkovic, P. A., Benkovic, S. J., Potts, R., Schleicher, E. and Floss, H. G. *Biochemistry* **16** (1977) 1093.
27. Schirch, L. and Gross, T. *J. Biol. Chem.* **243** (1968) 5651.
28. Taylor, W. and Weissbach, H. in *The Enzymes*, 3rd. edn. Boyer, P. (Ed.) Academic Press, New York **9** (1973) 121.
29. Tada, M., Miura, K., Okabe, M., Seki, S. and Mizukami, H. *Chem. Letters* **1981,** 33.
30. Krongelb, M., Smith T., and Abeles, R. *Biochem. Biophys. Acta* **167** (1968) 473.
31. Obata, Y. and Iimori, M. *J. Chem. Soc. Japan* **73** (1952) 832. *Chem. Abstr.* **47** (1953) 6093a.

32. Johnson, R., Guderian, R. H., Eden, F., Chilton, M. S., Gordon, M. P. and Nester, E. W. *Proc. Natl. Acad. Sci. USA* **71** (1974) 536.
33. Pinkus, L. and Meister, A., *J. Biol. Chem.* **247** (1972) 6119.
34. Powers, S. and Meister, A. *Proc. Natl. Acad. Sci. USA* **73** (1976) 3020.
35. Kjaer, A. and Larsen, P. O. in *Biosynthesis*, Specialist Periodical Reports, Chemical Society, London **3** (1973) 71.
36. Schütte, H. R. and Müller, P. *Biochem. Physiol. Pflanz.* **163** (1972) 528.
37. Maw, G. A. in *Sulfur in Organic and Inorganic Chemistry*, Senning, A. (Ed.) M. Dekker, New York **2** (1972) 113.
38. Okumura, S., Tsugawa, T., Tsunanda, T., and Motosaki, S. *J. Agric. Chem. Soc. Japan* **36** (1962) 599, 602.
39. Trainor, D. A., Parry, R. J. and Gitterman, A., *J. Am. Chem. Soc.* **102** (1980) 1467; Parry, R. J. and Naidu, M. V. *Tetrahedron Lett.* **1980**, 4783.
40. Crout, D. H. G. *J. Chem. Soc.* **1966**, 1968.
41. O'Donovan, D. G. and Long, D. J. *Proc. Roy. Irish Acad.* **75B** (1975) 465.
42. Cahill, R., Crout, D. G. H., Mitchell, M. B. and Müller, U.S. *J. Chem. Soc. Chem. Commun.* **1980**, 419.
43. Dewick, P. M. *Nat. Prod. Rep.* **1** (1984) 545; **11** (1994) 173.
44. Halkier, B. A. and Möller, B. L. *J. Biol. Chem.* **265** (1990) 21124; *Plant. Physiol.* **96** (1991)10.
45. Reay, P. F. and Conn, E. E. *J. Biol. Chem.* **249** (1974) 5826.
46. Duffey, S. S., Underhill, E. W. and Towers, G. H. N. *Comp. Biochem. Physiol.* **47B** (1974) 753.
47. Sammes, P. G. in *Prog. Chem. Org. Nat. Prod.* **32** (1975) 51.
48. Quilico, A. *Rec. Prog. Org. Med. Chem.* **1** (1964) 225.
49. Birch, A. J., Blance, G. E., David, S. and Smith, H. *J. Chem. Soc.* **1961**, 3128.
50. Casnati, G. and Pochini, A. *J. Chem. Soc. Chem. Commun.* **1970**, 1328.
51. Newbold, G. T., Sharp, W. and Spring, F. S. *J. Chem. Soc.* **1951**, 2679.
52. Baldwin, J. E. and Abraham, E. E. *Nat. Prod. Rep.* **5** (1988) 129.
53. Baldwin, J. E. J. *Heterocyclic Chem.* **27** (1990) 71.
54. Warren, S. C., Newton, G. G. F. and Abraham, E. P. *Biochem. J.* **103** (1967) 902.
55. Baldwin, J. E., Byford, M. F., Field, R. A., Shian, C.-Y., Sobey, W. J. and Schofield. C. J. *Tetrahedron* **49** (1993) 3221.
56. Bahadur, G., Baldwin, J. E., Wan, T., Jung, M., Abraham, E. P., Huddlestone, J. A. and White, R. L. *J. Chem. Soc. Chem. Commun.* **1981**, 1146.
57. Neuss, N., Nash, C. H., Baldwin, J. E., Lemke, P. A. and Grutzner, J. B. *J. Am. Chem. Soc.* **95** (1973) 3797.
58. Townsend, C. A. and Basak, R., *Tetrahedron* **47** (1991) 2591.
59. Baldwin, J. E., Adlington, R. M., Moroney, S. E., Field, L. D. and Ting, H.-H. *J. Chem. Soc. Chem. Comm.* **1984**, 984; Baldwin, J. E., Adlington, R. M., Aplin, R. T., Crouch, N. P. and Wilkinson, R. *Tetrahedron* **48** (1992) 6853.
60. Kita, Y., Shibata, N., Kawano, N., Tohjo T., Fujimori, C. and Ohishi, H. *J. Am. Chem. Soc.* **116** (1994) 5116.
61. Wieland, T. *Prog. Chem. Org. Nat. Prod.* **25** (1967) 214.
62. Nakai, K., Sasaki, T. and Hayashi, K. *Biochem. Biophys. Re. Commun.* **44** (1971) 893.

Chapter 8

The alkaloids

8.1 Introduction

The term alkaloids or alkali-like compounds was coined by Meissner, an apothecary in Halle, in 1819. Today it loosely defines naturally occurring basic compounds, apart from simple amines, derived from amino acids and the *N*-heterocycles of the pyrrole, pyrimidine, purine type, etc. There are exceptions, of course; alkaloids which are amides are not basic, e.g. colchicine. The classification is arbitrary and chemically artificial because there is no real need to separate the simple amines from alkaloids since both originate from amino acids as do the *N*-heteroaromatics. Nevertheless, the traditional classification is kept for some practical reasons. The function of the overwhelming majority of alkaloids is yet unknown, though we know that pyrimidine, purine, and pterin derivatives play an important role in the life processes. All alkaloids can be constructed of building blocks from the shikimic acid, polyketide, or mevalonic acid pathways in combination with an amino acid, a circumstance that automatically allows a consistent systemization of the numerous and structurally highly diverse compounds. In short, the amino acid component determines the character of the alkaloid and this classification harmonizes well with the classical system based on morphology. Alkaloids are therefore used as evolutionary or biogenetic markers.[1] Comparatively few amino acids are involved in the biosynthesis of alkaloids, e.g. glycine (in the *N*-heterocycles mentioned), glutamic acid, ornithine, lysine, phenylalanine, tyrosine, tryptophan, and anthranilic acid. The majority of alkaloids are found in the plant kingdom, from higher plants down to microorganisms. Few are found in the animal kingdom, and curiously enough, alkaloids are sparsely represented in the marine environment.

Since ancient times few compounds have been wrapped in as much mystery as the alkaloids. We find amongst them deadly poisons such as strychnine. It has been used as a rodenticide and vermin killer for centuries and as such it is responsible for the accidental death of many beloved pets. Since strychnine is very stable it has been detected in exhumed bodies several years after death; consequently, it is not the ideal homocidal agent. Coniine in *Conium maculatum* was used by the ancient Greeks for state executions, and here Socrates is certainly the most famous victim. Several hypnotics and hallucinogens are alkaloids including the opium group in *Papaver somniferum*, lysergic acid derivatives in *Claviceps purpurea*, the parasitic fungus on grain, causing convulsive "St. Anthony's fire" in medieval times, mescaline in the Indian peyote cactus *Lophophora williamsii*, and psilocybin from the Mexican mushroom, *Psilocybe mexicana*, used by the upper priesthood of the Mayas to gain transcendental spiritual contact with their ancient gods. Several alkaloids are used as valuable drugs in medicine, e.g. morphine as a pain reliever, reserpine in psychiatry as a tranquillizer, curare alkaloids in general anaesthesia—also utilized by South American indians as an arrow poison—atropine in eye surgery, ergonovine to induce or make childbirth easier, the indole alkaloids vincristine and vinblastine as antitumour agents, and quinine as an antimalarial.

8.2 Alkaloids derived from ornithine and lysine. The pyrrolidine and piperidine alkaloids

Ornithine is the immediate precursor of the pyrrolidine alkaloids. The biosynthesis of hyoscyamine, has been studied extensively. When *Datura stramonium* was fed with 2-^{14}C-labelled ornithine, these alkaloids were labelled at one bridgehead, which demonstrates asymmetric incorporation[2] of ornithine (Fig. 1). Therefore, the symmetric putrescine does not appear free on the pathway, even though feeding experiments show that it can serve as precursor. *N*-Methylated putrescine and δ-*N*-methylated, but not α-*N*-methylated ornithine are incorporated, and ^{15}N-labelling experiments demonstrate further that only the δ-N of ornithine is retained in the pyrrolidine ring.[3] We know that decarboxylation requires PLP and participation of this coenzyme in the biosynthesis of the pyrrolidine nucleus gives a rationale of all the data available. The asymmetric incorporation of ornithine is explained by the assumption that the hydrolysis of the Schiff's base to putrescine is slower than the cyclization and methylation. At the same time putrescine becomes acceptable as a potential alternative precursor. The Schiff's base blocks the α-N and this is specifically eliminated by cyclization.

The three-carbon bridge derives from two molecules of acetyl CoA.[4] Cyclization in combination with oxidation gives the cocaine skeleton. Reduction, methylation and benzoylation completes the biosynthesis of cocaine, which is the chief alkaloid of the South American coca bush, *Erythroxylum coca*, Fig. 1. It has anaesthetic properties but is nowadays replaced in medicine by other drugs since it causes addition.

Decarboxylation of the 2-pyrrolidine-3-oxobutanoate and subsequent cyclization give tropinone via hygrine. Reduction of the carbonyl group and esterfication with tropic acid produce (-)-hyoscyamine which by 6-hydroxylation and introduction of the epoxide function gives scopolamine.[5] It was quite a surprise to find that the oxygen atom at C-6 was retained which im-

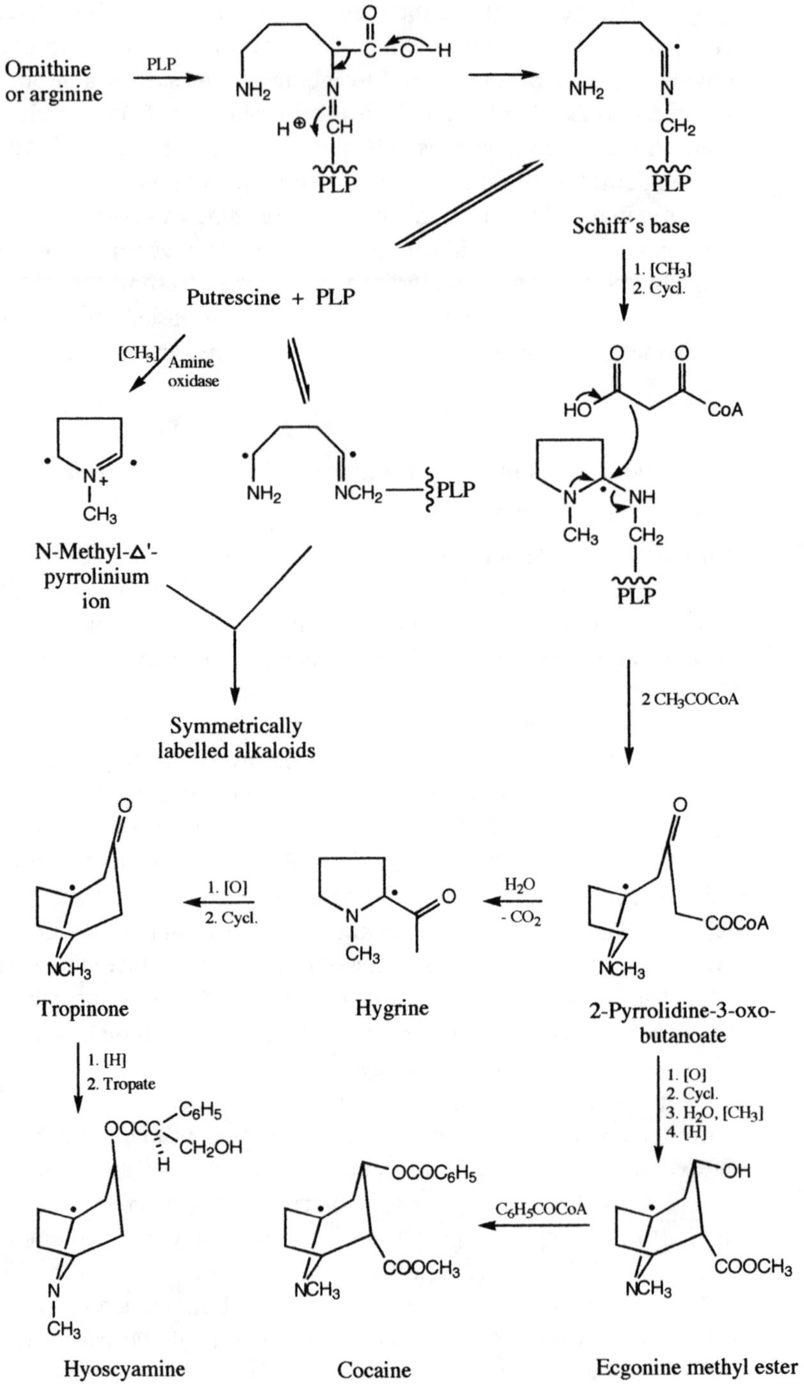

Fig. 1   Biosynthesis of tropane alkaloids

plies that the corresponding C-6,7 alkene is not an intermediate. The obvious is not always the trustworthy. The mechanism formulated in (1) involving molecular oxygen and a non-haem iron catalyst explains the direct introduction of the epoxide function into saturated hydrocarbons without formation of intermediary alkenes, cf. section 4.3.

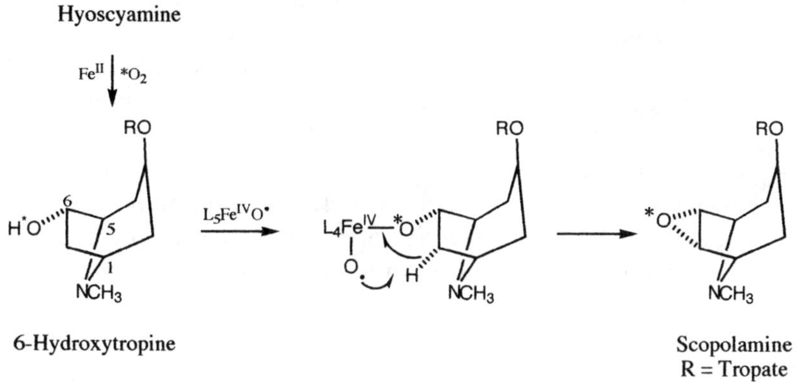

Hyoscyamine

(1)

6-Hydroxytropine

Scopolamine
R = Tropate

In the case where asymmetric incorporation does not occur the Schiff's base is rapidly hydrolysed to the free diamine. A considerable body of evidence proves that the pyrrolidine ring of nicotine, hyoscyamine and cocaine is formed via a symmetrical intermediate which *inter alia* is dependent on the organism.[6,7] The symmetrical incorporation of ornithine into hyoscyamine in *Hyoscyamus albus* has another explanation in that the additon of acetyl CoA to the pyrrolidinium ion gives an enantiomeric mixture of 2-pyrrolidine-3-oxobutanoate that likewise leads to equal distribution of label at the bridge heads of tropinone.[8]

Tropinone (Fig. 1) was elegantly synthesized by Robinson under virtually physiological conditions from succinaldehyde, methylamine and acetonedicarboxylic acid (2) and it was once thought—incorrectly—that this mild reaction mimicked the cellular reaction.

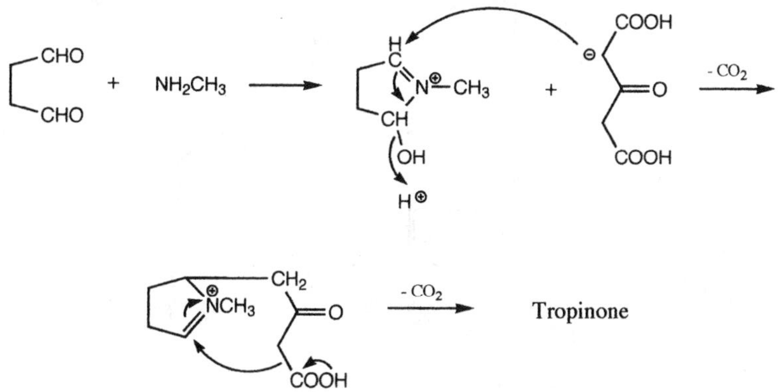

(2)

Symmetrical incorporation occurs in the pyrrolizidine bases which are formed from two molecules of ornithine.[9] In nature these bases occur as esters of necic acids, see section 7.6. By using isotopic labelling a detailed picture of the biosynthesis has emerged.[10] It has been shown that oxidative (diamine oxidases) condensation of two molecules of putrescine first gives the symmetrical triamine homospermidine, which via the pyrrolinium ion forms the pyrrolizidine skeleton, **1,2,** Fig. 2. As expected the label at C^1 of putrescine appears at $C^{3,5,8,9}$ in retronecine and by stereospecific hydrogen labelling at $C^{1,2}$ of putrescine it was established which one of the prochiral hydrogen atoms was involved in the redox and cyclization steps. Trachelanthamidine **1** is the preferred precursor for retronecine which forms dicrotaline with 3-hydroxy-3-methylglutaric acid. Isoretronecanol **2** gives rosmarinecine by hydroxylation at $C^{2,7}$ with normal retention of configuration. The hydrogens at $C^1\alpha$ and $C^2\beta$ of rosmarinecine originate from the 2-*pro-S* and 2-*pro-R* hydrogens, re-

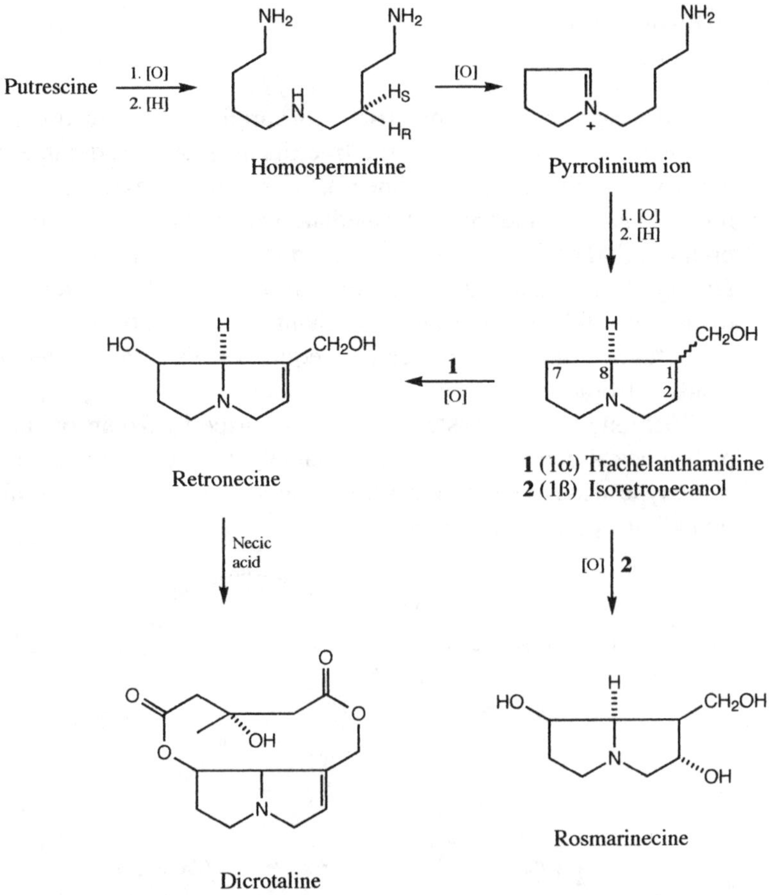

Fig. 2   Biosynthesis of pyrrolizidine alkaloids

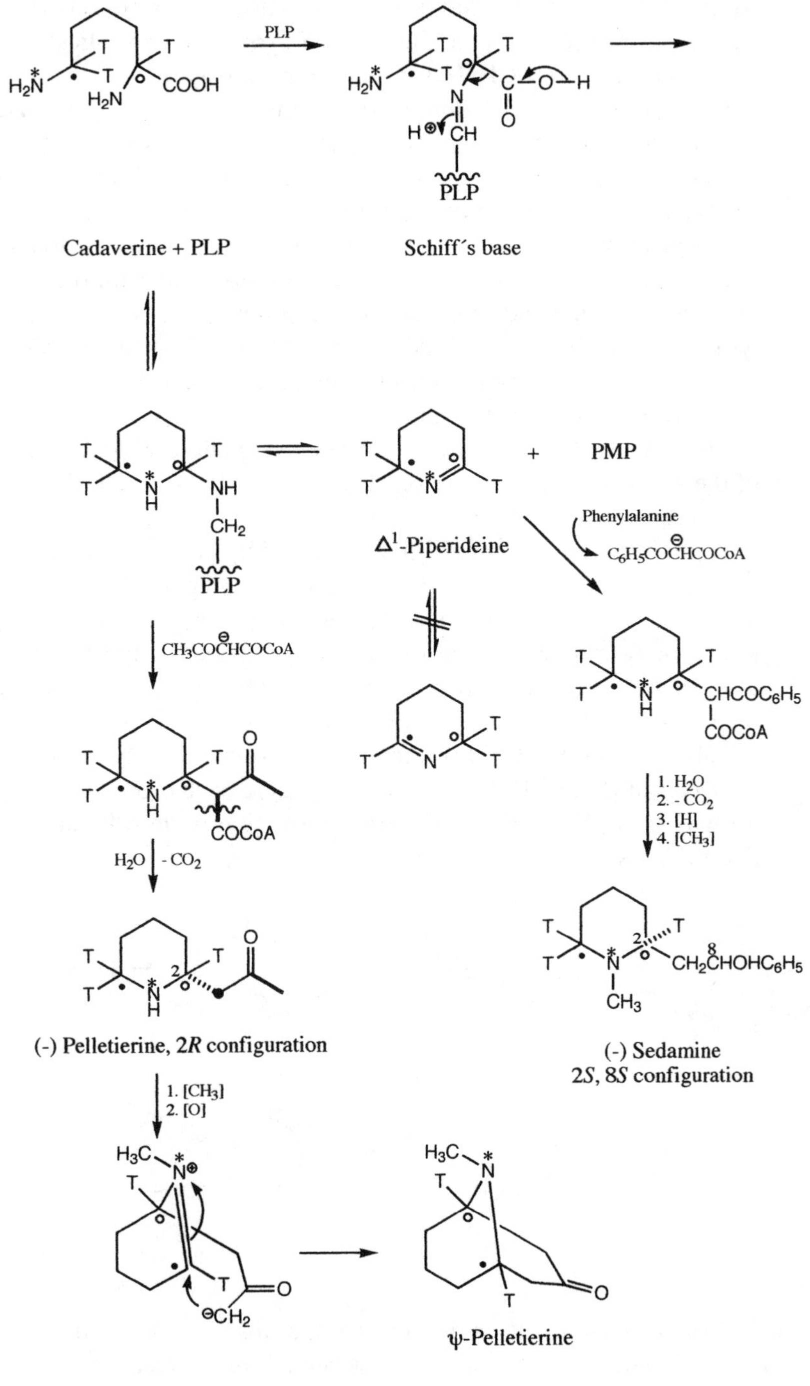

Fig. 3   Biosynthesis of sedamine, pelletierine, and ψ-pelletierine

spectively, in putrescine and the olefinic hydrogen at C^2 of retronecine comes from the 2-*pro-R* hydrogen. The 1-*pro-S* hydrogen is specifically removed in the oxidation of the –CH$_2$NH$_2$ group to the aldehyde.

Several genera of the *Boraginaceae* family are rich in pyrrolizidine alkaloids possessing hepatic toxicity which occasional'y have caused losses in pasturing livestock. Certain species of the plants have been used in traditional medicine and should be avoided for the same reason.

Lysine undergoes a series of similar reactions and gives rise to the six-membered piperidine ring alkaloids. Isotopic experiments with ψ-pelletierine, *N*-methylpelletierine and sedamine (Fig. 3) show that ε-N, C^6-H$_2$, and C^2–H of lysine are retained, and that lysine is incorporated unsymmetrically.[11–13] These results exclude α-amino-δ-formylpentanoic acid, α-keto-ε-aminohexanoic acid and free cadaverine as intermediates.

The *N*-methylation seems to be a late event and is therefore not the cause of the asymmetric incorporation. The C_3-side-chain in pelletierine originates from acetylacetic acid which is different to the previously discussed pathway to cocaine, cf. Fig. 1, as has been established by the mode of incorporation of $^{13}C_2$ acetate. Sedamine derives analogously from phenylalanine via benzoylacetic acid.[14] The asymmetry is thus maintained by the rather stable Schiff's base. Cadaverine, which acts as a precursor, can enter the system at the step of the Schiff's base formation either by oxidation with PLP or with a diamine oxidase to Δ^1–piperideine. Another mechanistic detail becomes evident from the observation that C^2-H is retained. The α-imino group must be generated by decarboxylation. But there are cases where exchange of C^2-H occurs. The formation of pipecolic acid demands formation of an α-ketocarboxylic acid equivalent, since the C^2-H* is lost (3).

(3)

L-(-)-Pipecolic acid

A final point worth mentioning is the absolute stereostructure at C^2 in (—)pelletierine and (—)sedamine which are of opposite configuration. That means that different faces of the piperideine ring are oriented towards the polyketide on the enzyme surface in the alkylation step in the two cases.

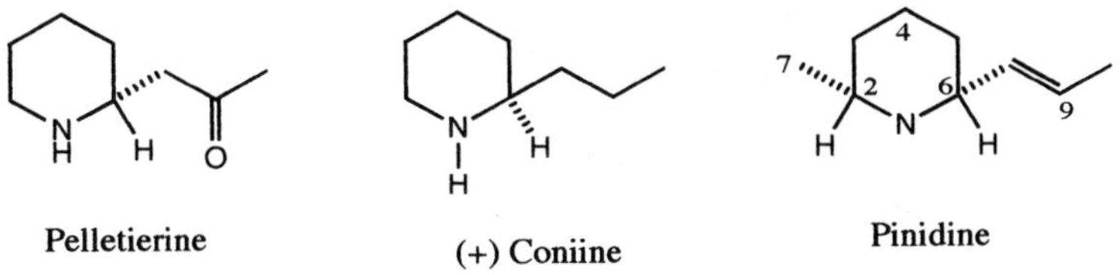

Pelletierine   (+) Coniine   Pinidine

Fig. 4   Structure of piperidine alkaloids of different biogenesis. Pelletierine comes from lysine, whereas coniine and pinidine come from acetate

The piperidine alkaloids give us another instructive example showing that the biosynthetic origin cannot be deduced simply by analogy and structural considerations. Pinidine in *Pinus* spp. and coniine in the *Conium maculatum* strongly resemble pelletierine in *Punica granatum* (Fig. 4), but surprisingly, tracer studies demonstrated their acetate origin. As to coniine octanoic acid, 5-keto-octanoic acid, the corresponding aldehyde, and coniceine act as specific precursors.[15, 16] These results suggest the sequence presented in Fig. 5. In analogy with coniine, pinidine is labelled by 1-^{14}C-acetate at alternate carbon atoms, $C^{2,4,6,9}$, suggesting the dioxodecanoic acids in Fig. 6 or derivatives as intermediates. The starter end was determined by feeding 1-^{14}C-malonic acid together with inactive acetic acid to *Pinus jeffreyi*. The activity of the derived pinidine was three times higher at C^9, than at C^2, from which it follows that $C^{2,7}$ is the starter.[17] However, feeding experiments with decanoic acid and the diketo derivative gave no incorporation, possibly because of poor exchange between the medium and the multifunctional enzyme system, section 5.2. Anatabine has a different origin in that it comes from two molecules of nicotinic acid, Fig. 8.

The biosynthetic path followed by the *Nicotiana* alkaloids turned out to be full of surprises. The aromatic pyridine ring is not derived from lysine by cyclization and dehydrogenation as one intuitively may think, nor is it derived from acetate. Its immediate precursor is nicotinic acid, the important pro-

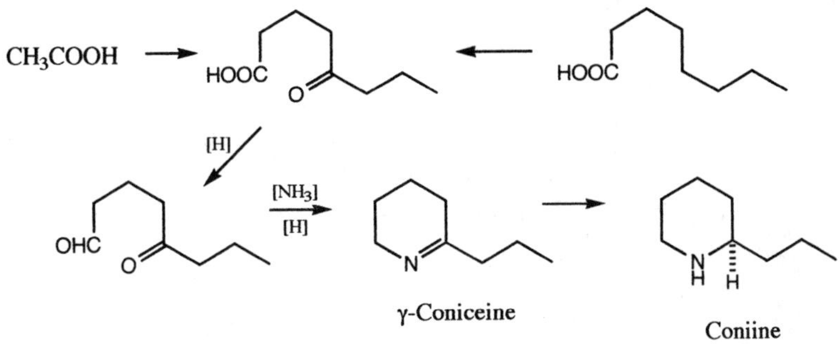

γ-Coniceine   Coniine

Fig. 5   Biosynthesis of coniine

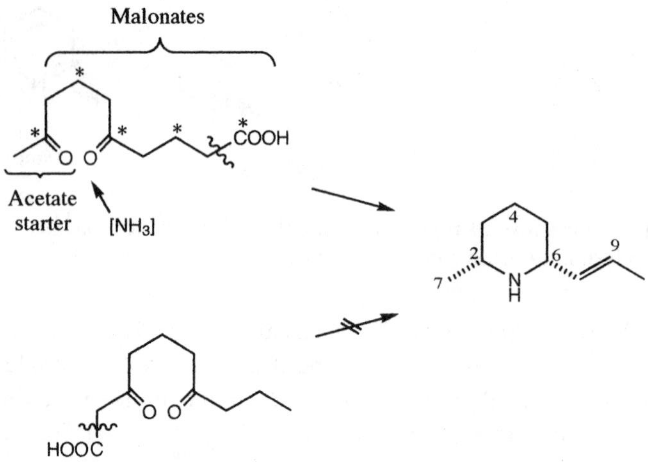

Fig. 6   Biosynthesis of pinidine

vitamin, trivially called niacin, the amide of which is a constituent of NADH. Deficiency of nicotinic acid leads to pellagra in humans and black tongue in dogs. It was first prepared by oxidation of nicotine with nitric acid, actually one of the degradations that once gave the structure of half of the nicotine molecule (4), and from which its name derives.

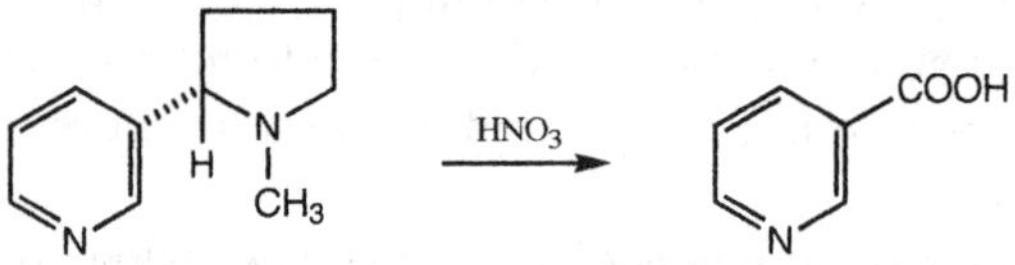

(4)

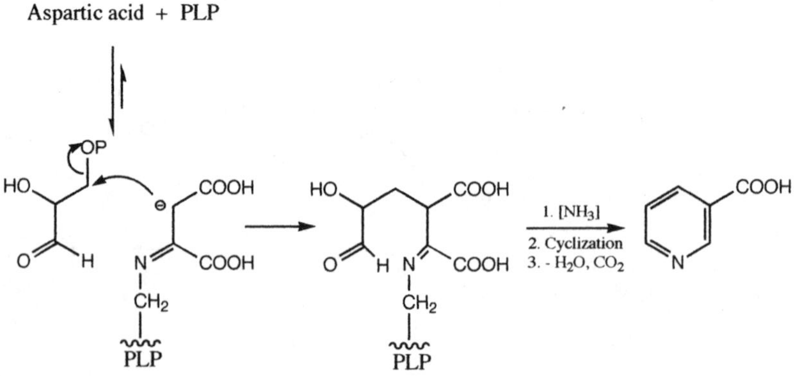

Nicotinic acid

Fig. 7   Suggested biosynthesis of nicotinic acid

Most animals can form nicotinic acid catabolically from tryptophan via kynurenine and 3-hydroxyanthranilic acid, a route that is not followed by *Nicotiana* spp. However, in a still not well understood pathway, plants are able to elaborate nicotinic acid from glyceraldehyde and aspartic acid or congeners, e.g. according to Fig. 7 as a number of tracer experiments have suggested. Incorporation studies with ^{15}N-labelled aspartic acid show that there is a considerable loss of ^{15}N in nicotinic acid.

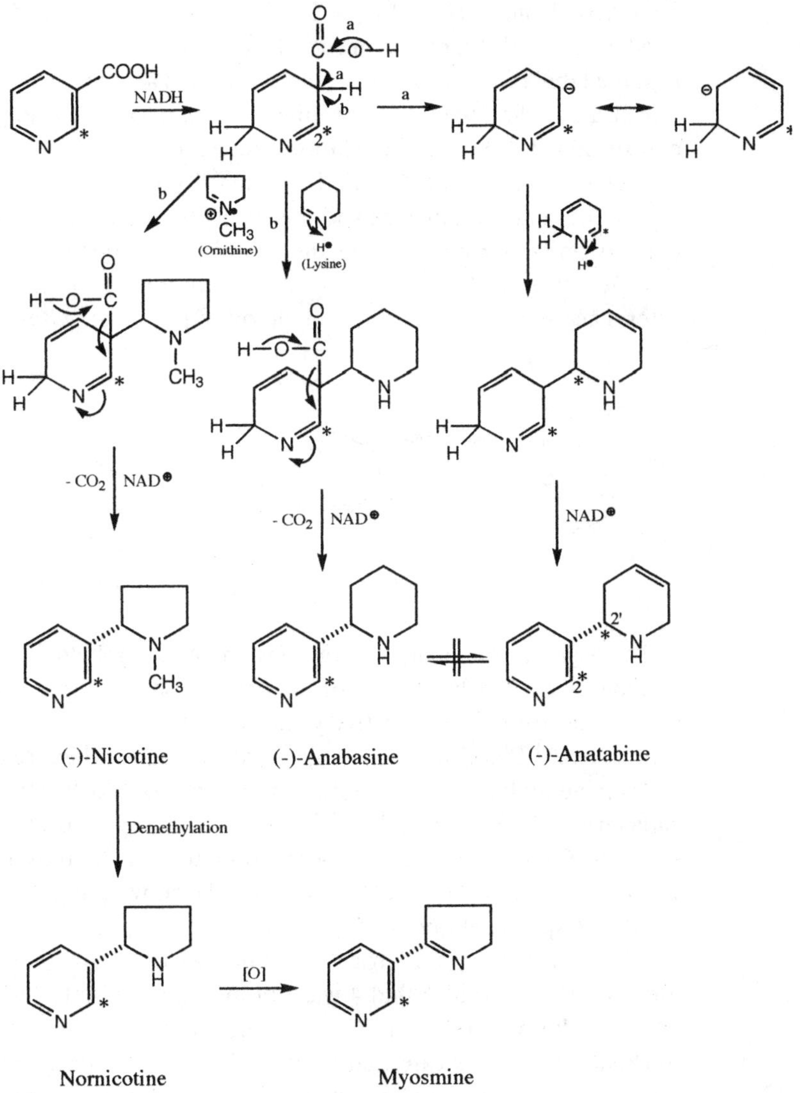

Fig. 8 Biosynthesis of *Nicotiana* alkaloids

The pyridine ring of nicotine,[18] anabasine[19] and quite unexpectedly both rings of anatabine[20] are derived from nicotinic acid by decarboxylation and specific coupling at C^3 (Fig. 8). 2-^{14}C-Labelled nicotinic acid gives rise to 2-[14]-C-labelled nicotine and anabasine and 2,2'-^{14}C-labelled anatabine. Decarboxylation is facilitated by reduction of nicotinic acid to 3,6-dihydronicotinic acid which contains the β-imino structure necessary for stabilizing an anion. In support of the 3,6-dihydronicotinic acid is an observation of a considerable loss of ^{3}H label in nicotine from 6-^{3}H-nicotinic acid as precursor compared with those from 2-[3]H-, 4-[3]H-, or 5-^{3}H-nicotinic acid. It is not definitely settled at which stage the carboxyl group of 3,6-dihydronicotinic acid is lost. The fact that the label of 2-^{14}C-nicotinic acid is retained at C^2 of nicotine, speaks in favour of a coupling with Δ^1-pyrroline or Δ^1-piperideine preceding or possibly concerted with the decarboxylation (Fig. 8, route b). A delocalized carbanion may give rise to 2,6-^{14}C-labelled nicotine. Lysine serves as precursor for the piperidine ring in anabasine and it is incorporated unsymmetrically. No interconversion between anabasine and anatabine has been observed.

1-Deoxy-D-xylulose and 4-hydroxy-L-threonine (possibly as phosphate esters) account for the pyridine skeleton of pyridoxol, vitamin B_6,(5).[21]

$$(5)$$

Vitamin B_6

The asymmetric incorporation of lysine is not upheld for the quinolizidine alkaloides, i.e. the hydrolysis of the Schiff's base is rapid and the symmetric cadaverine formed is effectively incorporated into lupinine and tetracyclic sparteine. Δ^1-Piperideine is a logical intermediate which reacts with another cadaverine molecule under redox conditions. Metabolic studies of 1-^{13}C, ^{15}N-cadaverine, **3**, (S)- and (R)-1-[2]H-cadaverines, (S)- and (R)-2-[2]H-cadaverine and 1,2-^{13}C-cadaverine gave a detailed picture of the biosynthesis.[22] When **3** was fed to a *Lupinus sp.* $C^{4,6,10,11}$ of lupinine showed equal enrichment in the ^{13}C NMR spectrum and a [13]C-^{15}N coupling was visible for C^6 but not for C^4 which excludes the symmetrical triamine **4** as an intermediate. Feeding experiments established also that 4 was not incorporated into lupinine which contrasts to the previously discussed route to the pyrrolizidine alkaloids. The thickened bonds indicate that intact 1-[13]C-^{15}N moieties in **3** have been incorporated. The 2-^{2}H label in cadaverine was lost at the bridgehead carbon atoms $C^{7,9}$ in sparteine because of keto-enol and imine-enamine equilibria at the final cyclization step, Fig. 9.

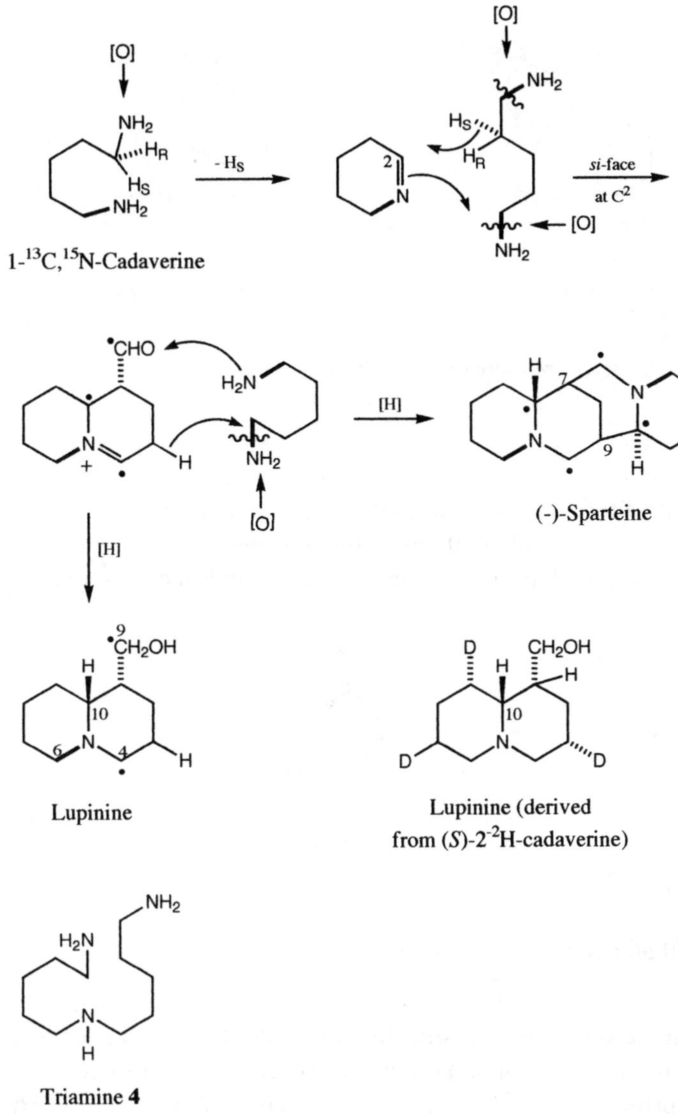

1-^{13}C,^{15}N-Cadaverine

(-)-Sparteine

Lupinine

Lupinine (derived from (S)-2-^{2}H-cadaverine)

Triamine **4**

Fig. 9   Biosynthesis of quinolizidine alkaloids

Plants of the *Euphorbiaceae* family synthesize alkaloids of quite diverse structures, such as the diterpenoid alkaloids of *Daphniohyllum* spp.,[23a] benzylisoquinolines of *Croton* spp. or the *Securinega*[23b] alkaloids derived from lysine and tyrosine in an unusual coupling reaction (Fig. 10), indicating the polyphyletic character of the family. *Ricinus communis* produces the toxic

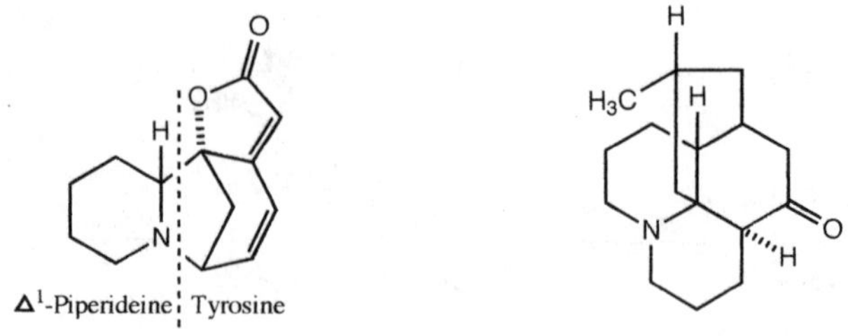

Δ¹-Piperideine ⋮ Tyrosine

Securinine
Securinega suffruticosa

Lycopodine
Lycopodium selago

Fig. 10   Structure of more complex, lysine derived alkaloids

alkaloid ricinine, derived from nicotinic acid. The carboxylic group is retained as nitrile, formed by dehydration of nicotinamide (6).

Lycopodium[23c] alkaloids from the *Lycopodaceae* family derive from two molecules of pelletierine but only one of them is incorporated intact.

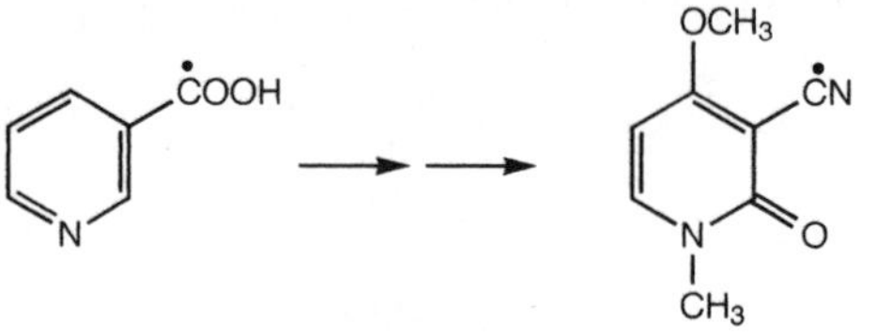

(6)

Ricinine

8.3 Alkaloids derived from tyrosine

The success of the biosynthetic principles can hardly be seen more clearly than in the systematic classification of alkaloids derived from aromatic amino acids, the largest group. Yet, in retrospect, with all respect to their success, these principles could not have been discovered without our experience derived from classical aromatic electrophilic substitution and from radical reactions. The theories of Robinson and Barton *et al.* and the extensive tracer work by Battersby and others gradually gave us a coherent biosynthetic picture. The various skeletons of, for example, simple isoquinolines, opium alkaloids, *Amaryllidaceae* alkaloids, *Ipecacuanha* alkaloids, *Erythrina* alkaloids, benzophenanthridine alkaloids, phenethylisoquinoline alkaloids and of the related indole alkaloids can be constructed by application of a few fundamental principles:

1. aromatic hydroxylation, *O*-methylation, and decarboxylation of the amino acid to form hydroxylated β-arylethylamines;

2. Pictet–Spengler condensation of β-arylethylamine with the appropriate carbonyl compound;
3. a phenol coupling.

Occasionally ring fission and recyclization occur, as for the benzophenanthridine and indole alkaloids.

(7)

Most tyrosine-based alkaloids possess additional hydroxyl groups in the aromatic ring. Some of them are methylated or have become part of a methylenedioxy ring. The latter is formed by further oxidation of a methoxy group located *ortho* to a hydroxyl followed by ring closure (7) rather than by direct condensation of *ortho*-substituted hydroxyls with a formaldehyde equivalent. It appears from feeding experiments in various plants that hydroxylation of tyrosine precedes decarboxylations, whereas methylation is often a later event. This is understandable in light of the electronic requirements of the electrophilic Pictet–Spengler condensation and the phenolic coupling which may follow. Free hydroxyls promote a much higher electron density in *ortho* and *para* positions than methoxyls as a result of dissociation, and the phenoxy radicals are formed by oxidation of the phenolate ion. Thus, at least one free hydroxyl is required *ortho* or *para* to the position at which ring closure occurs. In salsoline, 1-methyl-6-hydroxy-7-methoxytetrahydroisoquinoline, *O*-methylation could very well occur at any stage in the biosynthesis, but in lophocerine, the 6-methoxy isomer (Fig. 11) methylation presumably takes place at a late stage.

The isopropyl group of lophocerine arises independently, either from mevalonic acid or leucine[24,25] (Fig. 11). Two points in the scheme merit discussion. β-Phenylethylamine could conceivably first be alkylated by isopentenyl phosphate. Oxidation of the secondary amine to an imine followed by cyclization could then give the alkaloid. However, it is known from several other similar cases that the secondary amines act as poor precursors which makes this route unlikely. The carbonyl function is thus a prerequisite in the biosynthesis. The second point concerns the timing of the leucine decarboxylation. Does leucine or the derived α-keto acid form isopentanal as an intermediate or, alternatively, does the α-keto acid condense directly with the amine to a new amino acid which then decarboxylates? This possibility has not been examined in this particular case but is has to be considered because this pathway is followed in other cases, e.g. in the biosynthesis of anhalonidine[26] (Fig. 11)

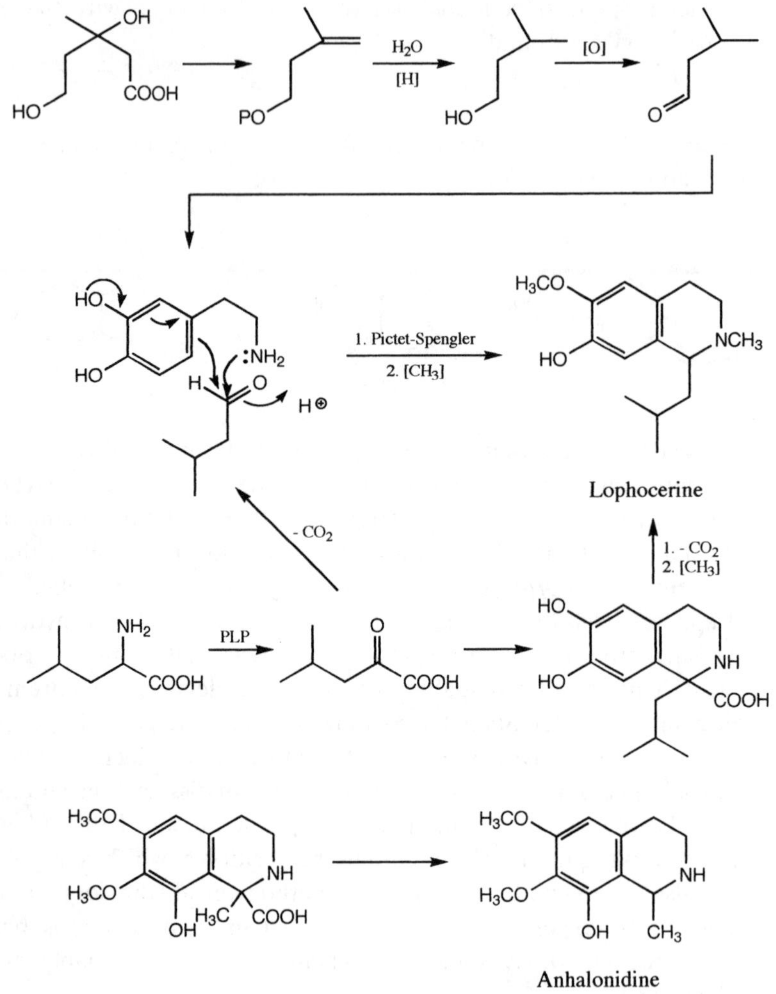

Fig. 11 Possible pathways to lophocerine and anhalonidine

from pyruvic acid, and it has been suggested that phenylpyruvic acid is directly involved in the cyclization step of the pivotal benzylisoquinoline reticuline.[27,28]

The 1-benzylisoquinoline skeleton arises from two molecules of tyrosine, one of which is oxidized and decarboxylated to give 3,4-dihydroxyphenylethylamine, dopamine. For a long time it was taken more or less for granted that the other tyrosine molecule gave 3,4-dihydroxyphenylacetaldehyde via regular PLP and thiamine mediated oxidation and decarboxylation, respectively, and that this aldehyde then condensed with the amine to form nor-

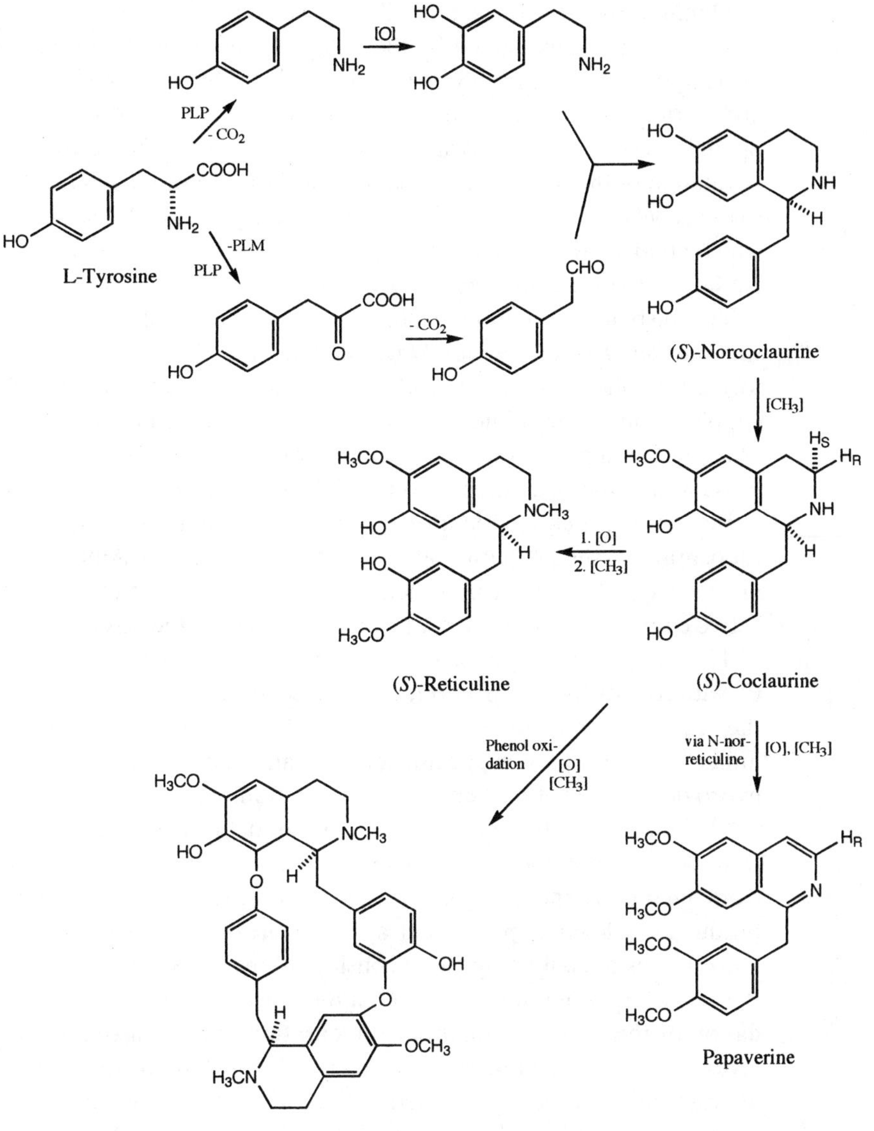

Fig. 12   Biosynthesis of (*S*)-reticuline, papaverine and (*S*,*S*)-bebeerine

laudanosoline. The latter reaction can in principle be carried out in the cell by
(*S*)-norlaudanosoline synthetase, but extensive tracer work has led to a re-
evaluation of the early steps of the biosynthesis of the key metabolite (*S*)-re-
ticuline, demonstrating that norlaudanosoline is not on the pathway after all.
It was found that dopa and dopamine, curiously enough, were incorporated

predominantly into the isoquinoline nucleus, the "upper part" of the molecule. This can be explained by their sluggishness to undergo transamination
into 3,4-dihydroxyphenylpyruvic acid and 3,4-dihydroxyphenylacetaldehyde,
and consequently they are unable to partake as the "lower part" in the Pictet–
Spengler cyclization. 4-Hydroxyphenylpyruvic acid, like tyrosine, is incorporated in both parts of the molecule and furthermore it was established that
(S)-norcoclaurine was the first formed benzyltetrahydroisoquinoline from
which an array of benzylisoquinolines originated. The sequence of reactions
leading to (S)-reticuline, bebeerine and papaverine is shown in Fig. 12.[29]

The bisbenzylisoquinoline alkaloids, distributed primarily in the *Meni-
spermaceae*, *Berberidaceae*, *Magnoliaceae* and *Monimiaceae* families, are
formed by phenol coupling. (S,S)-Bebeerine derives from (S)-coclaurine by
coupling the isoquinoline nucleus to the lower part of the other molecule.
Phenol couplings between the two upper or the two lower parts are also represented in the structures of some of the bisbenzylisoquinoline alkaloids.

Bebeerine (or chondodendrine, *SS* and *RR*-structures) occurs together with
tubocurarine, the N-methyl ammonium salt, in South American *Chondro-
dendron* spp. Tubocurarine which has strong curare-like effects, is the highly
active component of the arrow poison prepared by Indians of the Amazon.

The dehydrogenation of norreticuline to papaverine is stereospecific at
C^3 with loss of the *pro-S* H but non-stereospecific at C^4.[30] It is assumed that
the reaction starts with an enzyme controlled oxidation at C^3 to give the 2,3-
imine followed by a non-specific imine-enamine rearrangement to the 3,4-de-
hydro derivative without enzyme participation.

The important but rather toxic anti-amoebic drug, emetine, an *Ipeca-
chuanha* alkaloid contains a more complicated aliphatic part which derives
from the monoterpene, loganin (section 6.3). Loganin and its oxidative fission
product, secologanin, play a central role in the biosynthesis of the indole alkaloids. It was actually biogenetic considerations based on results gained from
the monoterpenoid field that guided the work on emetine towards the elucidation of the correct structure. Condensation of secologanin with dopamine
gives (—)-N-deacetylipecoside (Fig. 13). Hydrolysis of the glycosidic bond
liberates another aldehyde function that cyclizes with the amine to an immonium ion. Reduction, decarboxylation and a new Pictet–Spengler condensation completes the biosynthesis. Rather surprisingly, it turned out that the
precursor N-deacetylipecoside had the opposite of emetine's configuration at
C^5. Tracer work showed that deacetylipecoside incorporated well in emetine
but its C^5 epimer, N-deacetylisoipecoside, with the same configuration as
emetine not at all.[31] This implies that only ipecoside fits the enzyme surface
adequately for the glycolysis and the cyclization and at some stage an inversion must occur. It was demonstrated by ^{3}H-labelling of the aldehyde proton
in secologanin that it was retained at C^{11b} in emetine. This finding excludes an
equilibration of the immonium ion into conjugation with the aromatic ring.

But the immonium structure is well suited for another rearrangement which could lead to the wanted inversion. We have a "push-pull" situation with an electron donating hydroxyl *para* to the carbon carrying the immonium function, which can bring about a fission of the C–N bond and rearrangement to the epimer (Fig. 13).

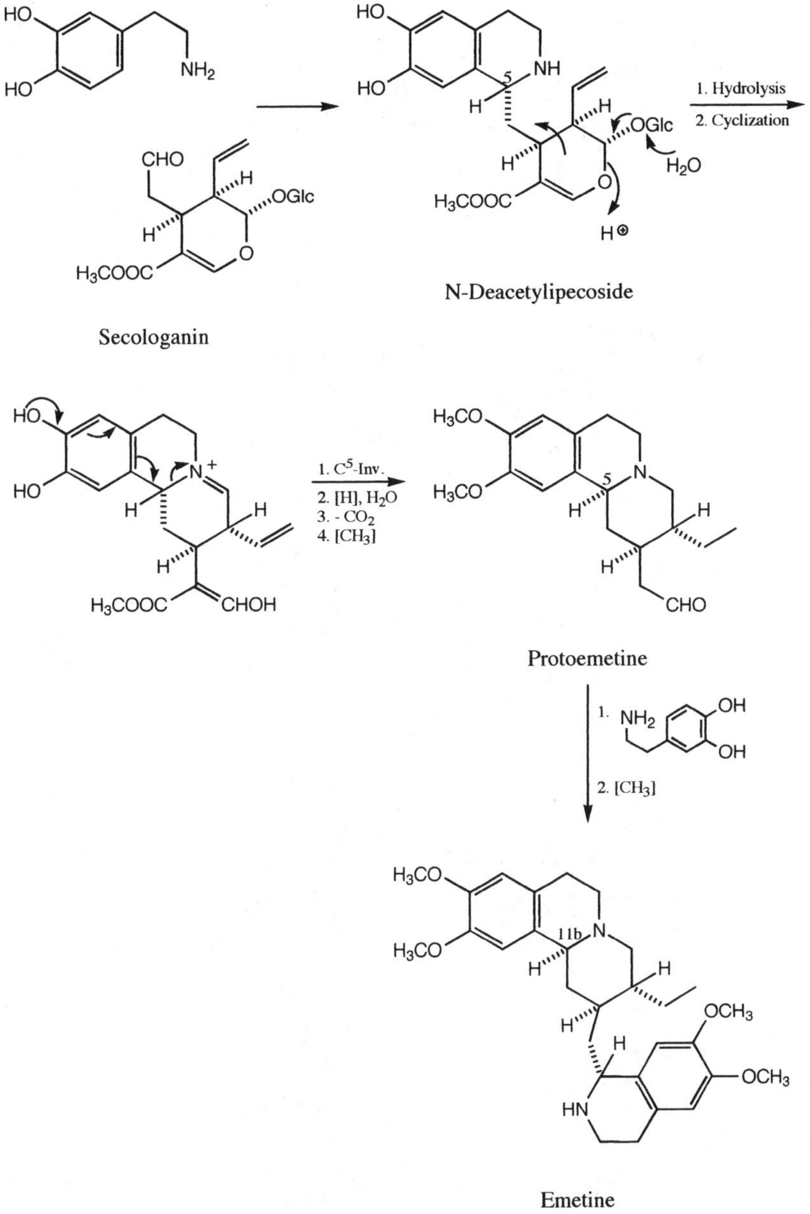

Fig. 13   Biosynthesis of emetine

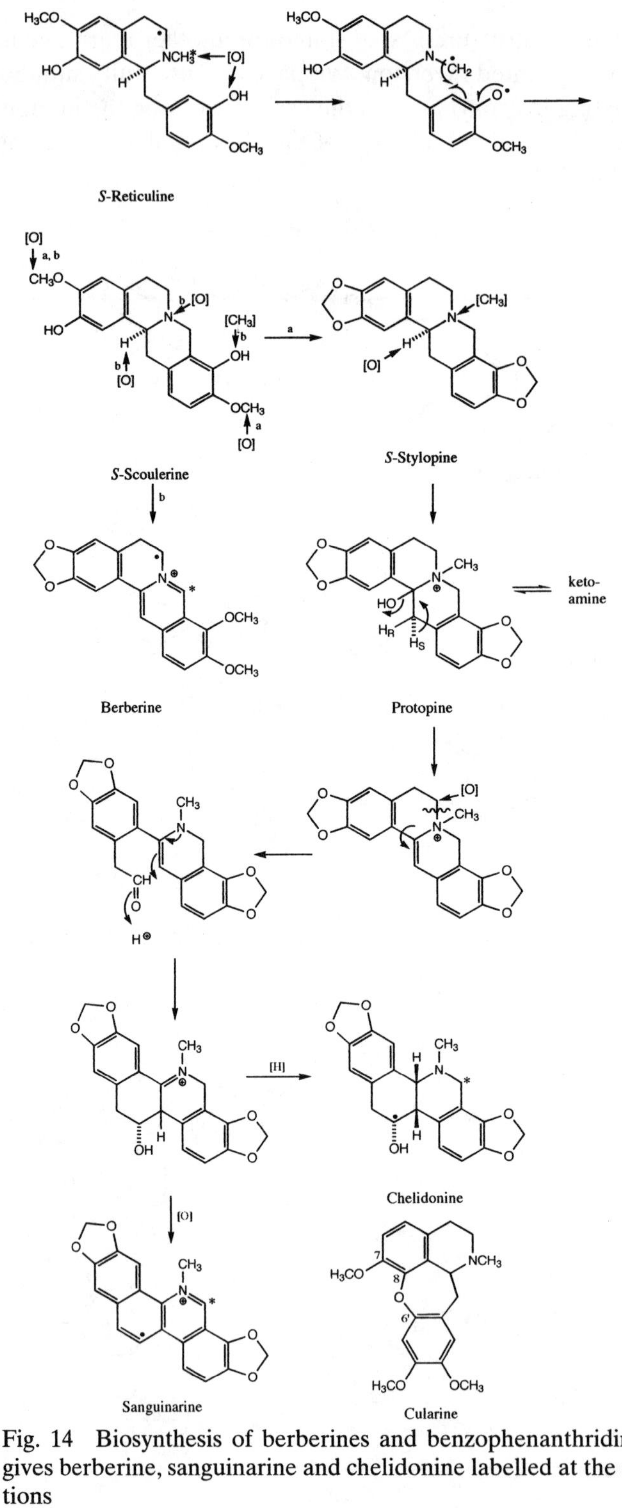

Fig. 14 Biosynthesis of berberines and benzophenanthridines. Labelled reticuline gives berberine, sanguinarine and chelidonine labelled at the starred and dotted positions

The berberine skeleton, Fig. 14, is formed by P450/NADPH/O_2-catalysed hydrogen abstraction[32] from the *N*-methyl group of reticuline and subsequent phenolic coupling to the aromatic ring—surprisingly in the crowded *ortho* position of the hydroxy group. Very few berberine alkaloids are known where the cyclization occurs in the sterically less hindered *para* position. Reactions carried out *in vitro* give both isomers. Feeding experiments support the pathways to the berberine and benzophenanthridine alkaloids shown in Fig. 14. A mechanistic interpretation is given in Fig. 15 applying the ferroxene formalism for hydroxylation with the 33d-form as oxidant, cf. section 4.3. This mechanism accounts for the attack at the *ortho* position and the inversion of configuration at the *N*-methyl group. The formation of the methylenedioxy ring proceeds analogously by radical abstraction of a hydrogen atom from the methoxy group and a subsequent inner sphere oxygen rebound step with retention of configuration.

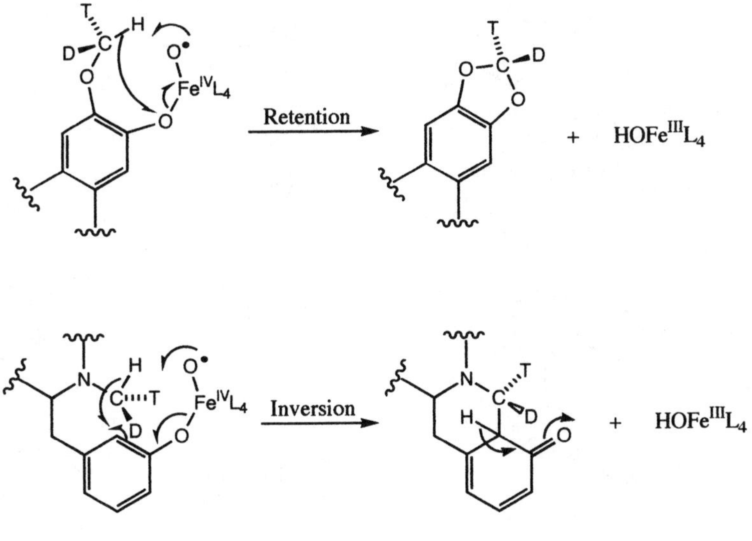

Fig. 15   Mechanisms proposed for the formation of the methylenedioxy ring and the berberine bridge

The cularine alkaloids are characterized typically by the 7,8-hydroxylation pattern in the quinoline part and the 6′,8-oxygen bridge formed by phenol oxidation involving the 8-oxygen atom. The structure of the parent compound cularine is given in Fig. 14. It is biosynthesized by condensation of dopamine (upper part) with 4-hydroxyphenylacetaldehyde and subsequent nuclear hydroxylation and *O*-methylation.[32a]

Appropriate twisting of the reticuline biradical gives aporphines and the

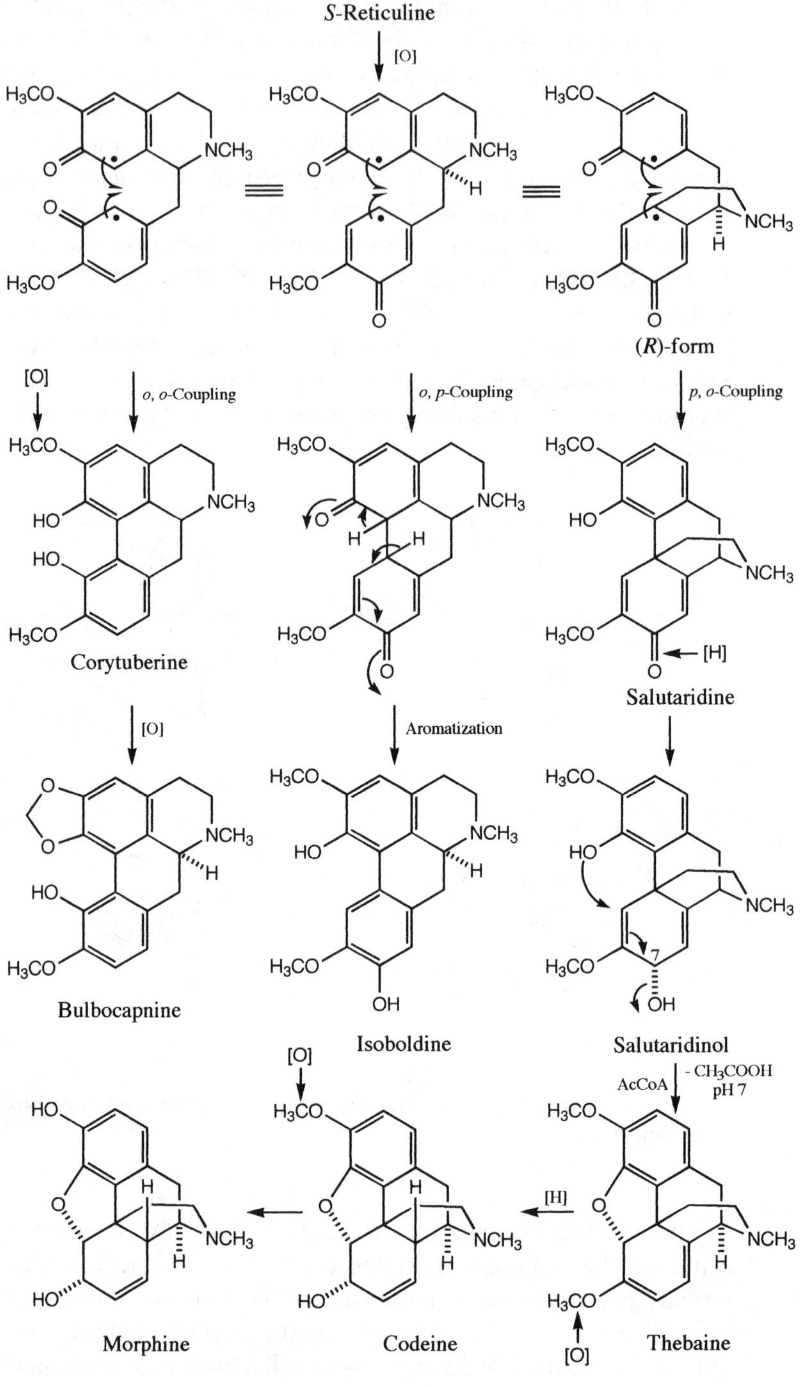

Fig. 16   Biosynthesis of aporphine and morphine type of alkaloids

bridged morphine skeleton by intramolecular C-C coupling in *ortho-ortho* and *ortho-para* positions to the free hydroxyl, Fig. 16. Formation of the morphine skeleton requires first conversion of (*S*)-reticuline into the (*R*)-isomer, which proceeds by way of an intermediary immonium ion. Phenol oxidation gives salutaridine, which is reduced to the alcohol salutaridinol.[33] The oxygen bridge forms spontaneously at pH 8 by acetylation of the C[7]-OH group with acetyl CoA. Final demethylation proceeds via radical oxygenation of the methyl group and not as expected by enol ether hydrolysis because it was shown that this reaction involved no loss of labelled oxygen. The compara-

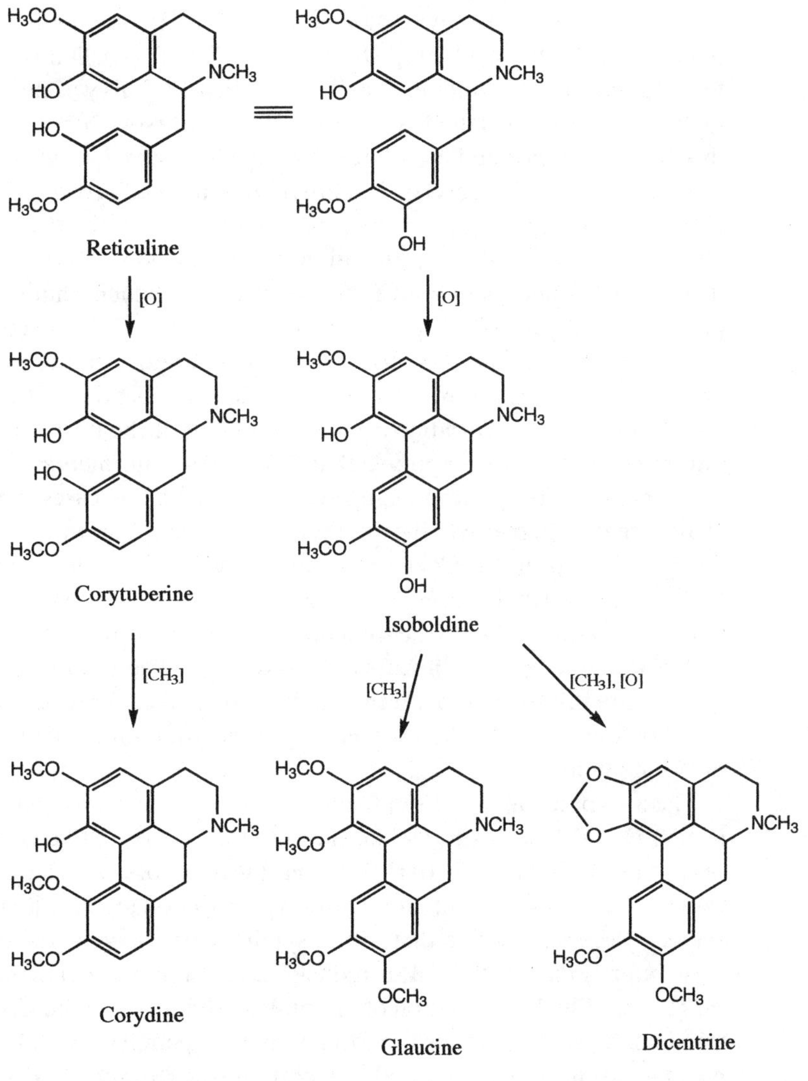

Fig. 17   Disproved routes to corydine, glaucine, and dicentrine in *Dicentra eximia*

tively complex morphine skeleton can thus be deduced in a simple way with correct oxygenation pattern. Most amazing is the observation that morphine also occurs in mammalian tissues and that enzyme systems there are capable of converting codeine into morphine.

We must never fall into the habit of routine thinking just because our scheme works beautifully in one case. The straightforward phenol oxidation in Fig. 16 was neatly proved by feeding experiments. Consequently the "obvious" routes to the closely related aporphine alkaloids, corydine, glaucine, and dicentrine, should, by analogy, be the ones depicted in Fig. 17 with reticuline as the precursor. However, we have no guarantee that nature has chosen the route we believe is "obvious". Failure to obtain the labelled aporphine alkaloids by incubating *Dicentra eximia* with labelled reticuline led to the discovery of the "unlikely" pathways,[34] Fig. 18. This was done by testing a number of possible labelled intermediate precursors for incorporation. It turned out that norlaudanosoline, its 4´-methoxy and 7,4´-dimethoxy derivative but no other *O*-methyl or *N*-methyl isoquinolines gave efficient incorporation. Double labelling proved that the 4´-methoxy derivative was incorporated intact because the isotopic ratio was unchanged in corydine. Phenol oxidation and *N*-methylation give the two dienones which undergo a dienone–phenol rearrangement to boldine, which was found to be efficiently incorporated into glaucine and dicentrine. Methylation and oxidation give the final products. One point, however, needs further clarification. The methylation pattern of ring A of corydine is opposite to that of the postulated aporphine intermediate. It is not yet proved that this is a true intermediate, but if so, how is the methylation pattern changed? From analogous cases it may be concluded that a demethylating–methylating sequence is more likely than a direct methyl transfer. Coclaurine labelled with ^{14}C in the *O*-methyl group lost part of its label during its transformation into crotonosine[35] in *Croton linearis* (8) and (±) 6-$O^{14}CH_3$-reticuline lost 64 per cent of its label when incorporated into boldine in *Litsea glutinosa* (9).[36] These findings are interesting because this plant apparently can use reticuline as a precursor for boldine via isoboldine, whereas *Dicentra eximia* cannot, but requires the isomeric norprotosinomenine.

The aporphine alkaloid stephanine shows an unusual oxygenation pattern in ring D which seems to violate the rule that the oxidative coupling occurs *ortho* or *para* to the hydroxyl function. Furthermore, it is unlikely that this ring would originate from an *o*-hydroxyphenylethylamine. The problem was solved by the supposition that the biosynthetic pathway passes through a dienone intermediate which, after reduction, undergoes a dienol–benzene rearrangement, Fig. 19. Oxidation of orientaline, reduction to the alcohol and water elimination account for the formation of stephanine. A number of dienone alkaloids have been isolated, e.g. crotonosine (8), supporting the pathway suggested.

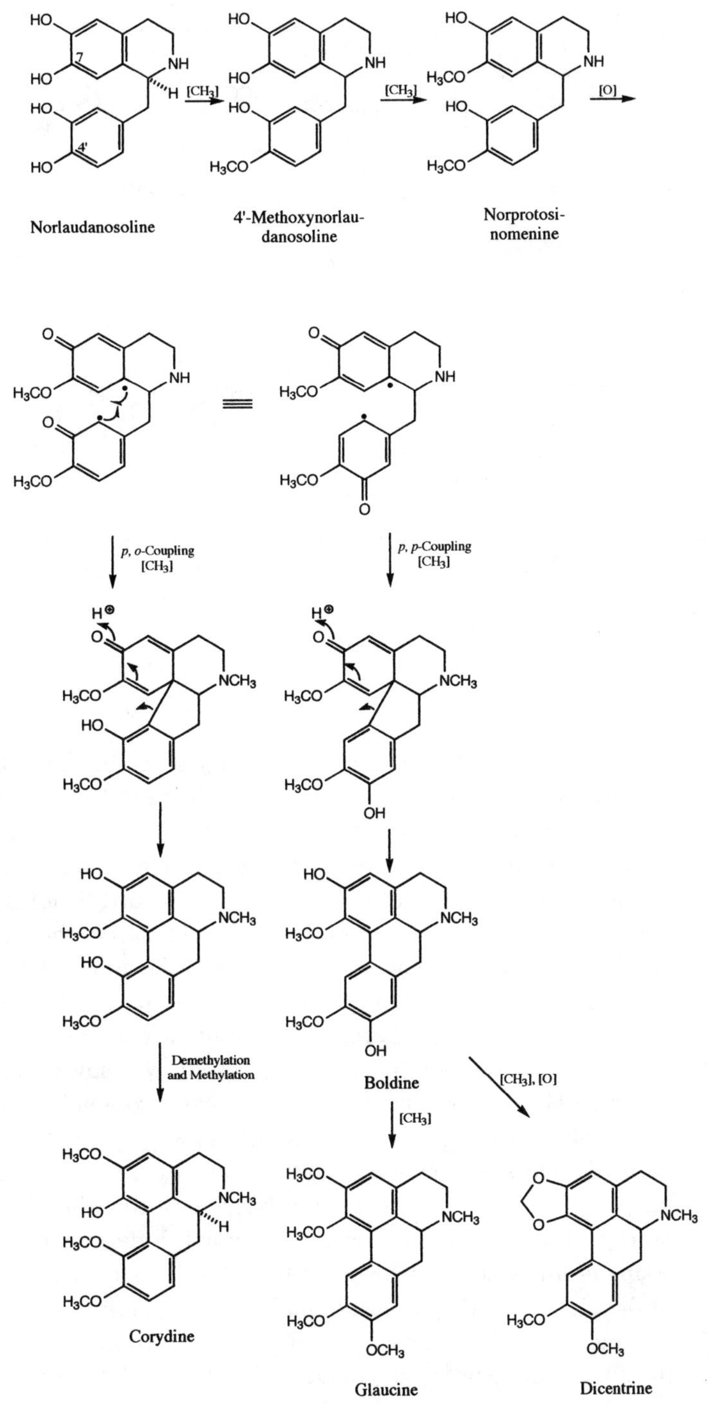

Fig. 18 Biosynthesis of corydine, glaucine and dicentrine

Coclaurine   Crotonosine

(8)

Isoboldine   Boldine

(9)

Several of the phenol couplings discussed can be mimicked in the laboratory by oxidation of phenolic tetrahydroisoquinolines with potassium ferricyanide. An early attempt to synthesize the aporphine skeleton by oxidation
of laudanosoline did not give the product, but instead a compound having the
dibenzopyrrocoline skeleton (10). At that time no alkaloids of that type were
known, but several years later alkaloids isolated from the bark of *Cryptocarya
bowiei* were shown to have this structure. If the amino function is blocked by
quaternization, the reaction leads to the aporphine skeleton in good yield.

The occurence of the *Erythrina* alkaloids is, with few exceptions, limited to
the genus *Erythrina* of the *Leguminosae* family. They are physiologically distinguished by a strong curare-like effect. Approximately 1 mg kg^{-1} of the most
potent alkaloids injected intravenously in a frog is sufficient to cause total
paralysis. The skeleton of the *Erythrina* alkaloids is another artful variation of
the phenol coupling theme. Erythraline and isococculidine, Fig. 20, stem from
S-norprotosinomenine—only the *S*-isomer is incorporated—by *para–para*
coupling followed by cleavage of the C^{Ar}–C^1 bond, supported by interaction
of the electron lone pair at the nitrogen. The early steps of the biosynthesis
are thus identical to those of the boldine biosynthesis, Fig. 18. Reduction of
the intermediate imine and oxidation of the *p,p*-dihydroxybiphenyl to a diphenoquinone give, after cyclization with the amino function, the *Erythrina*
alkaloid skeleton and by conventional reactions finally erythraline. Isococcu-

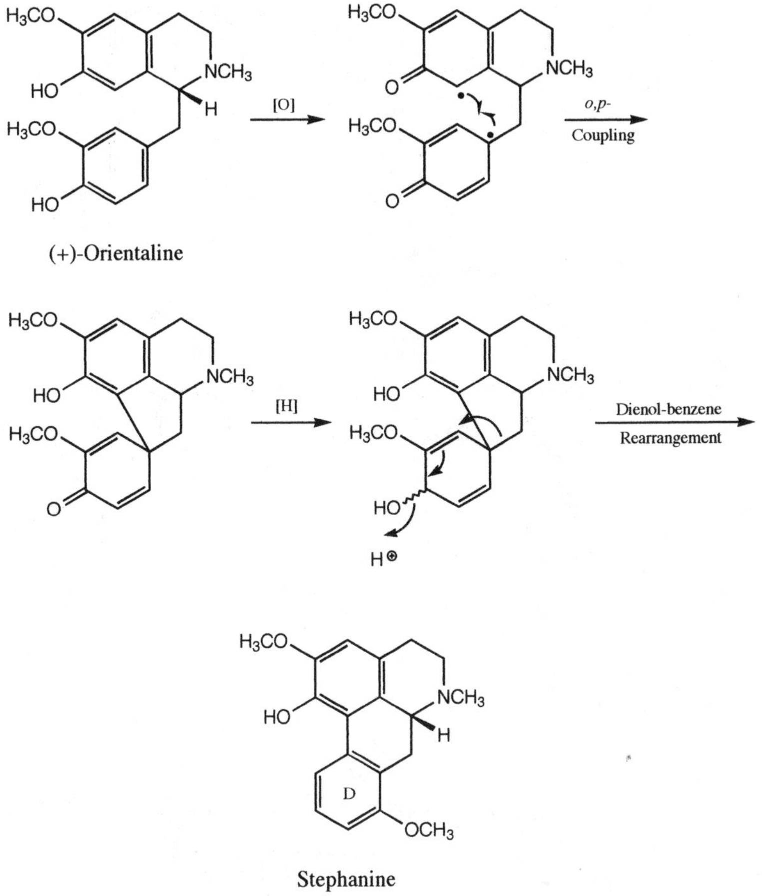

Fig. 19 Biosynthesis of stephanine

lidine lacks the oxygen function at C^{16}. This can in principle be removed by reductive elimination either at the first asymmetric quinonoid step or at the second quinonoid step, where asymmetry is lost. It was found that 7-$^{14}CH_3O$-norprotosinomenine diverted its label into both parts of isococculidine,[37] i.e. the reduction occurs at the symmetrical diphenoquinone stage.

All the tyrosine-derived alkaloids we have encountered so far, can be dissected into two C_6–C_2 groups. The *Amaryllidaceae* alkaloids are characterized having one C_6–C_2 and one C_6–C_1 unit originating from tyrosine and phenylalanine, respectively, Fig. 21. Although tyrosine and phenylalanine are metabolically closely related and the units in the alkaloids are highly oxygenated, very little randomization occurs due to selective enzymatic and compartmental effects. Phenylalanine is degraded to protocatechuic aldehyde

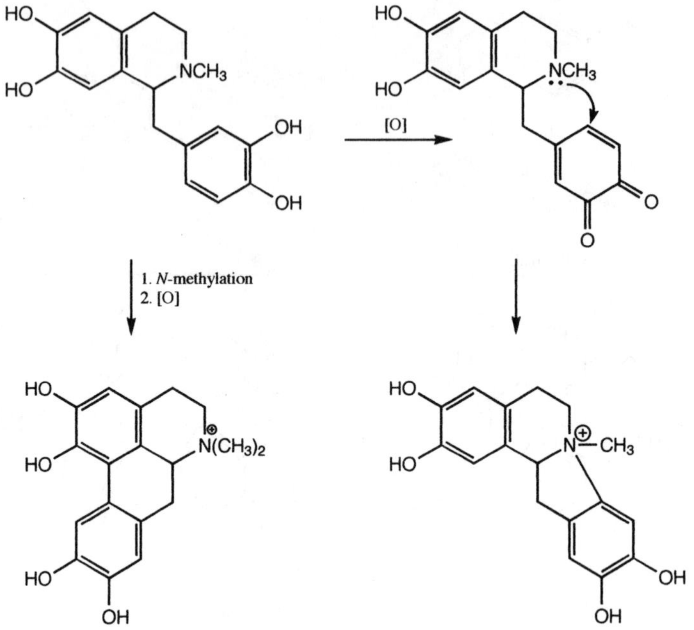

(10)

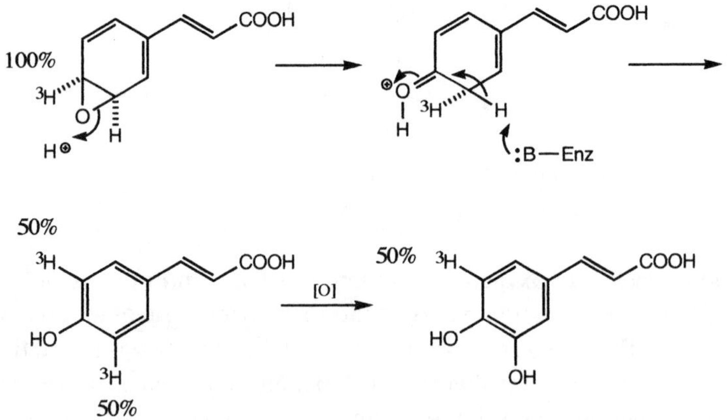

(11)

(section 4.4) by the sequence phenylalanine → cinnamic acid → coumaric acid → caffeic acid → protocatechuic aldehyde. It was established by ^{3}H-labelling that phenylalanine at some stage lost both hydrogens at C^3, which suggests that the fragmentation of the cinnamic acid cannot be formulated simply as a β-hydroxylation followed by a retroaldol condensation. At some

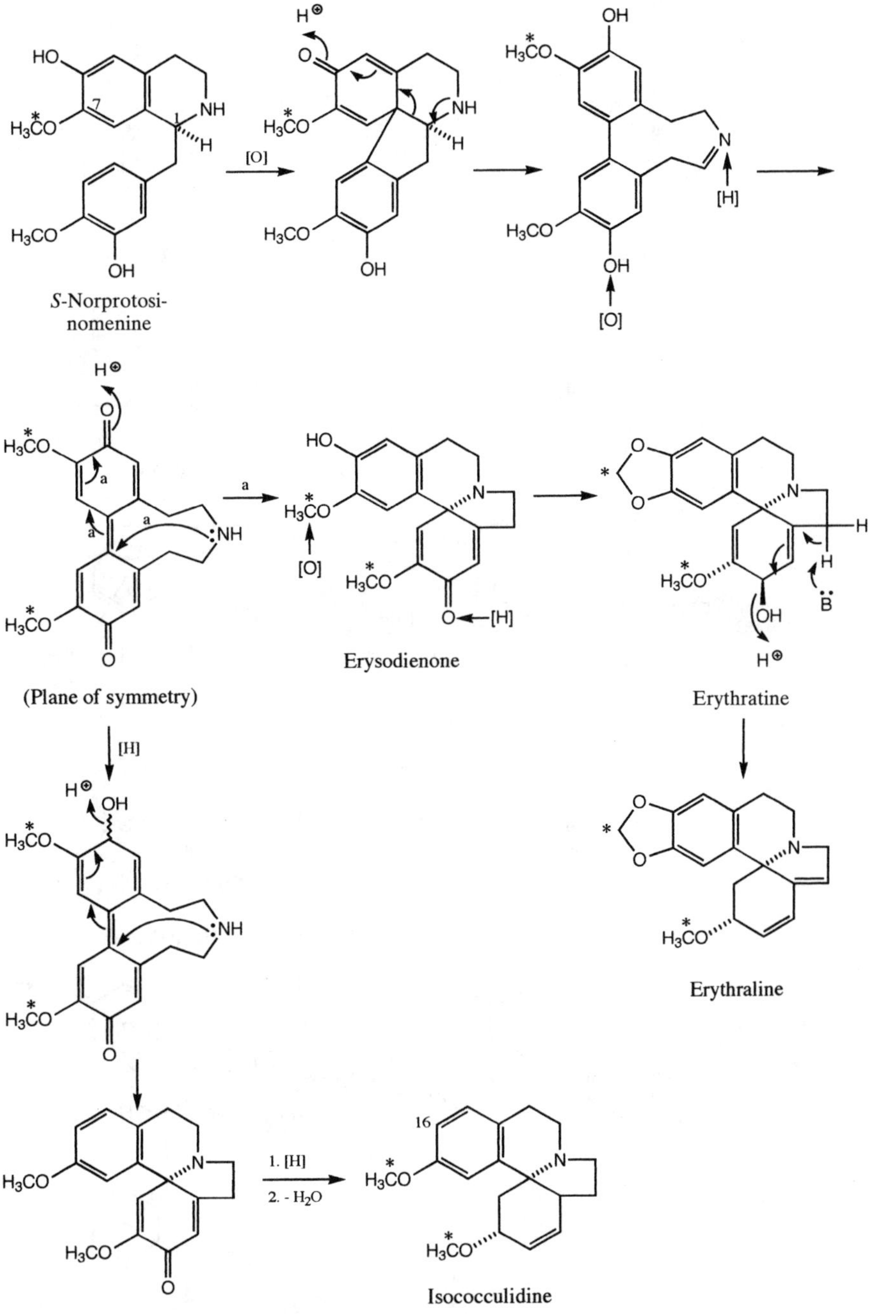

Fig. 20 Biosynthesis of *Erythrina* alkaloids

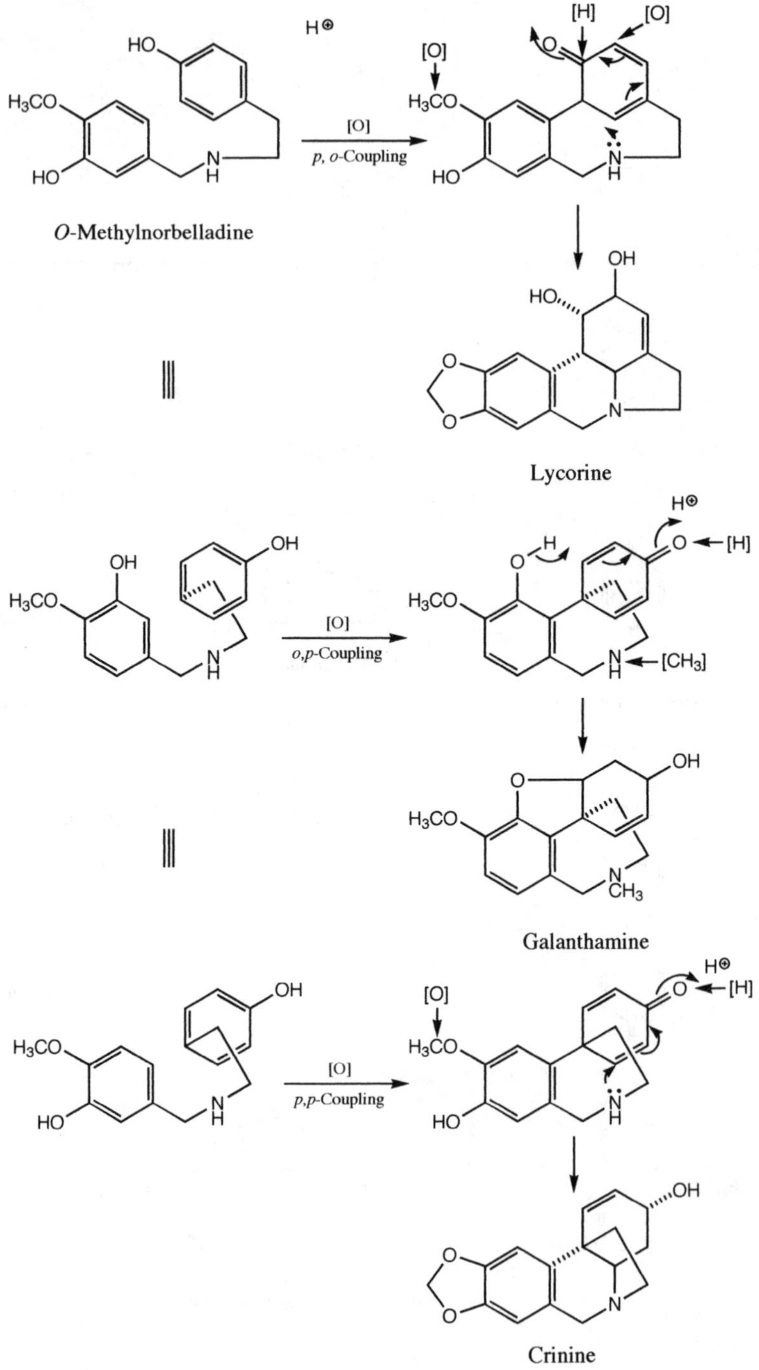

Fig. 21   Biosynthesis of *Amaryllidaceae* alkaloids

stage of the biosynthesis of the C_6–C_1 unit, C^3 of the amino acid is oxidized to the ketone level. The immediate precursor of protocatechuioc aldehyde is probably protocatechuoyl CoA. The biological hydroxylation of 4-^{3}H-cinnamic acid to *p*-coumaric acid involved complete migration and retention of tritium (NIH shift, section 4.3). The introduction of the *m*-hydroxy group proceeded with loss of 50 per cent of the tritium. The first complete retention could conceivably be explained by a large ^{3}H isotope effect but since incorporation of 3,5-^{2}H$_2$-4-^{3}H-cinnamic acid gave the same result, it is suggested that the stereospecific proton elimination is enzymatically controlled[30,31] (11). Tyrosine is incorporated via tyramine but dopa is not metabolized. Protocatechuic aldehyde condenses with tyramine and gives on reduction the pivotal norbelladine which, depending upon the folding, gives rise the basic *Amaryllidaceae* skeletons,[38,39] (Fig. 21).

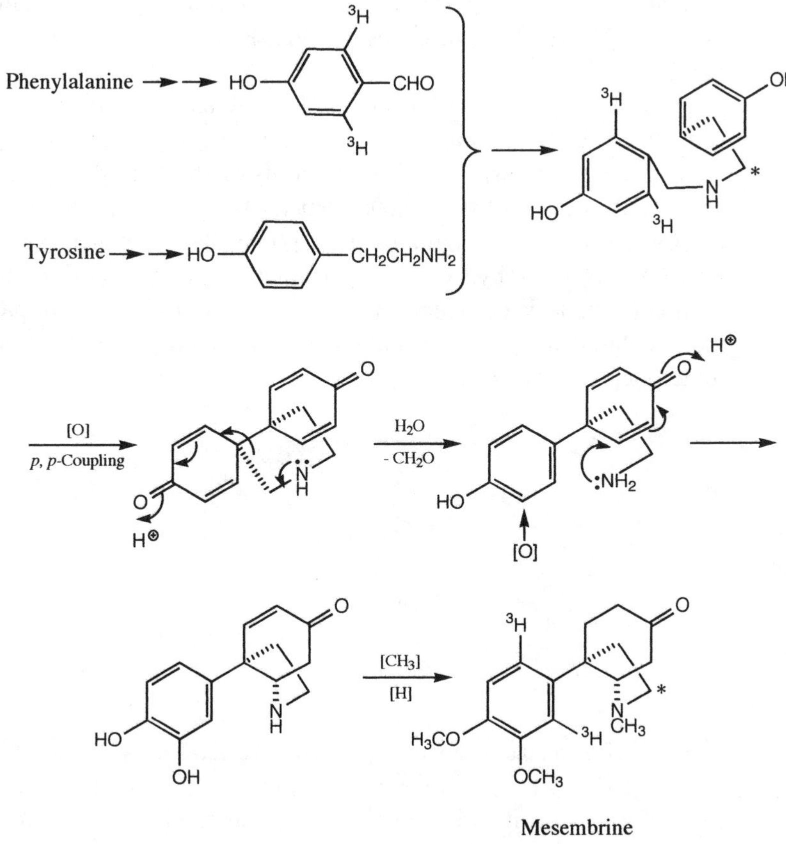

Fig. 22   Biosynthesis of mesembrine

The C_6–C_1–N–C_2–C_6 unit is uncommon outside *Amaryllidaceae.* Crypto-styline,[40] which belongs to a small group of alkaloids found in *Orchidaceae* is a 1-phenylisoquinoline derivative formed analogously to the 1-benzylisoquin-olines. Mesembrine, Fig. 22, found in *Sceletium strictum* of the *Aizoceae* fam-ily shows a strong structural kinship to the crinine group, but the C_1 carbon is missing. It is biosynthesized from one phenylalanine and one tyrosine unit. Tracer experiments show that:

1. methionine is the donor of *O*- and *N*-methyls;
2. 2,6-di-^{3}H-phenylalanine retains both tritiums in the aromatic ring of mesembrine;
3. phenylalanine with hydrogen isotopes in the side chain gives inactive alka-loids;
4. tyrosine is incorporated intact and is the precursor of the hydroindole frag-ment;
5. the second hydroxylation of the aromatic ring is a late event.[41]

The fact that both ^{3}H are retained in mesembrine excludes a crinine-like path-way, Fig. 21, which requires loss of one ^{3}H. The findings are rationalized in Fig. 22.

The phenethylisoquinoline alkaloids[42] isolated from six genera of *Liliaceae*, characteristically contain the C_6–C_3–N–C_2–C_6 unit. Their biosynthesis par-allels closely the corresponding alkaloids of the benzylisoquinoline groups. The C_6–C_2 fragment originates from tyrosine via dopa and dopamine and the C_6–C_3 fragment originates from phenylalanine which conventionally is transformed into 4-hydroxydihydrocinnamaldehyde according to (12).[43] The homoaporphine kreysigine is formed by condensation of dopamine with a methoxylated dihydrocinnamaldehyde to autumnaline which subsequently undergoes phenol oxidation, Fig. 23.

(12)

Colchicine, a structurally most intriguing alkaloid, is the active principle of several *Colchicum* spp. It is a toxic compound used in medicine in the treat-ment of gout and in biology for doubling the number of chromosomes during cell division in plants. The structure of colchicine, Fig. 24, was a long-standing problem for organic chemists. The clue to the structure came first with the

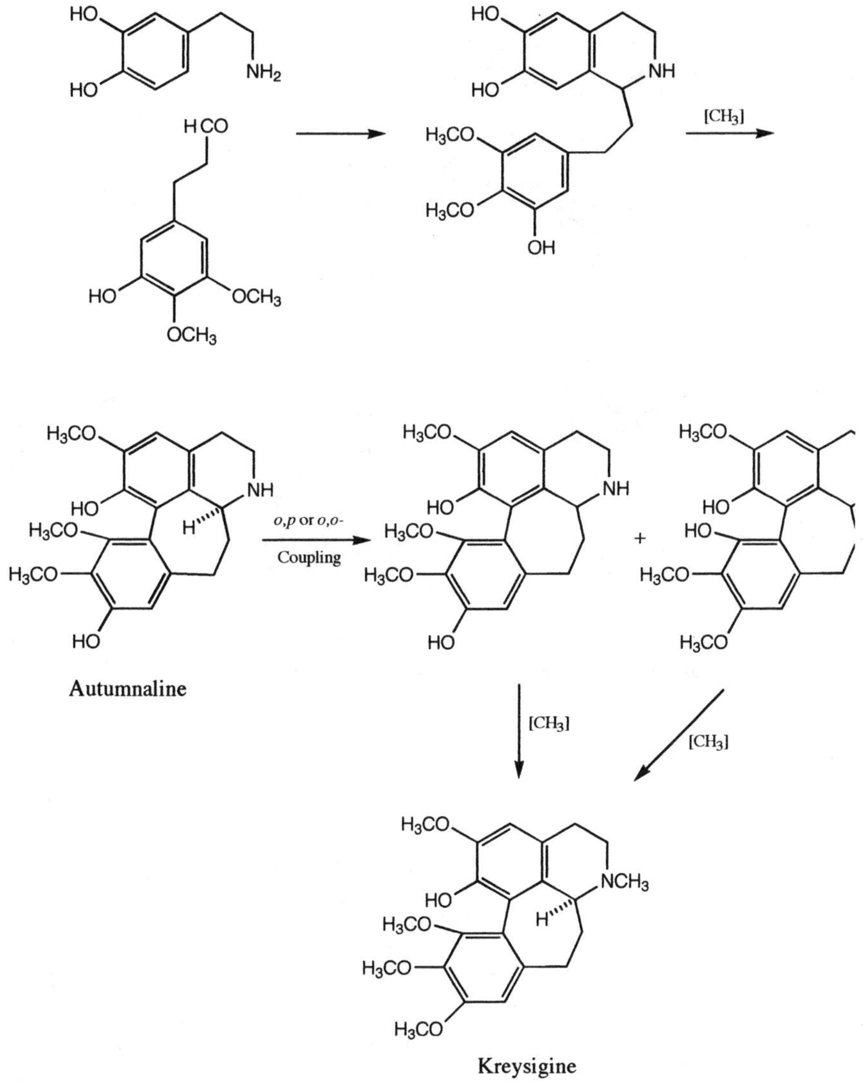

Fig. 23   Biosynthesis of kreysigine, a homoaporphine alkaloid

concept of tropolone aromaticity by Dewar. The position of the carbonyl group was finally determined by an X-ray diffraction study. Tracer studies in *Colchicum* spp. revealed that intact phenylalanine provided the C_6–C_3 fragment and the aromatic nucleus of tyrosine, and C^3 provided the seven membered tropolone system by ring enlargement. The nitrogen originates from tyrosine which was proved by a double labelling technique. The $^{14}C/^{15}N$ ratio of autumnaline was identical to the $^{14}C/^{15}N$ ratio in colchicine. Thus, it cannot,

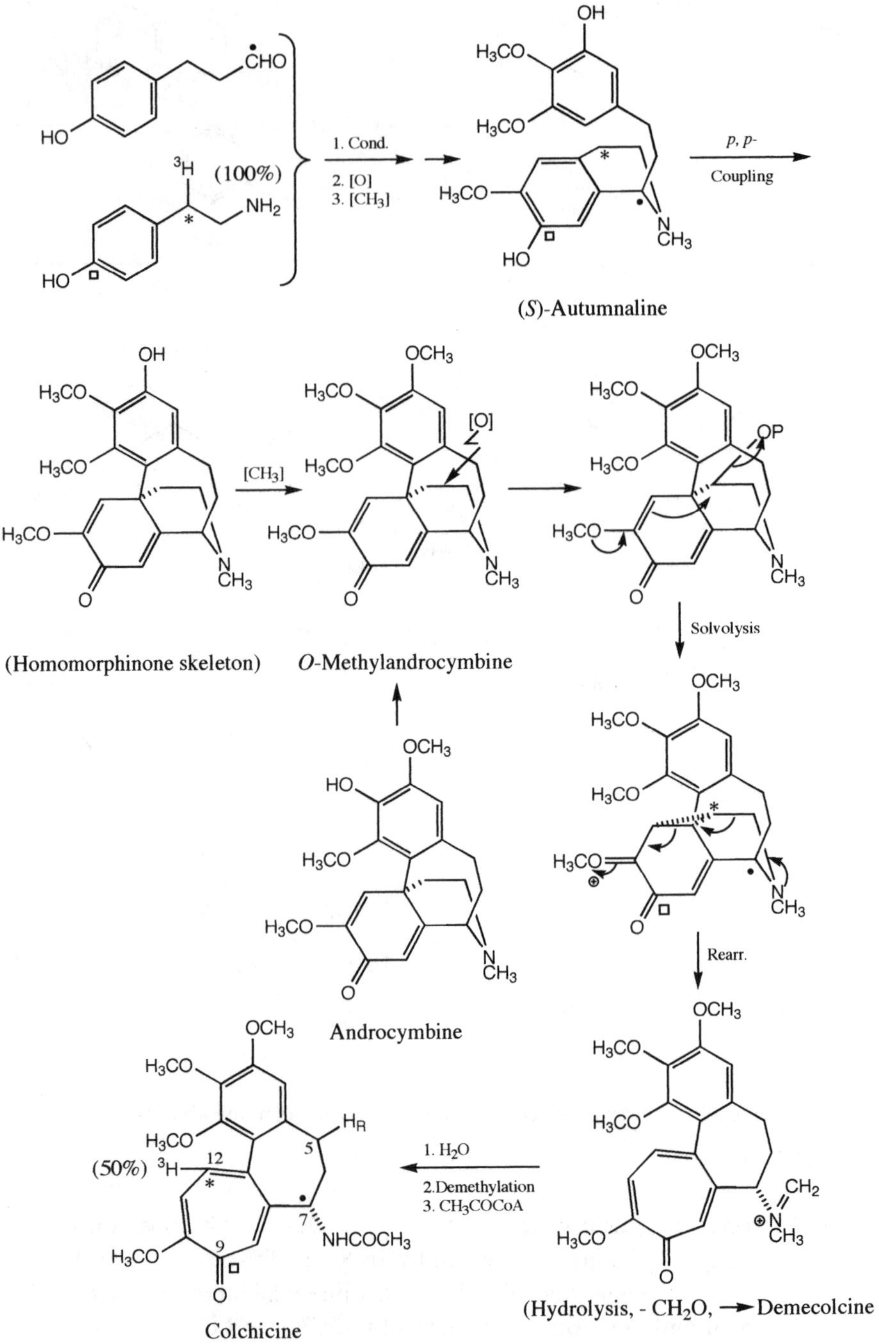

Fig. 24   Biosynthesis of colchicine

for example, be introduced by some late transamination reaction occurring in the condensed 6,7,7-membered tricyclic structure.

The lyase mediated elimination of ammonia involves loss of the pro-H_S at C^3 of phenylalanine, consequently, stereospecifically labelled H_R should be retained at C^5 in colchicine which was shown to be the case.[44] 1-^{14}C-Labelled phenylalanine, 3-^{14}C- and 4'-^{14}C-labelled tyrosine appear at C^7, C^9 and C^{12}, respectively. The specific labelling of colchicine proves that the metabolic paths of phenylalanine and tyrosine follow separate lines in the plant. Little by little a fascinating biosynthetic scheme emerged out of the results of numerous feeding experiments (Fig. 24). The two amino acids give first a phenethylisoquinoline derivative which undergoes a directed *p,p*-phenol coupling to give the homomorphinone skeleton. The presumed relay alkaloid autumnaline proved, in fact, to be efficiently incorporated, whereas phene-thylisoquinolines with other methoxylation patterns were incorporated poor-ly. It so happened that androcymbine, an isomer of the presumed, *p,p*-phenol coupling product from autumnaline, was isolated from *Androcymbium melanthioides*, a relation of *Colchicum autumnale*, and its *O*-methyl deriva-tive was found to be incorporated remarkably well.[43,44] Therefore, the next step is a methylation of the homomorphinone to *O*-methylandrocymbine. The insertion of C^{12} is best explained by enzymatic controlled hydroxylation because half of the tritium was removed from 3-^{3}H-labelled autumnaline. Non-specific hydroxylation would result in an isotope effect and oxidation to the carbonyl state with a complete loss of tritium. Solvolysis of the phosphate group supported by electron donation by the methoxy group leads to the pre-sumed cyclopropane derivative which eventually rearranges to the tropolone nucleus. C^2 of tyrosine is believed to be lost as formaldehyde. The intermedi-ate *N*-methyl derivative, demecolcine, is efficiently incorporated into colchi-cine. The ring expansion has, in fact, a laboratory analogy in reaction (13) which further strengthens the suggested biosynthetic route.

(13)

8.4 Alkaloids derived from tryptophan. The indole alkaloids

Apart from some simple derivatives of tryptophan, e.g. indolylalkylamines, such as gramine, bufotenine, and the hallucinogenic psilocybine extracted from a Mexican mushroom, physostigmines, and β-carbolines (Fig. 25), the formation of which have ample analogies in the tyrosine series, the over-whelming majority of the indole alkaloids is distinguished by a formidable structural variation and complexity. The tryptamine is visibly an invariant feature, but the C_9 or C_{10} aliphatic unit concealed for a long time its true or-igin. The indole alkaloids, which primarily are confined to three plant fam-ilies, *Apocynaceae, Loganiaceae* and *Rubiaceae*, attracted interest, partly because of their significant neurophysiological action, but not least because their structural elucidation, syntheses and biogenesis presented to the organic chemist a challenge on the highest intellectual level. In this area biosynthesis has celebrated triumphs in bringing sense and order to a bewildering variety of structures.

In a few instances the indole nucleus is modified to an isoquinoline nucleus, e.g. in the calycanthine and cinchonine alkaloids.

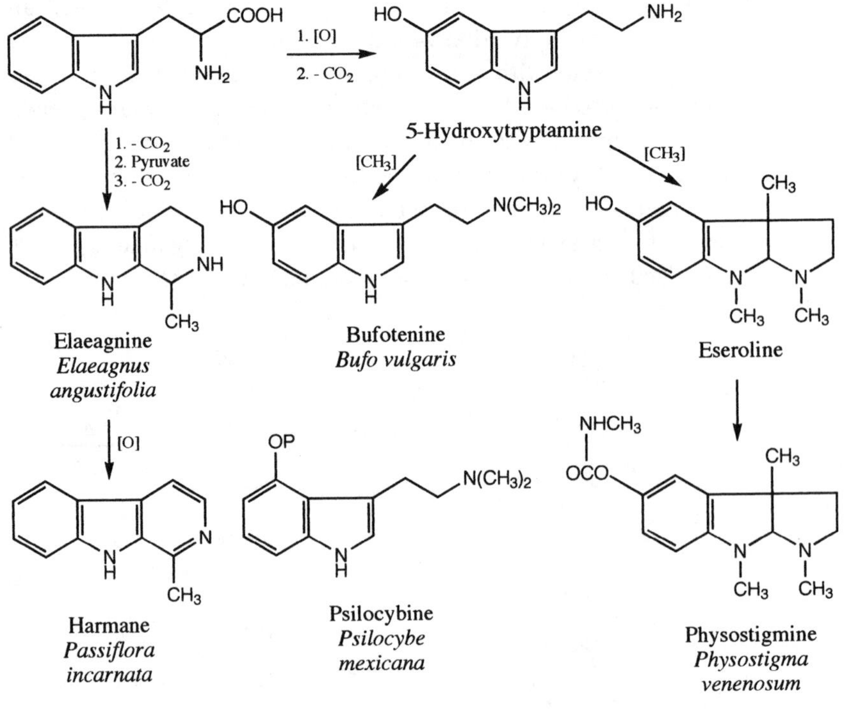

Fig. 25   Biosynthesis of some simple tryptophan alkaloids

There were wild speculations as to the origin of the aliphatic part until the monoterpene hypothesis was presented.[45,46] Subsequently it was shown to be correct by the incorporation of mevalonate or more efficiently, geranyl phosphate.[47] Specific incorporation was ensured by administering synthetic 2- and 4-^{14}C-geranyl phosphate to *Catharanthus roseus (Vinca rosea)* which is known to produce a variety of indole alkaloids of widely different structures (Fig. 26). Degradation showed that the label was entirely located at the marked positions. Thus, being confident of the origin of the non-tryptamine part, the next step was to identify the intermediates en route from geraniol. The monoterpene hypothesis suggested cyclopentene derivatives as likely candidates. Aid came from structural work on ipecoside (section 8.3) that was carried out at the same time.[48] It contains a monoterpenoid part which could be traced back to loganin as an attractive precursor which by cleavage of the five-membered ring gives secologanin (section 6.3). Condensation of secologanin with dopa gives *N*-deacetyl ipecoside. Labelled loganin as well as secologanin were subsequently shown to be efficiently incorporated into a number of indole alkaloids, and moreover, loganin was shown to be present in *Catharanthus roseus*. Hence, it is quite plausible that tryptamine and secologanin in the plant give rise to an analogous indole glycoside located somewhere on the path between loganin and the indole alkaloids (Fig. 27). Direct *in vitro* condensation of secologanin with tryptamine afforded two isomeric glycosides,

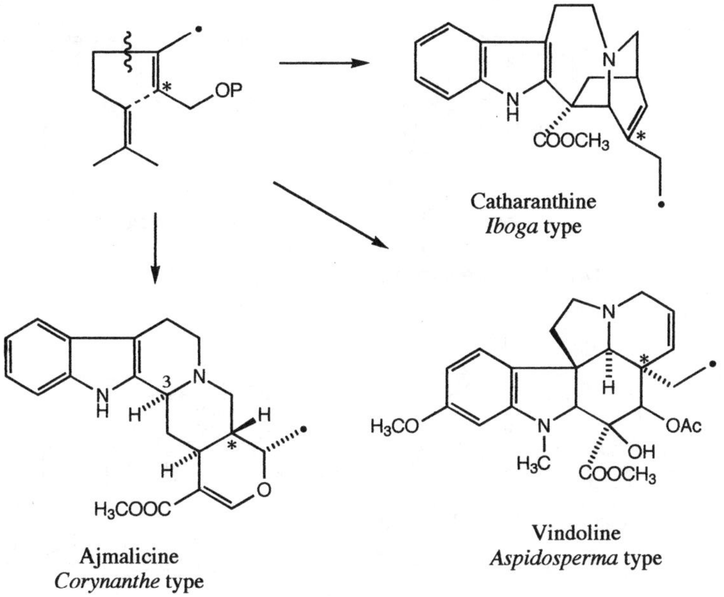

Catharanthine
Iboga type

Ajmalicine
Corynanthe type

Vindoline
Aspidosperma type

Fig. 26   Incorporation of 2- and 4-^{14}C-geraniol into indole alkaloids

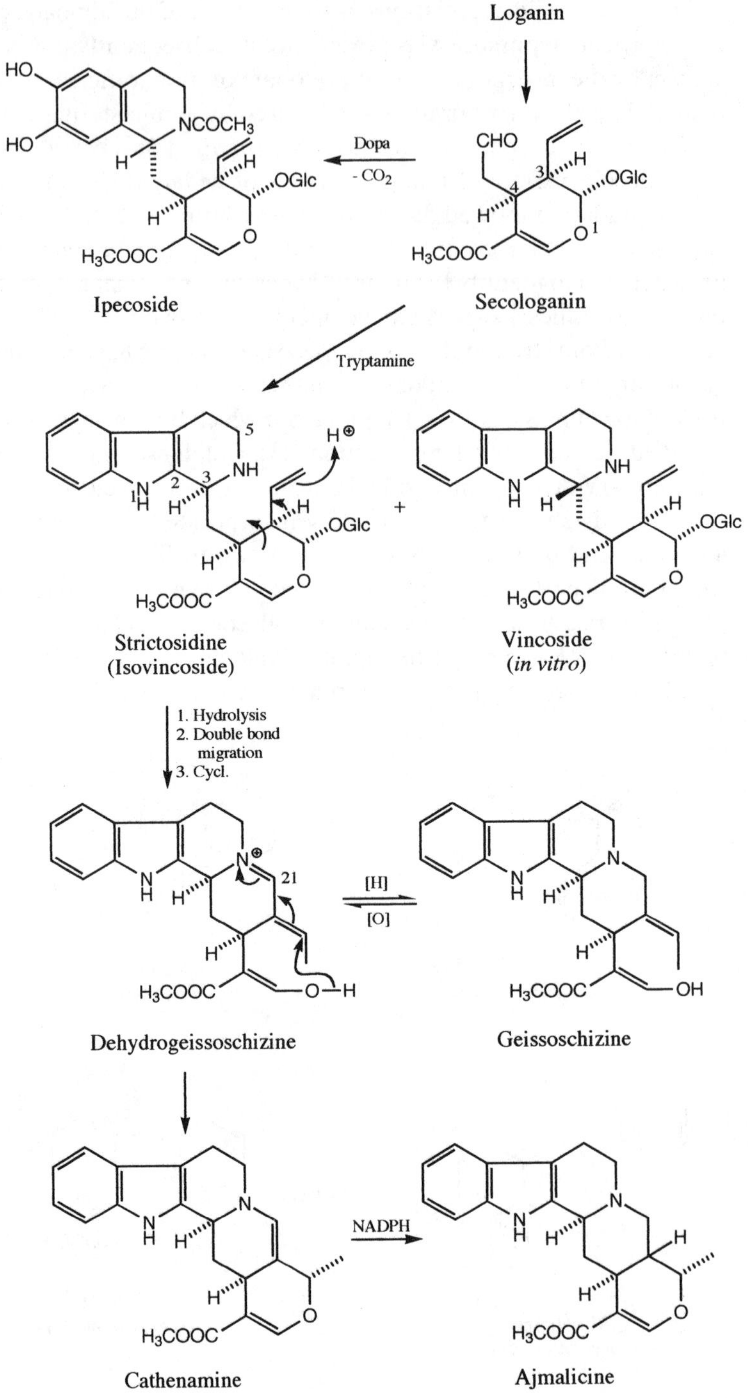

Fig. 27   Biosynthesis of ajmalicine

vincoside and isovincoside which by dilution analysis were proved to be present in *C. roseus.*[49] This technique is often used in cases where the presumed intermediate occurs in amounts so small that it cannot be isolated. The intermediate, the carrier, is therefore synthesized and added together with a radioactive precursor to the plant. After a suitable time it is again recovered and purified. The compound appears on the pathway if it has acquired radioactivity.

Strictosidine, an indole glycoside of the same gross structure, was simultaneously isolated from another plant *Rhazya stricta,*[50] and later proved to be identical to isovincoside.[51] Only this isomer with (*S*)-configuration at C_3 is incorporated into ajmalicine, vindoline, catharanthine, and other indole alkaloids. The biosynthesis of ajmalicine is illustrated in Fig. 27. If the scheme is correct, the C^3-H of secologanin is lost, but the C^4-H is retained during the transformations. This was also borne out in practice. Geissoschizine was proved by double labelling to be incorporated intact into ajmalicine and several other indole alkaloids. The intermediacy of cathenamine was proved by experiments in cell-free cultures. In the absence of NADPH the synthesis stopped at this stage but proceeded enzymatically to ajmalicine on addition of NADPH.

Strictosidine appears to be a universal intermediate in the biosynthesis of indole alkaloids.[52] The biosynthesis of the *Rauwolfia* alkaloids yohimbine, route b, and reserpine, route a, a tranquillizer used since ancient times in Indian folk medicine, is depicted in Fig. 28. The timing of the methoxylation and the epimerization is not known. Reserpine contains a carbocyclic E-ring and six asymmetric carbon atoms. The formidable task to elaborate the compound stereospecifically was overcome in the now classical synthesis by Woodward *et al.*[53]

So far, we have only considered condensations at C^2 of the indole nucleus but it is known to undergo electrophilic substitution at C^3 as well, reactions which in essence are reversible. Reactions of secologanin at C^3 leads to the strychnine skeleton which still has an intact monoterpenoid structure. Again, the initial C^3 condensation is also the prerequisite for the rearrangements leading to the *Aspidosperma* and *Iboga* alkaloids (Figs 29, 30).

Sequence analysis, i.e. determination of the appearance and disappearance of radioactivity in intermediates, after administration of 2-^{14}C-tryptophan in *C. roseus*, demonstrated that

1. geissoschizine and preakuammicine are dynamic intermediates but catharanthine, ajmalicine, or vindoline are end-products;
2. the *Corynanthe* skeleton is formed first and subsequently converted to the *Strychnos, Iboga*, and *Aspidosperma* types;
3. the Pictet-Spengler condensations at C^2 and C^3 of the indole nucleus are reversible.

The fragmentation of preakummicine (Fig. 29) leading to akuammicine can be regarded as a retroaldol condensation. In order to arrive at strychnine, the carboxyl group (Fig. 29) condenses with an acetate unit and the formyl group is oxidatively eliminated as carbon dioxide. The ketide is reduced to the alcohol and final cyclization gives strychnine, route a.

The bridgehead nitrogen in preakuammicine is in a position to support a fragmentation with re-establishment of the aromatic indole nucleus as an extra driving force (Fig. 30). This leads, after reduction of the intermediary immonium function, to stemadenine, another alkaloid isolated from *C. roseus* and shown to be incorporated intact into tabersonine, vindoline, and cathar-

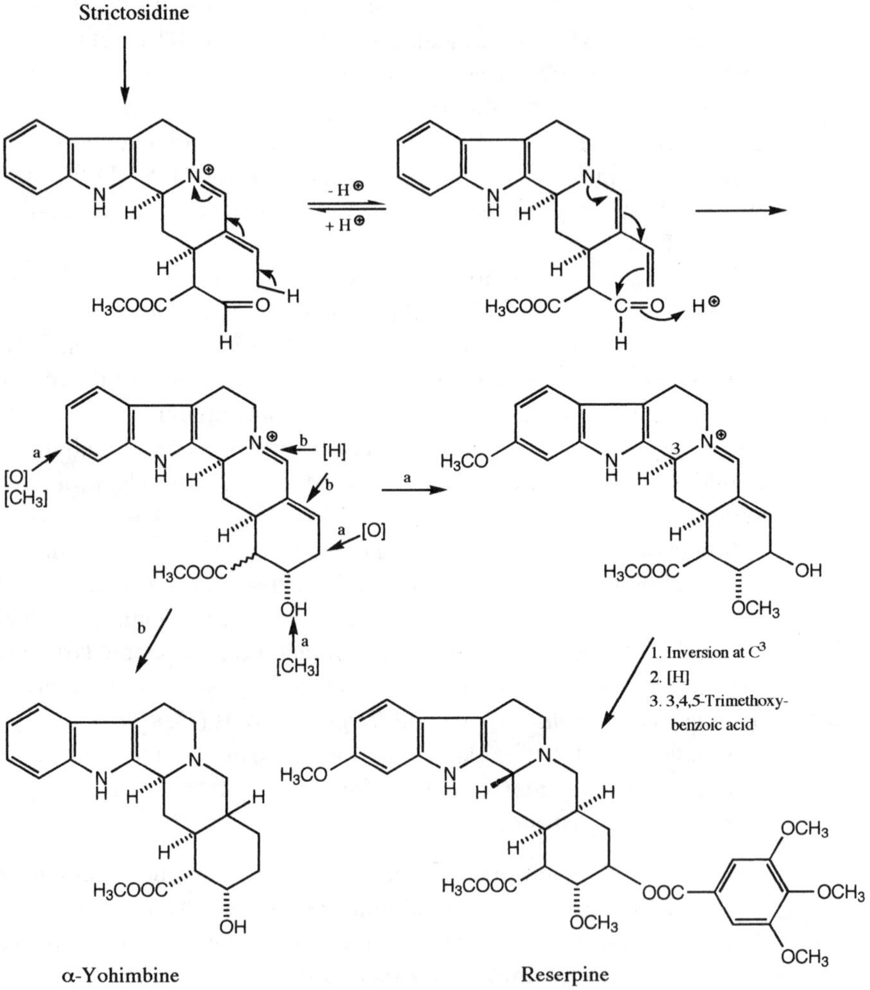

Fig. 28   Biosynthesis of reserpine and α-yohimbine

anthine.[54] The observation that tabersonine is also incorporated into catharanthine implies that these arrangements possess a high degree of reversibility.
The rearrangements proceed by way of a dihydropyridine–acrylic ester struc-

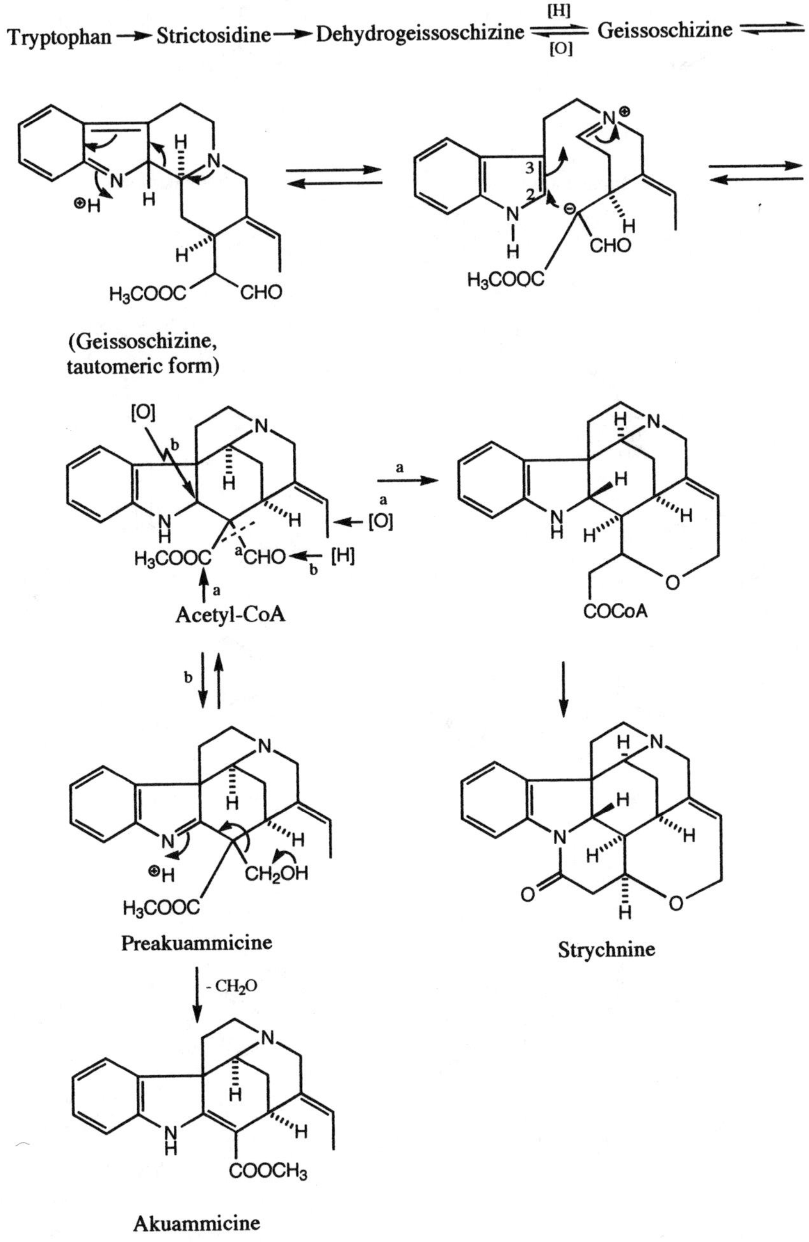

Fig. 29   Biosynthesis of akuammicine and strychnine

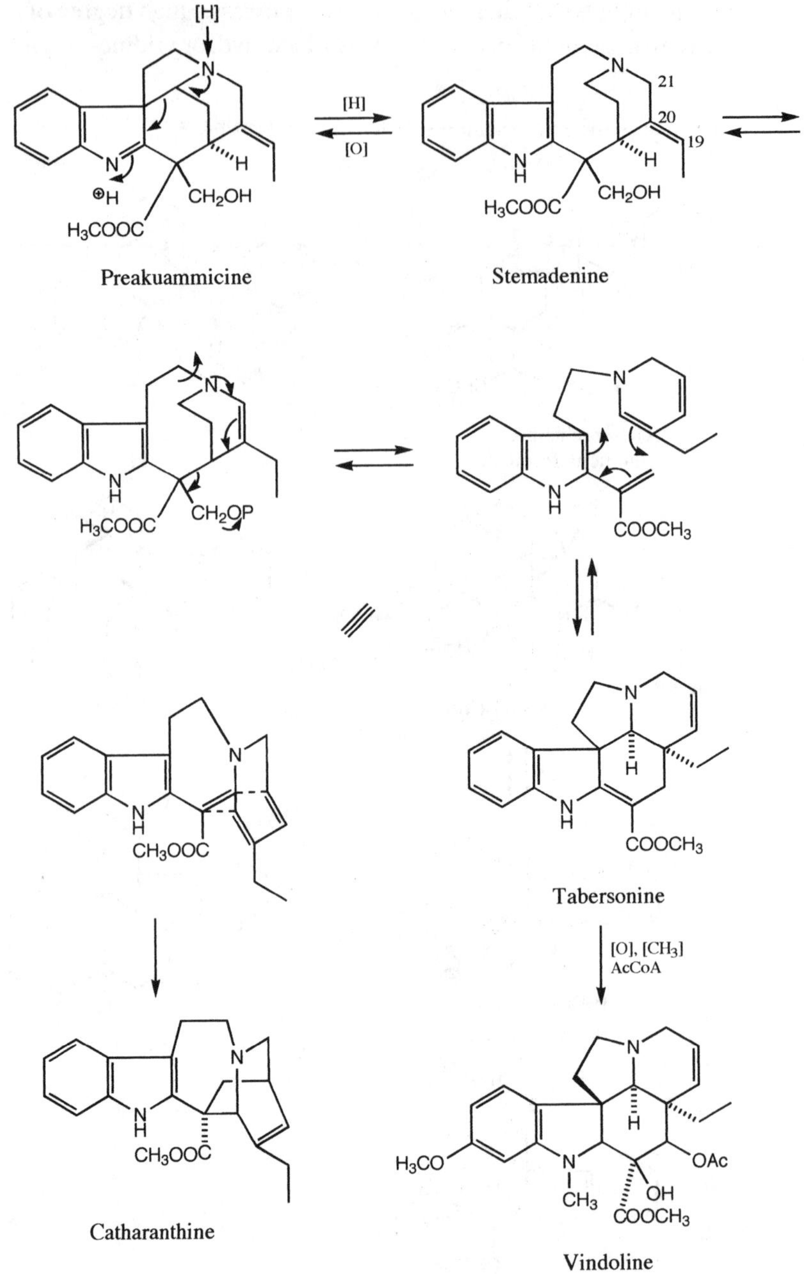

Fig. 30   Biosynthesis of *Aspidosperma* and *Iboga* alkaloids

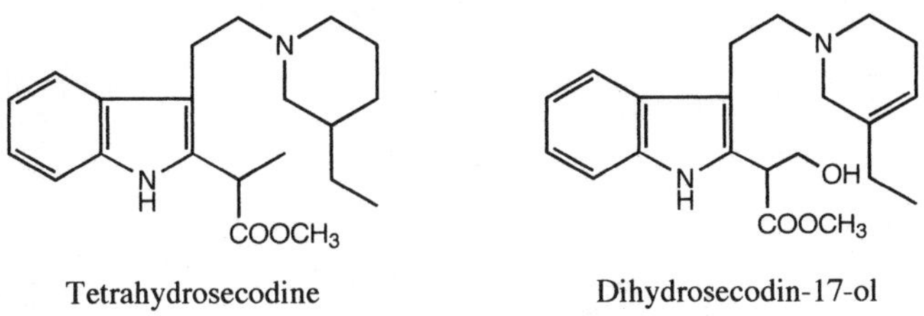

Tetrahydrosecodine   Dihydrosecodin-17-ol

Fig. 31   Structures of secodines

ture which at first sight may seem unlikely. However, strong support for such a formulation was obtained from isolation of the related secodines in *Rhazya* spp. (Fig. 31). The cyclization can be regarded either as an internal Diels–Alder reaction or stepwise Michael addition depending on the concertedness of the reaction. The reversible nature of these reactions speaks in favour of a Michael addition. The double bond at $C^{19,20}$ or $C^{20,21}$ (Fig. 30) is essential for the fragmentation to occur in conformity with the mechanism depicted; in fact, if the vinyl group of strictosidine is reduced the product does not incorporate.[55] It ought to be pointed out that alternative mechanistic schemes are still conceivable, simply because detailed experimentation is still lacking.

A number of bisindole alkaloids[56] are formed in the plant cell as, for example, the antileukaemic vinblastine derived from vindoline and catharanthine (14).

The ergot alkaloids are produced by the fungus *Claviceps purpurea* which is parasitic on rye and certain grasses. The metabolities attracted the interest

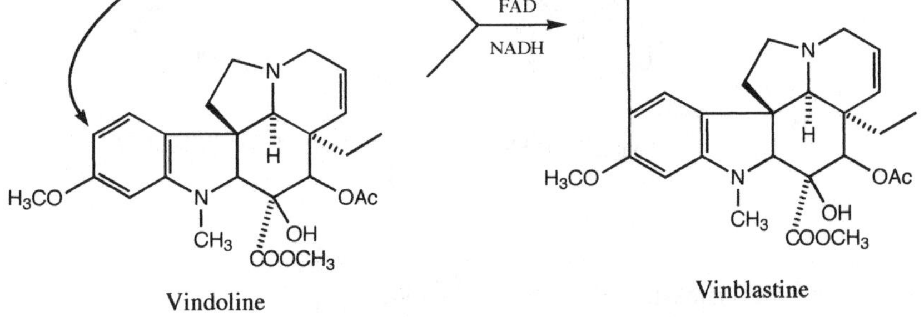

Catharanthine

(14)

FAD

NADH

Vindoline

Vinblastine

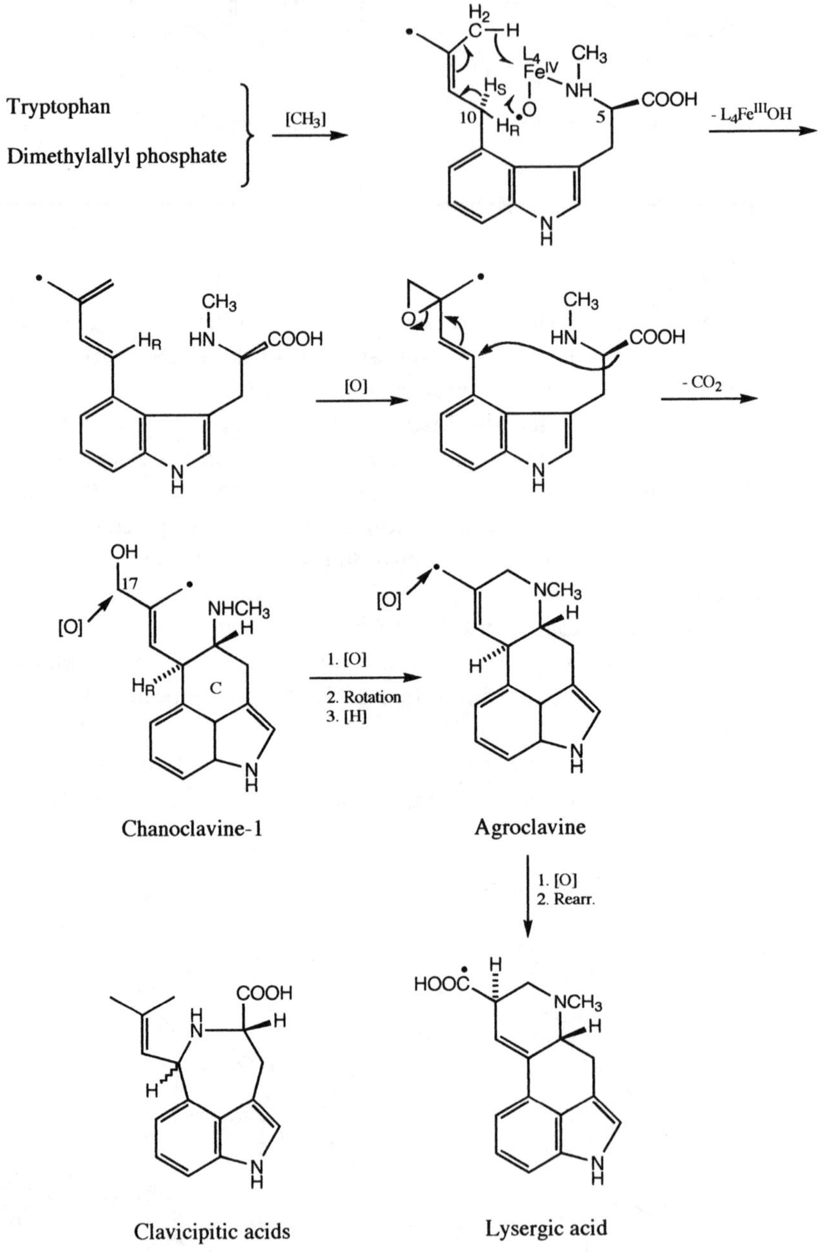

Fig. 32   Biosynthesis of ergot alkaloids

of chemists because of their strange and dramatic action on the human mind, not least after Hofmann's heroic experiments on himself in the 1940s. One of the most potent derivatives, LSD or lysergic acid diethylamide, is misused as a hallucinatory drug, often with unfortunate schizophrenic side effects. Several tons of ergot alkaloids are manufactured per year by fermentation or extraction of field cultivated *Claviceps* spp. Modified alkaloids are used in the treatment of hypertension, migraine and Parkinson's disease.

Biosynthetically, the ergot alkaloids can be dissected into tryptophan and an isoprene unit which has been verified experimentally although the details of ring C and D formation are still not well understood. Tryptophan is alkylated at C^4 by dimethylallyl pyrophosphate with inversion of configuration followed by *N*-methylation to give *N*-methyl-4-dimethylallyltryptophane. One puzzling feature about the ring closure is that two *cis-trans* isomerizations of the double bond occur as demonstrated by isotopic labelling, Fig. 32.[57] The oxygen atom of the hydroxymethyl group originates from molecular oxygen and is introduced prior to the cyclization. The prochiral H_S proton at C^{10} is lost in the dehydrogenation step forming dienyltryptophan as an intermediate which subsequently is epoxidized at the terminal double bond. The initial oxidation of C^{10} is supported by isolation of the clavicipitic acids, presumably formed in a nitrogen rebound reaction, Fig. 32. Decarboxylation, simultaneous cyclization involving retention of configuration at C^5, attack of the anion on the *re*-side of C^{10} and opening of the epoxide give chanoclavine-I. It was found that one of the hydrogens at C^{17} of chanoclavine was lost in the conversion into agroclavine which implies that an aldehyde was formed as an intermediate. The double-bond isomerization can be explained by an addition-elimination mechanism involving the enzyme, and the aldehyde is trapped in the *cis*-configuration as the imine and reduced. Agroclavine gives lysergic acid by oxidation of C^{17} to the carboxyl group and transposition of the double bond in conjugation with the aromatic ring.

8.5 Alkaloids derived from anthranilic acid

The structural unit pertaining to anthranilic acid is recognized in several alkaloids having the quinoline, acridine and quinazoline skeletons. They are frequently represented in the *Rutaceae* family. Their structures are comparatively simple but none the less all three metabolic main streams converge into their biosynthesis (Fig. 33). Coenzyme A activated anthranilic acid, which originates from shikimic acid, first undergoes chain extension with acetyl or malonyl CoA; cyclization then gives the quinoline or acridine nuclei. Isoprenylation of the quinoline skeleton leads to the furanoquinoline alkaloids. Tracer experiments established the isoprenoid origin of the furan ring.[58] Thus, C^5-labelled mevalonic acid gives skimmianine labelled at C^3. 4-Hydroxy-2-quinolone, 3-isoprenyl-4-hydroxy-2-quinolone and platysdesmine were found

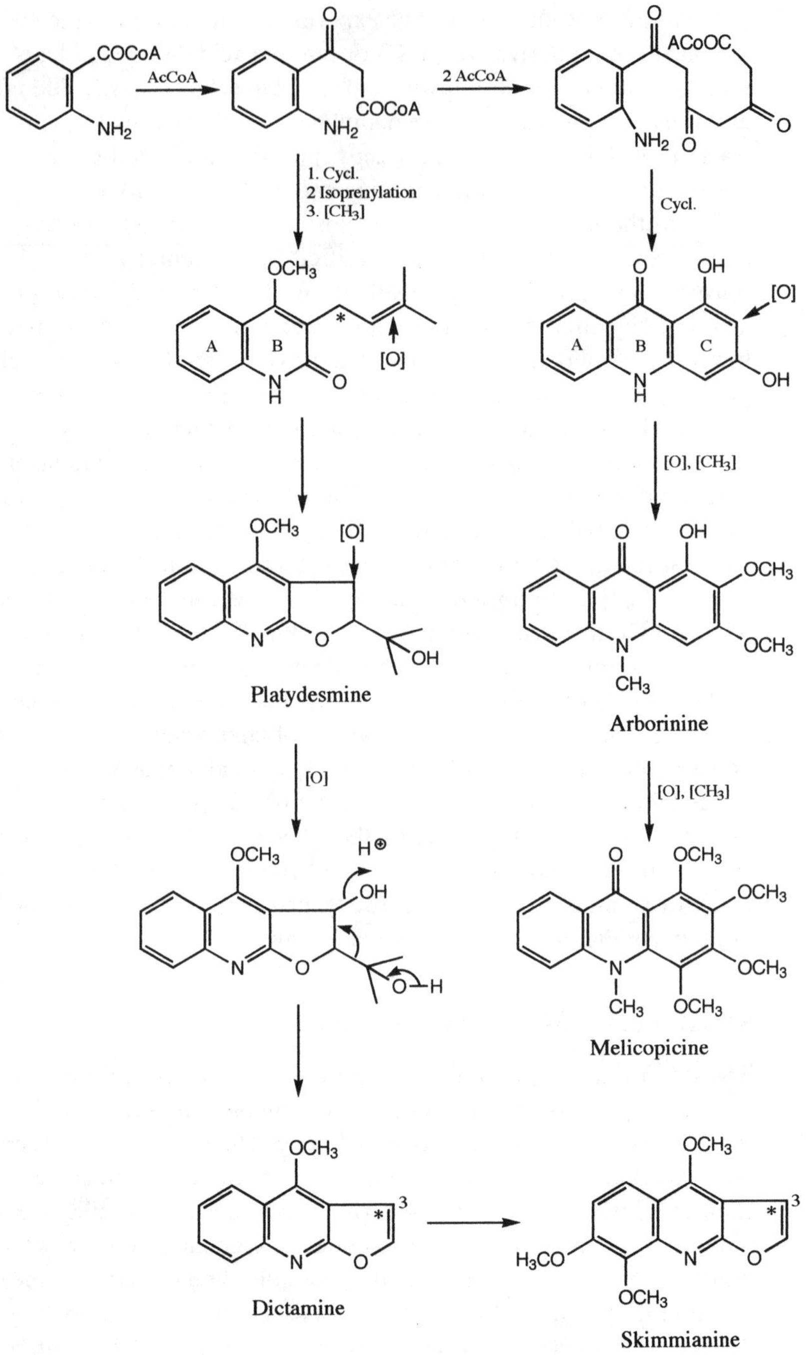

Fig. 33   Biosynthesis of skimmianine alkaloids derived from anthranilic acid

to be efficiently incorporated into dictamine[59,60]. The annelation of the furan ring corresponds closely to its formation in the furanocoumarins (section 4.5). Additional support for acetate as a precursor for the B and C rings came from tracer work on a quinoline alkaloid produced by *Pseudomonas aeruginosa*[61], (Fig. 34). The oxygenation patterns in the C ring of the simple acridine alkaloids are also in harmony with expectations.

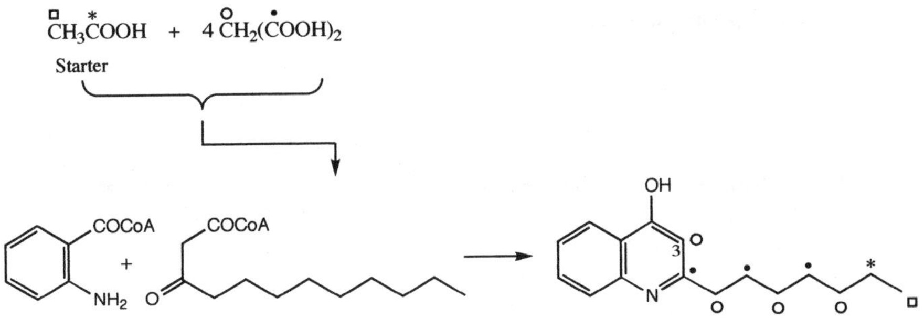

Fig. 34   Biosynthesis of a quinoline alkaloid in *Pseudomonas aeruginosa*. The alternate labelling pattern is consistent with the acetate hypothesis. C^3 is derived from the methylene group of malonate

The quinoline alkaloid, peganine, can be dissected into a pyrrolidine and an anthranilic acid unit. Fig. 35 shows a hypothetical pathway. The involvment of anthranilic acid is valid but the origin of the pyrrolidine is not quite settled yet. *Peganum harmala* of the *Zygophyllaceae* family seems to produce the pyrrolidine ring from ornithine[62] whereas *Adhatoda vasica* of the *Acanthaceae* family uses aspartic acid.[63] The use by plants of different enzymatic steps for the production of the same alkaloid is unique.

The phenazines constitute a small group of bacterial alkaloids which formally can be dissected into two molecules of anthranilic acid. They originate from shikimic acid but it is not settled whether diversion occurs at an intermediate stage or whether the precursor is anthranilic acid. Phenazine-1,6-di-

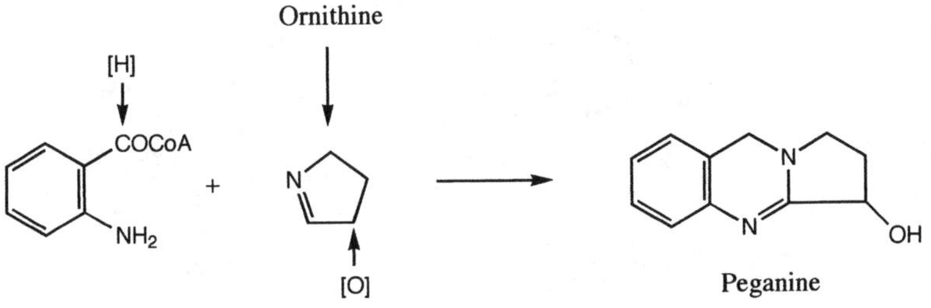

Fig. 35   Plausible pathway for biosynthesis of peganine in *Peganum harmala*

carboxylic acid is accumulated in mutants of *Pseudomonas phenazinium,*[64] and is found to be efficiently metabolized into other phenazines by ether pre-treated *P. aureofaciens* cells to facilitate transport across the cell wall.[65] It is suggested that chorismic acid is aminated by glutamine to give 2-amino-3-hydroxycyclohexan-4,6-dienoic acid, which is further oxidized and dimerized to phenazine-1,6-dicarboxylic acid (Fig. 36). Shikimic acid labelled at C^6 gives $C^{5a,10a}$-labelled iodinin,[66] and strong evidence indicates that anthranilic acid itself is on the pathway.[67]

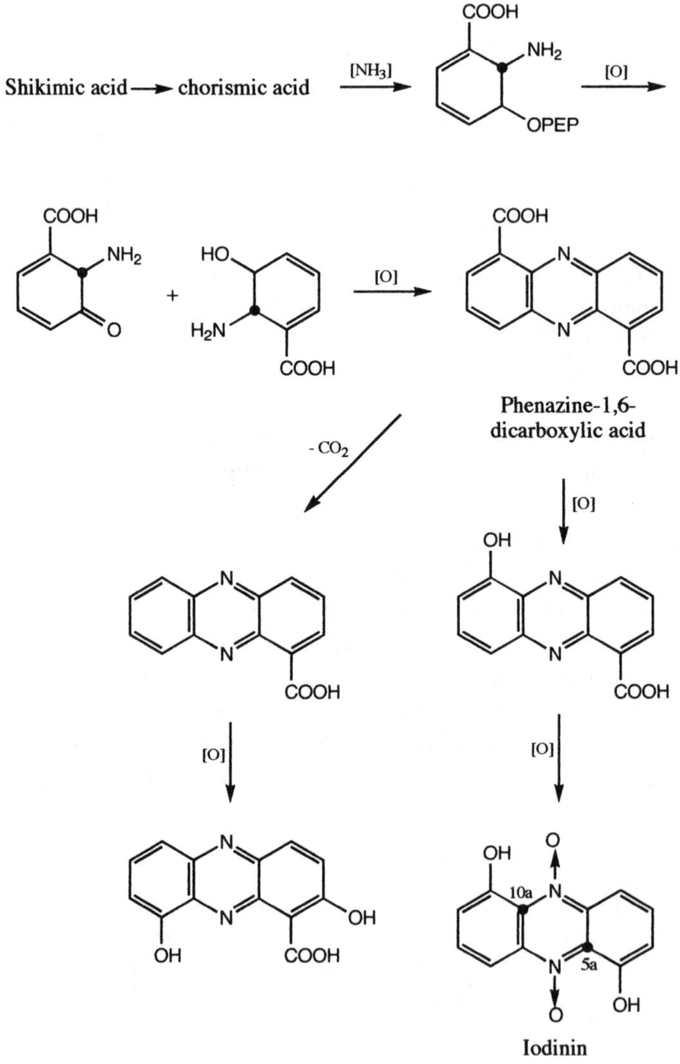

Fig. 36   Plausible route to phenazine alkaloids

8.6 Alkaloids derived by amination of terpenes

The biosynthesis of terpene alkaloids is largely a question of the biosynthesis of the parent compounds. Therefore they could equally well be categorized with them as functionalized derivatives in the same way as amino sugars are classified as carbohydrates. Diversification or structural modifications occur on the terpenoid level, i.e. amination is a late event. In this presentation we have chosen to let the basic character of the compound rule the systemization. The amination can take place according to a redox process (15), a substitution (16) or an addition (17). The substitution is expected to proceed with inversion. Loss of label at the α-position implies that the reaction proceeds via formation of a carbonyl group.

$$(15)$$

$$(16)$$

$$(17)$$

$$(18)$$

A hypothetical reaction is the iron catalysed nitrogen rebound reaction (18) which is analogous to sulfuration in the penicillin series or biological hydroxylation of saturated hydrocarbons.

The monoterpene alkaloids derive from iridoids of varying oxidation levels. Mevalonate, but not loganin or actinidine, is incorporated into the skytanthine-like alkaloids.[68] Amination occurs therefore at a stage prior to formation of loganin, and reduction of the pyridine ring does not take place.

As shown in section 6.3 ring closure occurs at the dial stage to give iridodial which subsequently reacts with ammonia, released from glutamine, either under reducing conditions to skytanthines or under oxidating conditions to pyridine derivatives. Reactions with tyramine lead to the quaternary *Valeriana* alkaloids[69], Fig. 37.

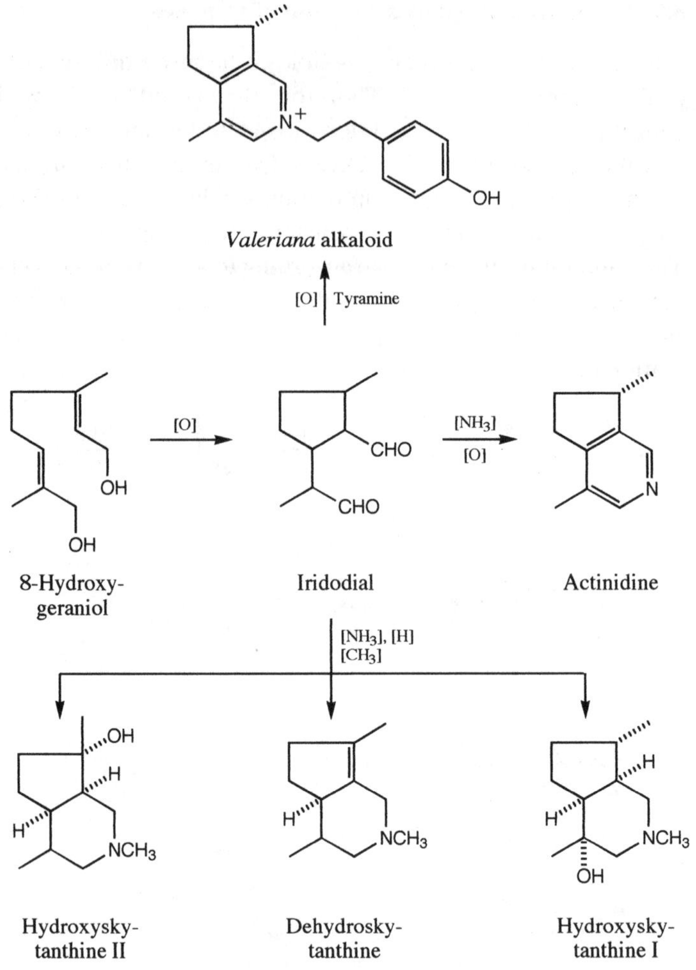

Fig. 37   Possible routes to monoterpene alkaloids

It is popularly believed that the somewhat musty smell of *Valeriana offici-nalis* attracts cats as do *Actinidia polygama* and *Nepeta cataria*. In fact, the plants contain similar monoterpenes and alkaloids. Actinidine has an effect on the EEG of the cat so there are chemical grounds for the old sayings.

Gentianine is a common monoterpene alkaloid. In several plants it beco-mes an artefact on treatment of the extracts with ammonia, a common work-up procedure for the isolation of alkaloids. It is known that the secoiridoid glycosides, gentiopicroside, and swertiamarin (section 6.3, Fig. 9) form genti-anine by treatment with ammonia *in vitro* (19). *Enicostema littorale*, known to contain swertiamarin, gave 0.18 per cent of gentianine in the presence of am-

(19)

Gentiopicroside   Gentianine

monia, whereas when processed without ammonia none was found.[70] However, gentianine has been isolated from several plants without use of ammonia in the work-up procedure.[71]

Sesquiterpene alkaloids are found in *Nymphaeaceae*, e.g. in the romantic water lily, *Nuphar* spp., and in *Orchidaceae*, e.g. in *Dendrobium nobile*. Castoramine[72] was isolated from the scent glands of the Canadian beaver, and most likely it is not endogenous, but accumulated and modified by the beaver feeding on *Nuphar* roots. The C^{15} *Nuphar* alkaloids can visually be segmented into three head-to-tail isoprene units. No biosynthetic work has been carried out as yet to investigate the mechanism of amination and cyclization. The dendrobine skeleton does not obey the first order isoprene rule but is generated via secondary fissions of a germacrene intermediate (Fig. 38). *trans–trans*-Farnesol, but not the *cis–trans*-isomer, is incorporated.[73] The label from 1-$^3H_2$-*trans–trans*-farnesol, fed to *Dendrobium nobile*, was localized to C^5 52 per cent and C^8 47 per cent, and when 1-$^3H(S)$-*trans–trans*-farnesol was administered, 85 per cent of the label was recovered at C^5, i.e. the 1,3H_R shift is a key step and it has a high degree of stereospecificity. A proposal involving two consecutive 1,2 shifts is thus eliminated. In this case one tritium would be located at C^4.

The diterpene alkaloids occur in the genera *Aconitum* and *Delphinium* of the *Ranunculaceae* family and in *Garrya* spp. of the *Cornaceae* family. *Aconitum* spp. are ubiquitously distributed plants cultivated in gardens and also are to be found growing tall in the fertile valleys of Sarek, the last European wilderness north of the arctic circle in Sweden. They contain a group of extremely poisonous alkaloids. Poisoning is diagnosed by a tingling sensation in the whole body followed by numbness. The skeletal elucidation turned out to be difficult with classical methods, and as in several other cases the key breakthrough came with an X-ray study which finally established the structure of lycoctonine[74] (Fig. 39). There is an obvious structural resemblance between garryfoline and the diterpene (—)kaur-16-en-15β-ol. Characteristic for the diterpene alkaloids are the *N*-ethyl or *N*-β-hydroxyethyl bridge across ring A.

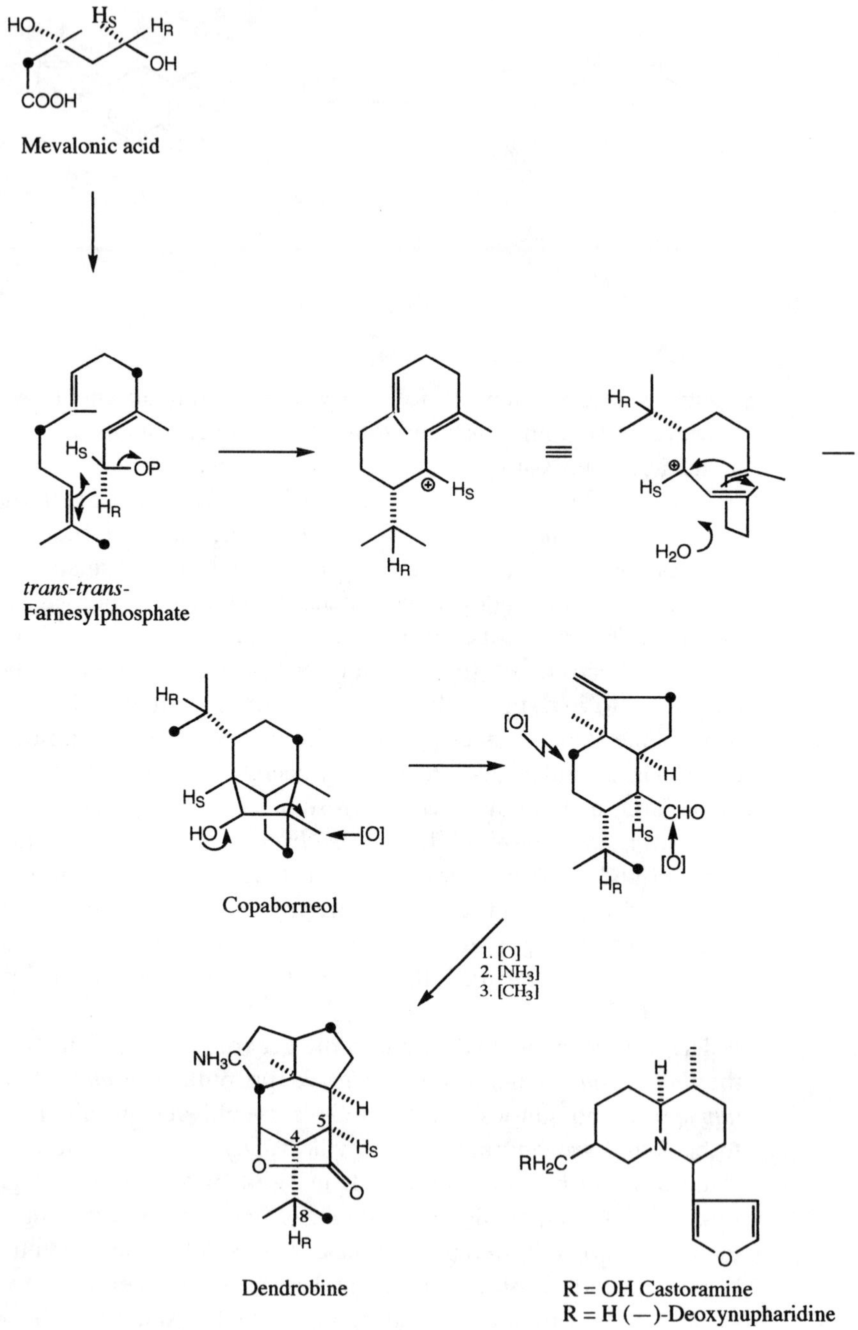
HO
Hs
HR
OH
COOH
Mevalonic acid
Hs
OP
HR
trans-trans-
Farnesylphosphate
Hs
HR
HR
Hs
H2O
HR
Hs
HO
[O]
Copaborneol
[O]
H
CHO
Hs
[O]
HR
1. [O]
2. [NH3]
3. [CH3]
NH3C
H
5
4
Hs
O
O
8
HR
Dendrobine
H
N
RH2C
O
R = OH Castoramine
R = H (—)-Deoxynupharidine

Fig. 38 Structure and biosynthesis of sesquiterpene alkaloids

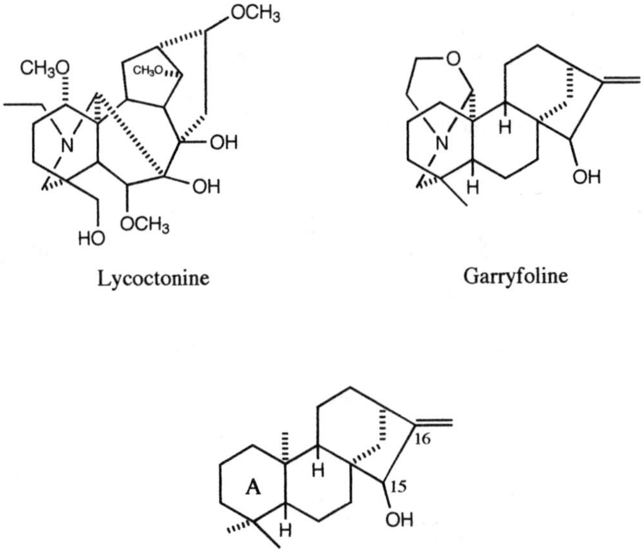

Lycoctonine

Garryfoline

(−) Kaur-16-en-15β-ol

Fig. 39   Structure of diterpene alkaloids

The steroid alkaloids are characteristic metabolites of the *Solanaceae, Liliaceae,* and *Buxaceae* families. *Zygadenus* spp. containing zygadenine (Fig. 40) are poisonous plants causing losses among livestock in North America. The *C*-nor-D-homo-steroid nucleus is formed via C^{12} hydroxylation which initiates the rearrangement (20).

Cholesterol, though incorporated with low efficiency into some plants, is thought to be a precursor for the steroid alkaloids. Solasodine and tomatidine have an intact steroid skeleton. As pointed out earlier, amino functionalization is a late event. Cholesterol is specifically hydroxylated at C^{26} or C^{27}, aminated and cyclized to the piperidine derivative (Fig. 40).[75-77] In tomatidine the C^{26} and in solasodine the C^{27} is derived from C^2 labelled mevalonic acid. The amination proceeds without loss of ^3H label at $C^{26/27}$ consistent with direct amination without intervention of a carbonyl group. On the other hand, formation of the piperidine ring proceeds via an imine. Introduction of a hydroxyl at C^{16} and formation of the tetrahydrofuran ring complete the biosynthesis.[78] The amination of $C^{26,27}$ is a case where mechanism (18) with advantage could be applied instead of substitution (16).

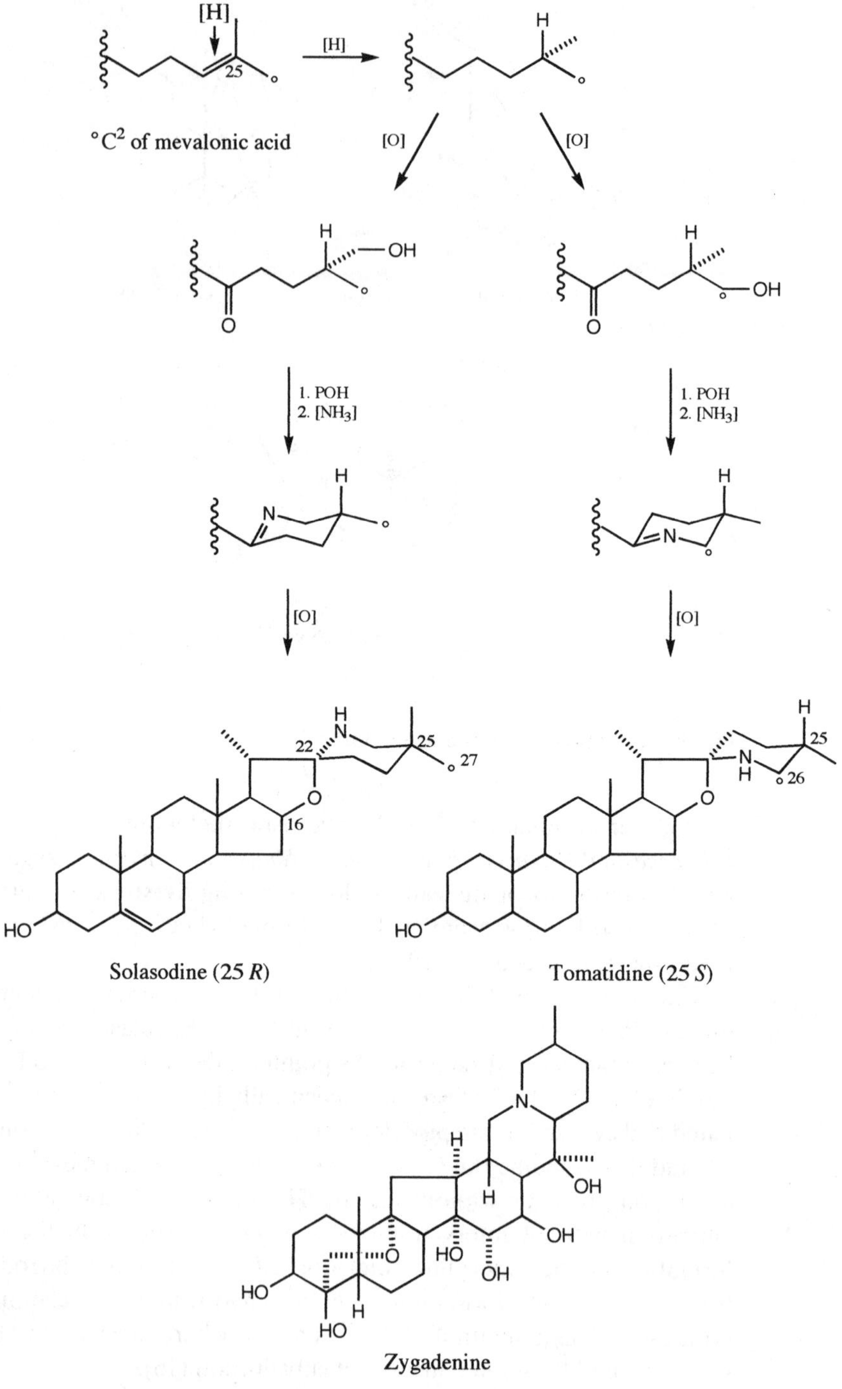

Fig. 40   Steroid alkaloids

(20)

Zygadenine ←[O]← Cevanine skeleton

8.7 Problems

8.1. (+)-Isothebaine is generated from (—)-orientaline. Suggest a mechanism for this transformation that accounts for the unusual oxygenation pattern.

(—)-Orientaline   (+)-Isothebaine

8.2 The formation of the *Amaryllis* alkaloid homolycorine can be explained by a sequence of reactions starting from norbelladine which undergoes phenolic coupling, hydroxylation, ring scission, methylations and finally ring closure. Formulate the intermediate stages.

Norbelladine   Homolycorine

8.3 Gramine is derived from tryptophan by loss of C^1 and C^2 but the α-nitrogen is most probably retained. Tracer experiments showed that the 3-^{14}C:3-^{3}H ratio of tryptophan is unchanged in gramine. Suggest a mechanism that accounts for these facts. (Gross, D., Nemeckova, A. and Schütte, H. R. *Z. Pflanzenphysiol.* **57** (1967) 60.)

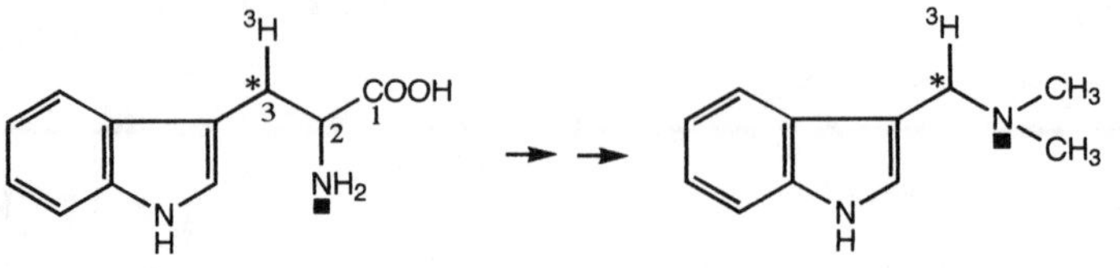

8.4 Suggest a plausible biosynthesis of ajmaline from strictosidine.

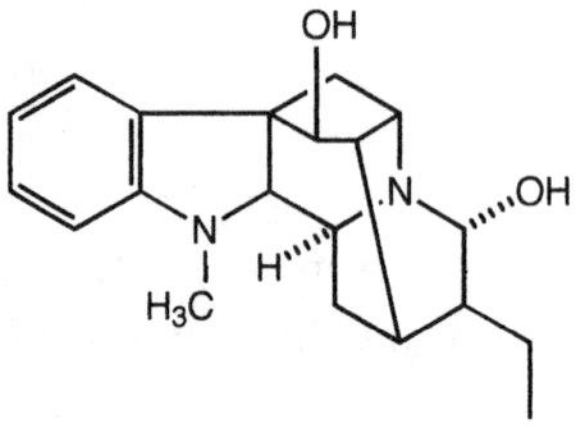

Ajmaline

8.5 Suppose that the biosynthesis of sedamine is not yet known and you wish to carry out such an investigation. Discuss a working hypothesis, feeding procedures, suitable precursors, spectroscopy, isolation procedures, etc. How would you solve the problem about the determination of the stereostructure of the compound. Do a literature search relevant to the problems and compare your strategies with those of the "professionals".

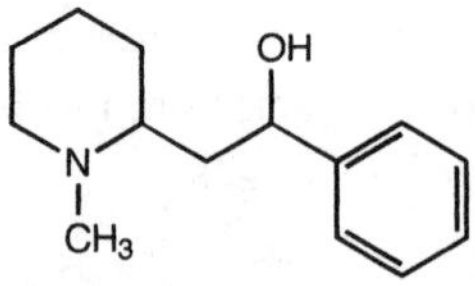

Sedamine
Sedum acre

8.6 Suggest a biosynthetic pathway for quinine from methoxygeissoschizine. The label from 2-^{14}C-tryptophan is located at the α-carbon of the quinoline ring in quinine.

Methoxygeissoschizine   Quinine

8.7 When 1-^{3}H$_2$-geraniol, 2-^{14}C-geraniol, and 5-^{3}H-loganin are fed to *Catharanthus roseus*, radioactive tabersonine (Fig. 30) and catharanthine (Fig. 26) can be isolated. Where are the alkaloids labelled?

8.8 Tylophorinine has its genesis in tyrosine, phenylalanine and ornithine. Phenacylpyrrolidine and its mono- and dihydroxylated derivatives are incorporated intact. Phenol oxidation occurs at one stage of the biosynthesis. Discuss a plausible pathway. (Herbert, R. B., Jackson, F. B. and Nicolson, I.T. *J. Chem. Soc. Chem. Commun.* **1976**, 450.)

Tylophorinine

Bibliography

1. Siegler, D. S. in *The Alkaloids*, Vol. XVI. p.1, Manske, R. H. F. (Ed.) Academic Press, New York, 1977.
2. Ahjmad, A and Leete, E. *Phytochemistry* **9** (1970) 2345; Hashimoto, T., Yamada, Y. and Leete, E. *J. Am. Chem. Soc.* **111** (1989) 1141.
3. Liebisch, H. W., Radwan, A. S. and Schütte, H. R. *Liebigs Ann.* **721** (1969) 163.
4. Leete, E., Bjorklund, J. A., Couladis, M. M. and Kim, S. H. *J. Am. Chem. Soc.* **113** (1991) 9286.
5. Hashimoto, T., Kohno, J. and Yamada, Y. *Plant Physiol.* **84** (1987) 144: *Phytochemistry* **28** (1989) 1077.
6. Leete, E. and Yu, M. L. *Phytochemistry* **19** (1980) 1093.

7. Herbert, R. B. *Nat. Prod. Chem.* **9** (1992) 507.
8. Sankawa, U., Noguchi, H., Hashimoto, T. and Yamada, Y. *Chem. Pharm. Bull.* **38** (1990) 2066.
9. Bottomley, W. and Geissman, T.A. *Phytochemistry* **3** (1964) 357.
10. Robins, D. J. *Chem. Soc. Rev.* **18** (1989) 375; Freer, I. K., Matheson, J. R., Rodgers, M., and Robins, D. J. *J. Chem. Res. (S)* **1991** 41; Denholm, A. A., Kelly. H. A. and Robins, D. J. *J. Chem. Soc, P. I.* **1991** 2003.
11. Gupta, R. N. and Spenser, I. D. *Phytochemistry* **9** (1970) 2329.
12. Keogh, M. F., and O'Donovan, D. G. *J. Chem. Soc. (C)* **1970**, 1792, 2470.
13. Leistner, E. and Spenser, I. D. *J. Am. Chem, Soc.* **95** (1973) 4715.
14. Hemscheidt, T. And Spenser, I. D. *J. Am. Chem. Soc.* **112** (1990) 6360.
15. Leete, E. and Olsen, J. O. *J. Am. Chem. Soc.* **94** (1972) 5472.
16. Roberts, M. F. *Phytochemistry* **14** (1975) 2393.
17. Leete, E., Leichleiter, J. C. and Carver, R. A. *Tetrahedron Lett.* **1975,** 3779.
18. Leete, E. and Liu, Y.–Y. *Phytochemistry* **12** (1973) 593.
19. Solt, M. L., Dawson, R. F. and Christman, D. R. *Plant. Physiol.* **35** (1960) 887.
20. Leete, E. and Slattery, S. A. *J. Am. Chem. Soc.* **98** (1976) 6326.
21. Kennedy, I. A., Hill, R. E., Paulosky, R. M., Sayer, B. G. and Spenser, I. D. *J. Am. Chem. Soc.* **117** (1995) 1661.
22. Rana, J. and Robins, D .J. *J. Chem. Soc. P. I* **1986** 1133; Golebiewski, W. M. and Spenser, I. D. *Can. J. Chem.* **63** (1985) 2707; *J. Am. Chem. Soc.* **106** (1984) 1441; Robins, D. J. and Sheldrake, G. N. *J. Chem. Soc. Chem. Commun.* **1994** 1331; *J. Chem. Res. (S)* **1987** 256.
23a. Yamamura, S. and Hirata, Y. in *The Alkaloids*, Manske, R. H. F. (Ed.), Academic Press, **15** (1975) 41. b. Snieckus, V. in *The Alkaloids*, Manske, R. H. F. (Ed.), **14** (1973) 425; Sankawa, U., Ebizuka, Y. and Yamasaki, Y. *Phytochemistry* **16** (1977) 561. c. MacLean, D. B. in *The Alkaloids*, Manske, R. H. F. (Ed.), **14** (1973) 347.
24. O'Donovan, D. G. and Barry, E. *J. Chem. Soc. Perkin I* **1974** 2528.
25. Schütte, H. R. and Seelig, G. *Liebigs Ann.* **730** (1970) 186.
26. Kapadia, G. J., Rao, G. S. and Leete, E. *J. Am. Chem. Soc.* **92** (1970) 6943.
27. Bhakuni, D. S., Singh, A. N., Tewari, S. and Kapil, R. S. *J. Chem. Soc. Perkin I* **1977** 1662.
28. Wilson, M. L. and Coscia, C. J. *J. Am. Chem. Soc.* **97** (1975) 431.
29. Zenk, M. H., Reuffer, M. Kutchan, T. M. and Galneder, E. in *Applications of Plant Cell and Tissue Culture*, J. Wiley, Chichester, 1988, p. 213; Stadler, R., Kutchan, T. M. and Zenk, M. H. *Phytochemistry* **28** (1989) 1083; Stadler, R. and Zenk, M. H. *Liebigs Ann.* **1990** 555.
30. Battersby, A. R., Sheldrake, P. W., Staunton, J. and Summers, M. C. *Bioorg. Chem.* **6** (1977) 43.
31. Battersby. A. R. and Parry, R. J. *J. Chem. Soc. Chem. Commun.* **1971** 901.
32. Tanahashi, T. and Zenk, M. H. *Phytochemistry* **29** (1990) 1113; Kobayashi, M., Frenzel, T., Lee. J. P., Zenk, M. H. and Floss, H. G. *J. Am. Chem. Soc.* **109** (1987) 6185; Frenzel, T., Beale, J. M., Kobayashi, M., Zenk, M. H. and Floss., H. G. *J. Am. Chem. Soc.* **110** (1988) 7878; Kutchan, T. M., Ditrich, H., Bracher, D. and Zenk, M. H. *Liebigs Ann.* **1993** 557.
32a. Mueller, M. J. and Zenk, M. H. *Liebigs Ann.* **1993** 557.
33. Battersby, A. R., Foulkes, D. M. and Binks, R. *J. Chem. Soc* **1965** 3323; Lotter, H., Gollwitzer, J. and Zenk, M. H. *Tetrahedron Lett.* **1992** 2443; Lenz, R. and Zenk, M. H. *Tetrahedron Lett.* **1994** 3897.
34. Battersby, A. R., McHugh, J. L., Staunton, J. and Todd, M. *J. Chem. Soc. Chem. Commun.* **1971,** 985.

35. Barton, D. H. R., Bhakuni, D. S., Chapman, G. M., Kirby, G. W., Haynes, L. J. and Stuart, K. L. *J. Chem. Soc. (C)* **1967,** 1295.
36. Bhakuni, D.S., Tewari, S. and Kapil, R.S. *J. Chem. Soc. Perkin I* **1977** 706.
37. Bhakuni, D. S. and Singh, A. N. *J. Chem. Soc. Perkin I* **1978,** 618.
38. Bowman, W. R., Bruce, I. T. and Kirby, G. W. *Chem. Commun.* **1969,** 1075.
39. Fuganti, C. in *The Alkaloids*, Vol. XV, p. 83, Manske, R. H. F. (Ed.), Academic Press, New York, 1975.
40. Agurell, S., Granelli, I., Leander, K. and Rosenblom, J. *Acta Chem. Scand.* **B28** (1974) 1175.
41. Jeffs, P. W., Karle, J. M. and Martin, N. H. *Phytochemistry* **17** (1978) 719.
42. Kametani, T. and Koizumi, M. in *The Alkaloids*, Vol. XIV, p. 265, Manske, R. H. F. (Ed.), Academic Press, New York, 1973.
43. Battersby, A. R., Herbert, R. B., McDonald, E., Ramage, R. and Clements, J. H. *J. Chem. Soc. P. I.* **1972** 1741; Herbert, R. B., Kattah, A. E. and Knagg, E. *Tetrahedron* **46** (1990) 7119.
44. Battersby, A. R., Herbert, R. B., Pijevska, L., Santavy, F. Sedmera, P. *J. Chem. Soc. P. I.* **1972** 1736.
45. Thomas, R. *Tetrahedron Lett.* **1961**, 544.
46. Wenkert, E. *J. Am. Chem. Soc.* **84** (1962) 98.
47. Battersby, A. R. in *Pure and Applied Chemistry* **14** (1967) 117, Butterworths, London, and in *The Alkaloids* **1** (1971) 31, Specialist Periodical Reports, The Chemical Society, Saxton, J. E. (Ed.). Scott, A. I. In *MTP Int. Rev. Sci.* **9** (1973) 105, Hey, D. H. and Wiesner, K. F. (Eds.), Butterworths, London.
48. Battersby, R. A., Gregory, B., Spenser, H., Turner, J. C., Janot, M. M., Potier, P., Francois, P. and Levisalles, J. *Chem. Commun.* **1967,** 219.
49. Battersby, A. R., Burnett, A. R. and Parsons, P. G. *J Chem Soc. (C)* **1969** 1187.
50. Smith, G. N. *Chem. Commun.* **1968,** 912.
51. Stöckigt, J. and Zenk, M. H. *J. Chem. Soc. Chem. Commun.* **1977,** 646.
52. Rueffer, M., Nagakara, N. and Zenk, M. H. *Tetrahedron Lett.* **1978,** 1593.
53. Woodward, R. B., Bader, F. E., Bickel, H., Frey, A. J. and Kierstad, R. W. *Tetrahedron* **2** (1958) 1.
54. Qureshi, A. A. and Scott, A. I. *J. Chem. Commun.* **1968** 948; Heimberger, S. I. and Scott, A. I. *J. Chem. Soc. Chem. Commun* **1973** 217; Furuya, T., Sakamoto, K., Iida, K., Asada, Y., Yoshikawa T., Sakai, S. and Aimi, N. *Phytochemistry*, **31** (1992) 3065.
55. Brown, R. T., Smith, G. F., Stapleford, K. S. J. and Taylor, D. A. *J. Chem. Soc. Chem. Commun.* **1970** 190.
56. Kutney, J. P. *Nat. Prod. Rep.* **7** (1990) 85.
57. Shibuya, M., Chou, H.-M., Fountoulakis, M., Hassam, S., Kim, S.-U., Kobayashi, K., Otsuka, H., Rogalska, E., Cassady, J. M. and Floss, H. G. *J. Am. Chem. Soc.* **112** (1990) 279; Kozikowski, A. P., Chen, C., Wu, J.-P., Shibuya, M., Kim, C.-G. and Floss, H. G. *J. Am. Chem. Soc.* **115** (1993) 2482.
58. Colonna, A.O. and Gros, E.G. *Chem. Commun.* **1970,** 674.
59. Grundon, M. F., Harrison, D. M. and Spyropoulos, C. G. *J. Chem. Soc. Chem. Commun.* **1974,** 51.
60. Collings, J. F., Donelly, W. J., Grundon, M. F. and James, K. J. *J. Chem. Soc. P. I.* **1974** 2177; Neville, C. F., Grundon, M. F., Ramachandran, V. R. Reisch, G. and Reisch, J. *J. Chem. Soc. P. I.* **1991** 2261.
61. Ritter, C. and Luckner, M. *European J. Biochem.* **18** (1971) 391.
62. Liljegren, D. R. *Phytochemistry* **10** (1971) 2661.
63. Johne, S., Waiblinger, K. and Gröger, D. *Pharmazie* **28** (1973) 403.

64. Byng, G. S. and Turner, J. M. *J. Gen. Microbiol.* **97** (1976) 57.
65. Buckland, P. R., Herbert, R. B. and Holliman, F. G. *Tetrahedron Lett.* **1981,** 595.
66. Hollstein, U., Mock, D. L., Sibbit, R. R., Roisch, U. and Lingens, F. *Tetrahedron Lett.* **1978,** 2987.
67. Essar, D. W., Eberly, L. Hadero, A. and Crawford, I. P. *J. Bacteriol.* **172** (1990) 884.
68. Gross, D., Berg, W. and Schütte, H. R. *Biochem. Physiol. Pflanz.* **163** (1972) 576.
69. Torssell, K. and Wahlberg, K. *Acta Chem. Scand.* **21** (1967) 53.
70. Govindachari, T. R., Sathe, S. S. and Viswanathan, N. *Indian J. Chem.* **4** (1966) 201.
71. Cordell, G. A. in *The Alkaloids* Vol. XVI, p. 431, Manske, R. H. F. (Ed.) Academic Press, New York, 1977.
72. Valenta, S. and Khaleque, A. *Tetrahedron Lett.* **12** (1959) 1.
73. Corbella, A., Gariboldi, P., Jommi, G. and Sisti, M. *J. Chem. Soc. Chem. Commun.* **1975,** 288.
74. Przybylska, M. and Marion , L. *Can. J. Chem.* **34** (1956) 185; Pelletier, S. W., Mody, N. V., Varughese, K. I., Maddry, J. A. and Desai, H. K. *J. Am. Chem. Soc.* **103** (1981) 6536.
75. Tschesche, R. and Hulpke, H. *Z. Naturforsch.* **21b** (1966) 893.
76. Guseva, A. R. and Paseshnichenko, V. A. *Biochemistry* (USSR) **27** (1962) 721.
77. Ronchetti, R. and Russo, G. *J. Chem. Soc. Chem. Commun.* **1974,** 785.
78. Kaneko, K., Seto, H., Motoki, C. and Mitsuhashi, H. *Phytochemistry* **14,** (1975) 1295.

Chapter 9

N-Heteroaromatics

9.1 Introduction

An account of secondary metabolism cannot be considered complete without a discussion of heteroaromatics such as pyrimidines, purines and porphyrins. Even though they are considered first order constituents participating in the fundamental chemistry of life and replication, this class of compounds contains several secondary metabolites. Furthermore, their mode of formation is of general chemical interest. Several *N*-heteroaromatics are vitamins and co-enzymes catalysing the most intriguing reactions—vitamin B_{12} in particular. We have already met representatives of the pteridine nucleus in the monoxygenase cofactors riboflavin, vitamin B_2, and the thiazole and pyrimidine nuclei in the cofactor thiamine, vitamin B_1. These and other nuclei will be discussed in this chapter. The redox reaction of the riboflavins is treated in section 4.3 in connection with biological hydroxylation.

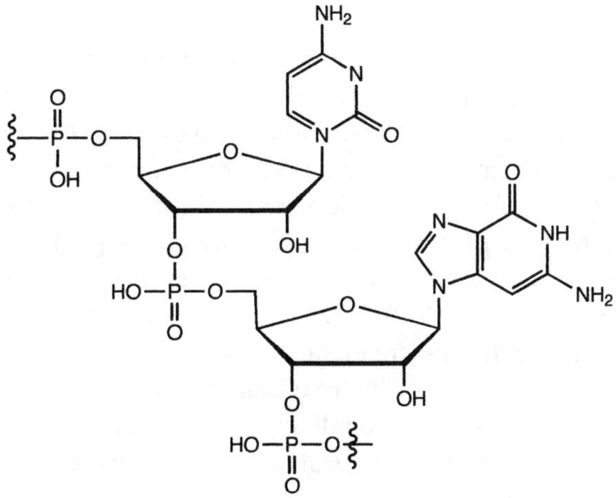

Fig. 1   Section of a ribonucleic acid (RNA)

9.2 Pyrimidines, purines and pteridines

The major bases in RNA (Fig. 1) are adenine, guanine, cytosine and uracil. In DNA, deoxyribonucleic acid, uracil is replaced by thymine (5-methyluracil). Like their ribosides the free bases are present in trace amounts in the cell as catabolic products from the nucleic acids or as intermediates in the biosynthesis of nucleic acids. In some fungi and sponges, nucleosides are produced *per se* in larger quantities having antibiotic properties, e.g. nebularine from *Clitocybe nebularis*[1] (Fig. 2). The N^6-prenylpurines and their ribosides, the so-called cytokinins, are growth hormones strongly promoting cell division, enlargement, and differentiation in plants.[2] These prenylated purine derivatives are also found as minor constituents in transfer RNA.[3] A hydrogenated purine skeleton is contained in the fatally poisonous saxitoxin which is produced by the marine dinoflagellate *Gonyaulax catanella* and also by the blue-green alga *Aphanizomenon flos-aque*, and occasionally accumulated in clams and mussels feeding on the dinoflagellates during blooms (red tide). When the toxin reaches man in the feeding chain, it acts by blocking the sodium ion channel and causes what is known as paralytic shellfish poisoning. The structure of saxitoxin is unusual and posed serious problems to the organic chemist. It was eventually solved by X-ray crystallography.[4] The purine derivatives caffeine and theobromine (3,7-dimethylxanthine) are widely used as stimulants. Caffeine is found in *Coffea* spp. (*Rubiaceae*), tea, *Thea sinensis* (*Theaceae*), in yerba maté from the evergreen shrub *Ilex paraguariensis* (*Aquifoliaceae*) and in the bark of sapindaceous jungle lianas in the Amazon, *Paullinia* spp. The fatty cacao seeds from *Theobroma cacao* (*Sterculiaceae*) contain *ca.* 1 per cent theobromine. The ancient Aztecs called the tree chocoatl, one of the few etymological legacies passed on to our modern culture from this once so powerful people. The British are the largest consumers of tea per head. Coffee became a popular beverage rather late and was once treated with suspicion. In Sweden during the eighteenth century it was forbidden by law to drink coffee; now the Swedes are among the largest coffee consumers in the world, whatever conclusion one can draw from that. Apart from functioning as a building block in nucleic acids, pyrimidines are rare molecules in nature.

The bicyclic pteridine (from Gr. *pteron*, wing) skeleton formed by a pyrimidine ring fused to a pyrazine ring is contained in the biologically important coenzymes tetrahydrofolic acid, biopterine and riboflavin. The pleasing, colourful and fluorescent pigments found in the wings of butterflies and in the skin of amphibians are hydroxy- and amino-substituted pteridines which on oxidation can form an extended quinonic system. These pigments play an important role in visual chemical communication in the terrestrial and marine environments.

Visual inspection of the structure is a misleading guide for chemical intuition concerning the biosynthesis of purines. Contrary to expectations the

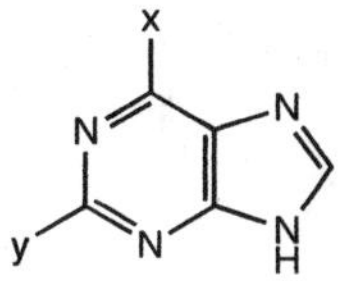

x = y = H   Purine
x = NH$_2$, y = H   Adenine
x = OH, y = H   Inosine
           (Hypoxanthine)
x = OH, y = OH   Xanthine
x = OH, y = NH$_2$ Guanine

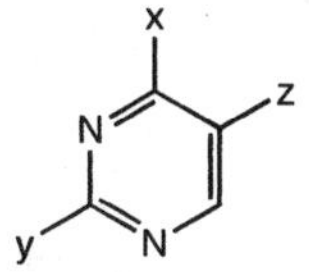

x = y = OH, z = H
           Uracil
x = NH$_2$, y = OH, z = H
           Cytosine
x = y = OH, z = CH$_3$
           Thymine

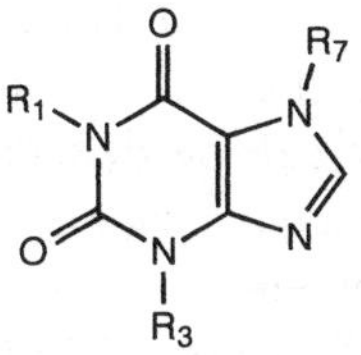

R$_1$ = R$_3$ = H, R$_7$ = CH$_3$ 7-CH$_3$-Xanthine
R$_1$ = R$_3$ = CH$_3$, R$_7$ = H   Theophylline
R$_3$ = R$_7$ = CH$_3$, R$_1$ = H   Theobromine
R$_1$ = R$_7$ = CH$_3$, R$_3$ = H   Paraxanthine
R$_1$ = R$_3$ = R$_7$ = CH$_3$ Caffeine

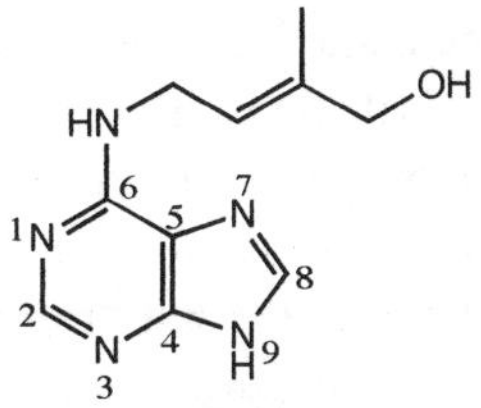

Zeatin
(N^6-(*trans*-4-hydroxy-3-methyl-
but-2-enyl)adenine)

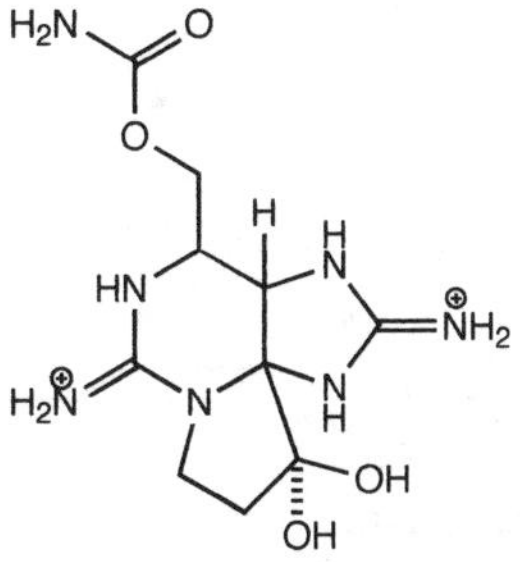

Saxitoxin

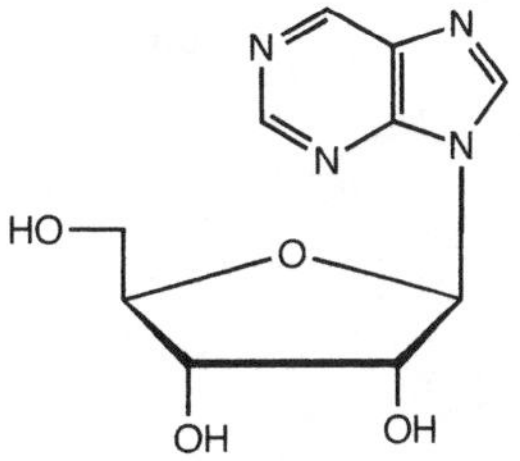

Nebularine
(N^9-ß-Ribosylpurine)

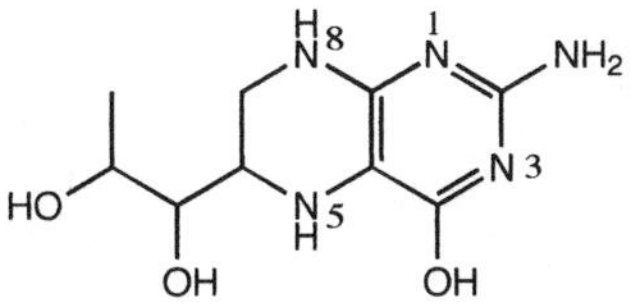

5,6,7,8-Tetrahydrobiopterin

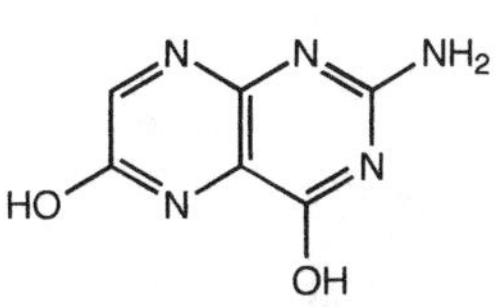

Xanthopterin

Fig. 2   Structure of purine and pteridine derivatives

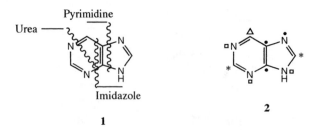

Fig. 3   **1**, Imaginable precursors of the purine ring; **2**, precursors of the purine ring
• Glycine, * Formate,   □ [NH₃], △ CO₂

nucleoside is not formed by ribosidation of a preformed purine derivative. Instead ribose is attached already at the start of the assembly of the purine ring
system. The condensed heteroaromatic can be dissected into a pyrimidine
ring, an imidazole ring and urea but no such derivatives are incorporated (Fig.
3, **1**). Feeding experiments with pigeons and analysis of uric acid (2,6,8-tri-
hydroxypurine) excreted, indicated that carbon dioxide, ammonia, formate
and intact glycine are built into the molecule. The biosynthetic scheme which
emerged, is as follows (Fig. 4).[5] C^1 of ribose-5-phosphate is activated by phosphorylation with ATP. The pyrophosphate group is displaced with "active"

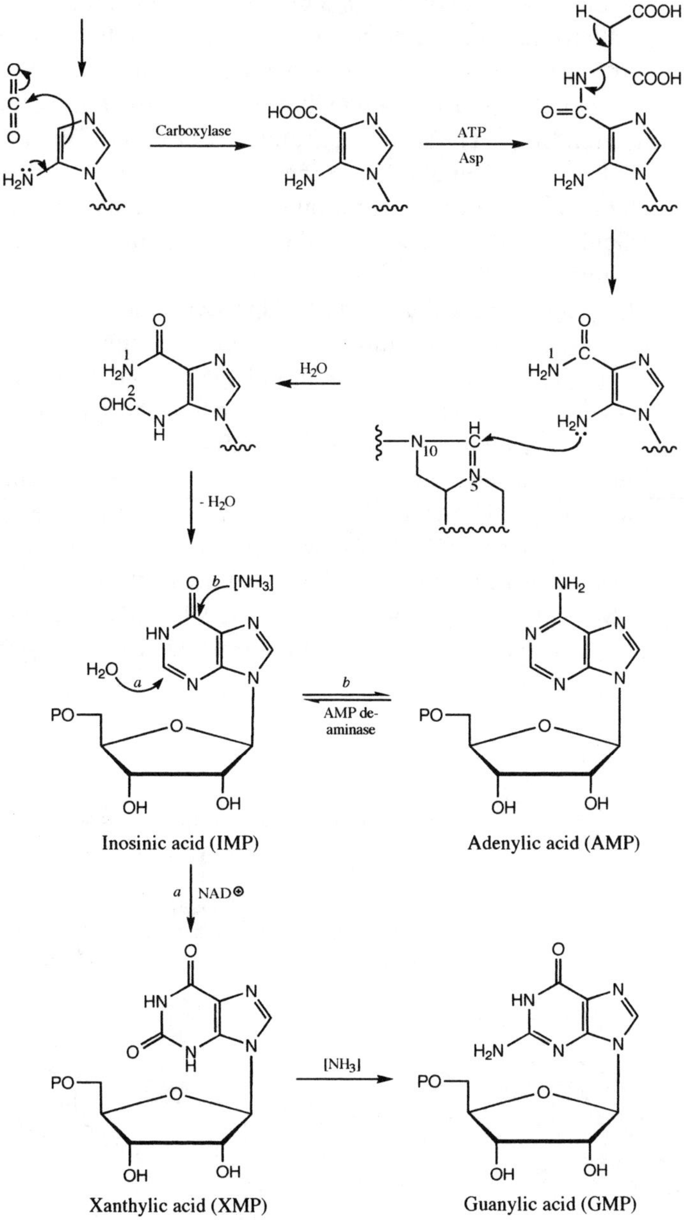

Fig. 4   Biosynthesis of purine nucleotides. The 5-phosphoribosyl moiety is omitted in some of the intermediates

ammonia, generated from glutamine, with inversion of configuarion at C^1 giving the β-N^9 atom of the final purine nucleotide. Glycine is activated by ATP and introduced intact to form an amide. In the following steps the atoms are introduced one by one. The free amino group of the glycine amide is formylated with N^5,N^{10}-formamidinyltetrahydrofolate, thus accomplishing the introduction of C^8. The next step before cyclization to the imidazole involves an ATP assisted amidination of the glycine amide by ammonia, generated from glutamine. The purine C^6 atom is introduced as carbon dioxide in an electrophilic substitution at C^4 of the imidazole nucleus. Nature uses yet another technique to introduce the N^1 atom. The carboxyl group of the imidazole is phosphorylated with ATP and reacted with aspartic acid. β-Elimination gives the carboxamide and fumaric acid. C^2 is finally introduced as formate in a second folate-assisted reaction and cyclization gives the nucleotide inosinic acid (IMP) which is a key intermediate for other purine nucleotides. The conversion of IMP to adenylic acid (AMP) is accomplished by amination of C^6 via the aspartic acid route (above) by assistance of Mg^{2+} and GTP. The biosynthesis of guanylic acid (GMP) is accomplished by oxidation of C^2 with $NAD^\oplus$ to xanthylic acid (XMP) followed by amination of C^2. Hydrolysis of the C-5 phosphate group gives the corresponding nucleoside.

By studying the incorporation of ^{14}C-labelled purine derivatives and the appearance and disappearance of intermediates it was possible to map the path-

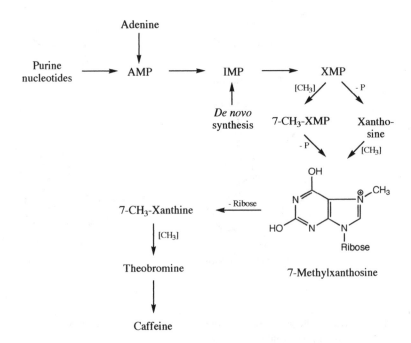

Fig. 5   Biosynthesis of purine alkaloids

way to the purine alkaloid caffeine.[6] The formation of the fused ring system of caffeine follows the classical nucleotide biosynthesis. It is postulated that caffeine is derived primarily from AMP in the nucleotide pool and by nucleic acid breakdown rather than by *de novo* purine biosynthesis. It is not settled whether the first methylation by adenosyl methionine occurs at the XMP or at the xanthosine stage, Fig. 5. Adenine is an excellent precursor for caffeine whereas hypoxanthine, 7-methylinosine, 1-methylxanthine, xanthine and guanine are poorly incorporated and consequently not on the pathway.

The pathway leading to pyrimidine nucleotides is much simpler than the pathway to purine nucleotides. A major difference between the two is that the pyrimidine nucleus is assembled prior to the attachment of ribose. Aspartic acid is carbamoylated with carbamoyl phosphate and cyclized to dihydroorotic acid (Fig. 6). The dehydrogenation to orotic acid is carried out by the flavin redox system, coupled with $NAD^{\oplus}$.[7] Subsequently ribosidation of N^1 by 5-phosphoribosyl-1-pyrophosphate (PRPP, Fig. 6) leads to the orotidylic acid (OMP) and decarboxylation gives uridylic acid (UMP).

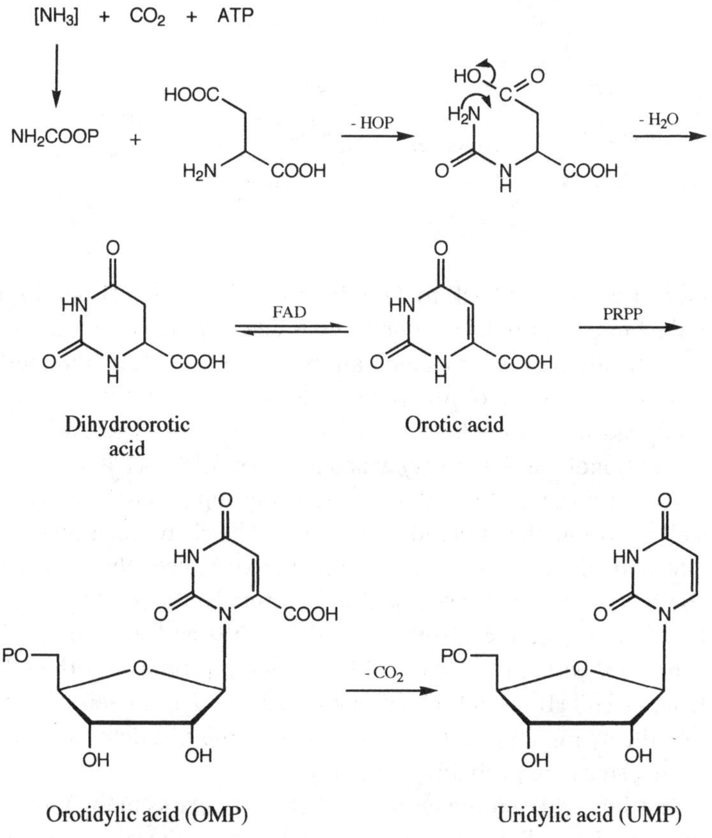

Fig. 6   Biosynthesis of pyrimidine nucleotides

The biosynthesis of thymine (5-methyluracil) is of mechanistic interest. It represents one of a few instances where methylation is carried out by 5,6,7,8-tetrahydrofolate (THF) rather than by methionine.[8] Uridylic acid (UMP) is the substrate but for the sake of clarity only the pyrimidine moiety is shown in Fig. 7. Formaldehyde originating from serine, is trapped by THF and forms the $N^{5,10}$-methylene bridge. THF represents the reduced partner of a quinonoid redox couple, the oxidized form of which can form two tautomers of dihydrofolic acid (DHF) (1). The 7,8-dihydrofolic acid is the most stable tauto-

5,6,7,8-Tetrahydrofolate
(THF, partial structure)

7,8-Dihydrofolate (DHF)

(1)

mer. It is reduced back to THF by NADH. The C^6–H bond is consequently rather weak. Uridylic acid undergoes electrophilic substitution at C^5 and rearomatization occurs by elimination of C^5–H, whereby the methylene bridge now moves over into the uridyl nucleus. The N^5-nitrogen serves as a leaving group assisted by the lone pair of uridine-N^3. Cleavage leads to an intermediate quinonoid uridyl moiety which picks up the C^6–H as a hydride (route a) or reacts with water to the hydroxymethyl group (route b). Route a can equally well be formulated as a radical reaction. The cleavage occurs homolytically or alternatively the quinonoid uridyl moiety oxidizes the THF ion which gives rise to two semiquinonoid radicals and finally, the uridyl radical abstracts the C^6–H atom (route c). Both the hydride shift and the radical abstraction are consistent with the finding that thymidine is produced with no loss of activity from 6-^{3}H-THF. A ^{3}H/^{1}H isotope effect of 5.2 has been observed indicating that the hydrogen transfer is taking part in the rate determining step. The preceding steps are probably equilibria.[9]

It is inferred that the enzyme which participates in the methylation of cytidine acts by adding across the 5,6 double bond thereby initiating the reaction with $N^{5,10}$-methylene-THF (2)[10] as shown by experiments with 5-fluorodeoxy-

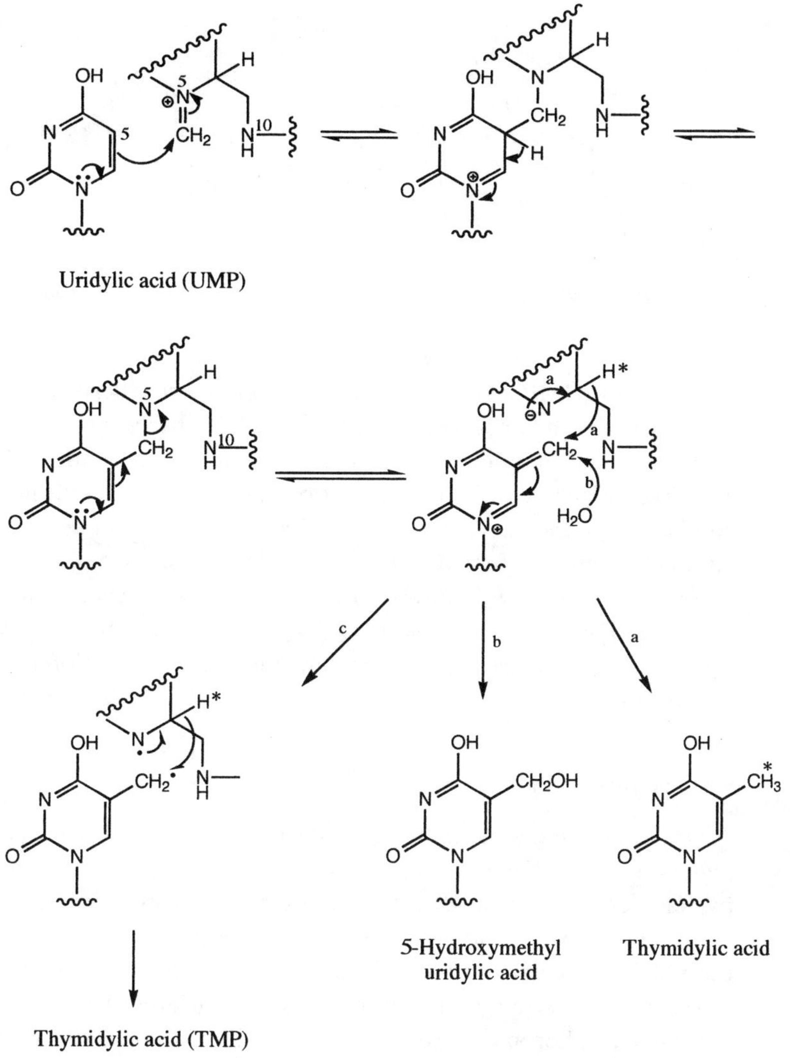

Fig. 7 Mechanism for methylation and hydroxymethylation of uridylic acid (UMP). Partial structure; 5-phosphoribose moiety is omitted

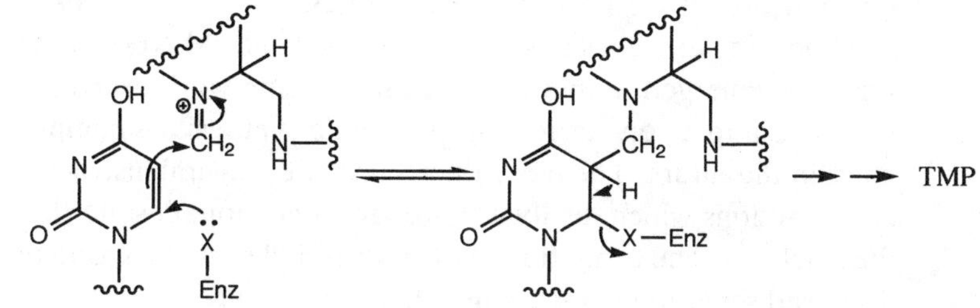

(2)

(3)

(F-dUMP)   Blocked enzyme

uridylate, F-dUMP, (3). It was found that the fluorine derivative causes rapid inactivation of the enzyme and that is loses UV absorbance due to loss of the pyrimidine chromophore on incubation with enzyme and $N^{5,10}$-methylene-THF in consistency with the formulation in (3), where fluorine now blocks the elimination of the enzyme. A slow exchange of 5-C-^{3}H in uracil was also noted which could be referred to as a reversible addition of enzyme across the 5,6 double bond. However, these observations can also be accommodated in the scheme in Fig. 7, if we assume that addition to the enzyme is a secondary process which takes place when rearomatization upon alkylation of uridylic acid cannot occur.

9.3 Pyrroles and porphyrins

Apart from being a building block in the porphyrins, corrins, and some gall pigments of common biogenesis, pyrrole derivatives are rare in nature. A few pyrrole derivatives (Fig. 8) are produced by microorganisms, sponges, higher plants and animals. The intensely red prodigiosins in *Serratia marcescen*s are derived from acetate and amino acids as established by MS and ^{13}C NMR.[11] However, this biosynthetic pathway is quite different from that of the porphyrins. The pheromone from the hair pencils of *Danaus* spp. originates from certain pyrrolizidine alkaloids produced by plants on which the butterfly thrives, and they are further metabolized by the insect.[12]

In view of their biological importance, the intensely coloured macrocyclic pigments of life have attracted special interest among scientists. By the combined efforts of workers in all fields of chemistry and across the borders of neighbouring branches of science a detailed and coherent picture of their biosynthesis emerged. Experiences from the team work exerted on vitamin B_{12} will have impact on strategies for organic synthesis of complex natural products in the future. The biosynthesis provides several instructive and remarkable reactions which justify a thorough discussion, despite the complexity of the stuctures. The formation of the corrin skeleton of B_{12} and its reactions are discussed separately in sections 9.4,5.

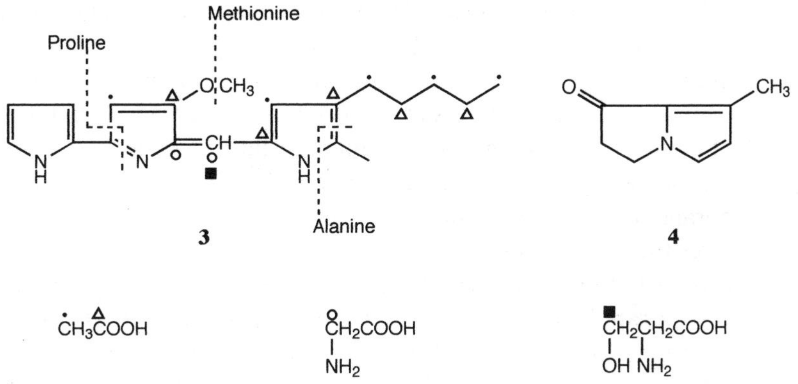

Fig. 8 Pyrrole derivates. **3**, Labelling pattern in prodigiosin observed from feeding experiments; **4**, a pheromone in *Danaus* spp. derived from a pyrrolizidine alkaloid

At the ring corners of the porphyrins we have four pyrrole units. The ^{14}C isotope technique was applied early by Shemin and colleagues[13] to trace the metabolic origin of these pyrrole units, which in animals and many micro-organisms are synthesized from glycine and succinyl CoA while in plants and some anaerobic bacteria the biosynthesis starts with glutamic acid. Glycine is activated by PLP and condensed in a Claisen-type reaction with succinyl CoA to 5-aminolevulinic acid (4). It is not clear whether 5-aminolevulinic acid is formed via 2-amino-3-ketoadipate (route a), i.e. PLP facilitates an α-H ionization of glycine, or by decarboxylative alkylation (route b), i.e. PLP catalyses the decarboxylation as in decarboxylation of amino acids. In any event, 2-

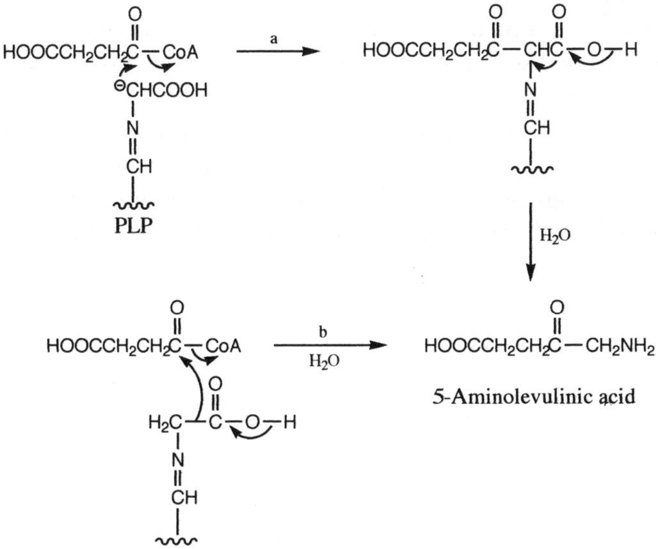

(4)

amino-3-ketoadipate is found to decarboxylate spontaneously in less than one minute at pH 7^{14} and its PLP derivative, i.e. the intermediate in route a, will most likely decarboxylate immediately. When 5-aminolevulinic acid was added to enzyme preparations from duck blood it served as an *in vitro* precursor of haem.[15] Two molecules of 5-aminolevulinic acid dimerize in a Knorr-type reaction to the pyrrole derivative porphobilinogen (PBG,5) which also incorporates well. Tetramerization of PBG, head to tail, catalysed by deaminase gives the open chain hydroxymethylbilane (HMB). Here comes the first surprise. At this stage a cosynthetase takes over, picks up released HMB, and diverts it rapidly to the rearranged uroporphyrinogen III via *ipso*-substitution, cleavage of the γ-*meso*-α-D-pyrrole bond, rotation of the D-unit and recyclization at the other α-pyrrole position, Fig. 9. Chemical cyclization or further cyclization with deaminase of HMB gives chiefly the biologically inactive uroporphyrogen I with alternating arrangement of acetate and propionate groups. Cosynthetase is unable to convert uroporphyrogen I into the isomer III and it was also shown that the assembly of the four rings started with ring A followed by sequential addition of rings B, C and D. On oxidation, e.g. by O_2, the uroporphyrogens form the strongly coloured conjugated 18-π-systems, the uroporphyrins I and III. The deaminase catalysed cyclization was followed by ^{13}C NMR of 5-amino-5-^{13}C-levulinic acid, enriched at 90 atom per cent, which gives rise to 2,11-labelled PBG and symmetrically labelled uroporphyrin I. The ^{13}C signal from the equivalent *meso* carbons, appears characteristically as a double doublet, rel. intensity *ca.* 90 per cent, $^1J_{13C\ 13C}$ 72 Hz, $^3J_{13C\ N\ C\ 13C}$ 5 Hz centred on a broad singlet, rel. intensity *ca.* 10 per cent at δ 97.5. The four pyrrole α-carbon atoms give a similar pattern at δ 143.5. This is in agreement with a straightforward cyclization.

In the biologically active porphyrins, A–C have alternating arrangements of the acetic and propionic acid groups, as in uroporphyrinogen I, but ring D is "turned around" so that C and D now have adjacent propionic acid groups as in uroporphyrinogen III.

The ^{13}C NMR spectrum of uroporphyrin III, produced by the combined enzyme complex, obtained either from avian blood or from the alga *Euglena gracilis*, was in agreement with the labelling pattern in Fig. 9. By dilution of the double-labelled (90 per cent) PBG with four parts of unlabelled PBG it is

(5)

Porphobilinogen (PBG)

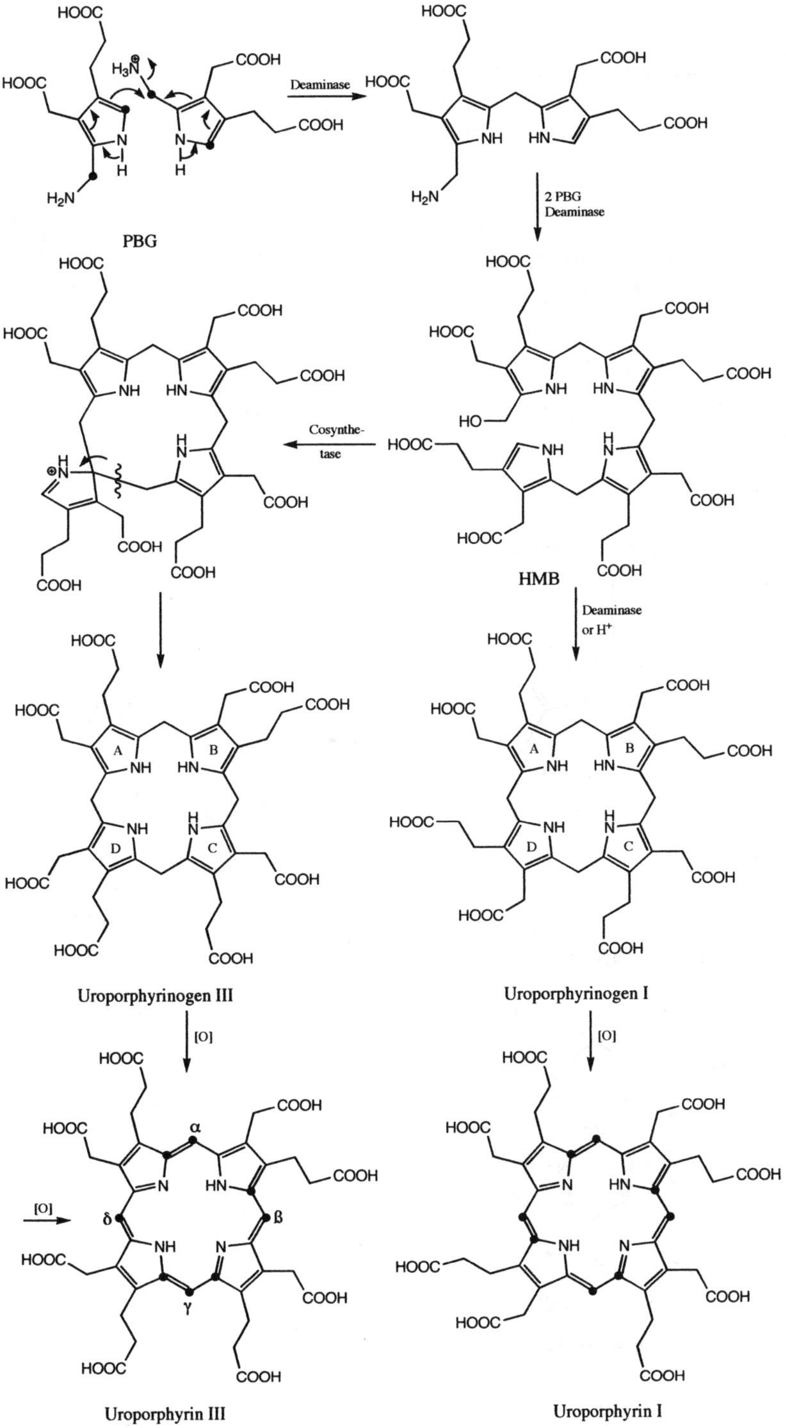

Fig. 9   Biosynthesis of uroporphyrinogens

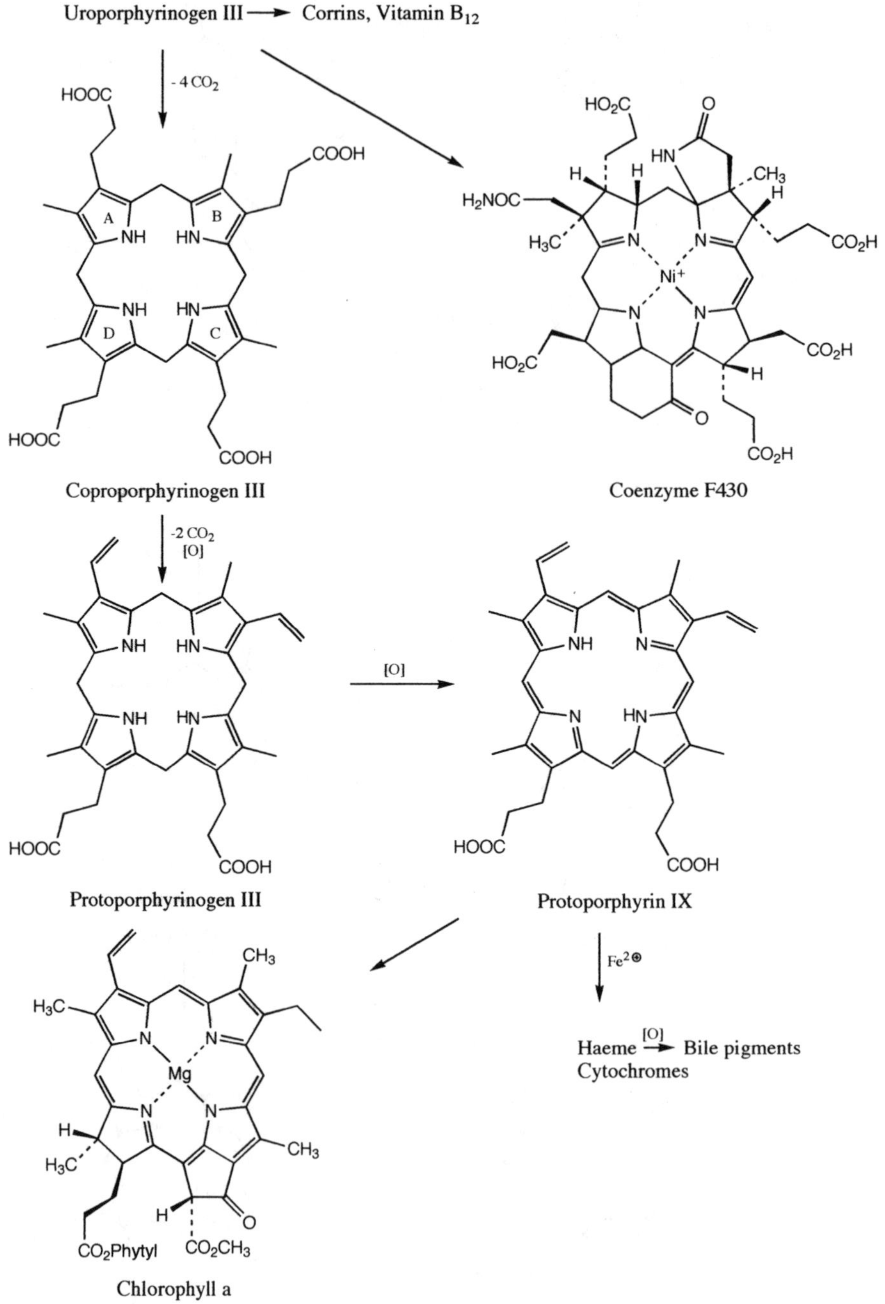

Fig. 10   Biosynthesis of chlorophyll a and haem

arranged so that the majority of the pyrrole units have only two interacting ^{13}C atoms which simplifies the fine structure of the proton decoupled ^{13}C NMR spectrum. The α-, β- and δ-carbons show primarily a fine doublet, $^3J_{13}$ CNC$_{13C}$ 5 Hz, proving that the A, B and C units are incorporated intact. The γ carbon signal is a doublet, $^1J_{13C\ 13C}$ 72 Hz, indicating an intramolecular rearrangement of the side chain from one α-carbon to the other in ring D.[16]

Initial formation of uroporphyrogen III is common for all pigments of life.[17,18] From here the pathways to haem and chlorophyll divert by decarboxylation of uroporphyrogen III, partly involving oxygen. A less well known porphyrin is the reduced and methylated nickel porphyrin, coenzyme F 430, Fig. 10, which catalyses the formation of methane by certain bacteria. The other route leads to vitamin B_{12} by a series of peripheral C-methylations. No net oxidation occurs in this case but redox reactions and ring contraction are involved. Certain bacteria are able to produce B_{12} anaerobically, section 9.4.

Uroporphyrogen III is converted to coproporphyrogen III by stepwise decarboxylation of the four acetic acid side chains to methyl groups starting with the acetic acid residue on ring D and continuing in a "clockwise" fashion with those on rings A, B, and C.[19] The decarboxylation proceeds stereospecifically with retention of configuration (6).[20]

This was demonstrated by incubation of haemolysed avian erythrocytes with chiral succinate, isolation of the biosynthesized haem and analysis of the acetic acid formed by oxidative degradation of the haem by the malate-fumarate procedure (section 6.2). It turns out that (2R)-succinic acid gives (2S)-acetic acid. The oxidative decarboxylation of the propionic acid side chains is also regiochemically controlled. The decarboxylation takes place first at ring A and subsequently at ring B.[21] The steric course of the decarboxylation of the propionic acid side chain was followed by investigation of

(6)

2-*R*-Succinic acid   Uroporphyrinogen III

Coproporphyrinogen III (section)   2-*S*-Acetic acid

$$(7)$$

the fate of PBG specifically deuterated in the side chain (7). The ^{1}H NMR signal of H_R in the vinyl group of protoporphyrin IX appears as a doublet, J_{H-H_R} 18 Hz, characteristic for a *trans* coupling. Earlier work with specifically labelled succinate has shown that only the pro-*S*-hydrogen is removed.[22] Hence, the vinyl groups are formed by antiperiplanar elimination of a hydrogen and carbon dioxide.[23] The mechanism of this reaction is still unknown but it is suggestive of the copper-catalysed decarboxylation of aliphatic acids leading to olefins which is of radical nature.[24] Several bacteriochlorophylls have ethyl side chains formed biosynthetically by reduction of the vinyl groups. The steric course of this reduction was determined in the following way. Specifically labelled (2*S*)-2-^{2}H, ^{3}H-ALA was metabolized by *Rhodopseudomonas spheroides* into bacteriochlorophyll a (8). The ethyl side chain was converted by oxidative degradation into acetate which by analysis via the maleate–fumarate procedure was shown to have the *R*-configuration. This shows that the vinyl group is reduced from the *si*-face.[25]

$$(8)$$

9.4 Biosynthesis of the corrin skeleton

The isolation, structure and action of vitamin B_{12}, which in its center contains a cobalt atom, is an exciting story starting with the discovery by Minot and Murphy in 1926 that liver extracts cure pernicious anaemia. The active factor, called vitamin B_{12}, was, after painstaking efforts, isolated in crystalline form as cyanide twenty years later.[26,27] The compound is present in minute quantities in liver and a test on patients suffering from pernicious anaemia was the only assay method accessible. The structural elucidation proved to be an extremely difficult problem which ultimately was solved by X-ray crystallography.[28] A year later a considerable stir was caused when it was found that the cyanide, now called cyanocobalamine, $[Co^{3\oplus}]$-CN was an artefact produced in the work-up.[29]

The active natural derivative has a 5´-deoxyadenosyl group as sixth ligand (Fig. 11). Thus, cyanide displaces the adenosyl during the work-up but if

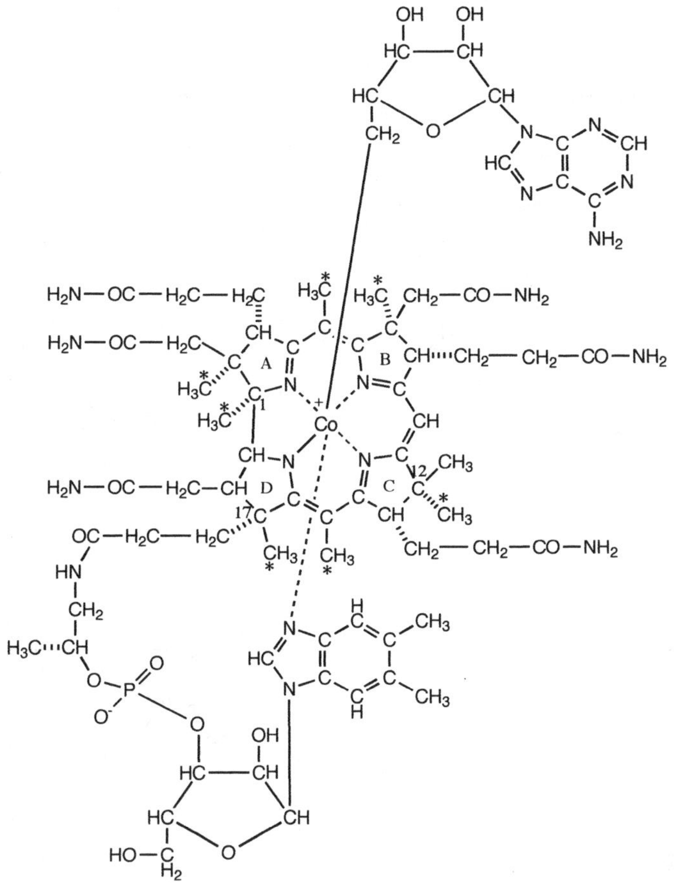

Fig. 11   Structure of vitamin B_{12}

cyanocobalamine is administered to the organism, the adenosyl derivative is formed again in the cell.

As a result of 40 years' dedicated work by English, French, Swiss and American scientists the complete story of the B_{12} biogenesis[17,18] can now be told, but its reactions as a cofactor are still a riddle, difficult to reconcile with conventional reaction mechanisms. This is partly because a descriptive formalism is lacking to handle the reactions of organo-transition-metallics suitably, which as a rule have no σ–C-metal bonds and are unstable unless their d-orbitals are tied up by N- or O-ligands, and partly because analogous rearrangements in organic chemistry are unknown.

The discussion is restricted to the biosynthesis of the corrin part of the molecule. Three major problems stand out: a. the C-methylation sequence, b. the ring contraction and c. the mechanisms of the redox reactions.

Experiments with the anaerobic *Propionibacterium shermanii* show that its

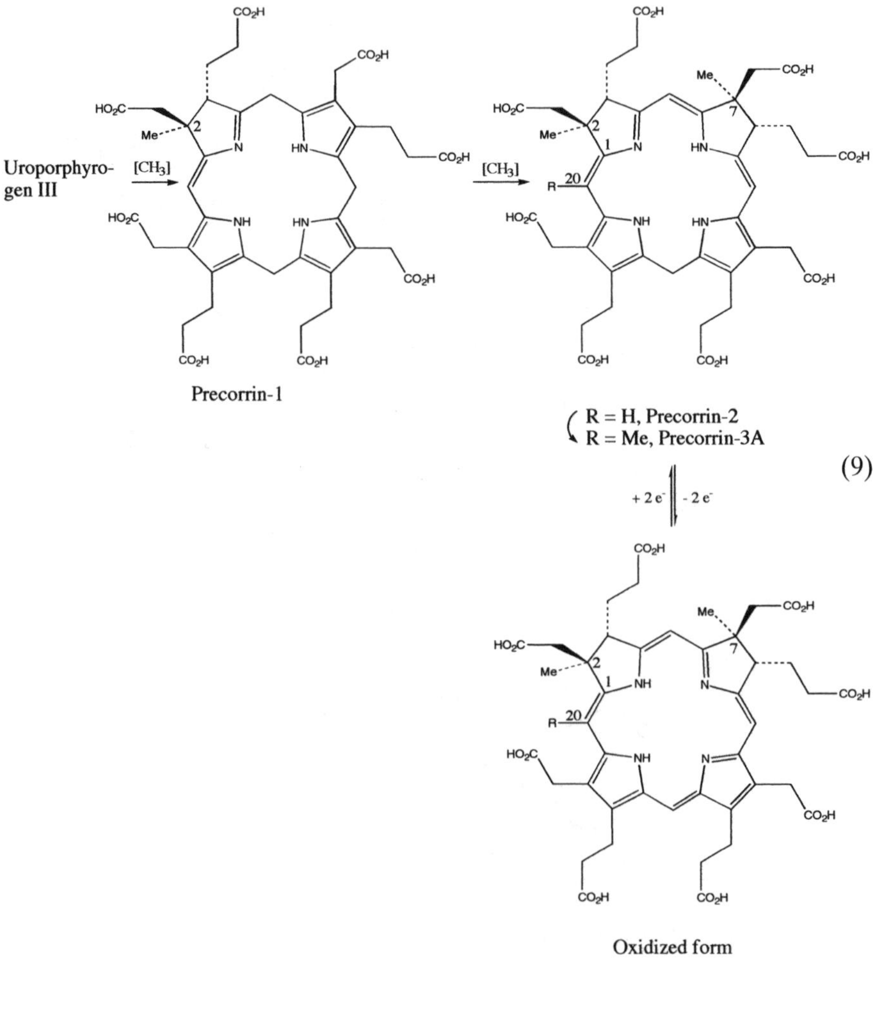

Uroporphyrogen III

Precorrin-1

$\left\{ \begin{array}{l} R = H, \text{Precorrin-2} \\ R = Me, \text{Precorrin-3A} \end{array} \right.$

(9)

$+2e^- \| -2e^-$

Oxidized form

pathway branches away at uroporphyrogen III by a series of electrophilic methylations with S-adenosylmethionine (SAM). In the absence of cobolt the biosynthesis is interrupted at precorrin-3 and it was possible to identify three new intermediates which all incorporate well into the corrins implying that they are located on the pathway. The name was introduced to denote intermediates preceding the fully built corrin macrocycle and the number denotes how many C-methyl groups have been introduced. The first methylation takes place at C^2 to give precorrin-1 which is further methylated at C^7 by the same enzyme system. The third methylation which occurs at C^{20} requires a second methylase enzyme. A reasonable supposition is that cobolt is inserted at this stage. The precorrins 1–3 are easily aerially oxidized to their highly conjugated forms (9).

For several years it was not possible to isolate any new intermediates between the 3A stage and cobyrinic acid, but in the meantime another important discovery was made that somewhere along that pathway the CH_3-C^{20} fragment was lost as acetic acid with C^{20} providing the carboxylic group (11). In this process the direct link between rings A and D is formed. The wild-type strains of the bacteria produce a complete set of enzymes in low concentration capable of metabolizing the substrates into final products leaving extraordinarily small amounts of methylated and rearranged intermediates behind.

Further advancement required new strategies and the breakthrough came, not quite unexpectedly[14], from research in the field of molecular genetics. The fundamental work was done on genetically engineered strains of aerobic *Pseudomonas denitrificans* mutants. Genes involved in the biosynthesis of cobyrinic acid were identified and overexpressed, which simply means that the individual essential enzymes required for the biosynthesis were overproduced. Higher yields of intermediates can in this way be prepared by incubating the substrates with cofactors and individual enzyme systems in high concentration. Alternatively, gene deletion was practised to accumulate an intermediate by interrupting the biosynthesis at a certain stage. The first new product isolated by incubating various ^{13}C labelled precorrin-3A with a selected set of enzymes and SAM ($^{13}CH_3$) as cofactor – but in the absence of NADPH – was precorrin-6A, (10). The molecule contained three new methyl groups at C^1, C^{11} and C^{17} and moreover ring contraction had occurred with loss of the CH_3-C^{20} fragment. The structure was solved by studying the ^1H, ^{13}C connectivity pattern around the molecule by NMR spectroscopy. One striking structural feature was the α-methyl substitution at C^{11}, which foreshadowed a suprafacial C^{11}-C^{12} methyl shift at a later stage preceded by decarboxylation of the C^{12}-acetate, on the basis that the vinylic β-imino structure is expected to facilitate the decarboxylation in comparison with decarboxylation at an sp3-centre. Restoring NADPH unveiled the next intermediate, precorrin-6B, which contained a saturated $C^{18,19}$ bond (10). It was shown by using tritium labelled NADPH that the purified reductase specifically directed the 4-H_R-

hydride of the cofactor to C^{19}, and C^{18} was protonated as expected. The reduction brings the molecule back to the oxidation level of precorrin-3A implying that a corresponding two electron oxidation must take place in the 3A-6A sequence both in the anaerobic *P. shermanii* and in the aerobic *Ps. denitrificans.*

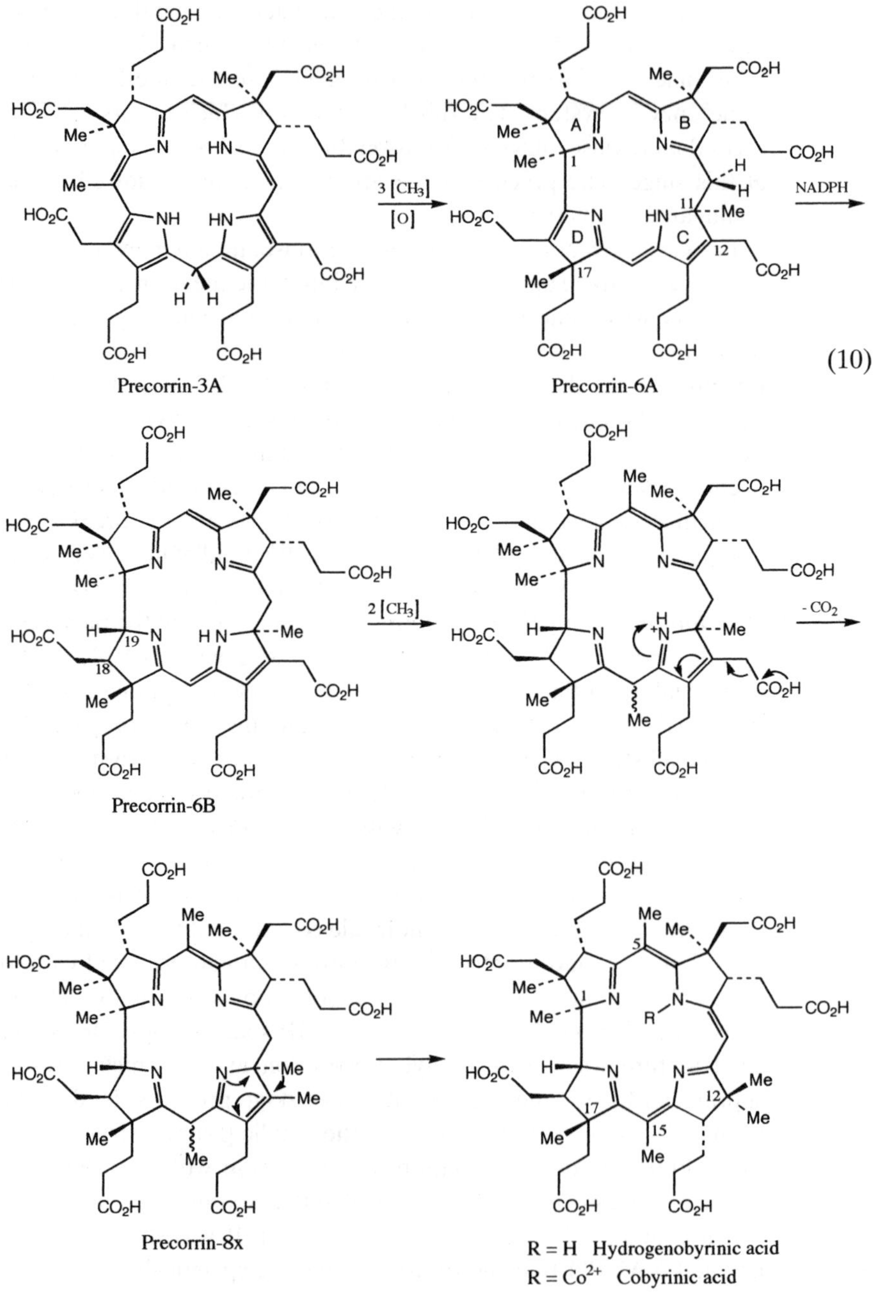
3 [CH3]
[O]
NADPH
2 [CH3]
- CO2
Precorrin-3A
Precorrin-6A
Precorrin-6B
Precorrin-8x
R = H Hydrogenobyrinic acid
R = Co2+ Cobyrinic acid

(10)

The major biosynthetic problems of B_{12} have gathered in the gap between precorrin-3A and 6A and here again genetic engineering shows its power. By deletion of a gene from a certain strain of *Ps. denitrificans*, a blocked enzyme system was prepared which converted precorrin-3A to the tetramethylated intermediate, precorrin-4. The fourth methyl group was located at C^{17}, but it turned out that the molecule also had been oxidized and that ring contraction had occurred with formation of an acetyl group at C^1. With access to specific enzyme preparations and admittance of oxygen as cofactor it was shown that precorrin-3A first gave an oxidized intermediate with intact skeleton, precorrin-3B, which contained a γ-lactone, IR: 1799 cm^{-1}, (11). This shows that the oxidation step is not the immediate cause of the ring contraction. Use of $^{18}O_2$ as oxidant yielded an oxygen labelled acetyl group in accordance with electrophilic attack of $L_5Fe^v=^{18}O$ (iron-sulphur cluster catalyst) at C^{20}, γ-lactonization and pinacol-type rearrangement as shown in (11). The carboxy group of the C^2-acetate did not suffer any loss of label in agreement with the mechanism presented. Incubation of precorrin-3B with 17-methyltransferase and SAM gave the known precorrin-4. At this stage ring C is *ipso*-methylated at C^{11} to give precorrin-5. One of its tautomers is suited for elimination of acetic acid in a reversed Claisen condensation with formation of a resonance stabilized anion which is methylated at C^1 from the α-side to give precorrin-6A (11).

The oxidation step in aerobic *Ps. denitrificans* is different from that in anaerobic *P. shermanii*. The interesting observation has been made that in

(11)

P. shermanii the C^2-acetate loses one oxygen atom which is transferred to the acetyl group at C^1. It is suggested that Co-ligated precorrin-3A undergoes a two electron oxidation followed by a nucleophilic C^2-acetate substitution at C^{20} with δ-lactone formation and subsequent rearrangement, Fig. 12.

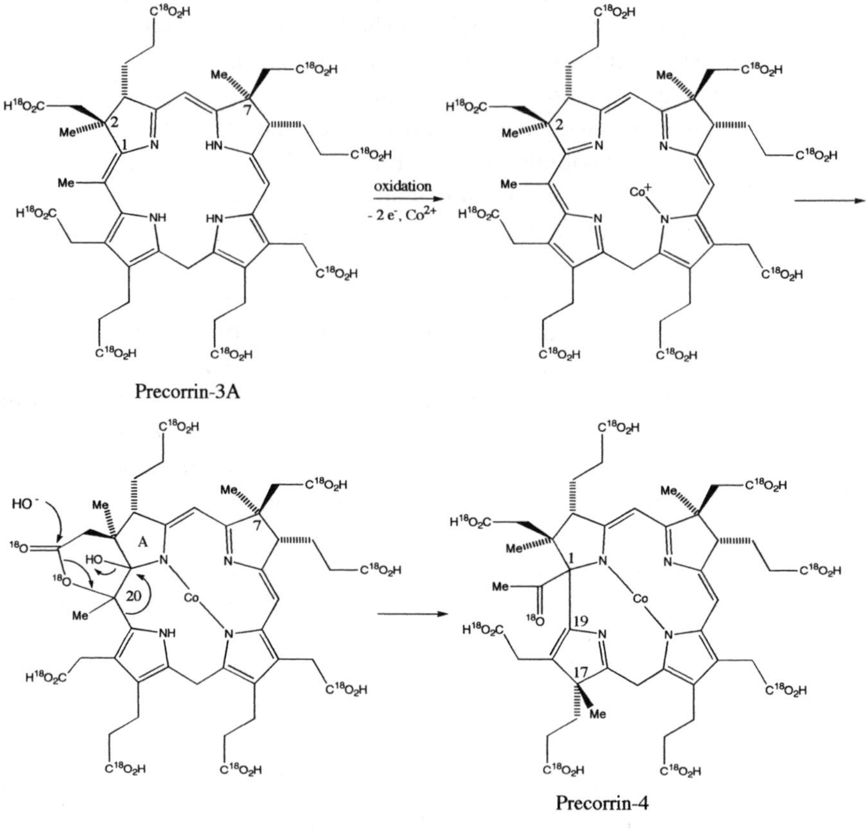

Fig. 12   Suggested reaction mechanism for oxidation of precorrin-3A in *Propionibacterium shermanii*

Thus, in the absence of a reducing agent the methylation stops at the 6A stage. Precorrin-8x could be conveniently prepared from precorrin-3A by using the basic set of enzymes together with both cofactors SAM and NADPH but in presence of a tenfold excess of hydrogenobyrinic acid (Co^{2+} replaced by H^+), which affects the final methyl rearrangement into hydrogenocobyrinic acid by product inhibition. Precorrin-6B was also converted into precorrin-8x by using a purified enzyme from an engineered strain of *Ps. denitrificans* in the presence of SAM. This enzyme complex also catalysed the decarboxylation of the acetate group which proceeded with retention of configuration.[30] The last two methyl groups are introduced at C^5 and C^{15} but the succession is unknown at present. Circumstantial evidence indicates that at least one of these methylations must occur before the decarboxylation takes place. The last step is the enzyme catalysed C^{11}-C^{12} methyl shift which exposes the extended conjugated system hydrogenobyrinic acid (10).

Synthesis of hydrogenobyrinic acid from ALA requires twelve enzymes all of which have been overproduced by genetic engineering and isolated and they have also demonstrated their excellence as a tool for elucidating biosynthetic pathways. One thrilling experiment had to be carried out as a coronation of the work. The proposition was now: chiral hydrogenobyrinic acid should be formed if the substrate ALA is incubated with a cocktail of the enzymes and cofactors. Indeed this was shown to be the case. The genetically engineered multi-enzyme synthesis has bright prospects in total chiral synthesis, especially for more complex natural products of practical significance. The end of the inspiring Woodwardian era seems to be approaching. The advice to the organic chemist working on total synthesis or in the biosynthetic field on structurally complex systems is to cooperate with molecular biologists.

9.5 Reactons of vitamin B$_{12}$

Vitamin B_{12}, $[Co^{3\oplus}]$-R, is reduced under controlled potential to $[Co^{2\oplus}]$ and by $NaBH_4$ to $[Co^{\oplus}]$ (12), formally corresponding to homolytic and heterolytic cleavages of the Co–C bond (13, 14). The monovalent complex $[Co^{\oplus}]$ is an efficient nucleophile and reductant, reminiscent of magnesium in its properties.

$$[Co^{3\oplus}]\text{-R} \overset{e^{\ominus}}{\rightleftharpoons} [Co^{2\oplus}] \overset{e^{\ominus}}{\rightleftharpoons} [Co^{\oplus}] \tag{12}$$

$$[Co^{3\oplus}]\text{-R} \rightleftharpoons [Co^{2\oplus}] + R^{\bullet} \tag{13}$$

$$[Co^{3\oplus}]\text{-R} \rightleftharpoons [Co^{\oplus}] + R^{\oplus} \tag{14}$$

It reacts with alkyl halides, amines and phosphates with formation of alkylcobalamines, thus providing a facile route to organocobalt compounds (15).

$$[Co^{\oplus}] + RX \rightleftharpoons [Co^{3\oplus}]\text{-R} + X^{\ominus} \tag{15}$$

Vitamin B_{12} is formed from $[Co^{\oplus}]$ and ATP by expulsion of triphosphate.

The methylation of cobalamine by N^5-methyltetrahydrofolate activated by N^5-protonation represents the first part of the methyl transfer catalysed by cobalamin-dependent methionine synthetase in mammalian tissues.[31] This corresponds to R=CH$_3$ and X=H$^+$N^5-tetrahydrofolate in eqn (15). In a model reaction[32] it has been shown that the cobalt atom in $Co^\oplus$-cobalamine is methyl-ated by $PhN(CH_3)_3^+$. In the second step (17), actually the reversal of (15), the sulfide ion of homocysteine reacts nucleophilically with methylcobalamine to give methionine. Both methyl transfers proceed with inversion of configuration, i.e. the methyl group transfers with net retention of configuration.[33] The enzymatic reaction is accompanied by *ca.* 50 % racemization.

In the biomethylation of mercuric ions (16) and other metal salts a transient methyl carbanion is formed. Thus, alkylcobalamines undergo electrophilic as well as nucleophilic substitution but the homolytic equilibrium (13) seems to be the best base of support for exploration and explanation of vitamin B_{12} mediated reactions, categorized as 1,2-shifts (18), where X=alkyl, vinyl, carboxyl, hydroxyl and amino groups, here exemplified by the methyl-malonyl-succinyl rearrangement (19).

$$[Co^{3\oplus}]\text{-}CH_3 + Hg^{2\oplus} \rightarrow [Co^{3\oplus}] + CH_3Hg^\oplus \tag{16}$$

$$RS^\ominus + [Co^{3\oplus}]\text{-}CH_3 \rightarrow RSCH_3 + [Co^\oplus] \tag{17}$$

$$\tag{18}$$

$$CH_3HC\begin{smallmatrix}COOH\\COCoA\end{smallmatrix} \rightleftharpoons HOOCCH_2CH_2COCoA \tag{19}$$

A radical mechanism has been advanced featuring initial homolysis of the cobalt-adenosyl bond ($Co\text{-}CH_2Ad$) and a hydrogen abstraction from the ligated malonylmethyl group by $AdCH_2^\cdot$ (20). The substrate radical rearranges within the ligand sphere of Co^{2+}, a "cage reaction", to the product radical, which finally abstracts a hydrogen from CH_3Ad thus regenerating the coenzyme for a new cycle. This mechanism excludes hydrogen exchange with the solvent, accounts for the isotope effect and the facts that both the coenzyme and the two succinyl methylene groups become labelled by processing methylmalonate labelled in the methyl group. Activation of C-H and C-C bonds is a characteristic effect of transition metal cations and it is central to our understanding of their catalytic activity. This effect was previously demonstrated for the iron catalysed biological hydroxylation of hydrocarbons, section 4.3. The reaction of Fe^+ with ethane which is representative of the transition metal cation activation of hydrocarbon bonds has been both experimentally and theoretically studied.[34] Fe^+ forms an iron-ethane complex which is 27 kcal mol^{-1} (calcd.) below the reactants. The bond dissociation energy for $Fe\text{-}C_2H_5+$ is 56 kcal mol^{-1}, and most interestingly the energy of Fe^+-insertion into the C-H bond of ethane, $CH_3CH_2\text{-}Fe^+\text{-}H$, is calculated to be *ca.* -6 kcal mol^{-1}. The dissociation energy for the $AdCH_2\text{-}Co^{2+}$ bond of B_{12} is low[35], 30-31 kcal mol^{-1} and is further weakened by the enzyme. The thermodynamical data

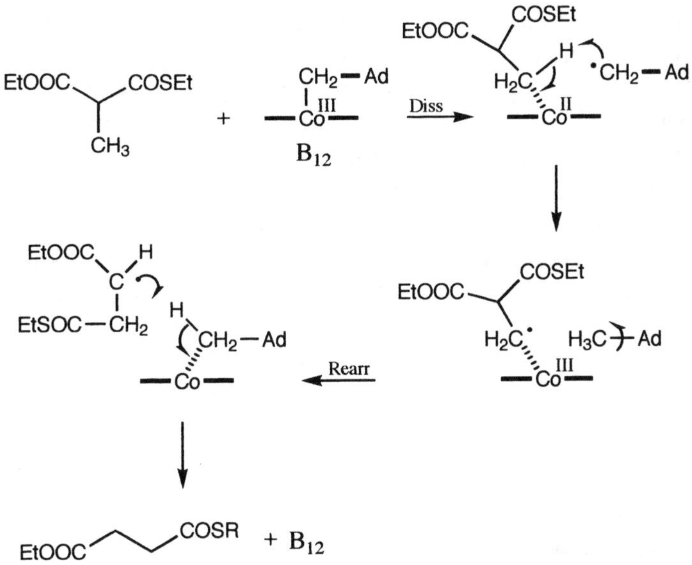

(20)

imply that Co^{2+} has a considerable activating effect on the C-H bonds of ligated hydrocarbons. Consequently the rearranged product radical is able to abstract a hydrogen from the less reactive CH_3Ad – but not necessarily the same labelled hydrogen. The nucleoside 5'-deoxyadenosine has been isolated from the reaction mixture, but it has not been possible to incorporate added nucleoside probably because it is strongly bound to the enzyme. As expected, 5'-C^2H_2-labelled coenzyme yields a labelled rearranged product.

The acid catalysed diol-one rearrangement (21) represents a conceivable model reaction but it fails to account for the unexpected findings that the tritium transfer is intermolecular in (22) and that the coenzyme also becomes tritiated implicating that vitamin B_{12} serves as a hydrogen carrier.[36]

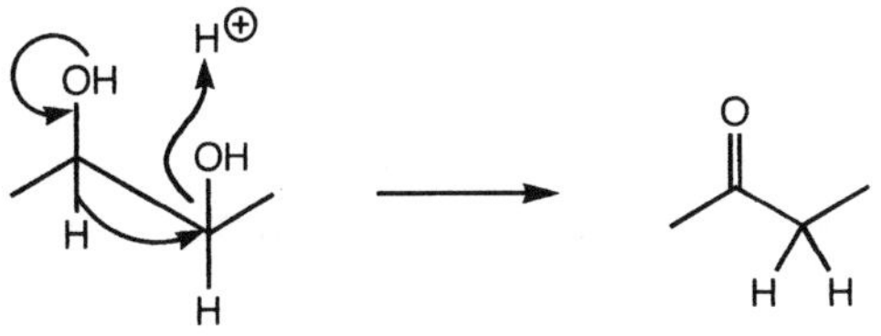

(21)

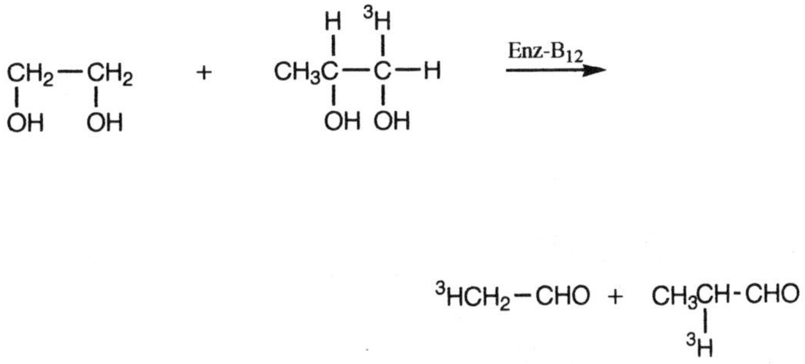

(22)

The radical mechanism for the B_{12} reactions is well documented experimentally: a. no incorporation of protons from the solvent is observed, b. observation of isotope effects, c. observation of radical intermediates by ESR spectroscopy,[37] d. inhibition by radical scavengers, e. probing the radical mechanism by carrying out model rearrangements, and f. evidence for spontaneous radical 1,2-rearrangements.[38]

The rearrangements shown in reaction (23) are illustrative. The radical intermediate gives acyl migration (23a,c), whereas the cationic intermediate gives phenyl migration (23b)[39]. Anion intermediates are expected to incorporate protons from the solvent which in fact is not observed.

The rearrangement depicted in (20) has been mimicked in a non-enzymic model reaction assisted by the cobalamin cofactor (24). In absence of the B_{12}-

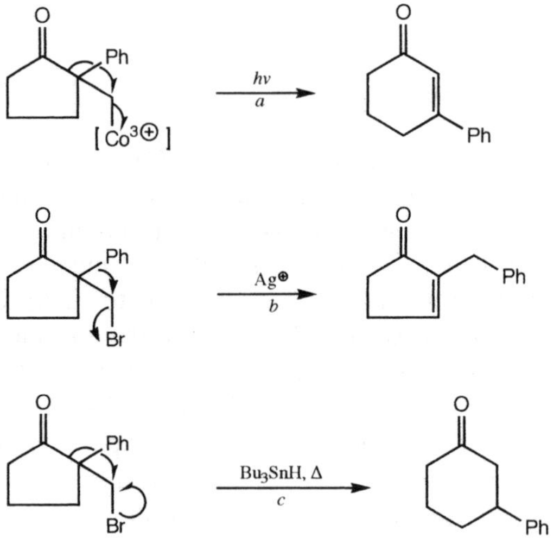

(23)

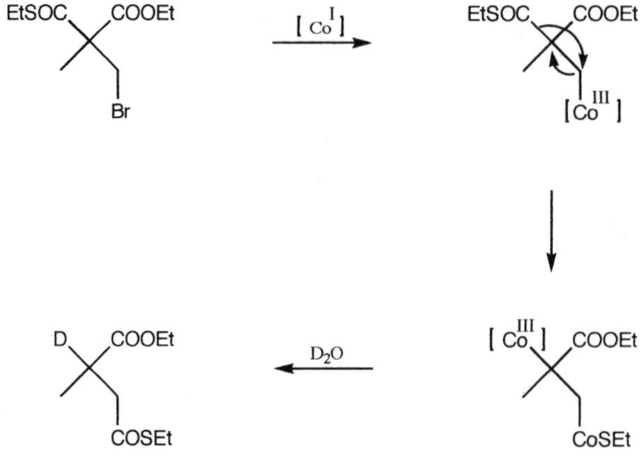

(24)

adenosyl group the rearranged organo cobalt compound is hydrolysed and deuterated at the methin position as anticipated provided that the reaction is run in a deuterated solvent.[40]

Coenzyme B_{12} dependent diol dehydratase is supposed to promote a radical 1,2-shift involving OH and NH_2 groups. These reactions have no precedence in organic chemistry and are thermodynamically unlikely. The reaction has one analogue in the hydroxy radical abstraction of a hydrogen atom from glycol yielding the acetaldehyde radical which was observed by ESR (25).[41] The enzymatic reaction with propanediol can be formulated similarly thus avoiding HO· migration (26)[42]. Examination of the stereospecificity of the rearrangement shows that 1-pro-(R)-H and 1-pro-(S)-H migrate selectively in (2R) and (2S)-propanediol with inversion at C^2. Another interesting feature of the reaction is that (2R) and (2S) propanediol labelled with ^{18}O at C^1 retain 8 per cent and 88 per cent of the label in propanal, respectively. This shows that the rearrangement stereospecifically passes via a 1,1-geminal diol (26), the hydrolysis of which is enzymatically controlled.[43]

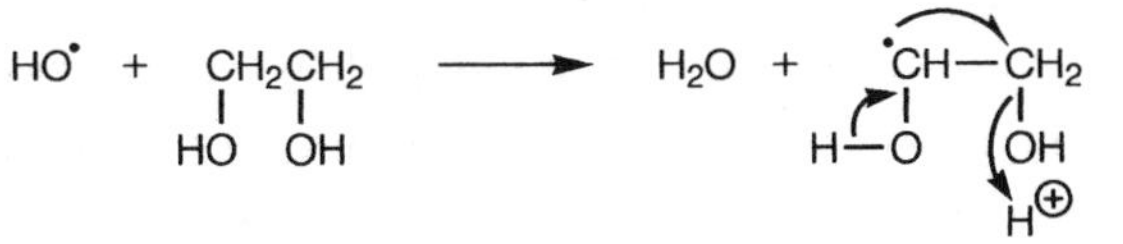

(25)

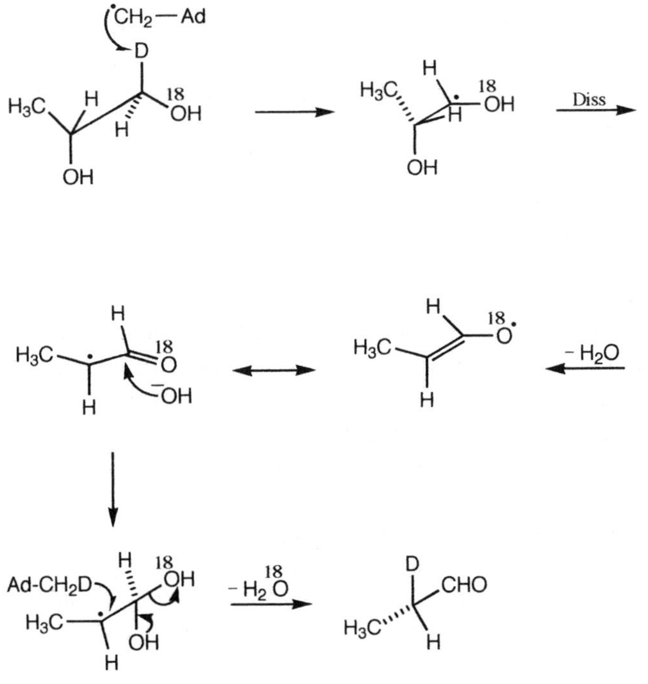

(26)

9.6 Retrospect and prospect

We expect a student of physics to have a knowledge of the theoretical background for a mathematical formula describing a physical phenomenon. The same should be applied to the study of natural products. A knowledge of the structures of a number of compounds is not sufficient; more essential is the knowledge of how they are formed. It is evident that natural product chemistry made its greatest leap forward with the elucidation of the biosynthetic mainstreams, which brought system to a seemingly chaotic network of processes. Considering the diversity and complexity of naturally occurring compounds one is amazed by the relatively small number of reaction types and agents nature utilizes for the synthesis of its products. In the laboratory we are aided by at least a hundred reducing or oxidizing agents to perform various redox reactions, while nature accomplishes its task with just a few. Nature has already made its entry into the laboratory in more than one way. Microorganisms and purified enzymes are routinely used in industry to perform stereospecific transformations in practically all classes of compounds, natural as well as synthetic. Redox reactions, epoxidations, hydroxylations, dehydrations, condensations (formations of C–C bonds), cleavages, rearrangements etc., are now performed with the aid of certain strains of microorganisms as single steps in a synthetic sequence and of special advantage is the introduction of chirality in such a process.

A milestone has been passed in this area with the complete elucidation of the biosynthesis of vitamin B_{12} which has given us new perspectives on how to combine genetic engineering with organic chemistry. Identification of genes and overexpression of the corresponding enzymes made it possible step by step to follow the detailed synthesis of B_{12}. The same principle could be applied to large scale production of intermediates for further use in synthesis. Traditional total synthesis of complex natural products will be outdated and replaced by identification and overexpression of the operating genes. A substrate incubated with a suitable set of enzymes will then be transformed to the desired chiral product. Many enzymes are not substrate specific and may accept substrates of some structural variation.

Another astonishing feature is that, essentially, biosynthesis is identical in the simplest prokaryote, in the highest plant, and in the human being, implying that the refined synthetic machinery arose very early in biological evolution.

The main biosynthetic pathways are well known today but still many connecting paths have to be explored and the enzymatic reaction mechanisms themselves are still imperfectly known.

Much is to be learned from the synthetic skill of nature. Thus, nature has in photosynthesis found an efficient method to convert the energy of light into chemical energy, thereby producing reactive ATP and splitting water into oxygen and the equivalent of hydrogen, NADPH. Man has not yet been successful in efficiently solving this long-standing significant problem for utilizing the energy of the sun.

As discussed in the chapter on ecological chemistry our knowledge of the biological and ecological functions of secondary metabolites is embryonic and a large field of important research is ahead of us. This will give us a deeper understanding of the conditions for coexistence of species on earth and will teach human beings to treat nature kindly. Up until now we can say that man's activities have only been a threat to nature and our own existence.

The plant kingdom is our major renewable resource both for nutrients, fibres, various chemicals such as drugs, essential oils, and resins, and for raw material for various chemical processes. The importance of drugs in human affairs appears from the fact that in 1984 *ca.* 23 per cent of all prescriptions (hormones, cardiovascular agents, analgesics, etc.) in the United States contained secondary natural products of plant origin, vitamin pills excluded. Another 20 per cent are of microbial (antibiotics) and animal origin. Considering that the majority of plant species, especially in the tropics, have not been investigated yet, the prospects for the development of natural product chemistry are good.

Higher plants continue to be an important source of medicinal agents, model compounds for structure-activity investigations and as inspiration for scientific endeavour. Tropical forests contain a disproportionate share of the earth's flora and fauna with interesting pharmacologically active constituents. Ethnopharmacology is a branch of science which essentialy concentrates on investigations of plants "screened" by local curanderos for their medicinal effects. It would be a catastrophe for mankind if we by greed and ignorance destroy these vulnerable parts of our common heritage.

Structural elucidation will always be a necessary prerequisite for research in this field but is perhaps no longer considered prestigious as in the earlier days. Traditionally chemists have turned to terrestrial flora and fauna. Technological advances now make it easier to collect marine organisms at various depths and recent research on marine natural products has indeed given interesting results, widening our perspectives at the interface between chemistry and biology. The aquatic environment has an influence on the structural pattern of marine metabolites. A number of unusual structures have been solved spreading enthusiasm amongst workers in the field.

Secondary metabolites play a distinguished role in all manifestations of human behaviour. Utilization of narcotics and stimulants is pervading all cultures. Poisons have always been feared and used for various purposes. Exotic perfumes and essential oils are more harmless but none the less of pleasant significance, whenever people gather. Nor should we forget the exciting spices stimulating our appetite and giving new dimensions to the pleasures of eating. And there is no prospect of substituting these subtle products with synthetic analogues. Evidently, natural product chemistry will continue to be part of the age in which we live.

Bibliography

1. Löfgren, N. and Lüning, B. *Acta Chem. Scand.* **7** (1953) 225.
2. Miller, C. O. *Science* **157** (1967) 1055.
3. Burrows, W. J., Armstrong, D. J., Kaminek, M., Skoog, F., Bock, R. M., Hecht, S. M., Dammann, L. G., Leonard, N. J. and Occolowitz, J. *Biochemistry* **9** (1970) 1867.
4. Bordner, J., Thiessen, W. E., Bates, H. A. and Rapoport, H. *J. Am. Chem. Soc.* **97** (1975) 6008.
5. Buchanan, J. M. and Hartman, S. C. *Adv. Enzymol.* **21** (1959) 199.
6. Suzuki, T., Ashihara, H. and Waller, G. R. *Phytochemistry* **31** (1992) 2575; Schulthess, B. H., Morath, P. and Baumann, T. W. *Phytochemistry* **41** (1996) 169.
7. Kondo, H., Friedmann, H. C. and Vennesland, B. *J. Biol. Chem.* **235** (1960) 1533.
8. Friedkin, M. in *Adv. Enzymology* **38** (1973) 235, Meister, A. (Ed.), J. Wiley, New York.
9. Sigman, D. and Mooser, G. *Ann. Revs. Biochem.* **44** (1975) 895.
10. Santi, D. and McHenry, C. *Proc. Natl. Acad. Sci. USA* **69** (1977) 1855.
11. Wasserman, H. H., Sykes, R. J., Peverada, P., Shaw, C. K., Cushley, R. J. and Lipsky, S. R. *J. Am. Chem. Soc.* **95** (1973) 6874.
12. Schneider, D., Boppre, M., Schneider, H., Thompson, W. R., Boriack, C. J., Petty, R. L. and Meinwald, J. *J. Comp. Physiol.* **97** (1975) 245.
13. Shemin, D. and Rittenberg, D. *J. Biol. Chem.* **166** (1946) 621, 637; **192** (1951) 315.
14. Lawer, W. G., Neuberger, A. and Scott, J. J. *J. Chem. Soc.* **1959,** 1474.
15. Shemin, D. and Russel, D. S. *J. Am. Chem. Soc.* **75** (1953) 4873.
16. Battersby, A. R., Hunt, E. and McDonald, E. *J. Chem. Soc. Chem. Commun.* **1973,** 442.
17. Blanche, F., Cameron, B., Crouzet, J., Debusshe, L., Thibaut, D., Vuilhorgne, M., Leeper, F. J. and Battersby, A. R. *Angew. Chem. Int. Ed.* **34** (1995) 383.
18. Scott, A. I. *Tetrahedron* **50** (1994) 13315.
19. Jackson, A. H., Sankovich, H. A., Ferramola, A. M., Evans, N., Games, D. E., Matlin, S. A., Elder, G. H. and Smith, S. G. *Phil. Trans. R. Soc. B.* **273** (1976) 191.
20. Barnard, G. F. and Akhtar, M. *J. Chem. Soc. Chem. Commun.* **1975,** 494.
21. Jackson, A. H., Jones. D. M., Philip. G., Lash, T. D. and Batlle, A. M. del C. *Int. J. Biochem.* **12** (1980) 681; Cavalerio. J. A. S., Kenner, G. W. and Smith, K M. *J. Chem. Soc. Chem. Commun.* **1973,** 183.
22. Zaman, Z., Abboud, M. M. and Akhtar, M. *J. Chem. Soc. Chem. Commun.* **1972,** 1263.
23. Battersby, A. R., McDonald, E., Wurzier, H. K. W. and James, K. J. *J. Chem. Soc. Chem. Commun.* **1975,** 493.
24. Sheldon, R. A. and Kochi, J. K. *Organic Reactions* **19** (1972) 279.
25. Battersby, A. R., Gutman, A. L., Fookes, C. J. R., Günther, H. and Simon, H. *J. Chem. Soc. Chem. Commun.* **1981,** 645.
26. Rickes, E. L., Brink, N. G., Koniuszy, F. R., Wood, T. R. and Folkers, K. *Science* **107** (1948) 396.
27. Smith, E. L. and Parker, L. F. J. *Biochem. J.* **43** (1948) viii.
28. Hodgkin, D. C., Kamper, J., Lindsey, J., MacKay, M., Pickworth, J., Robertson, J. H., Shoemaker, C. B., White, J. G., Prosen, R. J. and Trueblood, K. N. *Proc. Roy. Soc.* **A242** (1957) 228.
29. Barker, H. A., Weissbach, H. and Smyth, R. D. *Proc. Natl. Acad. Sci. USA* **44** (1958) 1093.

30. Battersby, A. R., Deutscher, K. R. and Martinoni, B. *J. Chem. Soc. Chem. Commun.* **1983**, 699.
31. Matthews, R. G. and Drummond, J. T. *Chem. Revs.* **90** (1990) 1275.
32. Pratt, J. M., Norris, P. R., Mamsa, M. S. A. and Bolton, R. *J. Chem. Soc. Chem. Commun.* **1994**, 1333.
33. Zydowsky, T. M., Courtney, L. F., Fraska, V., Kobayashi, K., Shimizu, H., Yuen, L.-D., Matthews, R. G., Benkovich, S. J., and Floss, H. G. *J. Am. Chem. Soc.* **108** (1986) 3152.
34. Holthausen, M. C., Fiedler, A., Schwarz, H. and Koch, W. *Angew. Chem. Int. Ed.* **34** (1995) 2282.
35. Hay, B. P. and Finke, R. G. *J. Am. Chem. Soc.* **109** (1987) 8012.
36. Frey, P. A., Eisenberg, M. K. and Abeles, R. H. *J. Biol. Chem.* **242** (1967) 5369.
37. Zhao, Y., Such, P. and Rétey, *J. Angew. Chem. Int. Ed.* **31** (1992) 215.
38. Wollowitz, S. and Halpern, J. *J. Am. Chem. Soc.* **110** (1988) 3112.
39. Tada, M., Miura, K., Okabe, M., Seki, S. and Mitzukami, H. *Chem. Letters Japan* **1981**, 33.
40. Dowd, P. and Trivedi, B. K. *J. Org. Chem.* **50** (1985) 206.
41. Buley, A. L., Norman. R. O. C. and Pritchett, R. J. *J. Chem. Soc. (B)* **1966**, 849.
42. Finke, R. G. and Schiraldi, D. A. *J. Am. Chem. Soc.* **105** (1983) 7605.
43. Rétey, J., Umani-Ronchi, A., and Arigoni, D. *Experientia* **22** (1966) 502; Abeles, R. H. and Dolphin, D. *Acc. Chem. Res.* **9** (1976) 114.

Answers to problems

Chapter 3

3.1

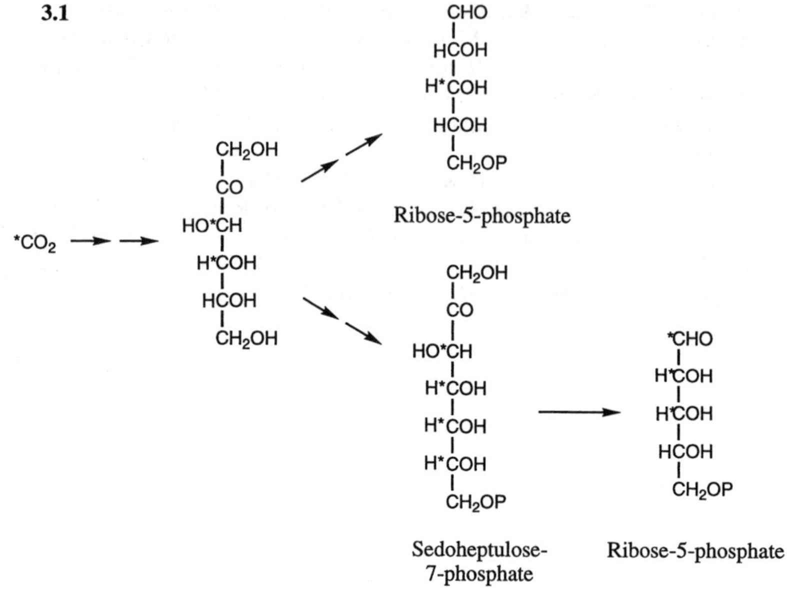

Ribose-5-phosphate

Sedoheptulose-
7-phosphate

Ribose-5-phosphate

3.2

2,3,6-tri-*O*-
methylglucose

2,3,4-tri-*O*-
methylgalactose

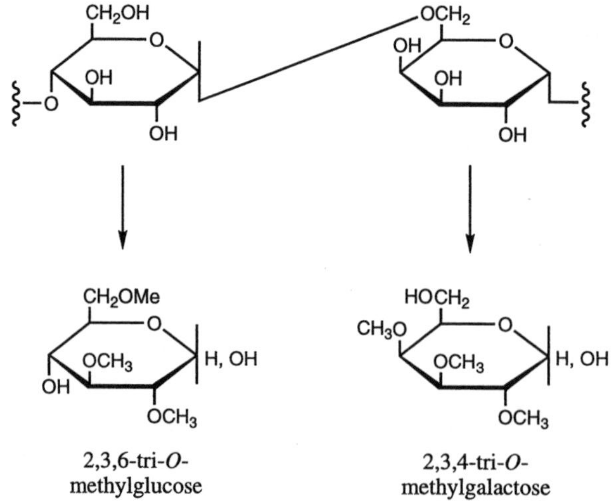

The chain contains *ca.* 2.6 x 10⁴ residues

3.3

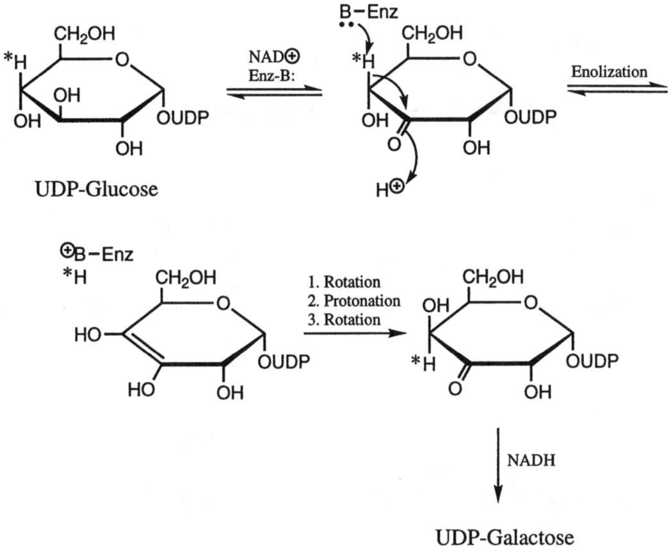

NAD⊕
Enz-B:
Enolization
B–Enz
UDP-Glucose
⊕B–Enz
1. Rotation
2. Protonation
3. Rotation
NADH
UDP-Galactose

The sugar must rotate twice around the C^1–OP axis to account for the steric outcome. It has also been argued that change of conformation of the sugar (chair-boat) would compensate for rotation or the proton could be transferred to a second base properly situated at the opposite phase of the molecule. This sequence is regarded as more unlikely than the route via 4-keto-UDP-glucose because fast proton exchange is expected at the basic site.

3.4

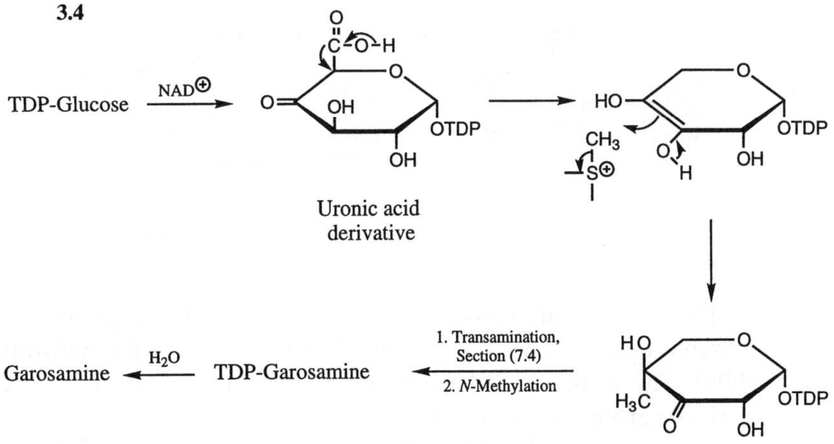

TDP-Glucose
NAD⊕
Uronic acid derivative
1. Transamination, Section (7.4)
2. N-Methylation
TDP-Garosamine
H2O
Garosamine

3.5 CF_3CH_2OH $\xrightarrow{NAD^+}$ $CF_3\overset{O}{\overset{\|}{C}H}$ + (methyl vinyl ketone)

$\xrightarrow{\text{Thiamine}}$ $CF_3COCH_2CH_2COCH_3$

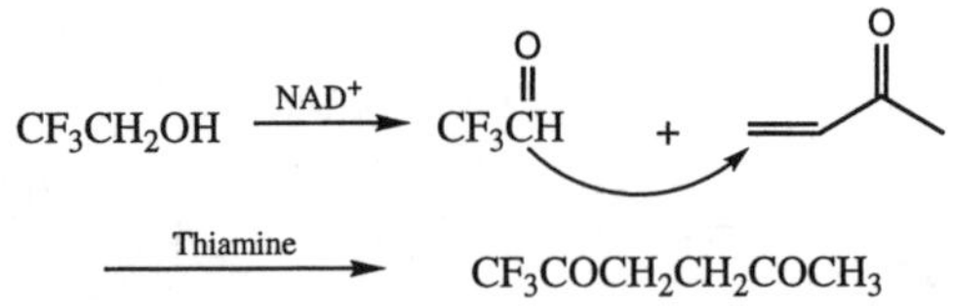

3.6

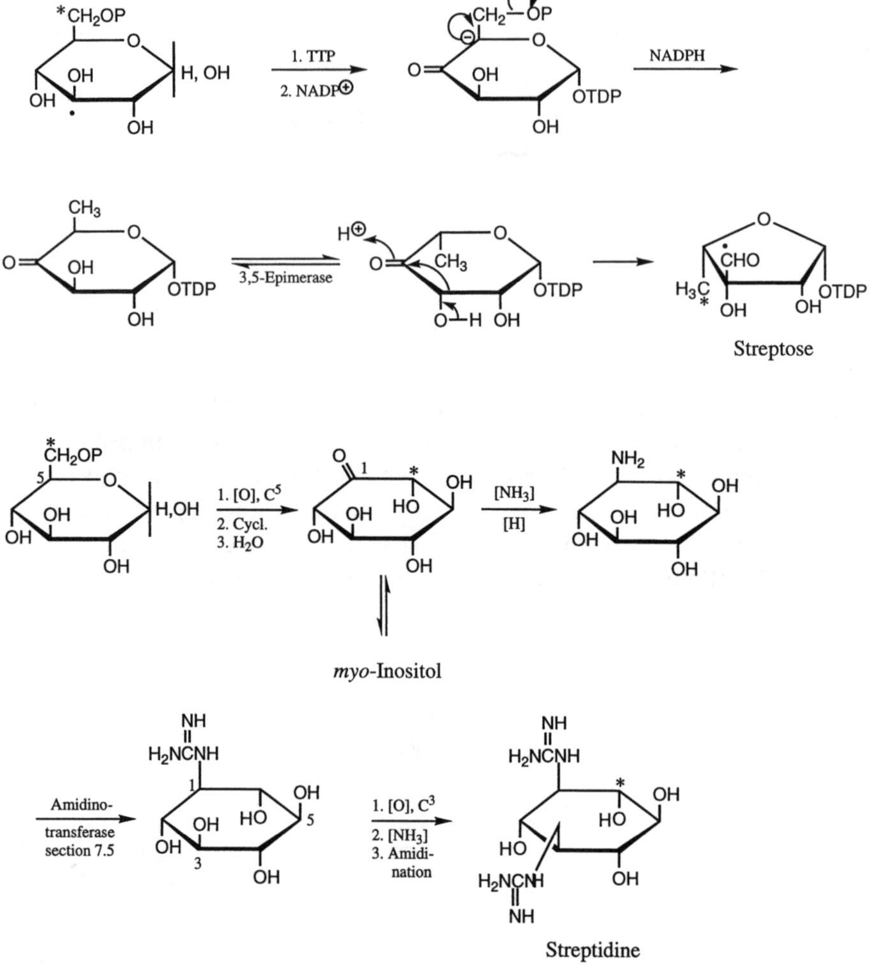

Streptose

myo-Inositol

Streptidine

The intermediate monoaminocyclitol has a plane of symmetry. The chiral enzyme oxidizes only C^3 but not C^5 as indicated by the absorption at 72.4 PPM. The peaks at 13.4 PPM and 61.2 PPM originate from the CH_3 and CH_2OH groups, respectively.

Since D-glucose (but not L-glucose) is incorporated, all chiral centres have to be inverted during the biosynthesis of 2-deoxy-2methylamino-L-glucose. The details of this epimerization are not known. The methylamino group is introduced by transamination followed by methylation with adenosylmethionine.

The streptomycin molecule is assembled by condensation of the nucleotide sugars with streptidine-6-phosphate in an S_N2 fashion.

Chapter 4

4.1

$$PLP + NH_3 + CH_3COCOOH$$

4.2

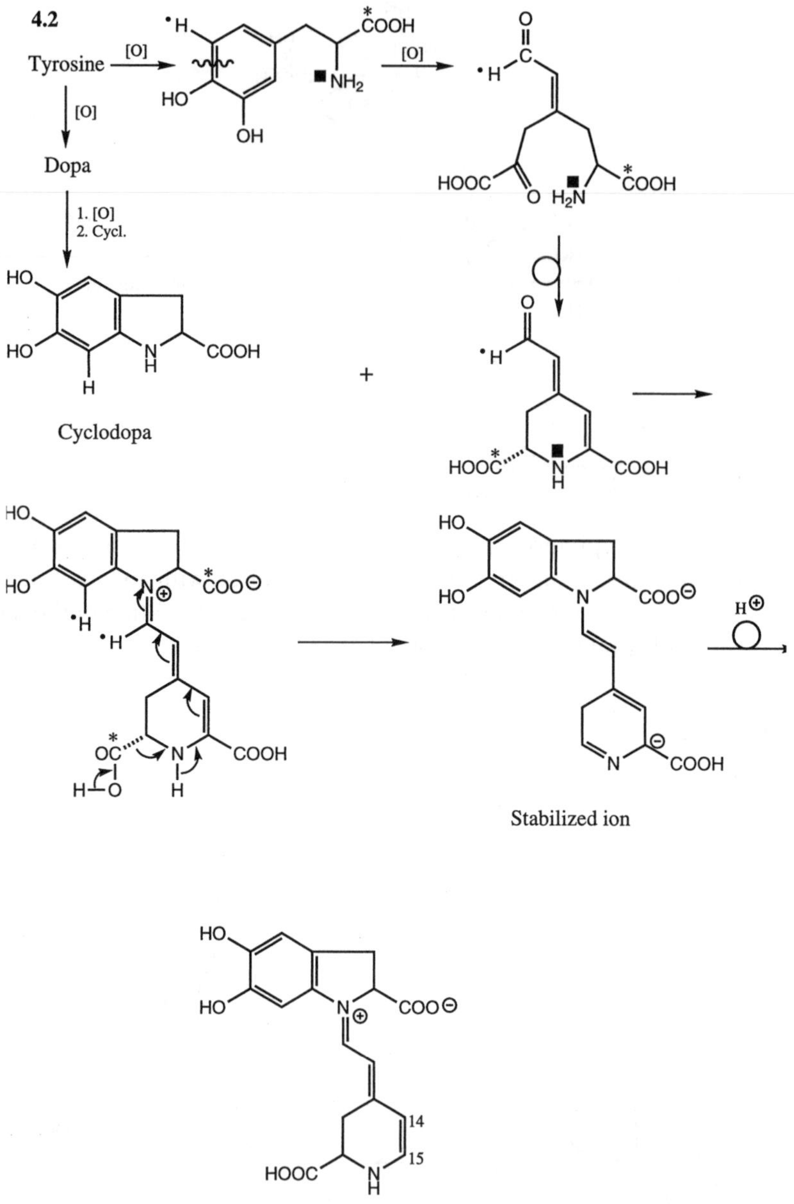

Tyrosine $\xrightarrow{[O]}$

Tyrosine $\xrightarrow{[O]}$ Dopa

Dopa $\xrightarrow{\begin{array}{l}1.\ [O]\\2.\ Cycl.\end{array}}$

Cyclodopa

+

Stabilized ion

4.3

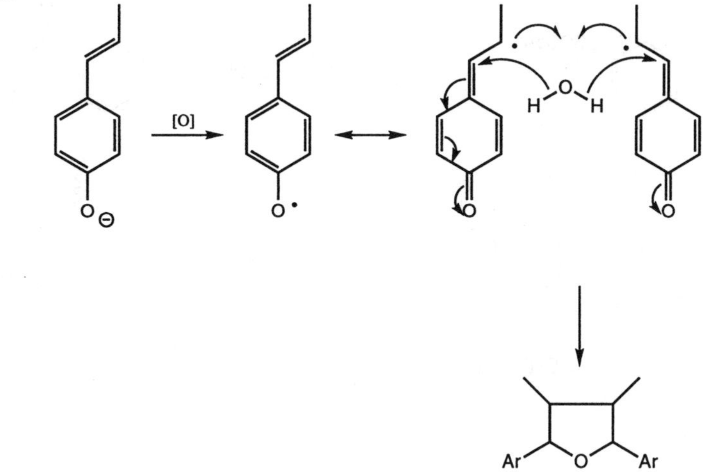

I

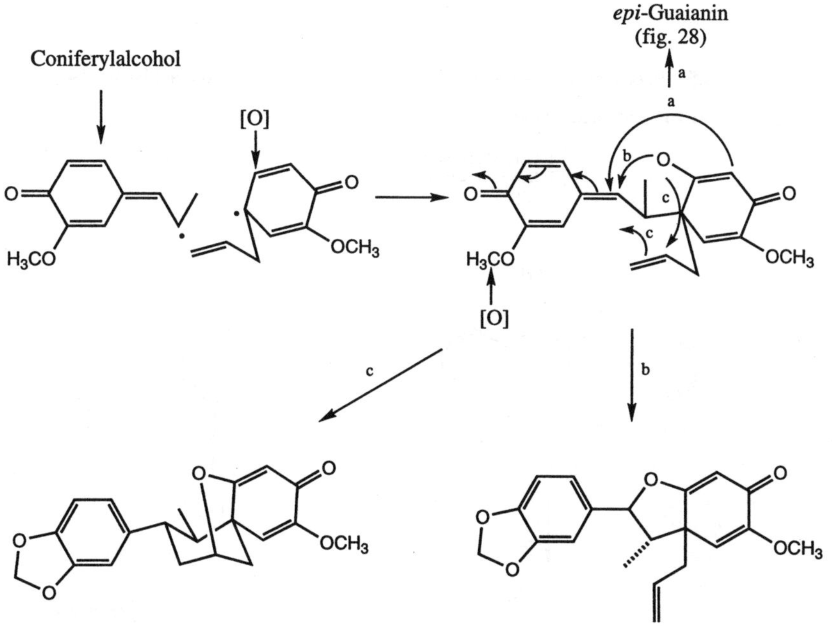

Coniferylalcohol

epi-Guaianin
(fig. 28)

Futoenone II

epi-Bursellin III

4.4

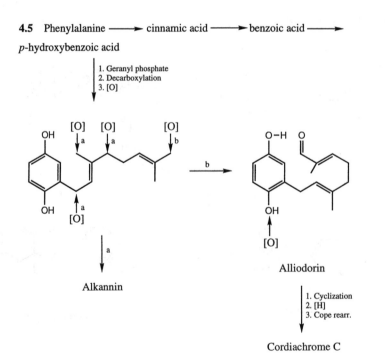

Cathinone

Cathinone is co-occurring with the dione in *Catha edulis*. It is possible that reduction precedes *trans*-amination.

4.5 Phenylalanine ⟶ cinnamic acid ⟶ benzoic acid ⟶
p-hydroxybenzoic acid

1. Geranyl phosphate
2. Decarboxylation
3. [O]

Alkannin

Alliodorin

1. Cyclization
2. [H]
3. Cope rearr.

Cordiachrome C

The cordiachrome pathway is supported by isolation of aldehyde alliodorin.

4.6

(a)

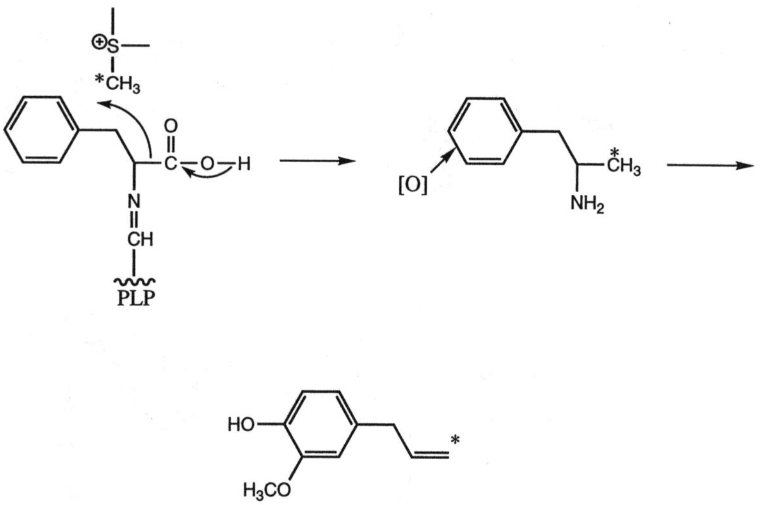

Eugenol

(b)

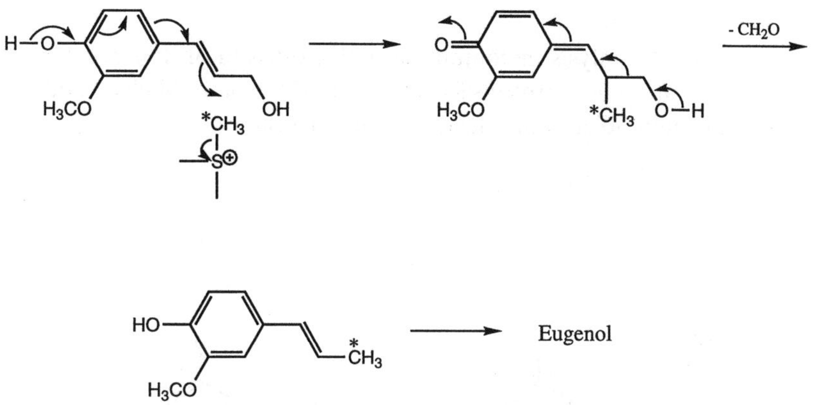

Eugenol

4.7

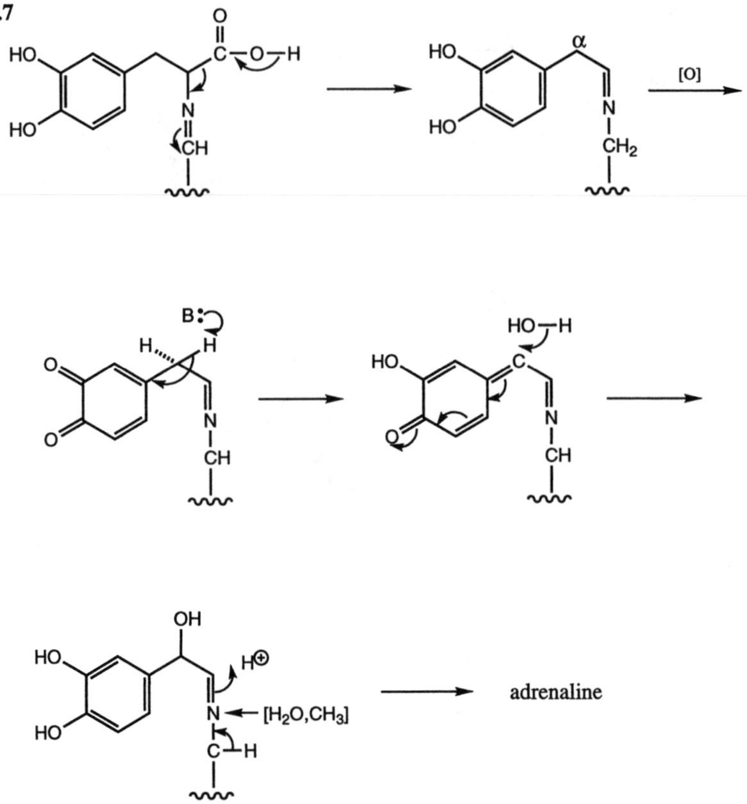

PLP catalysis in conjuntion with oxidation of dopamine to the *o*-quinone activates the α-protons for abstraction by a base. Michael addition of water to the quinoid structure leads to α-hydroxylation.

4.8

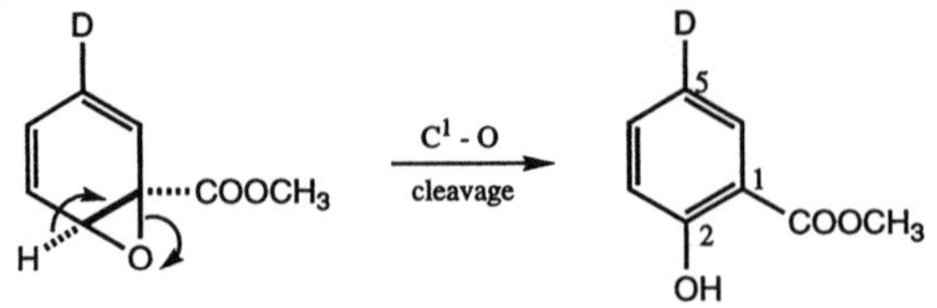

^{2}H-Labelling at C^5 gives three different products.

Chapter 5

5.1

Test, for example, incorporation of labelled decanoic acid and octanoic acid which undergo β-oxidation.

5.2

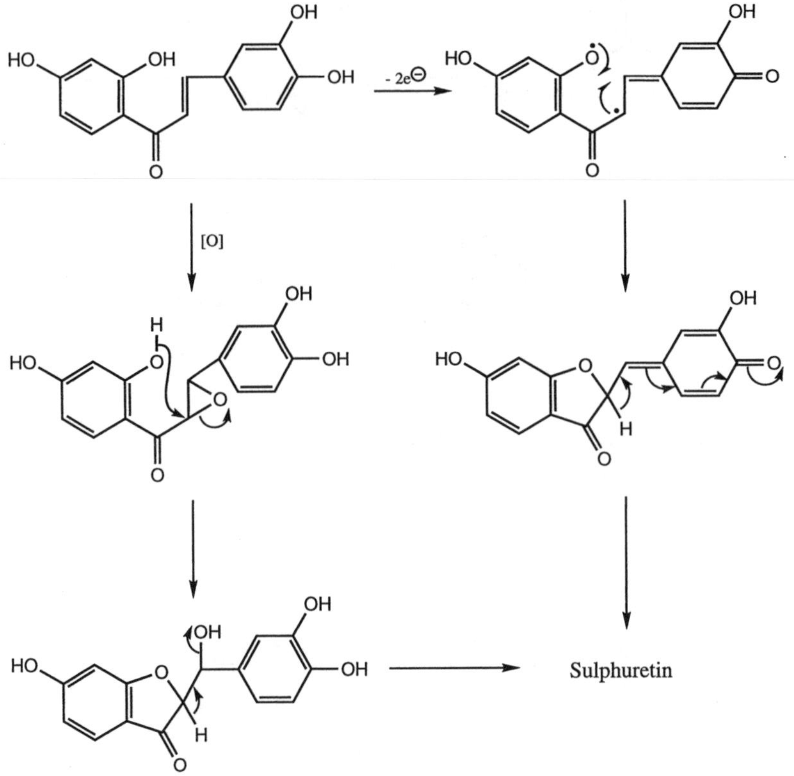

5.3

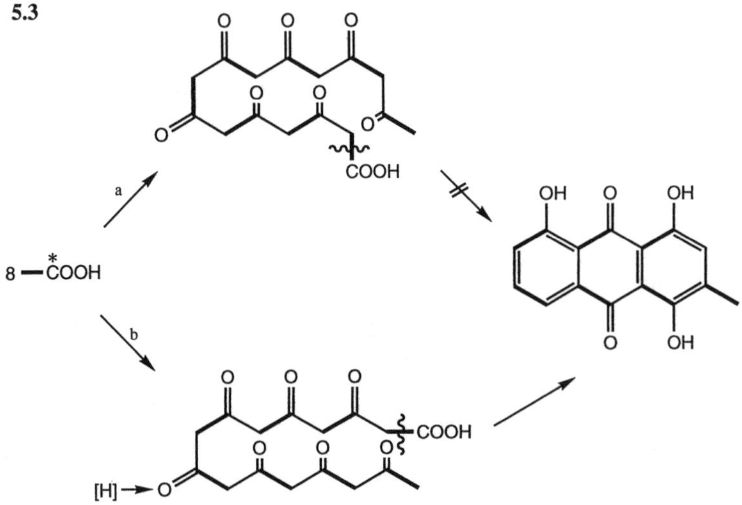

The $^{13}C_2$-acetate method gives different labelling patterns for routes a and b. The $J^{13}C-^{13}C$ data turned out to be compatible only in the route b. The ^{13}C shifts are assigned by incorporation of monolabelled acetate and by off-resonance decoupling of ^{1}H. The residual $^1J_{^{13}C-^1H}$ give information on the number of hydrogens attached to a certain carbon atom.

5.4

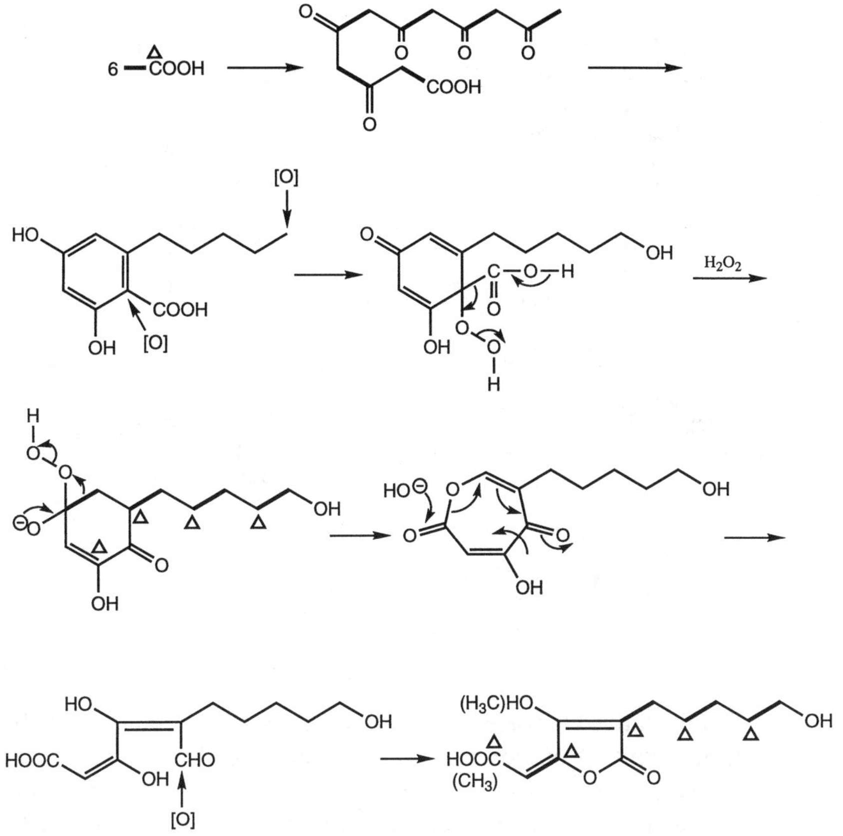

5.5

(a) Separate chain condensation

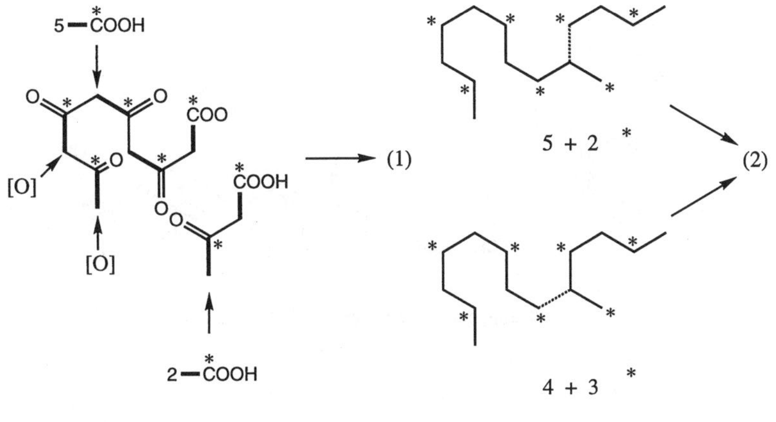

(b) Single chain condensation and cleavage

The oxidative cleavage of the fusarubin precursor must occur between two acetate units. Rotation and recyclization give (1) and (2). The nearly equal enrichment of label indicates that one single chain precursor is more likely. A two chain assembly should conceivably give rise to higher enrichment in the starter units.

5.6

1 = acetate   2 = butyrate   3 = propionate

Formation of the benzene ring:

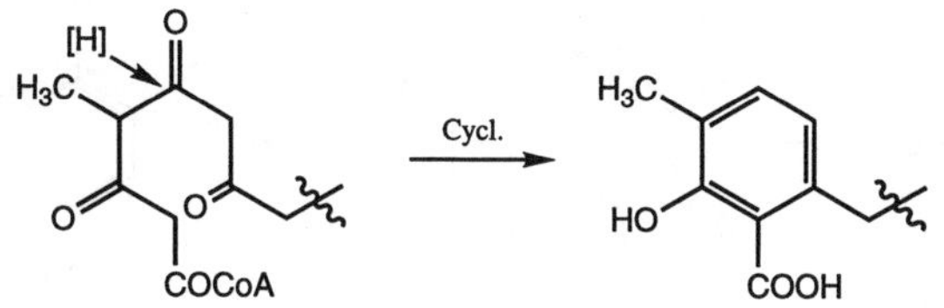

Formation of the tetrahydrofuran ring:

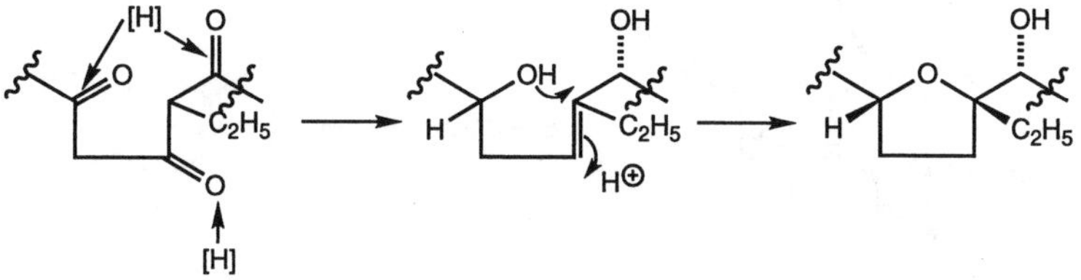

Another possibility: (*cf.* Cane, D. E., Liang, T.-C. and Hasler, H. *J. Am. Chem. Soc.* **103** (1981) 5962; Hutchinson, C. R., Sherman, M. M., McInnes, A. G., Walter, J. A. and Vederas, J. C. *J. Am. Chem. Soc.* **103** (1981) 5956) the vicinal hydroxy groups are derived from molecular oxygen via the olefins and epoxides. Cf. the biosynthesis of monensin, section 5.7.

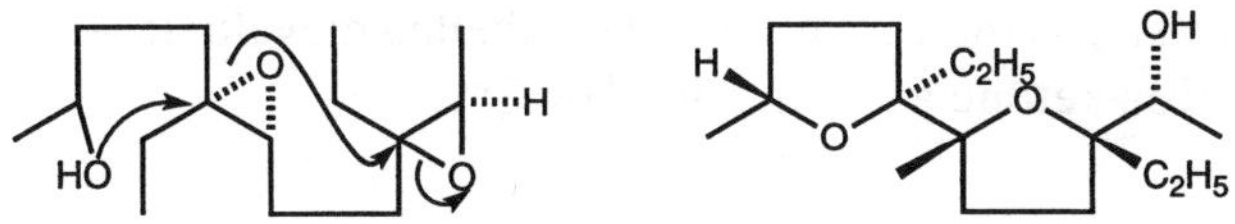

The methyl groups could conceivably come from methionine, but labelling experiments show that this is not the case.

5.7

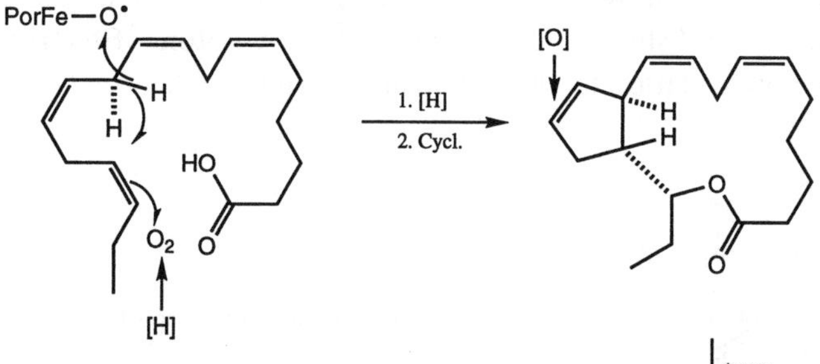

Octadeca-6,9,12,15-
tetraenoic acid
Attack at C^{11}

trans-
hydroxy-
lation

Ecklonia lactone

5.8

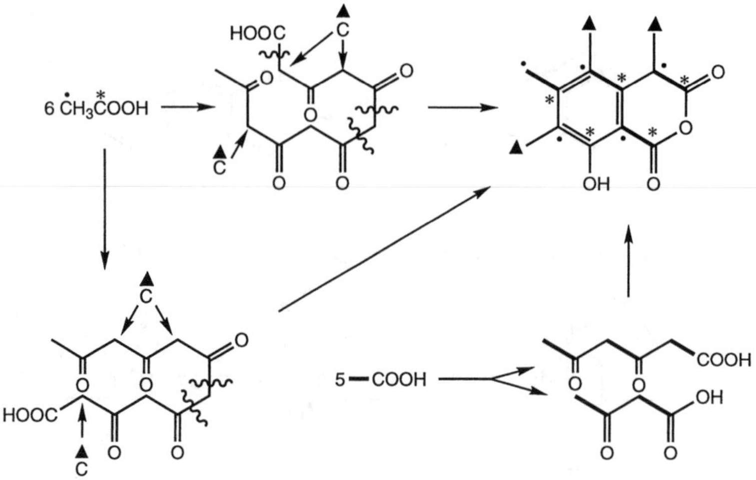

The labelling pattern from 1,2-$^{13}C_2$-acetate gives the correct mode of folding. It turns out that only the biosynthesis from two chains is compatible with the ^{13}C NMR spectroscopic results ($^1J_{13_C13_C}$).

Chapter 6

6.1 The key step is the 1,3 shift. C^5–$^2H_2$-labelled mevalonic acid is expected to label artemisia ketone at C^1 and at C^9 or C^{10}.

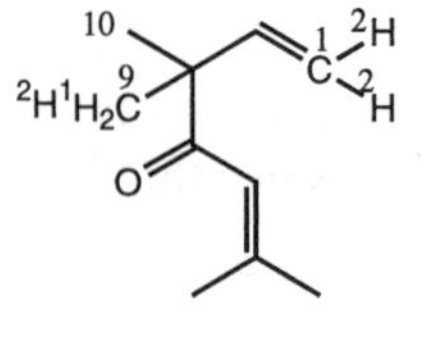

^1H NMR should show a triplet (1: 1: 1) for the methylene protons of C^9. Artemisia alcohol should be a precursor for the ketone. Electrophilic aliphatic C^4–C^2 condensation leads to the chrysanthemyl skeleton (section 5.3).

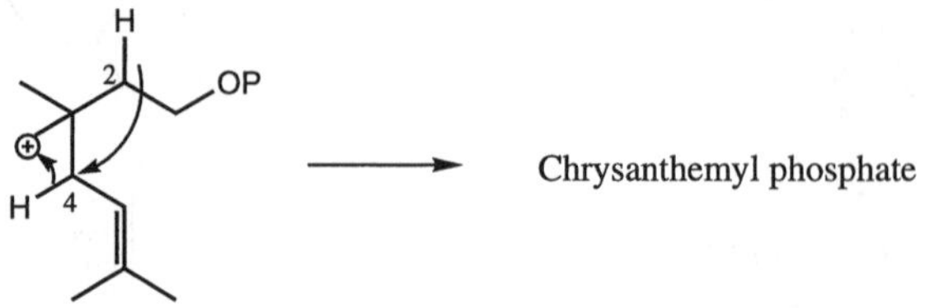

Chrysanthemyl phosphate

6.2

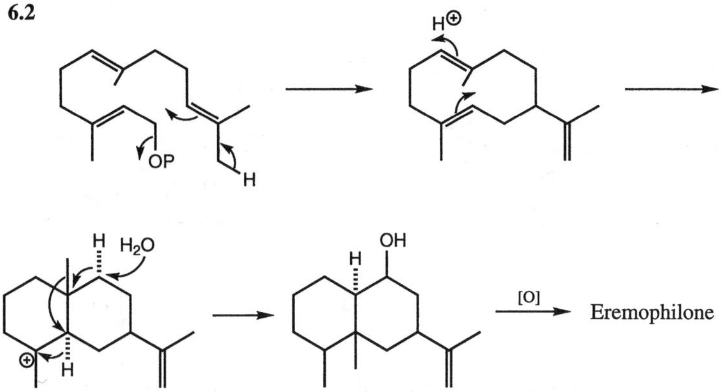

6.3

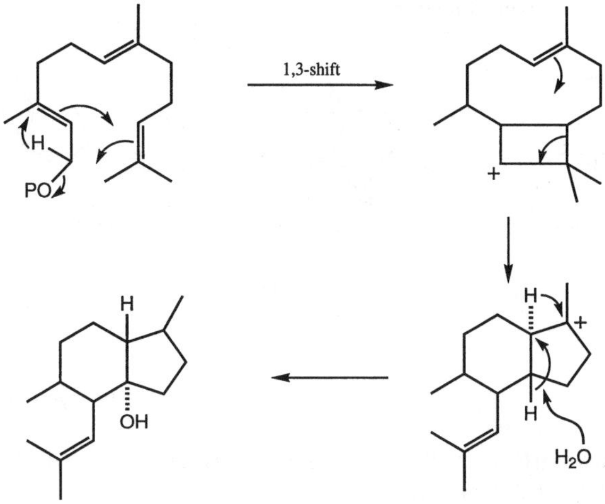

6.4

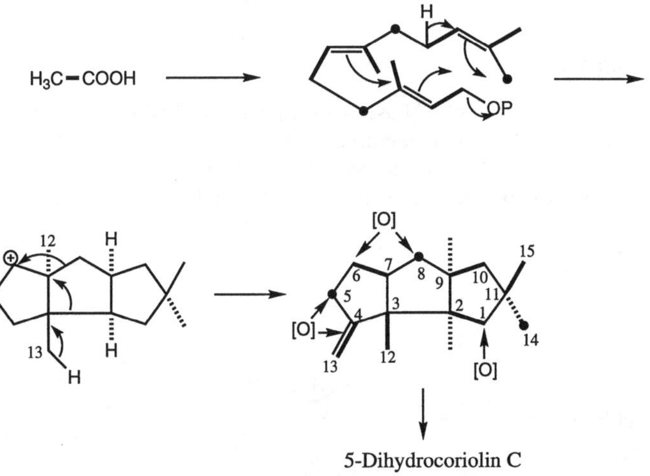

$H_3C-COOH$

5-Dihydrocoriolin C

If the $C^{12,13}$ methyl group undergo Wagner-Meerwein shifts, two acetate units will be cleaved.

6.5

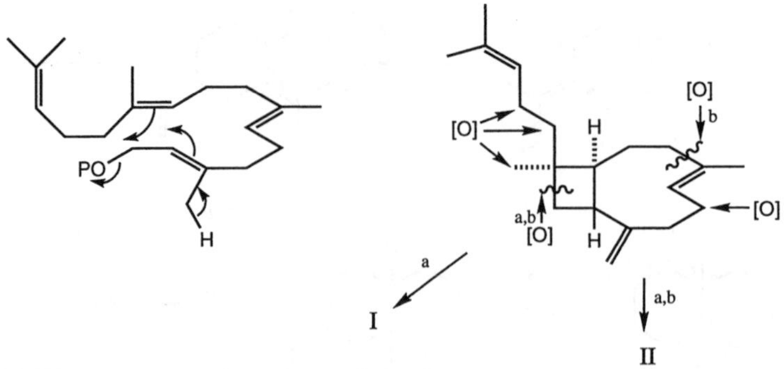

6.6 There are several ways to coil the C_{20} geranylgeranyl phosphate chain to obtain this cembrene derived diterpene, e.g.

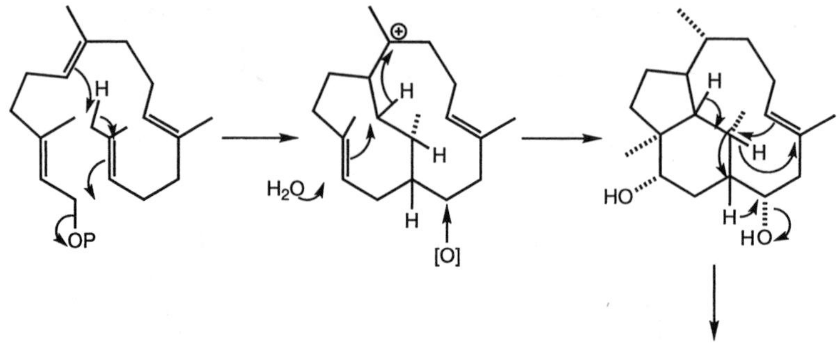

3α-Hydroxy-15-rippertene

This sequence starts from all-*trans*-geranylgeranyl phosphate which undergoes two 1,3 H shifts and one 1,2 methyl shift.

Chapter 7

7.1 PLP gives a Schiff base with aspartic acid and rearranges to the ketimine. This structure stabilizes a negative charge on C^3, i.e. it facilitates the cleavage of the C^3–COOH bond. Protonation of C^3 and C^2 gives alanine after hydrolysis. Protonation of C^3 and the pyridoxamine moiety followed by hydrolysis gives pyruvic acid and PMP.

7.1

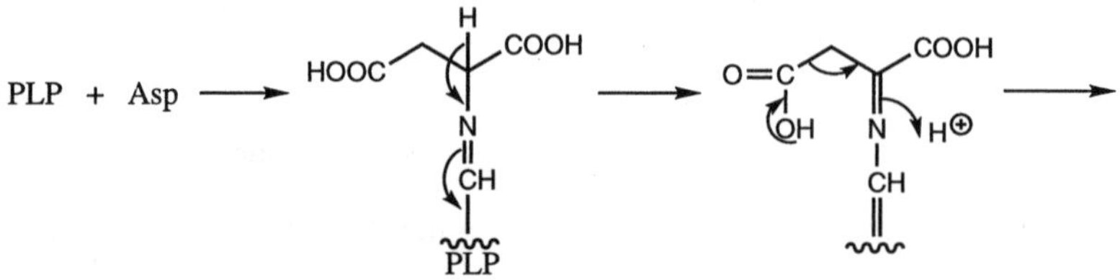

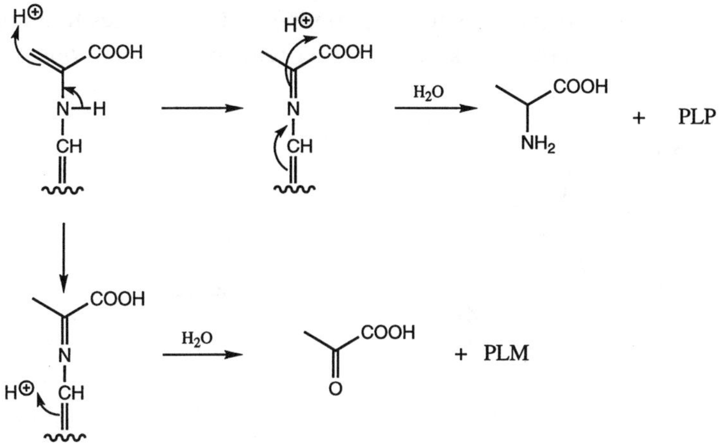

7.2

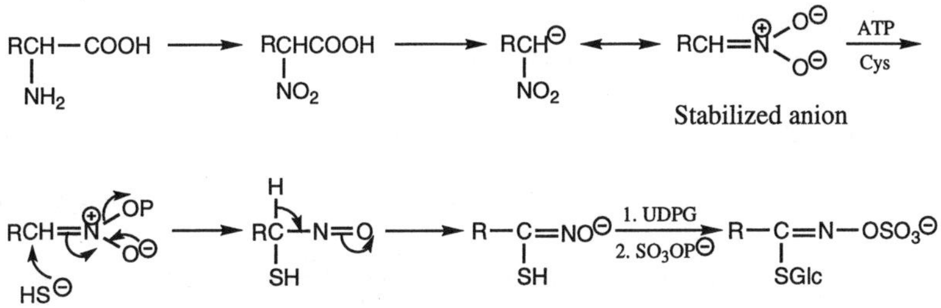

Stabilized anion

7.3

Gabaculine 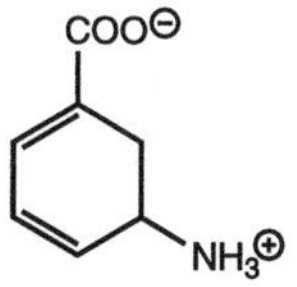 competes with

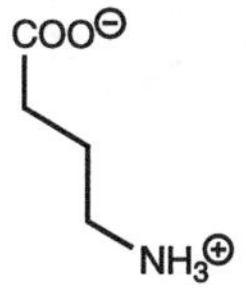

at the active site of the transaminase forming

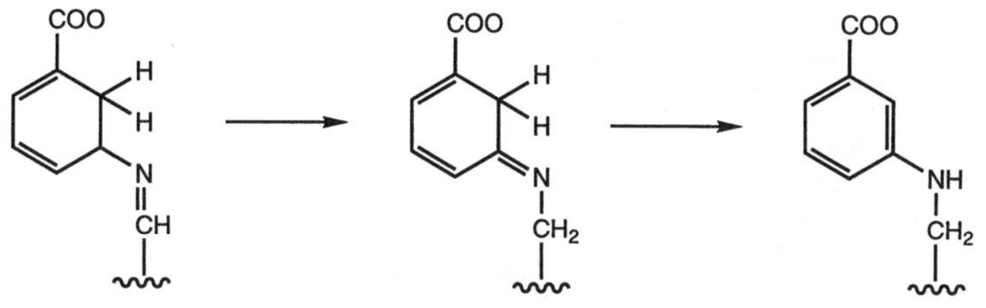

| PLP-aldimine | PLP-ketimine | Stable aromatic and blocked PLP-derivative |

PLP-aldimine rearranges to the ketimine and aromatizes to a non-hydrolysable PLP derivative blocking its further reactions as coenzyme.

7.4

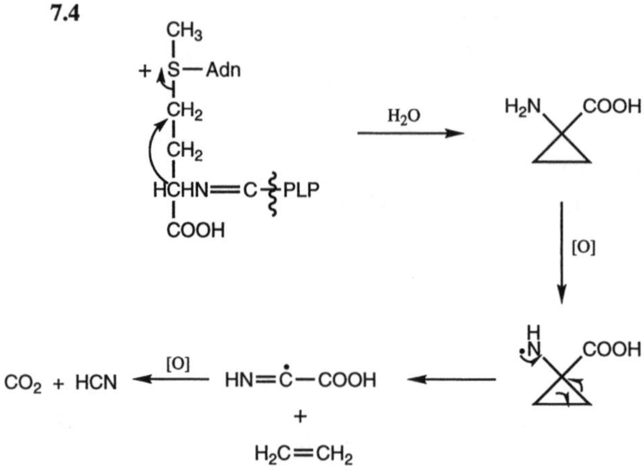

7.5

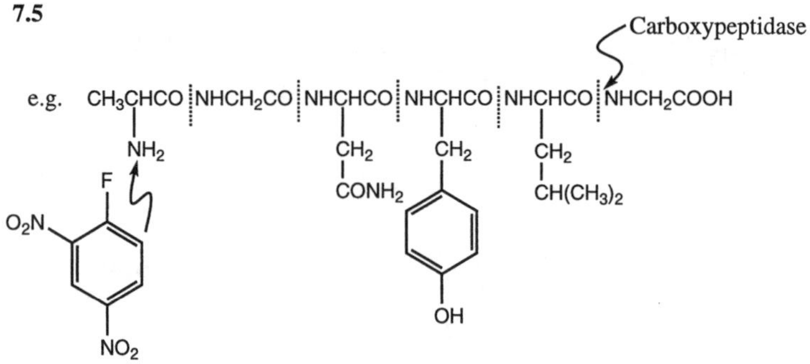

7.6

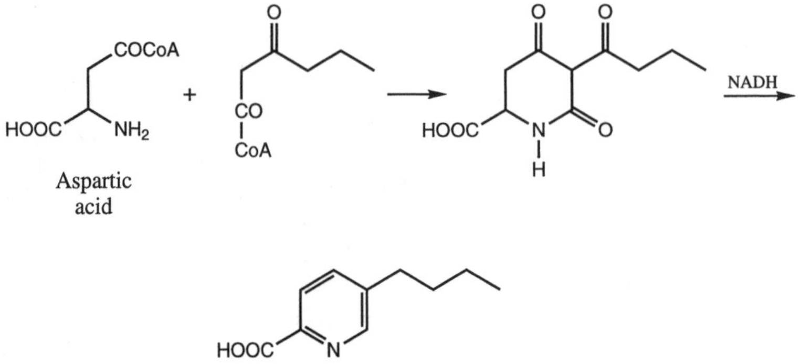

7.7   A working hypothesis:

Cystine

Other reasonable precursors: Cysteine, serine. ^{13}C-, ^{3}H-, ^{2}H-labelled or doubly label-led precursors are given to the culture medium. It is of interest to investigate whether C_α–H is retained, *cf.* the biosynthesis of penicillins. The metabolite is analysed by NMR and MS spectroscopy. Suggestion: carry out a literature search to see if the problem is solved.

7.8 Anthramycin can by visual inspection be dissected into an anthranilic acid deriva-tive and a substituted proline derivative, the latter originating from tyrosine by ring cleavage.

The order of events is still unknown but methylation of the proline part occurs late since one ^{3}H is retained.

7.9 Dehydrogenation of 2-methyl propionic acid, Michael addition of cysteine follow-ed by elimination of the amino acid fragment introduces the thio group at C–3. The re-actions are repeated at C–4. In this process ^{3}H-2 is lost but ^{3}H-3,4 are retained because of the isotope effect.

7.10

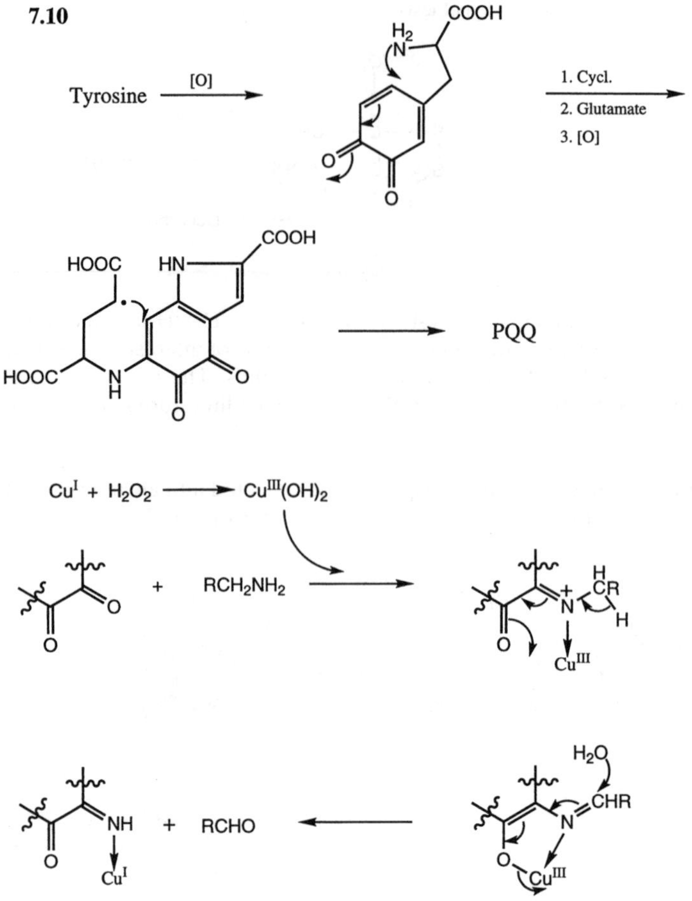
Tyrosine
[O]
COOH
1. Cycl.
2. Glutamate
3. [O]
HOOC
COOH
PQQ
Cu^I + H2O2
Cu^III(OH)2
RCH2NH2
Cu^III
H2O
CHR
NH
RCHO
Cu^I

7.11

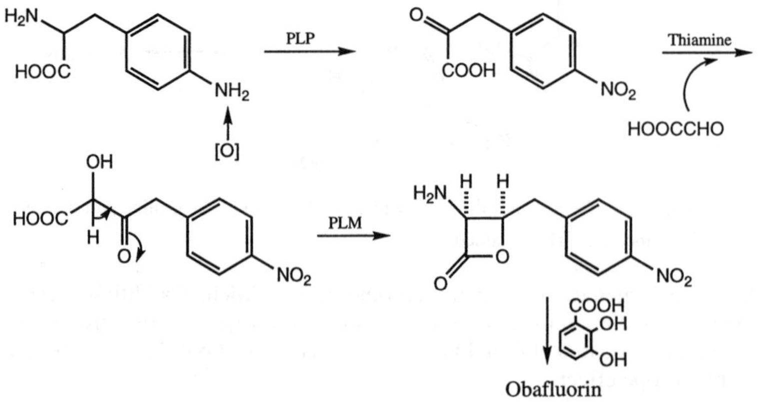
H2N
HOOC
NH2
[O]
PLP
COOH
NO2
Thiamine
HOOCCHO
OH
PLM
COOH
Obafluorin

Chapter 8

8.1

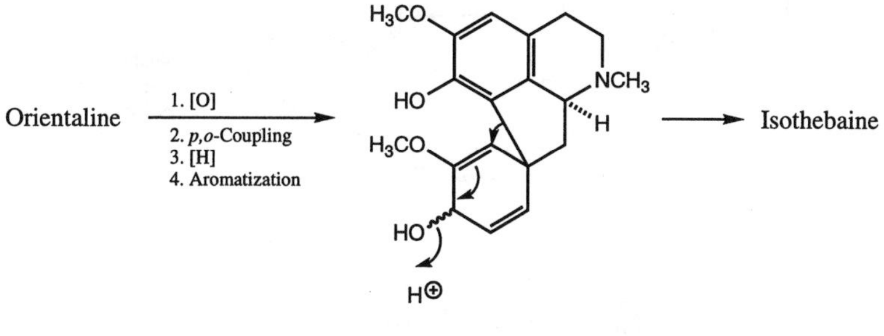
Orientaline
1. [O]
2. *p,o*-Coupling
3. [H]
4. Aromatization
H3CO
HO
H3CO
HO
NCH3
H
H⊕
Isothebaine

8.2

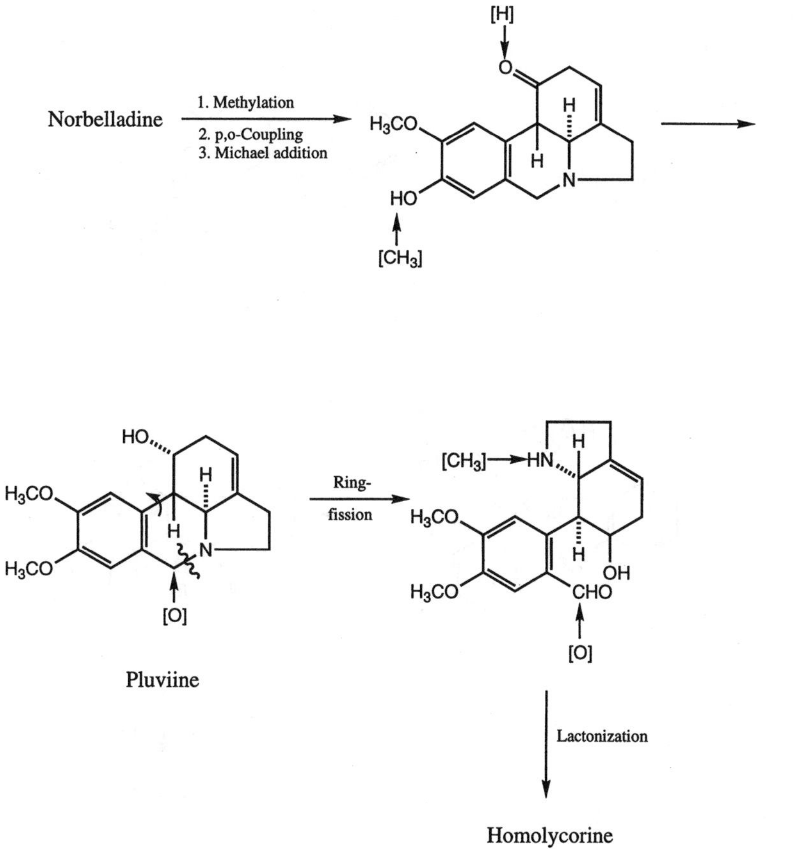
Norbelladine
1. Methylation
2. p,o-Coupling
3. Michael addition
[H]
O
H
H
H3CO
HO
N
[CH3]
HO
H3CO
H3CO
H
H
N
[O]
Pluviine
Ring-
fission
[CH3]
HN
H
H3CO
H3CO
H
OH
CHO
[O]
Lactonization
Homolycorine

8.3

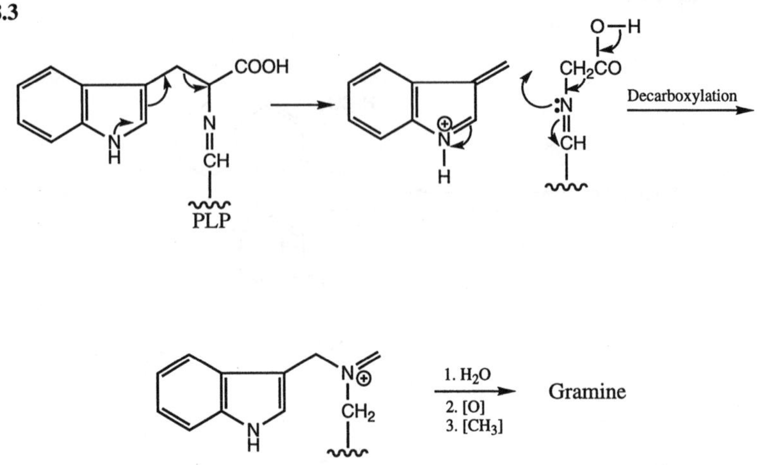
COOH
PLP
Decarboxylation
1. H_2O
2. [O]
3. [CH_3]
Gramine

8.4

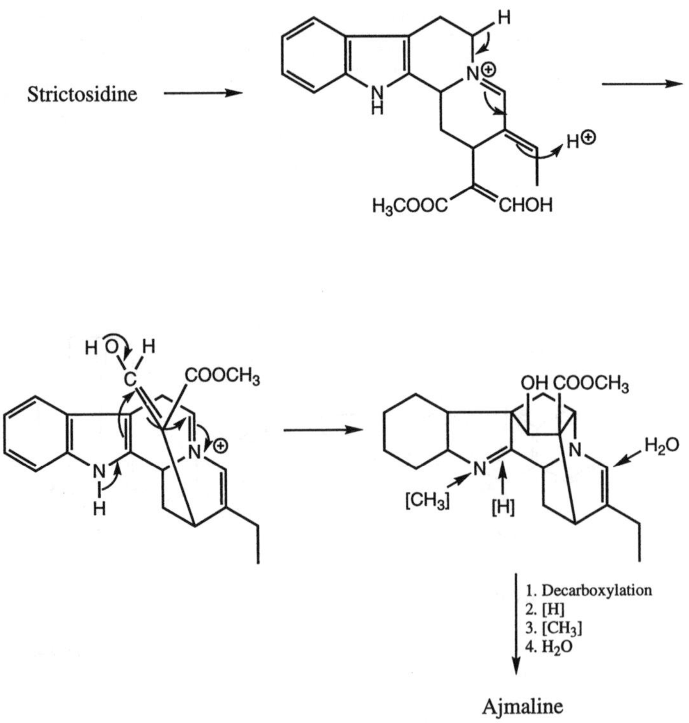
Strictosidine
H_3COOC
CHOH
$COOCH_3$
OH $COOCH_3$
H_2O
[CH_3]
[H]
1. Decarboxylation
2. [H]
3. [CH_3]
4. H_2O
Ajmaline

8.5 This problem could be discussed at a group seminar with your professor.

8.6

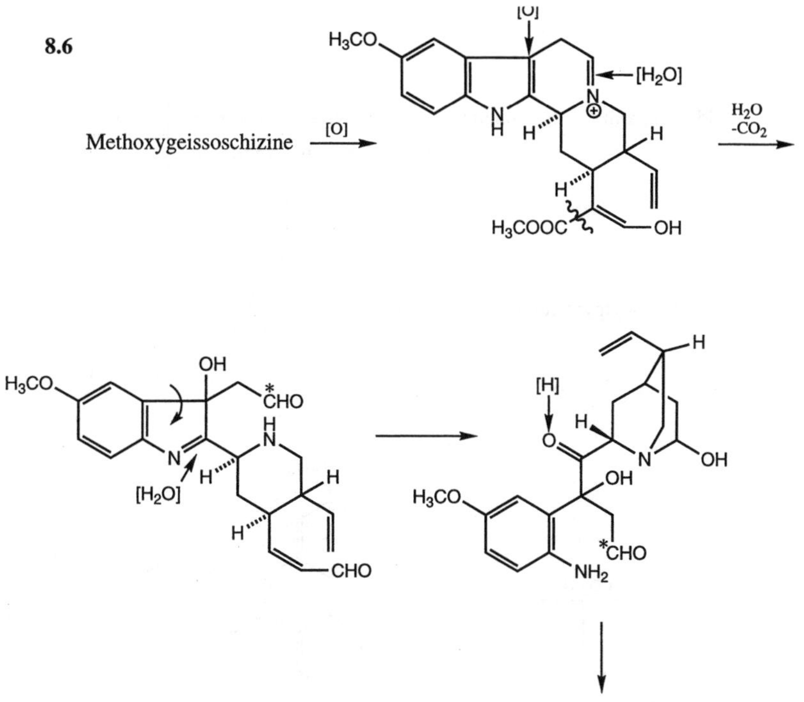

Methoxygeissoschizine $\xrightarrow{[O]}$

Quinine

8.7

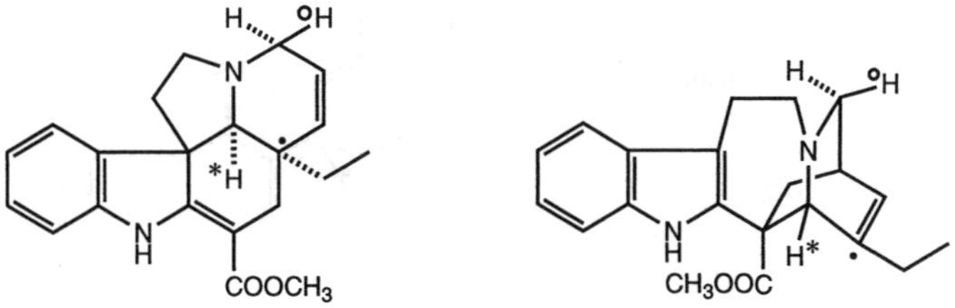

Tabersonine

Catharanthine

* $1\text{-}^3H_2\text{-Geraniol}$
• $2\text{-}^{14}C\text{-Geraniol}$
○ $5\text{-}^3H\text{-Loganin}$

8.8

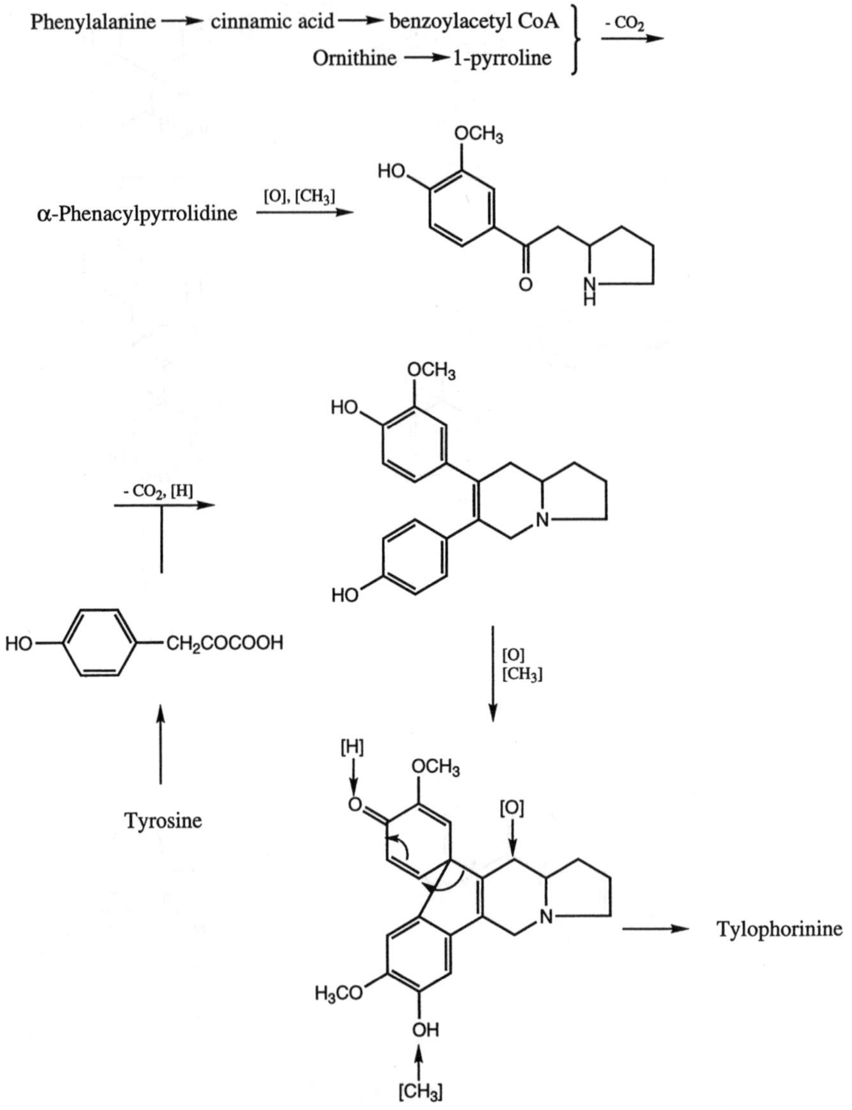

Phenylalanine ⟶ cinnamic acid ⟶ benzoylacetyl CoA ⎫ - CO_2
                                                      ⎬ ⟶
Ornithine ⟶ 1-pyrroline   ⎭

α-Phenacylpyrrolidine $\xrightarrow{\text{[O], [CH}_3\text{]}}$

$\xrightarrow{\text{- CO}_2, \text{[H]}}$

HO—⟨ ⟩—$CH_2COCOOH$

Tyrosine

[O]
[CH₃]

⟶ Tylophorinine

$[CH_3]$

Subject Index